Cross-Sectional Atlas of Rhesus Monkey Head

Jin Seo Park

Cross-Sectional Atlas of Rhesus Monkey Head

with 0.024-mm pixel size color images

Springer

Jin Seo Park
Department of Anatomy
Dongguk University School of Medicine
Gyeongju, Korea (Republic of)

ISBN 978-981-19-3253-3 ISBN 978-981-19-3254-0 (eBook)
https://doi.org/10.1007/978-981-19-3254-0

This Springer imprint is published by the registered company Springer Nature Singapore Pte Ltd.
The registered company address is: 152 Beach Road, #21-01/04 Gateway East, Singapore 189721, Singapore

Contents

Introduction

1

1 Introduction

This superb atlas sets a new standard in cross-sectional neuroanatomy based on the detailed sectioned images of rhesus monkey head with 0.05 mm intervals and 0.024 mm × 0.024 mm sized pixel. Based on the publication experience of the cross-sectional atlas of human head [1], I wrote this atlas by considering absolutely three points.

First is image quality. I already made color and high-resolution sectioned images of four human cadavers [2–5], a dog [6], and a cat [7] in Visible Korean Project. Thanks to the knowledge and know-how at that time, I could produce the superb sectioned images of rhesus monkey with true color and high resolution [8].

Second is reference system. A new reference system employed for the human head atlas of 7.0 Tesla MRI and the sectioned images is clearly explained by center of anterior commissure (AC) and posterior commissure (PC) [9]. In this atlas, I used the modified reference system of center of AC-PC, and fully annotated structures in the axial, coronal, and sagittal planes.

Third is anatomical terminology. In human anatomy, the terminology was already standardized and commonly used in most textbooks, atlases, and journal papers of anatomy. Although this is monkey atlas, not human, I used anatomical terminology of human by Terminologia Anatomica [10].

2 Producing Sectioned Images of Rhesus Monkey

Rhesus monkey (length, 758 mm; weight, 4.3 kg; age at sacrifice, 93 months, female) was donated from the National Primate Research Center of Korea. After embedding and freezing the monkey, the monkey was milled serially using cryomacrotome. Sectioned surface was observed with the naked eye, and the sectioned surface was trimmed manually. A well-trimmed surface was photographed, and the photograph was verified to make a sectioned image. This work was performed from head to toe of the monkey. As a result, we produced the axial sectioned images of monkey's whole body.

For this atlas, I selected axial sectioned images from crown to cervical vertebra IV of monkey head. Also, I chose coronal and sagittal images by reconstruction of axial sectioned images on a personal computer using in-house developed software [8].

The study protocol was reviewed and approved by the Institutional Review Board (IRB) of National Primate Research Center of Korea Research Institute of Bioscience and Biotechnology (IRB No. KRIBB-AEC-18087).

J. S. Park, *Cross-Sectional Atlas of Rhesus Monkey Head*, https://doi.org/10.1007/978-981-19-3254-0_1

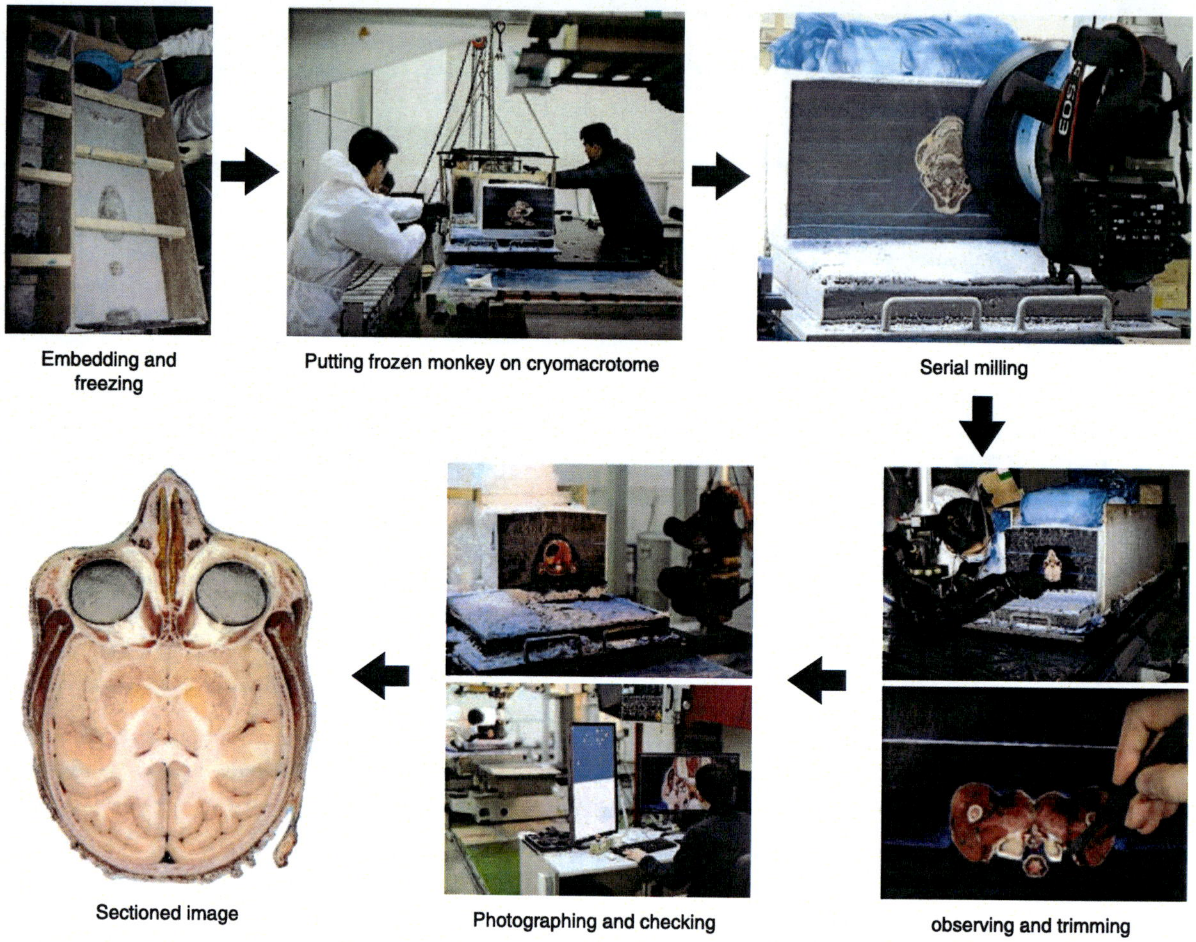

3 Reference System

A human head, especially brain, obeys absolutely the reference system including both the center of AC-PC and the AC-PC line (A) when MRI and CT are scanned for not only diagnosis but also research [9]. However, I could not use the typical axial plane of human head in this atlas because it was sectioned images of monkey head. So, the next best way was to use axial plane that passes only through the center of AC-PC and not follow AC-PC line (B). I applied the reference system of this next best way to axial, coronal, and sagittal sectioned images in this atlas.

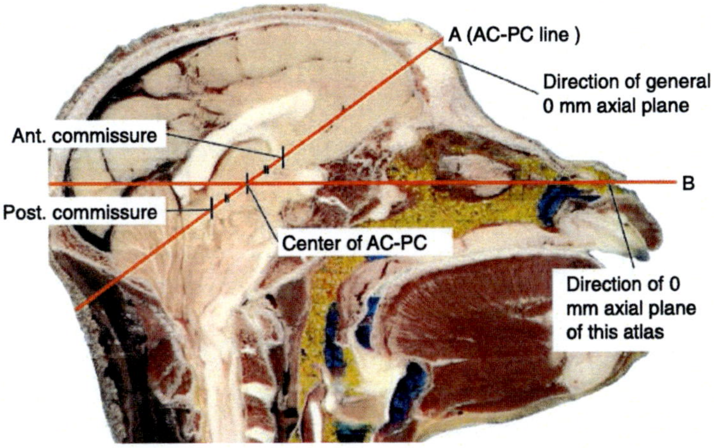

Typical axial plane (A) and modified axial plane of this atlas (B)

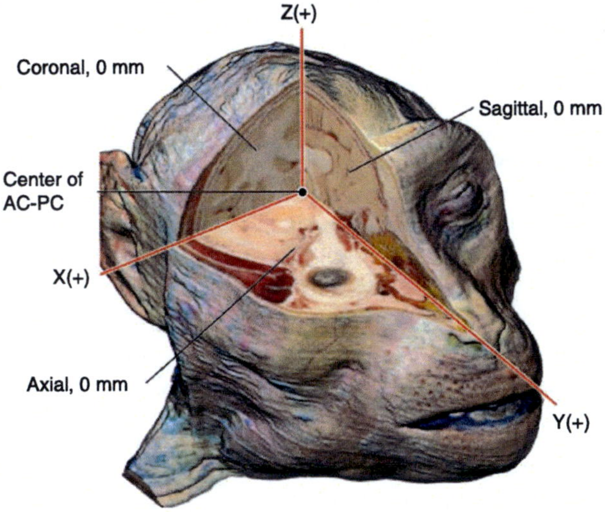

Axial, coronal, and sagittal planes of this atlas

4 Terminology and Labelling

Accurate anatomical terminology is essential to neuroanatomy atlas. Therefore, all anatomical terms in this atlas were used based on the authorized terms, Terminologia Anatomica [10].

It was difficult to identify directly each structure in only a sectional plane. Therefore, I identified each structure by using organically axial, coronal, and sagittal planes that can show three planes simultaneously on the volume model of monkey head in MRIcroGL software [11]. On the other hand, when I identified each structure for this atlas, I could not label various structures of brainstem in the sectioned images because shape of brainstem was crushed by escaping it to foramen magnum in these images.

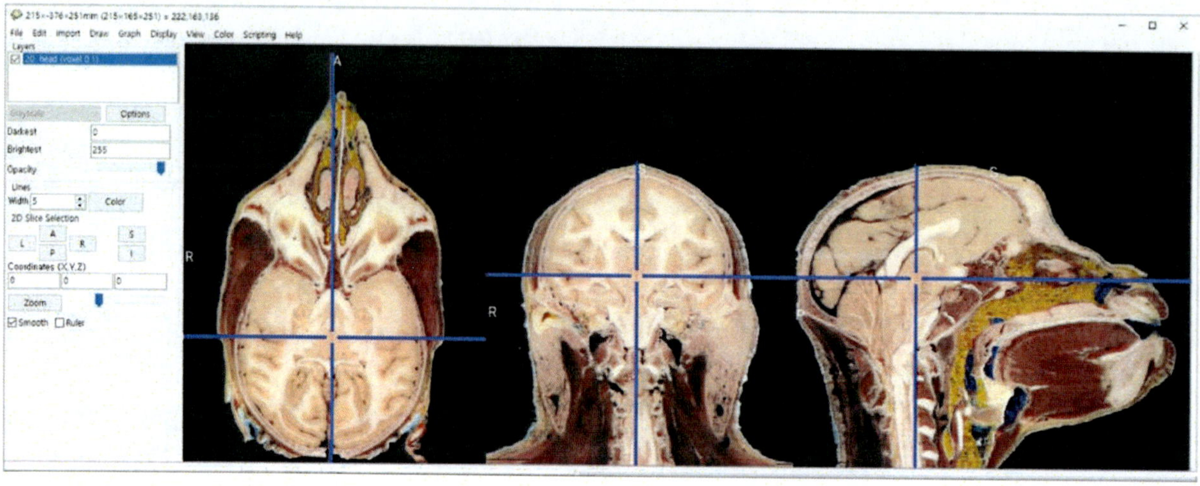

Volume model of rhesus monkey on MRIcroGL that any axial, coronal, and sagittal planes can be observed

5 Conclusion

The exquisite sectioned images contained within this atlas will be invaluable in providing both researchers and clinicians; therefore, this atlas will help in education and research of not only veterinary but also medicine. This is first edition of atlas for monkey sectional anatomy, and I will be upgrading continuously this head atlas and be making new sectional atlas of whole body of rhesus monkey. If there are missing and wrong labelled structures, please inform me using email (park93@dongguk.ac.kr).

Acknowledgments This research was also supported by a Basic Science Research Program through the National Research Foundation of Korea (NRF) funded by the Ministry of Education (NRF-021R1F1A1063044).

References

1. Park, J. S. (2018). *Cross-sectional atlas of the human head: With 0.1-mm pixel size color images*. Springer.
2. Park, J. S., et al. (2005). Visible Korean human: Improved serially sectioned images of the entire body. *IEEE Transactions on Medical Imaging, 24*(3), 352–360.
3. Park, H. S., Choi, D. H., & Park, J. S. (2015). Improved sectioned images and surface models of the whole female body. *International Journal of Morphology, 33*, 1323–1332.
4. Park, J. S., et al. (2009). Sectioned images of the cadaver head including the brain and correspondences with ultrahigh field 7.0 T MRIs. *Proceedings of the IEEE, 97*(12), 1988–1996.
5. You, Y., et al. (2022). Advanced-sectioned images obtained by microsectioning of the entire male body. *Clinical Anatomy, 35*(1), 79–86.
6. Park, H. S., et al. (2014). Improved sectioned images and surface models of the whole dog body. *Annals of Anatomy, 196*(5), 352–359.
7. Chung, B. S., et al. (2018). Sectioned images of a cat head to contribute to learning of its sectional anatomy. *International Journal of Morphology, 36*, 537–543.
8. Chung, B. S., et al. (2019). Rise of the Visible Monkey: Sectioned images of rhesus monkey. *Journal of Korean Medical Science, 34*(8), e66.
9. Park, J. S., et al. (2010). A proposal of new reference system for the standard axial, sagittal, coronal planes of brain based on the serially-sectioned images. *Journal of Korean Medical Science, 25*(1), 135–141.
10. Drake, R. L. (2011). *Terminologia anatomica: International anatomical terminology by FIPAT* (2nd ed.). Thieme.
11. Kim, C. Y., et al. (2020). Dawn of the Visible Monkey: Segmentation of the rhesus monkey for 2D and 3D applications. *Journal of Korean Medical Science, 35*(15), e100.

Part I

Whole Head

Abbreviations

a.	Artery
ant.	Anterior
b.	Bone
CS	Cervical segment of spinal cord
CV	Cervical vertebra
ext.	External
inf.	Inferior
int.	Internal
lat.	Lateral
lig.	Ligament
m.	Muscle
med.	Medial
n.	Nerve
nu.	Nucleus
post.	Posterior
sup.	Superior
v.	Vein

2 Axial

All images, 0.024 mm × 0.024 mm sized pixel

+27 mm ~ +19 mm, 2 mm intervals
+19 mm ~ −43 mm, 1 mm intervals

J. S. Park, *Cross-Sectional Atlas of Rhesus Monkey Head*, https://doi.org/10.1007/978-981-19-3254-0_2

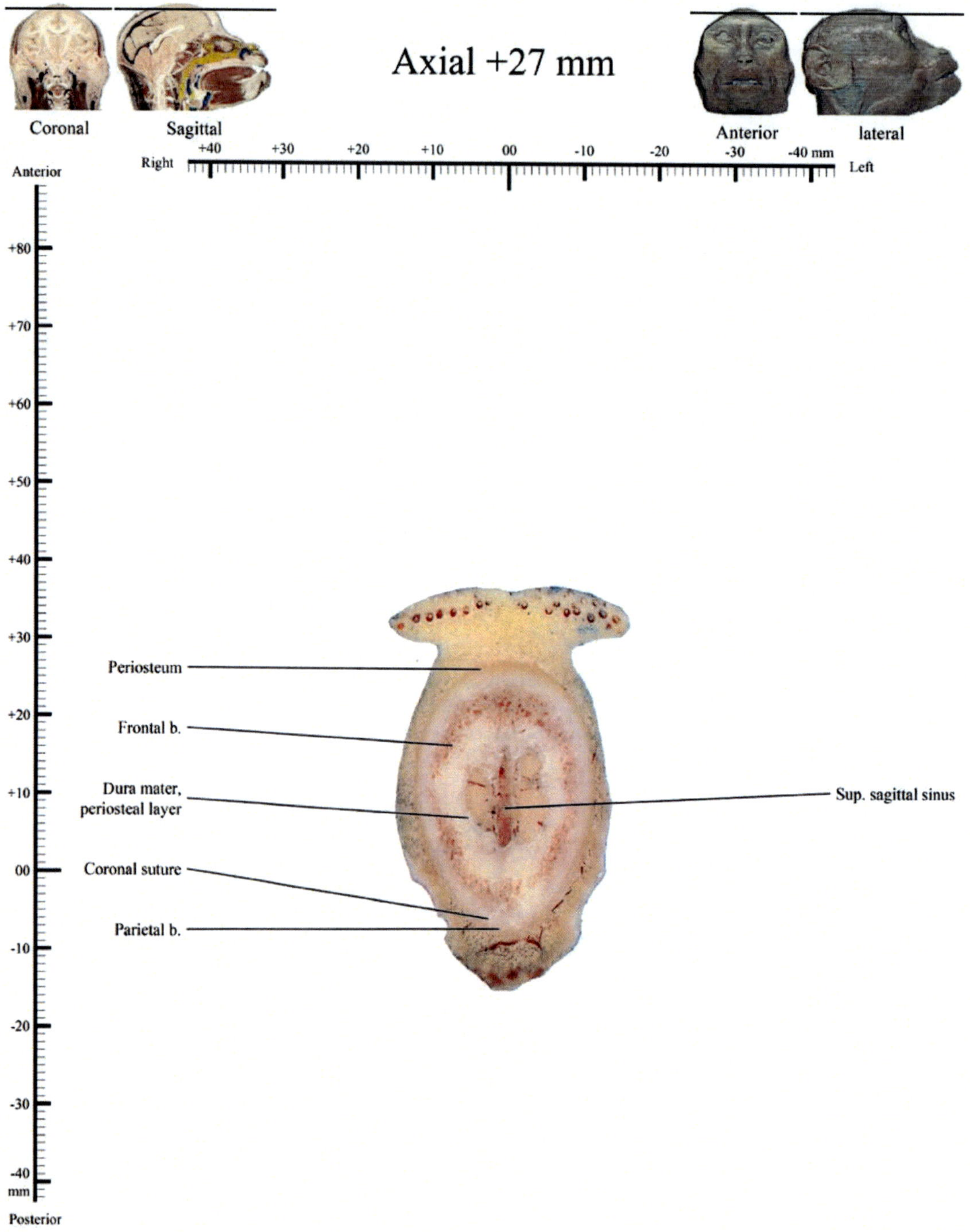
Axial +27 mm
Coronal
Sagittal
Anterior
lateral
Right
+40
+30
+20
+10
00
-10
-20
-30
-40 mm
Left
Anterior
+80
+70
+60
+50
+40
+30
+20
+10
00
-10
-20
-30
-40
mm
Posterior
Periosteum
Frontal b.
Dura mater, periosteal layer
Coronal suture
Parietal b.
Sup. sagittal sinus

Axial +25 mm

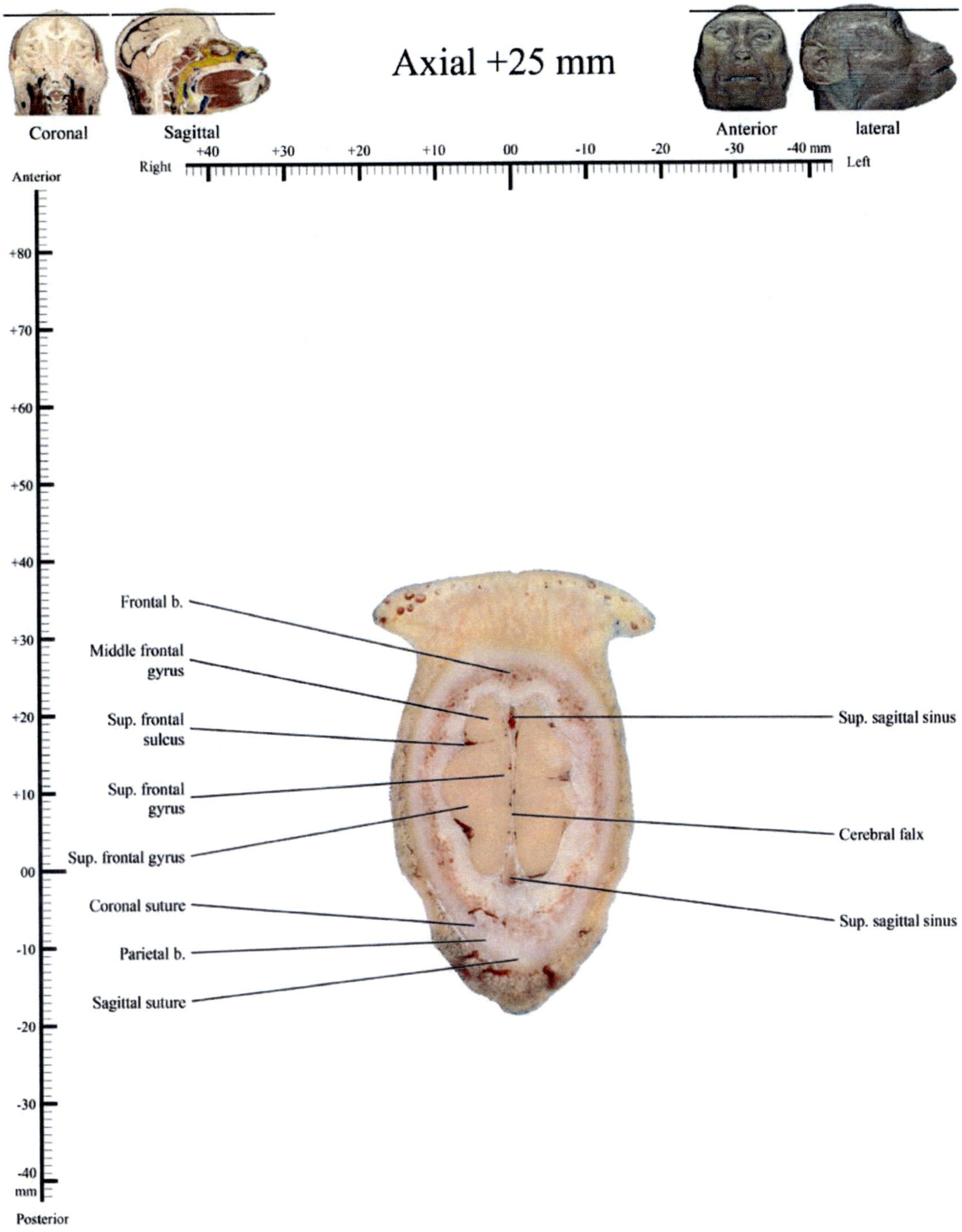
Coronal
Sagittal
Anterior
lateral
Right
+40
+30
+20
+10
00
-10
-20
-30
-40 mm
Left
Anterior
+80
+70
+60
+50
+40
+30
+20
+10
00
-10
-20
-30
-40
mm
Posterior
Frontal b.
Middle frontal gyrus
Sup. frontal sulcus
Sup. frontal gyrus
Sup. frontal gyrus
Coronal suture
Parietal b.
Sagittal suture
Sup. sagittal sinus
Cerebral falx
Sup. sagittal sinus

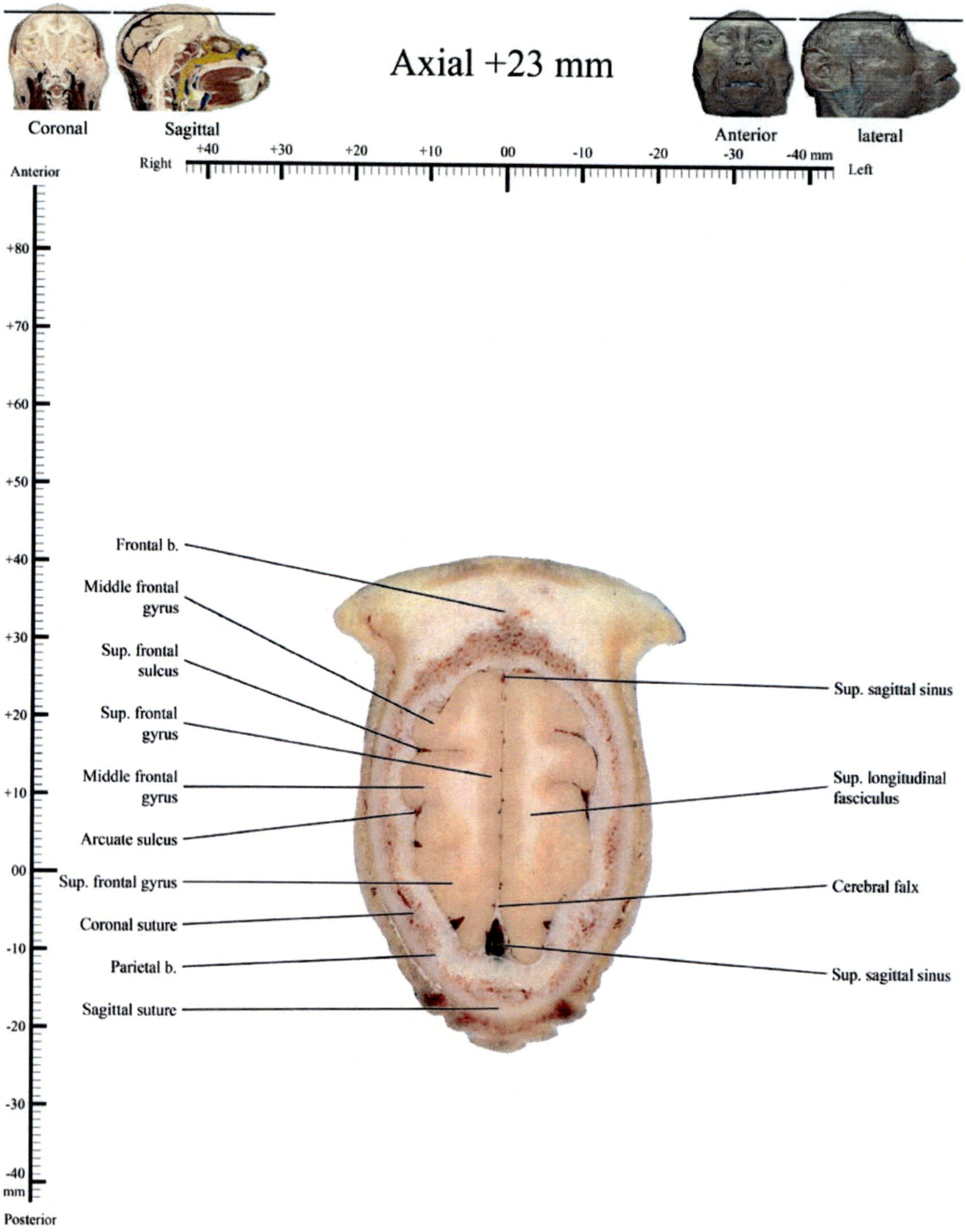
Axial +23 mm
Coronal
Sagittal
Anterior
lateral
Right
+40
+30
+20
+10
00
-10
-20
-30
-40 mm
Left
Anterior
+80
+70
+60
+50
+40
+30
+20
+10
00
-10
-20
-30
-40
mm
Posterior
Frontal b.
Middle frontal gyrus
Sup. frontal sulcus
Sup. frontal gyrus
Middle frontal gyrus
Arcuate sulcus
Sup. frontal gyrus
Coronal suture
Parietal b.
Sagittal suture
Sup. sagittal sinus
Sup. longitudinal fasciculus
Cerebral falx
Sup. sagittal sinus

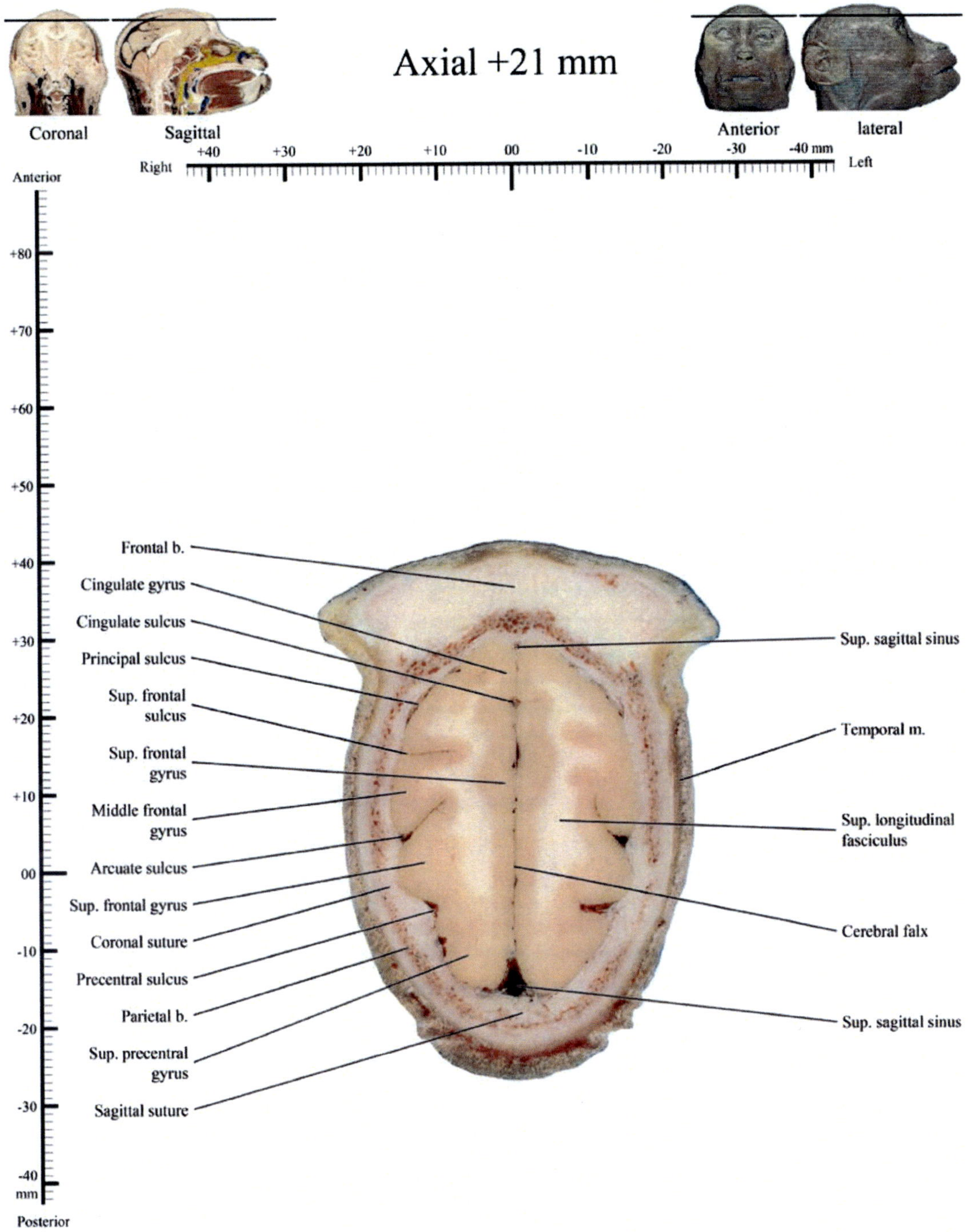
Axial +21 mm
Coronal
Sagittal
Anterior
lateral
Right
+40
+30
+20
+10
00
-10
-20
-30
-40 mm
Left
Anterior
+80
+70
+60
+50
+40
+30
+20
+10
00
-10
-20
-30
-40
mm
Posterior
Frontal b.
Cingulate gyrus
Cingulate sulcus
Principal sulcus
Sup. frontal sulcus
Sup. frontal gyrus
Middle frontal gyrus
Arcuate sulcus
Sup. frontal gyrus
Coronal suture
Precentral sulcus
Parietal b.
Sup. precentral gyrus
Sagittal suture
Sup. sagittal sinus
Temporal m.
Sup. longitudinal fasciculus
Cerebral falx
Sup. sagittal sinus

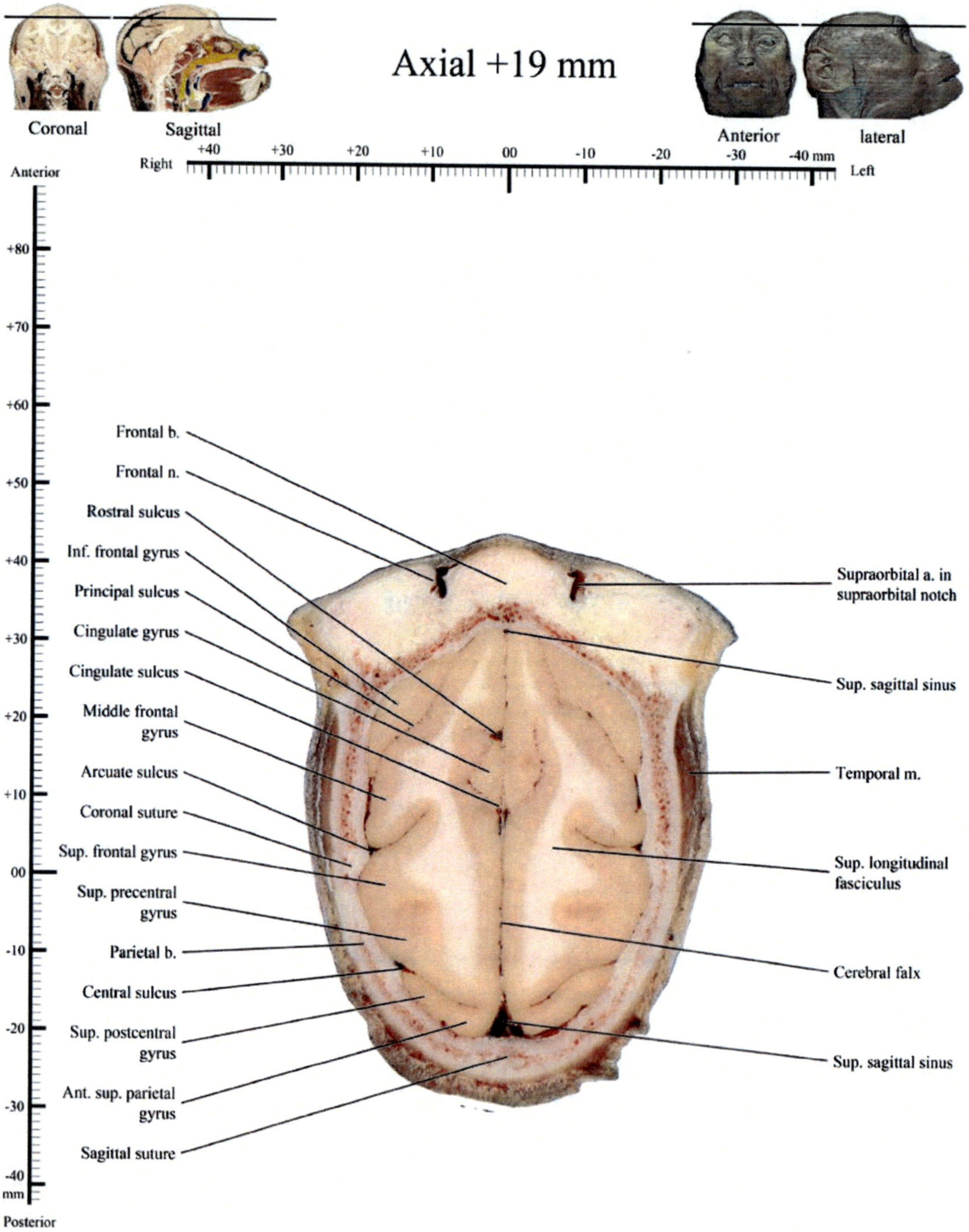
Coronal
Sagittal
Axial +19 mm
Anterior
lateral
Right
+40
+30
+20
+10
00
-10
-20
-30
-40 mm
Left
Anterior
+80
+70
+60
+50
+40
+30
+20
+10
00
-10
-20
-30
-40
mm
Posterior
Frontal b.
Frontal n.
Rostral sulcus
Inf. frontal gyrus
Principal sulcus
Cingulate gyrus
Cingulate sulcus
Middle frontal gyrus
Arcuate sulcus
Coronal suture
Sup. frontal gyrus
Sup. precentral gyrus
Parietal b.
Central sulcus
Sup. postcentral gyrus
Ant. sup. parietal gyrus
Sagittal suture
Supraorbital a. in supraorbital notch
Sup. sagittal sinus
Temporal m.
Sup. longitudinal fasciculus
Cerebral falx
Sup. sagittal sinus

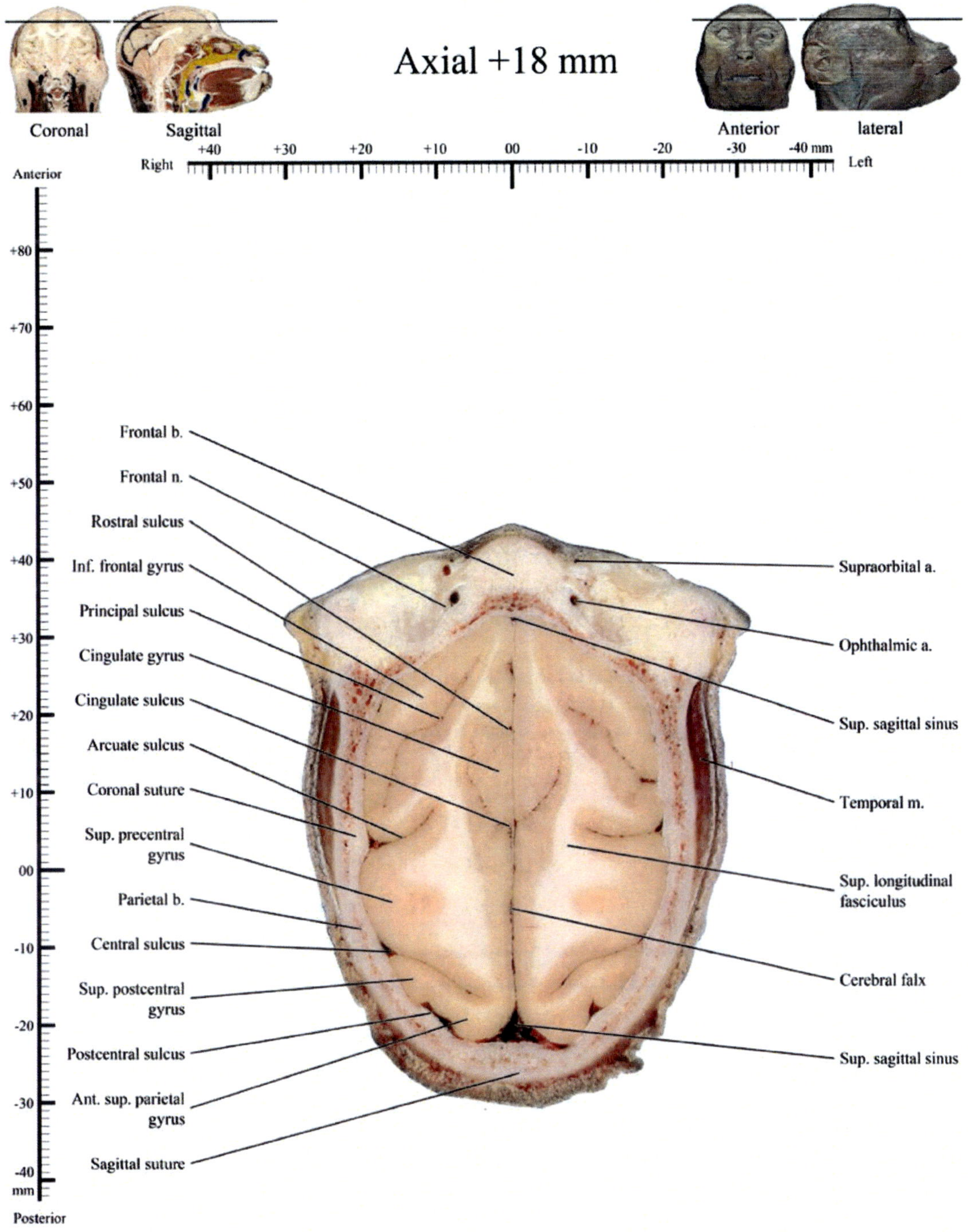
Axial +18 mm
Coronal
Sagittal
Anterior
lateral
Right
+40
+30
+20
+10
00
-10
-20
-30
-40 mm
Left
Anterior
+80
+70
+60
+50
+40
+30
+20
+10
00
-10
-20
-30
-40
mm
Posterior
Frontal b.
Frontal n.
Rostral sulcus
Inf. frontal gyrus
Principal sulcus
Cingulate gyrus
Cingulate sulcus
Arcuate sulcus
Coronal suture
Sup. precentral gyrus
Parietal b.
Central sulcus
Sup. postcentral gyrus
Postcentral sulcus
Ant. sup. parietal gyrus
Sagittal suture
Supraorbital a.
Ophthalmic a.
Sup. sagittal sinus
Temporal m.
Sup. longitudinal fasciculus
Cerebral falx
Sup. sagittal sinus

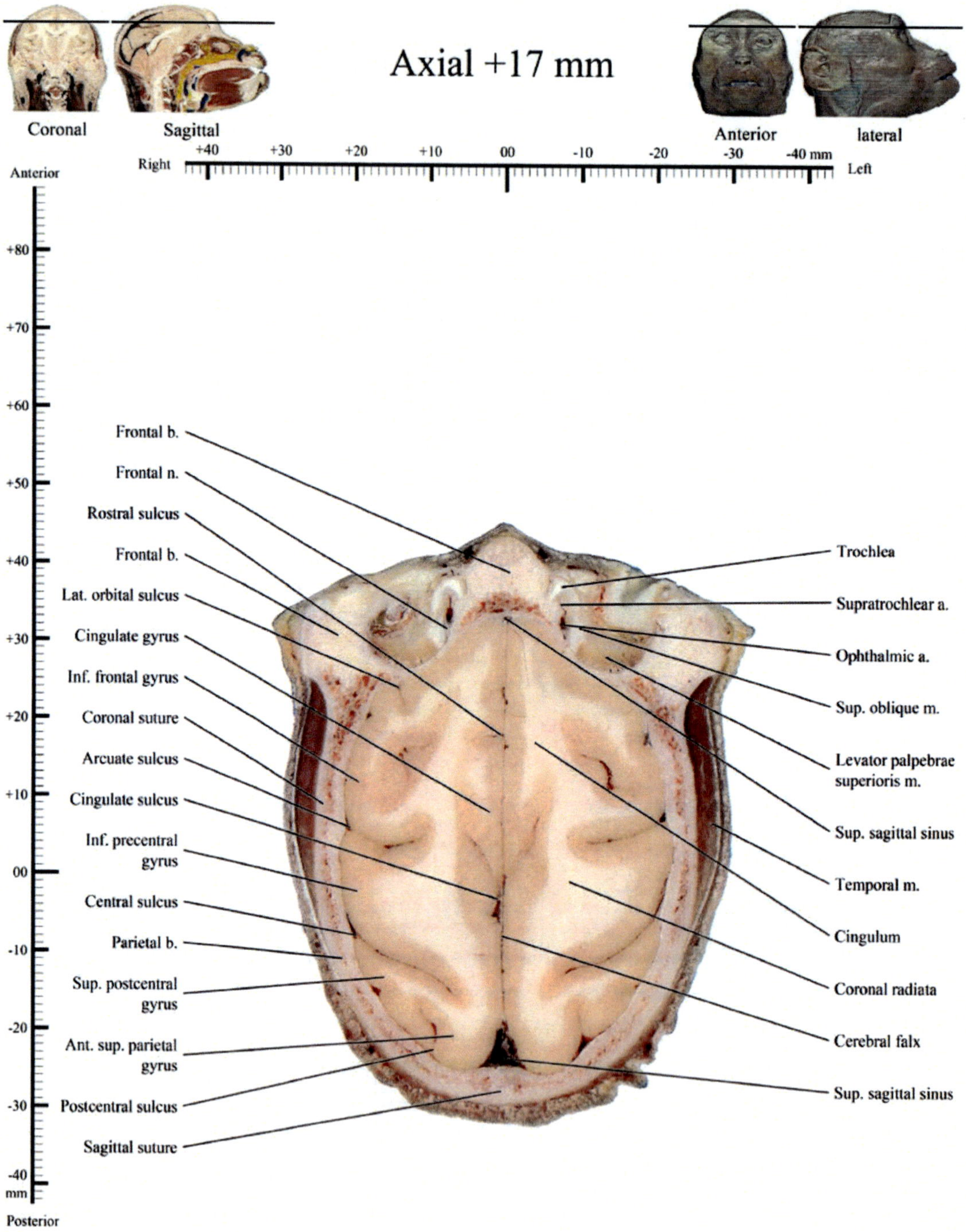
Coronal
Sagittal
Axial +17 mm
Anterior
lateral
Right
+40
+30
+20
+10
00
-10
-20
-30
-40 mm
Left
Anterior
+80
+70
+60
+50
+40
+30
+20
+10
00
-10
-20
-30
-40
mm
Posterior
Frontal b.
Frontal n.
Rostral sulcus
Frontal b.
Lat. orbital sulcus
Cingulate gyrus
Inf. frontal gyrus
Coronal suture
Arcuate sulcus
Cingulate sulcus
Inf. precentral gyrus
Central sulcus
Parietal b.
Sup. postcentral gyrus
Ant. sup. parietal gyrus
Postcentral sulcus
Sagittal suture
Trochlea
Supratrochlear a.
Ophthalmic a.
Sup. oblique m.
Levator palpebrae superioris m.
Sup. sagittal sinus
Temporal m.
Cingulum
Coronal radiata
Cerebral falx
Sup. sagittal sinus

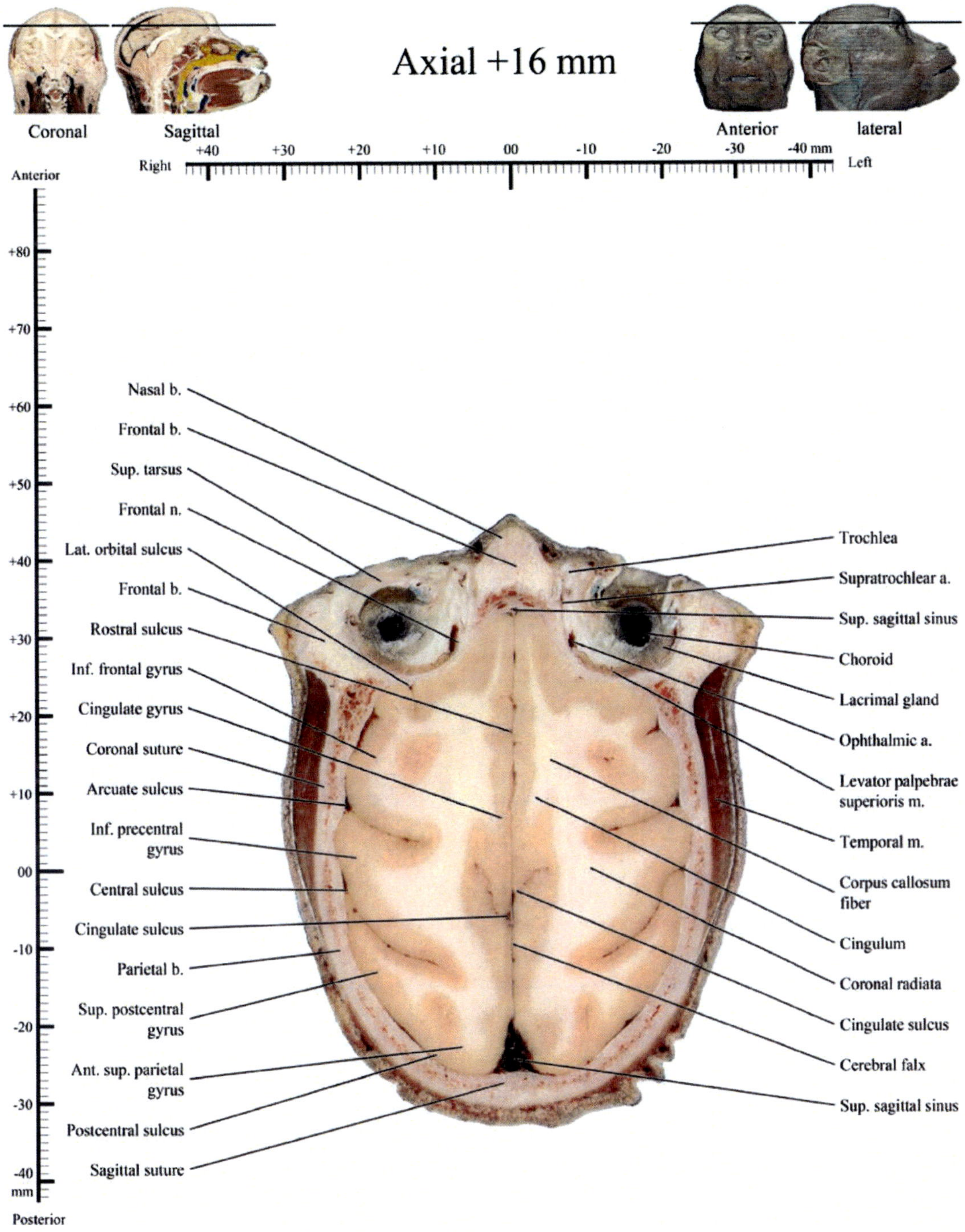

Coronal
Sagittal
Axial +16 mm
Anterior
lateral
+40
+30
+20
+10
00
-10
-20
-30
-40 mm
Right
Left
Anterior
+80
+70
+60
+50
+40
+30
+20
+10
00
-10
-20
-30
-40
mm
Posterior
Nasal b.
Frontal b.
Sup. tarsus
Frontal n.
Lat. orbital sulcus
Frontal b.
Rostral sulcus
Inf. frontal gyrus
Cingulate gyrus
Coronal suture
Arcuate sulcus
Inf. precentral gyrus
Central sulcus
Cingulate sulcus
Parietal b.
Sup. postcentral gyrus
Ant. sup. parietal gyrus
Postcentral sulcus
Sagittal suture
Trochlea
Supratrochlear a.
Sup. sagittal sinus
Choroid
Lacrimal gland
Ophthalmic a.
Levator palpebrae superioris m.
Temporal m.
Corpus callosum fiber
Cingulum
Coronal radiata
Cingulate sulcus
Cerebral falx
Sup. sagittal sinus

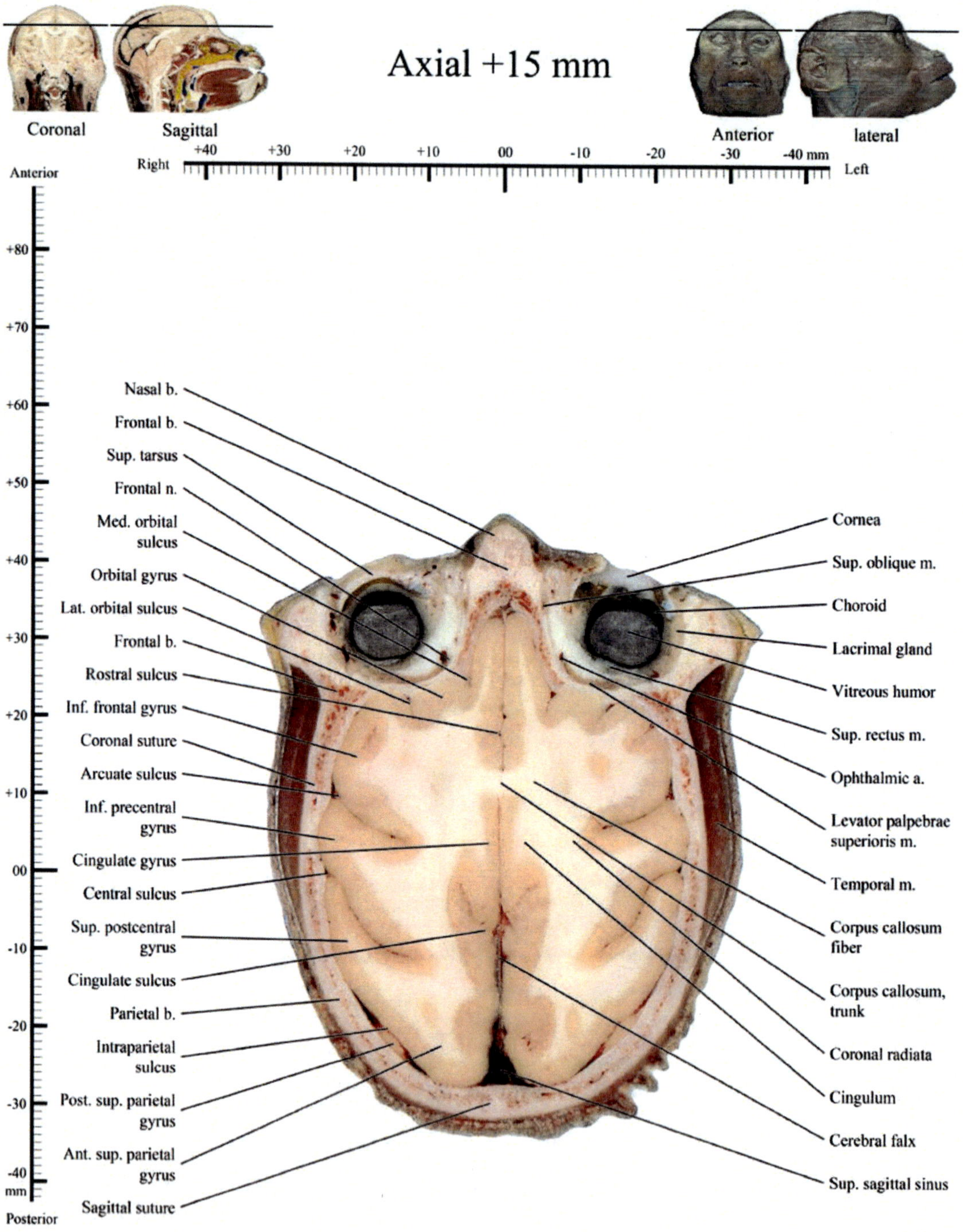
Axial +15 mm
Coronal
Sagittal
Anterior
lateral
Right
Left
+40
+30
+20
+10
00
-10
-20
-30
-40 mm
Anterior
+80
+70
+60
+50
+40
+30
+20
+10
00
-10
-20
-30
-40
mm
Posterior
Nasal b.
Frontal b.
Sup. tarsus
Frontal n.
Med. orbital sulcus
Orbital gyrus
Lat. orbital sulcus
Frontal b.
Rostral sulcus
Inf. frontal gyrus
Coronal suture
Arcuate sulcus
Inf. precentral gyrus
Cingulate gyrus
Central sulcus
Sup. postcentral gyrus
Cingulate sulcus
Parietal b.
Intraparietal sulcus
Post. sup. parietal gyrus
Ant. sup. parietal gyrus
Sagittal suture
Cornea
Sup. oblique m.
Choroid
Lacrimal gland
Vitreous humor
Sup. rectus m.
Ophthalmic a.
Levator palpebrae superioris m.
Temporal m.
Corpus callosum fiber
Corpus callosum, trunk
Coronal radiata
Cingulum
Cerebral falx
Sup. sagittal sinus

Axial +14 mm

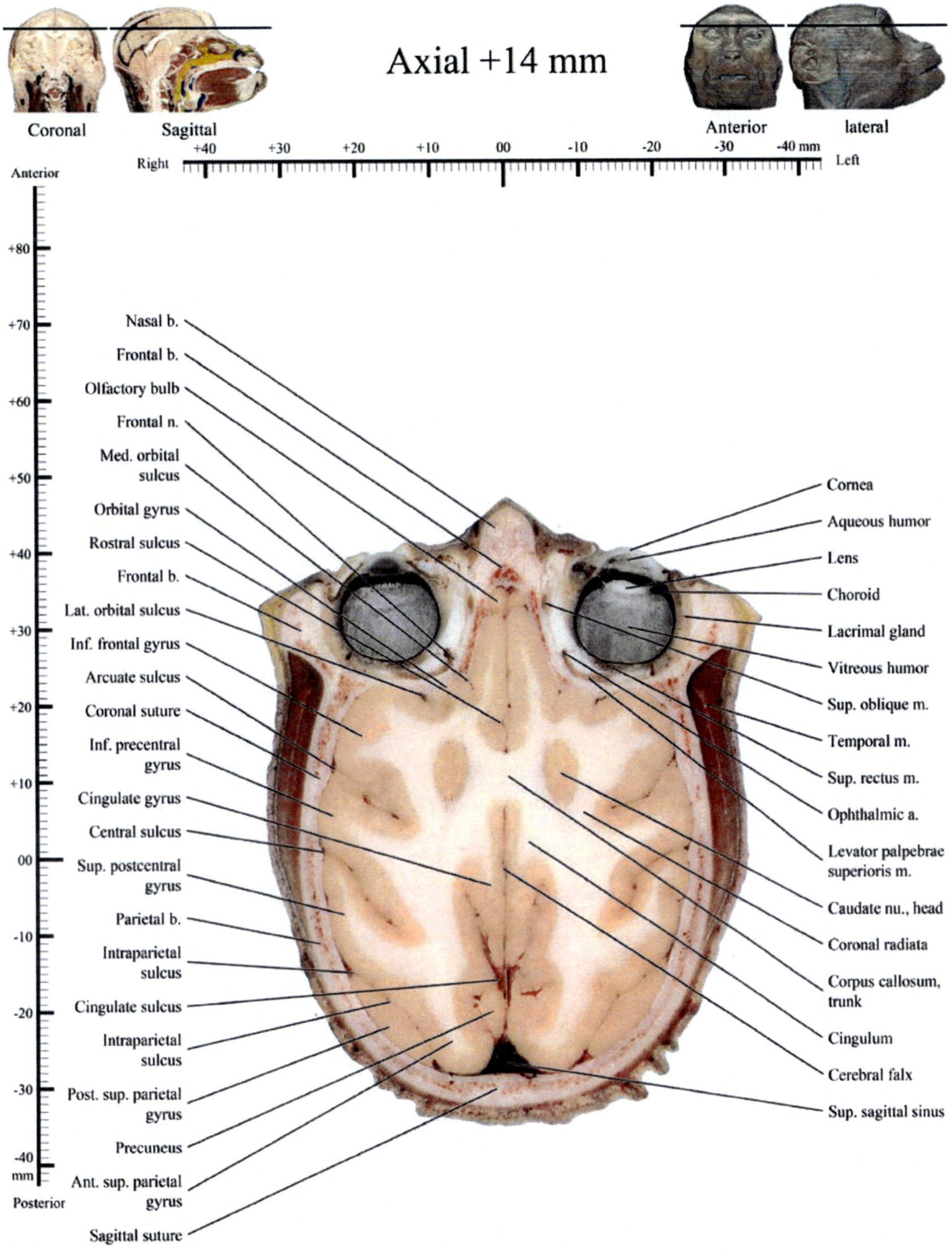

Coronal
Sagittal
Anterior
lateral
Right
Left
+40
+30
+20
+10
00
-10
-20
-30
-40 mm
Anterior
+80
+70
+60
+50
+40
+30
+20
+10
00
-10
-20
-30
-40
mm
Posterior
Nasal b.
Frontal b.
Olfactory bulb
Frontal n.
Med. orbital sulcus
Orbital gyrus
Rostral sulcus
Frontal b.
Lat. orbital sulcus
Inf. frontal gyrus
Arcuate sulcus
Coronal suture
Inf. precentral gyrus
Cingulate gyrus
Central sulcus
Sup. postcentral gyrus
Parietal b.
Intraparietal sulcus
Cingulate sulcus
Intraparietal sulcus
Post. sup. parietal gyrus
Precuneus
Ant. sup. parietal gyrus
Sagittal suture
Cornea
Aqueous humor
Lens
Choroid
Lacrimal gland
Vitreous humor
Sup. oblique m.
Temporal m.
Sup. rectus m.
Ophthalmic a.
Levator palpebrae superioris m.
Caudate nu., head
Coronal radiata
Corpus callosum, trunk
Cingulum
Cerebral falx
Sup. sagittal sinus

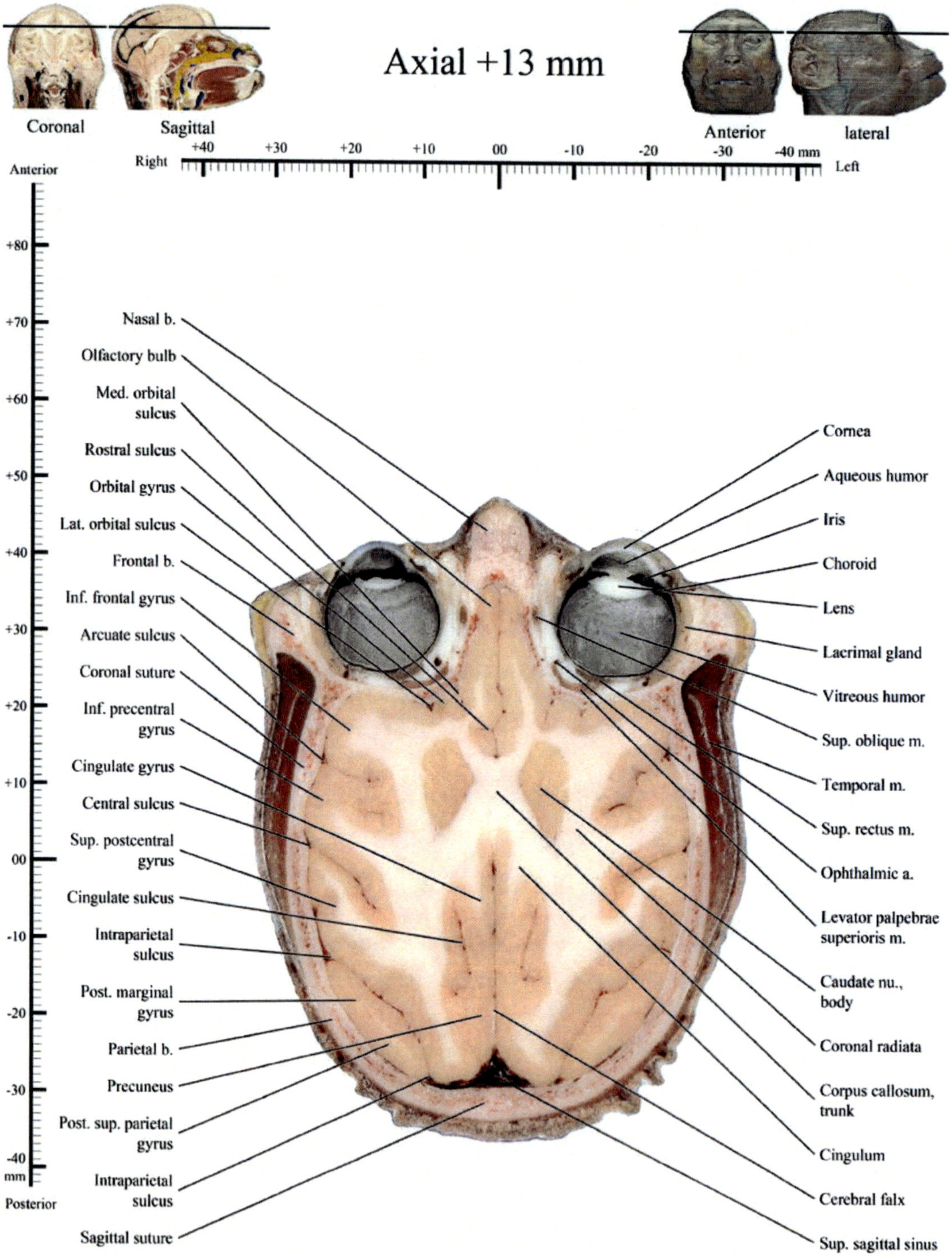
Axial +13 mm
Coronal
Sagittal
Anterior
lateral
Right
Left
+40
+30
+20
+10
00
-10
-20
-30
-40 mm
Anterior
+80
+70
+60
+50
+40
+30
+20
+10
00
-10
-20
-30
-40
mm
Posterior
Nasal b.
Olfactory bulb
Med. orbital sulcus
Rostral sulcus
Orbital gyrus
Lat. orbital sulcus
Frontal b.
Inf. frontal gyrus
Arcuate sulcus
Coronal suture
Inf. precentral gyrus
Cingulate gyrus
Central sulcus
Sup. postcentral gyrus
Cingulate sulcus
Intraparietal sulcus
Post. marginal gyrus
Parietal b.
Precuneus
Post. sup. parietal gyrus
Intraparietal sulcus
Sagittal suture
Cornea
Aqueous humor
Iris
Choroid
Lens
Lacrimal gland
Vitreous humor
Sup. oblique m.
Temporal m.
Sup. rectus m.
Ophthalmic a.
Levator palpebrae superioris m.
Caudate nu., body
Coronal radiata
Corpus callosum, trunk
Cingulum
Cerebral falx
Sup. sagittal sinus

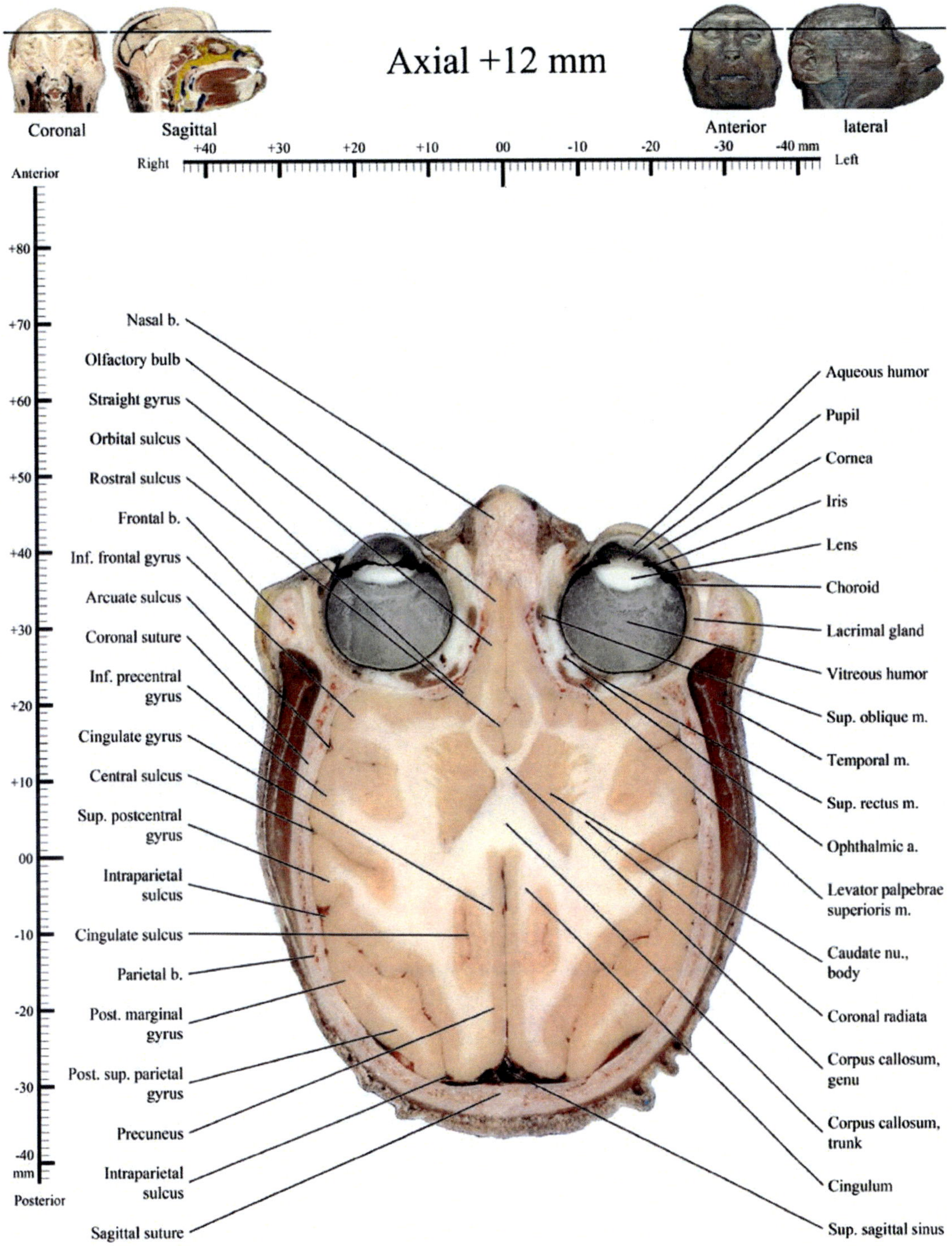
Axial +12 mm
Coronal
Sagittal
Anterior
lateral
Right
Left
+40 +30 +20 +10 00 -10 -20 -30 -40 mm
Anterior
+80 +70 +60 +50 +40 +30 +20 +10 00 -10 -20 -30 -40 mm
Posterior
Nasal b.
Olfactory bulb
Straight gyrus
Orbital sulcus
Rostral sulcus
Frontal b.
Inf. frontal gyrus
Arcuate sulcus
Coronal suture
Inf. precentral gyrus
Cingulate gyrus
Central sulcus
Sup. postcentral gyrus
Intraparietal sulcus
Cingulate sulcus
Parietal b.
Post. marginal gyrus
Post. sup. parietal gyrus
Precuneus
Intraparietal sulcus
Sagittal suture
Aqueous humor
Pupil
Cornea
Iris
Lens
Choroid
Lacrimal gland
Vitreous humor
Sup. oblique m.
Temporal m.
Sup. rectus m.
Ophthalmic a.
Levator palpebrae superioris m.
Caudate nu., body
Coronal radiata
Corpus callosum, genu
Corpus callosum, trunk
Cingulum
Sup. sagittal sinus

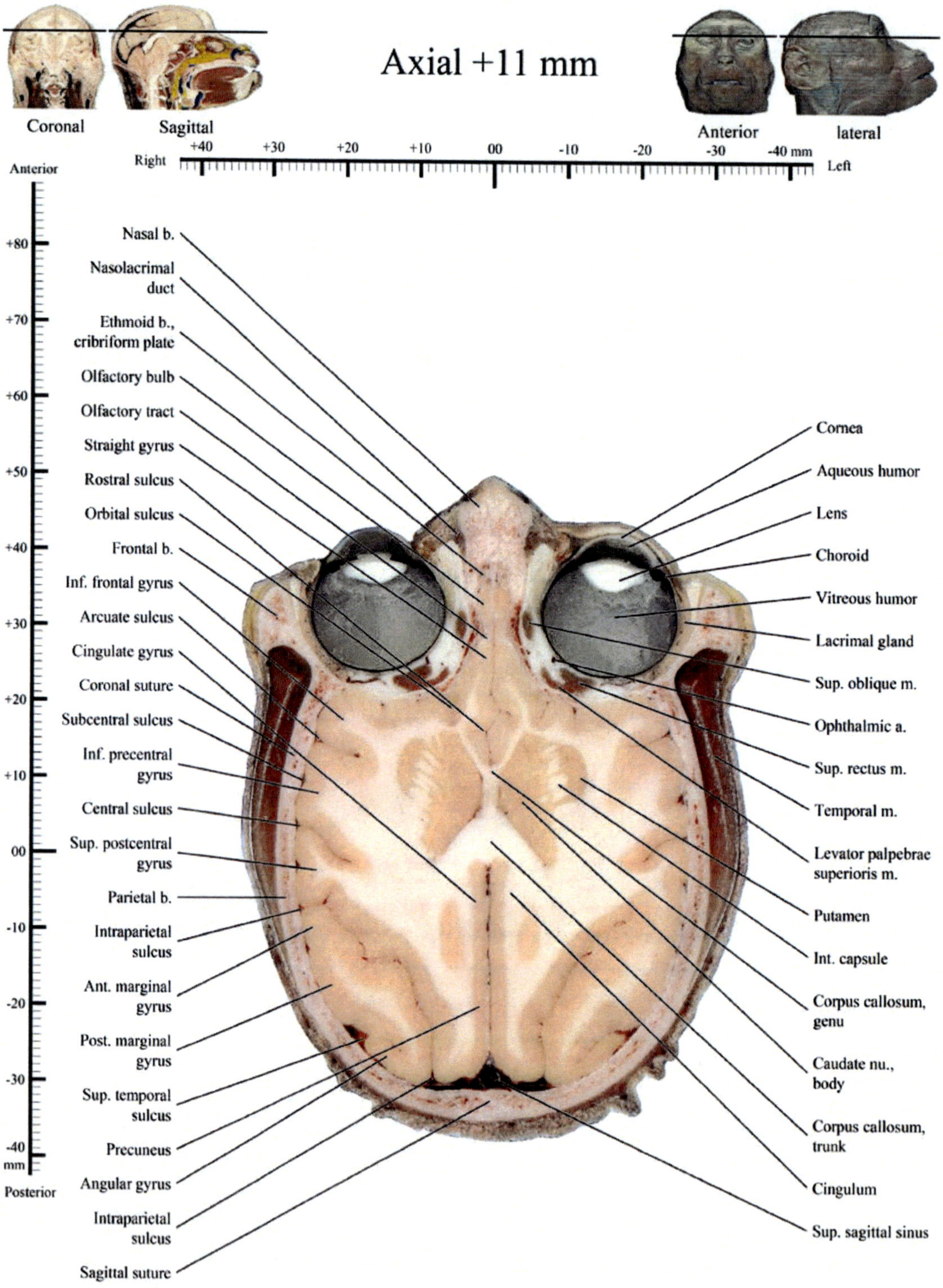
Coronal
Sagittal
Axial +11 mm
Anterior
lateral
Right
+40
+30
+20
+10
00
-10
-20
-30
-40 mm
Left
Anterior
+80
+70
+60
+50
+40
+30
+20
+10
00
-10
-20
-30
-40 mm
Posterior
Nasal b.
Nasolacrimal duct
Ethmoid b., cribriform plate
Olfactory bulb
Olfactory tract
Straight gyrus
Rostral sulcus
Orbital sulcus
Frontal b.
Inf. frontal gyrus
Arcuate sulcus
Cingulate gyrus
Coronal suture
Subcentral sulcus
Inf. precentral gyrus
Central sulcus
Sup. postcentral gyrus
Parietal b.
Intraparietal sulcus
Ant. marginal gyrus
Post. marginal gyrus
Sup. temporal sulcus
Precuneus
Angular gyrus
Intraparietal sulcus
Sagittal suture
Cornea
Aqueous humor
Lens
Choroid
Vitreous humor
Lacrimal gland
Sup. oblique m.
Ophthalmic a.
Sup. rectus m.
Temporal m.
Levator palpebrae superioris m.
Putamen
Int. capsule
Corpus callosum, genu
Caudate nu., body
Corpus callosum, trunk
Cingulum
Sup. sagittal sinus

Axial +10 mm

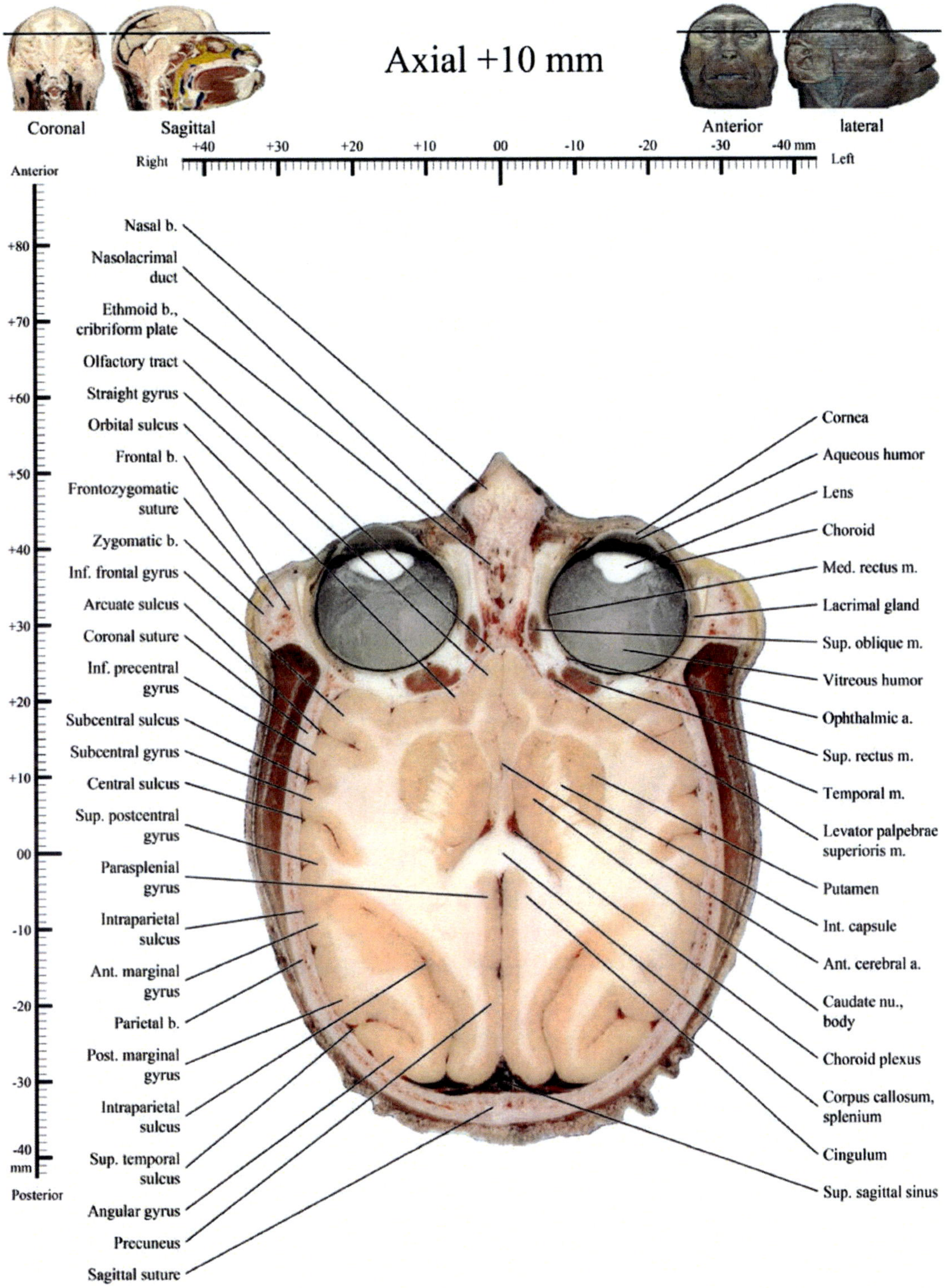
Coronal
Sagittal
Anterior
lateral
Right
Left
+40
+30
+20
+10
00
-10
-20
-30
-40 mm
Anterior
+80
+70
+60
+50
+40
+30
+20
+10
00
-10
-20
-30
-40 mm
Posterior
Nasal b.
Nasolacrimal duct
Ethmoid b., cribriform plate
Olfactory tract
Straight gyrus
Orbital sulcus
Frontal b.
Frontozygomatic suture
Zygomatic b.
Inf. frontal gyrus
Arcuate sulcus
Coronal suture
Inf. precentral gyrus
Subcentral sulcus
Subcentral gyrus
Central sulcus
Sup. postcentral gyrus
Parasplenial gyrus
Intraparietal sulcus
Ant. marginal gyrus
Parietal b.
Post. marginal gyrus
Intraparietal sulcus
Sup. temporal sulcus
Angular gyrus
Precuneus
Sagittal suture
Cornea
Aqueous humor
Lens
Choroid
Med. rectus m.
Lacrimal gland
Sup. oblique m.
Vitreous humor
Ophthalmic a.
Sup. rectus m.
Temporal m.
Levator palpebrae superioris m.
Putamen
Int. capsule
Ant. cerebral a.
Caudate nu., body
Choroid plexus
Corpus callosum, splenium
Cingulum
Sup. sagittal sinus

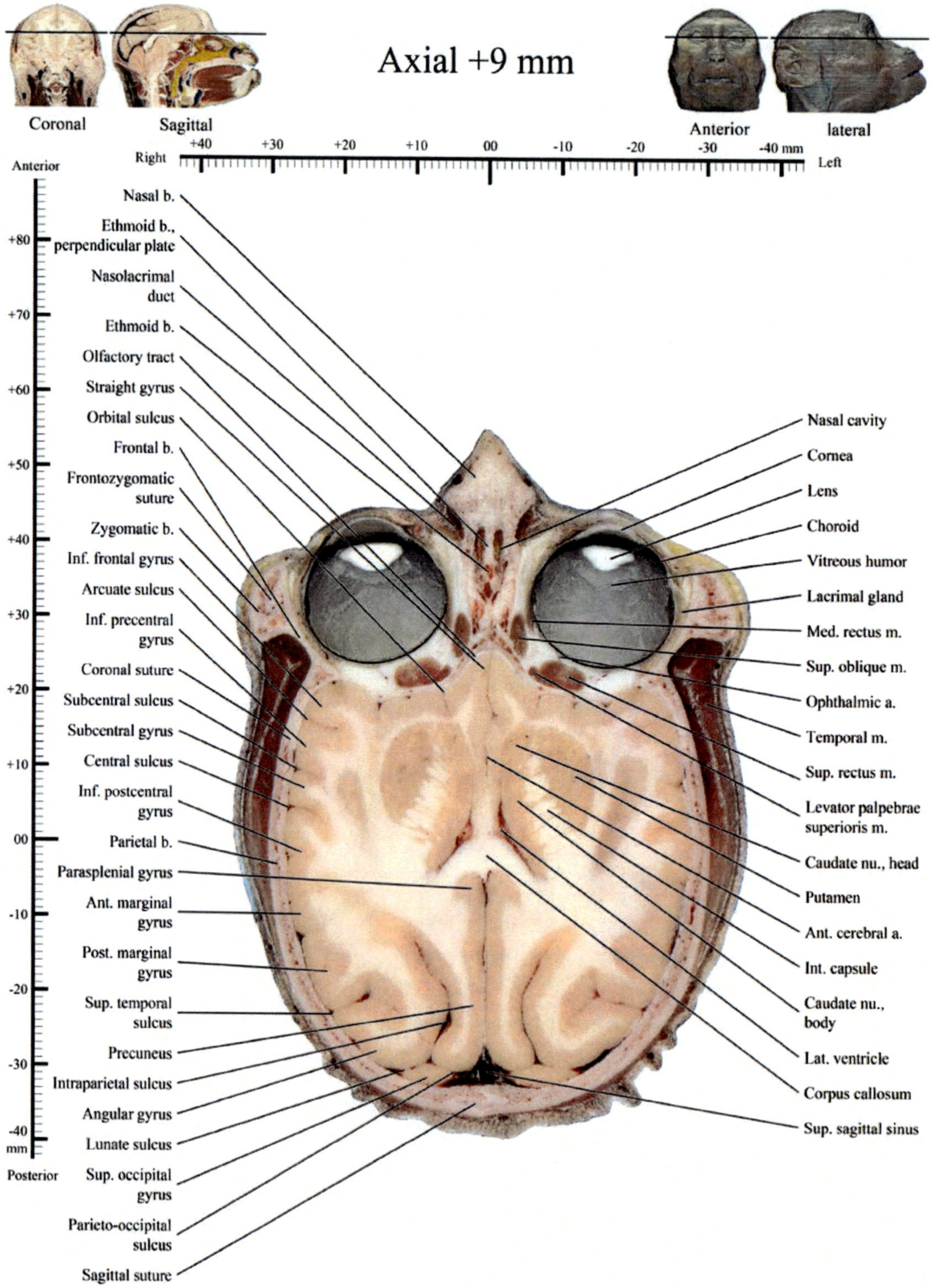
Coronal
Sagittal
Axial +9 mm
Anterior
lateral
Right
+40
+30
+20
+10
00
-10
-20
-30
-40 mm
Left
Anterior
+80
+70
+60
+50
+40
+30
+20
+10
00
-10
-20
-30
-40
mm
Posterior
Nasal b.
Ethmoid b., perpendicular plate
Nasolacrimal duct
Ethmoid b.
Olfactory tract
Straight gyrus
Orbital sulcus
Frontal b.
Frontozygomatic suture
Zygomatic b.
Inf. frontal gyrus
Arcuate sulcus
Inf. precentral gyrus
Coronal suture
Subcentral sulcus
Subcentral gyrus
Central sulcus
Inf. postcentral gyrus
Parietal b.
Parasplenial gyrus
Ant. marginal gyrus
Post. marginal gyrus
Sup. temporal sulcus
Precuneus
Intraparietal sulcus
Angular gyrus
Lunate sulcus
Sup. occipital gyrus
Parieto-occipital sulcus
Sagittal suture
Nasal cavity
Cornea
Lens
Choroid
Vitreous humor
Lacrimal gland
Med. rectus m.
Sup. oblique m.
Ophthalmic a.
Temporal m.
Sup. rectus m.
Levator palpebrae superioris m.
Caudate nu., head
Putamen
Ant. cerebral a.
Int. capsule
Caudate nu., body
Lat. ventricle
Corpus callosum
Sup. sagittal sinus

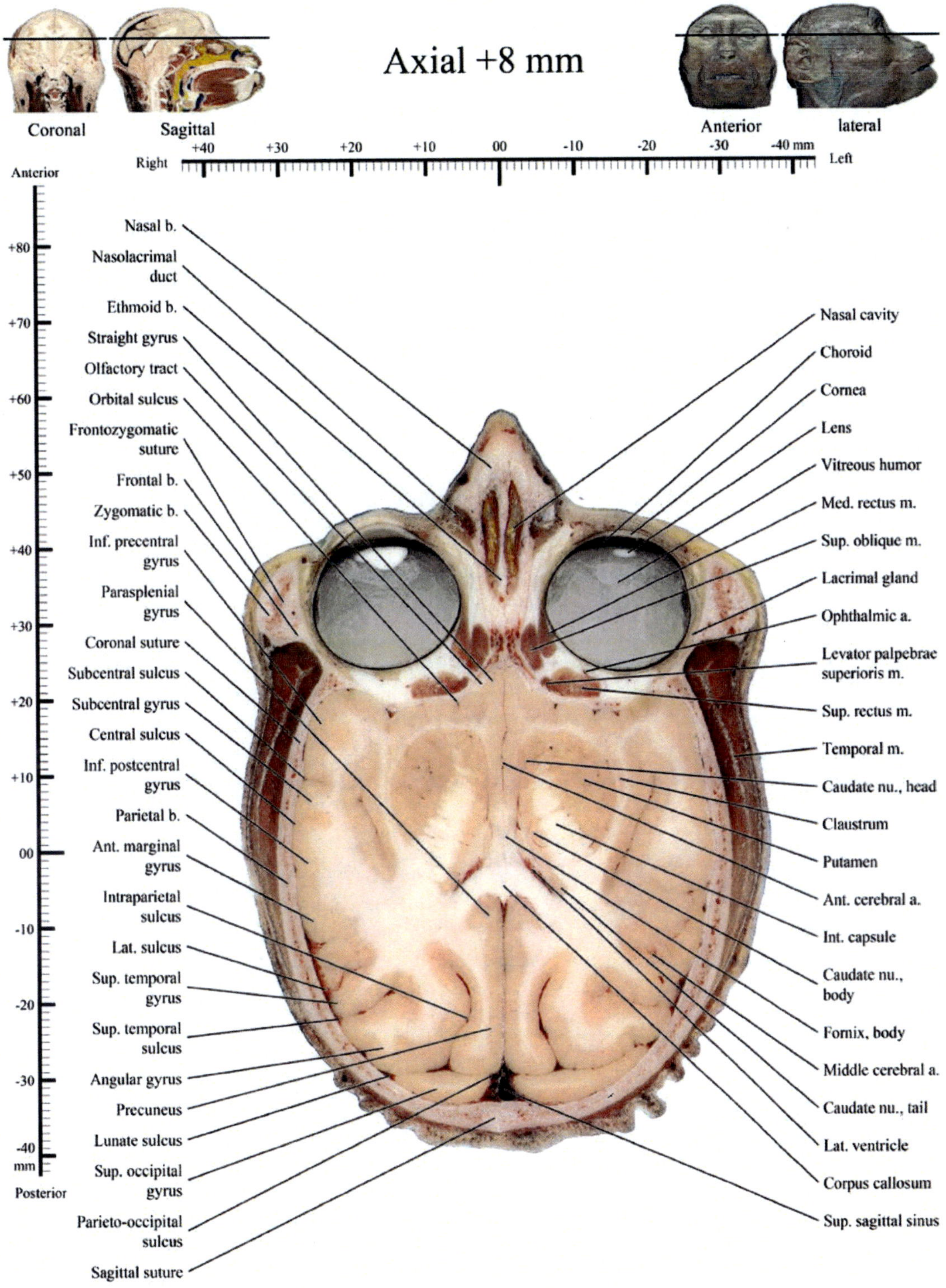
Coronal
Sagittal
Axial +8 mm
Anterior
lateral
Right
+40 +30 +20 +10 00 -10 -20 -30 -40 mm
Left
Anterior
+80 +70 +60 +50 +40 +30 +20 +10 00 -10 -20 -30 -40 mm
Posterior
Nasal b.
Nasolacrimal duct
Ethmoid b.
Straight gyrus
Olfactory tract
Orbital sulcus
Frontozygomatic suture
Frontal b.
Zygomatic b.
Inf. precentral gyrus
Parasplenial gyrus
Coronal suture
Subcentral sulcus
Subcentral gyrus
Central sulcus
Inf. postcentral gyrus
Parietal b.
Ant. marginal gyrus
Intraparietal sulcus
Lat. sulcus
Sup. temporal gyrus
Sup. temporal sulcus
Angular gyrus
Precuneus
Lunate sulcus
Sup. occipital gyrus
Parieto-occipital sulcus
Sagittal suture
Nasal cavity
Choroid
Cornea
Lens
Vitreous humor
Med. rectus m.
Sup. oblique m.
Lacrimal gland
Ophthalmic a.
Levator palpebrae superioris m.
Sup. rectus m.
Temporal m.
Caudate nu., head
Claustrum
Putamen
Ant. cerebral a.
Int. capsule
Caudate nu., body
Fornix, body
Middle cerebral a.
Caudate nu., tail
Lat. ventricle
Corpus callosum
Sup. sagittal sinus

Axial +7 mm

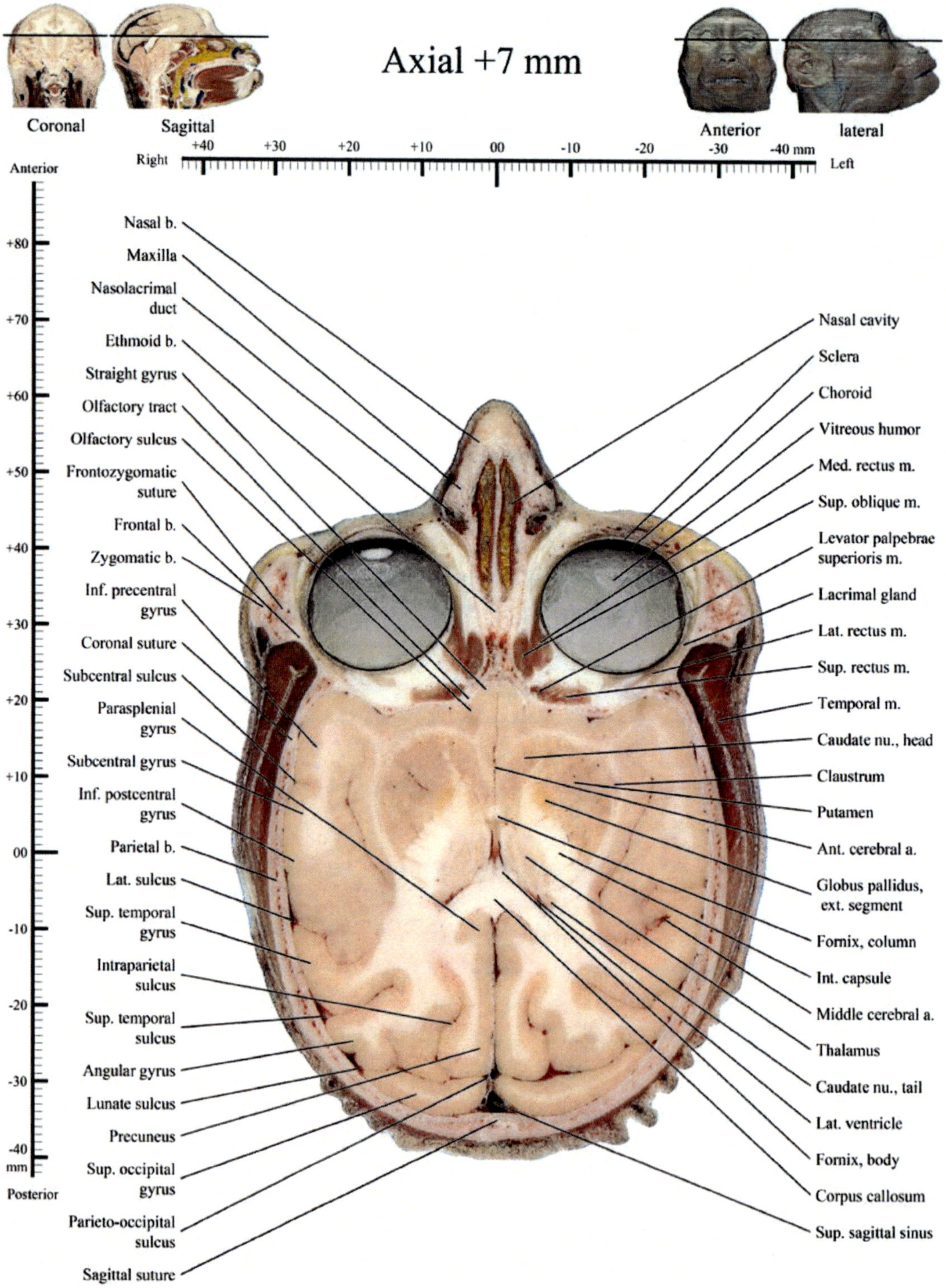
Coronal
Sagittal
Anterior
lateral
Right
+40
+30
+20
+10
00
-10
-20
-30
-40 mm
Left
Anterior
+80
+70
+60
+50
+40
+30
+20
+10
00
-10
-20
-30
-40
mm
Posterior
Nasal b.
Maxilla
Nasolacrimal duct
Ethmoid b.
Straight gyrus
Olfactory tract
Olfactory sulcus
Frontozygomatic suture
Frontal b.
Zygomatic b.
Inf. precentral gyrus
Coronal suture
Subcentral sulcus
Parasplenial gyrus
Subcentral gyrus
Inf. postcentral gyrus
Parietal b.
Lat. sulcus
Sup. temporal gyrus
Intraparietal sulcus
Sup. temporal sulcus
Angular gyrus
Lunate sulcus
Precuneus
Sup. occipital gyrus
Parieto-occipital sulcus
Sagittal suture
Nasal cavity
Sclera
Choroid
Vitreous humor
Med. rectus m.
Sup. oblique m.
Levator palpebrae superioris m.
Lacrimal gland
Lat. rectus m.
Sup. rectus m.
Temporal m.
Caudate nu., head
Claustrum
Putamen
Ant. cerebral a.
Globus pallidus, ext. segment
Fornix, column
Int. capsule
Middle cerebral a.
Thalamus
Caudate nu., tail
Lat. ventricle
Fornix, body
Corpus callosum
Sup. sagittal sinus

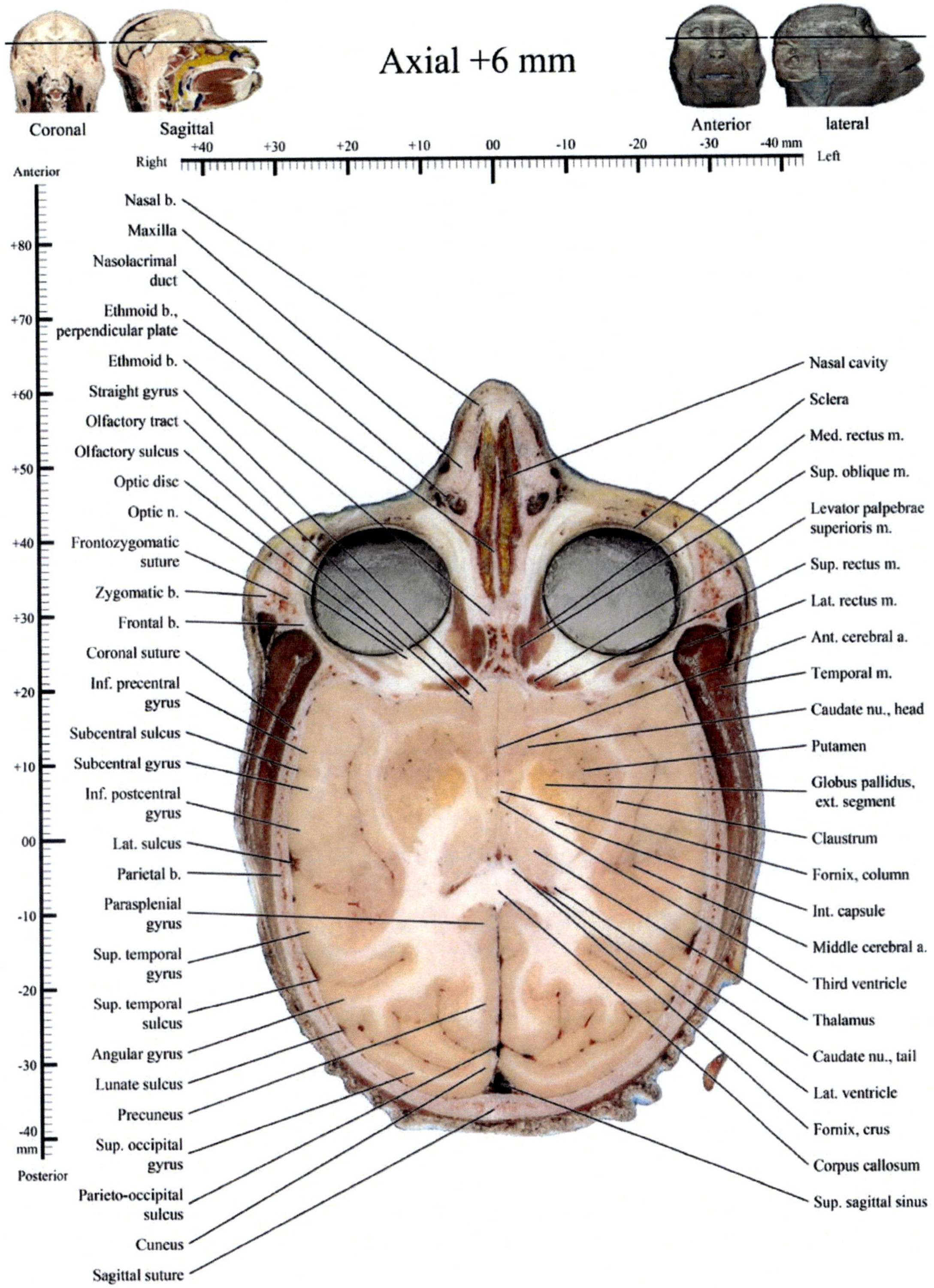
Axial +6 mm
Coronal
Sagittal
Anterior
lateral
Right
Left
+40
+30
+20
+10
00
-10
-20
-30
-40 mm
Anterior
+80
+70
+60
+50
+40
+30
+20
+10
00
-10
-20
-30
-40 mm
Posterior
Nasal b.
Maxilla
Nasolacrimal duct
Ethmoid b., perpendicular plate
Ethmoid b.
Straight gyrus
Olfactory tract
Olfactory sulcus
Optic disc
Optic n.
Frontozygomatic suture
Zygomatic b.
Frontal b.
Coronal suture
Inf. precentral gyrus
Subcentral sulcus
Subcentral gyrus
Inf. postcentral gyrus
Lat. sulcus
Parietal b.
Parasplenial gyrus
Sup. temporal gyrus
Sup. temporal sulcus
Angular gyrus
Lunate sulcus
Precuneus
Sup. occipital gyrus
Parieto-occipital sulcus
Cuneus
Sagittal suture
Nasal cavity
Sclera
Med. rectus m.
Sup. oblique m.
Levator palpebrae superioris m.
Sup. rectus m.
Lat. rectus m.
Ant. cerebral a.
Temporal m.
Caudate nu., head
Putamen
Globus pallidus, ext. segment
Claustrum
Fornix, column
Int. capsule
Middle cerebral a.
Third ventricle
Thalamus
Caudate nu., tail
Lat. ventricle
Fornix, crus
Corpus callosum
Sup. sagittal sinus

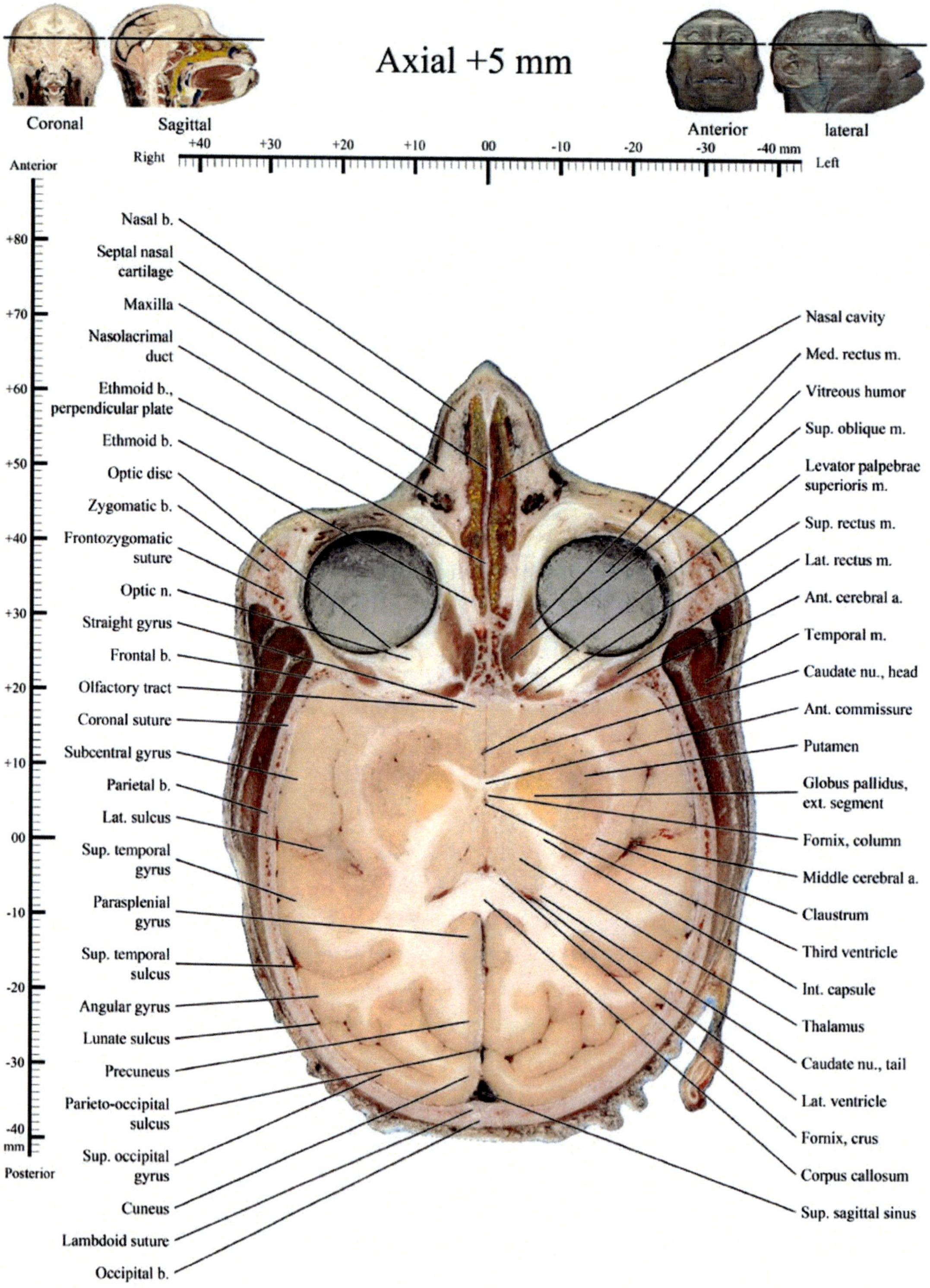
Axial +5 mm
Coronal
Sagittal
Anterior
lateral
Right
Left
+40
+30
+20
+10
00
-10
-20
-30
-40 mm
Anterior
+80
+70
+60
+50
+40
+30
+20
+10
00
-10
-20
-30
-40
mm
Posterior
Nasal b.
Septal nasal cartilage
Maxilla
Nasolacrimal duct
Ethmoid b., perpendicular plate
Ethmoid b.
Optic disc
Zygomatic b.
Frontozygomatic suture
Optic n.
Straight gyrus
Frontal b.
Olfactory tract
Coronal suture
Subcentral gyrus
Parietal b.
Lat. sulcus
Sup. temporal gyrus
Parasplenial gyrus
Sup. temporal sulcus
Angular gyrus
Lunate sulcus
Precuneus
Parieto-occipital sulcus
Sup. occipital gyrus
Cuneus
Lambdoid suture
Occipital b.
Nasal cavity
Med. rectus m.
Vitreous humor
Sup. oblique m.
Levator palpebrae superioris m.
Sup. rectus m.
Lat. rectus m.
Ant. cerebral a.
Temporal m.
Caudate nu., head
Ant. commissure
Putamen
Globus pallidus, ext. segment
Fornix, column
Middle cerebral a.
Claustrum
Third ventricle
Int. capsule
Thalamus
Caudate nu., tail
Lat. ventricle
Fornix, crus
Corpus callosum
Sup. sagittal sinus

Axial +4 mm

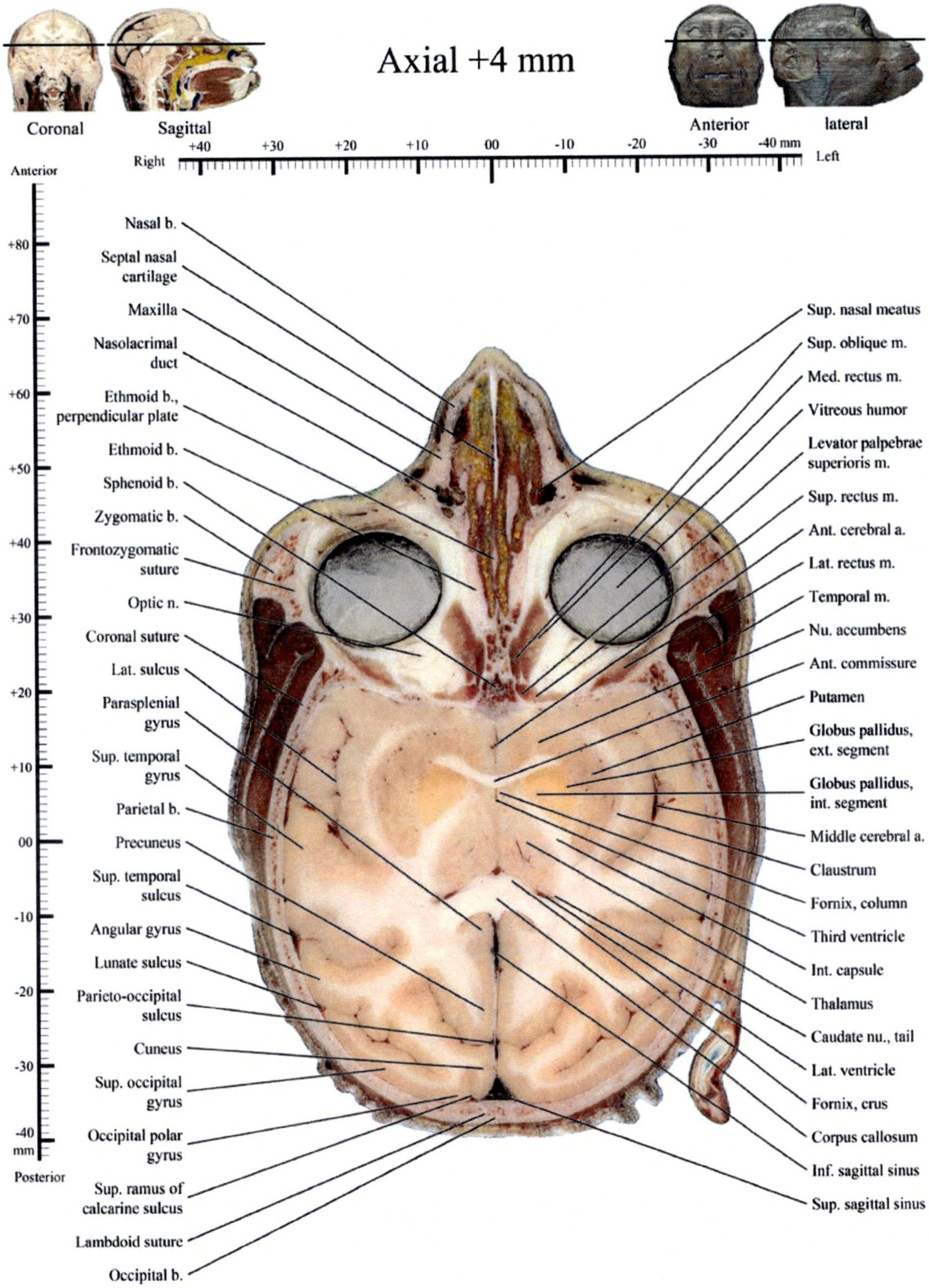

Coronal
Sagittal
Anterior
lateral
Right
Left
+40
+30
+20
+10
00
-10
-20
-30
-40 mm
Anterior
+80
+70
+60
+50
+40
+30
+20
+10
00
-10
-20
-30
-40
mm
Posterior
Nasal b.
Septal nasal cartilage
Maxilla
Nasolacrimal duct
Ethmoid b., perpendicular plate
Ethmoid b.
Sphenoid b.
Zygomatic b.
Frontozygomatic suture
Optic n.
Coronal suture
Lat. sulcus
Parasplenial gyrus
Sup. temporal gyrus
Parietal b.
Precuneus
Sup. temporal sulcus
Angular gyrus
Lunate sulcus
Parieto-occipital sulcus
Cuneus
Sup. occipital gyrus
Occipital polar gyrus
Sup. ramus of calcarine sulcus
Lambdoid suture
Occipital b.
Sup. nasal meatus
Sup. oblique m.
Med. rectus m.
Vitreous humor
Levator palpebrae superioris m.
Sup. rectus m.
Ant. cerebral a.
Lat. rectus m.
Temporal m.
Nu. accumbens
Ant. commissure
Putamen
Globus pallidus, ext. segment
Globus pallidus, int. segment
Middle cerebral a.
Claustrum
Fornix, column
Third ventricle
Int. capsule
Thalamus
Caudate nu., tail
Lat. ventricle
Fornix, crus
Corpus callosum
Inf. sagittal sinus
Sup. sagittal sinus

Axial +3 mm

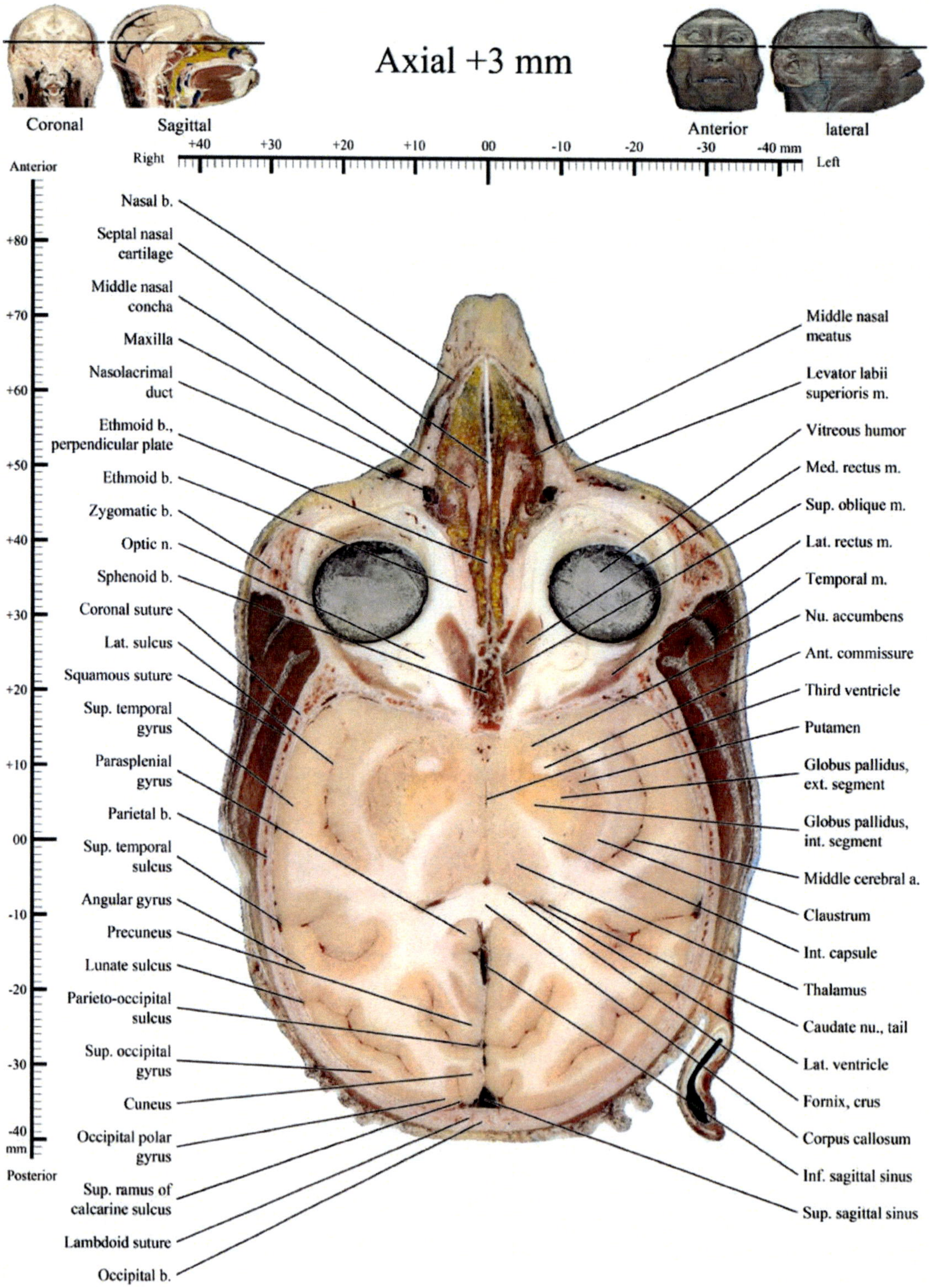
Coronal
Sagittal
Anterior
lateral
Right
Left
+40
+30
+20
+10
00
-10
-20
-30
-40 mm
Anterior
+80
+70
+60
+50
+40
+30
+20
+10
00
-10
-20
-30
-40
mm
Posterior
Nasal b.
Septal nasal cartilage
Middle nasal concha
Maxilla
Nasolacrimal duct
Ethmoid b., perpendicular plate
Ethmoid b.
Zygomatic b.
Optic n.
Sphenoid b.
Coronal suture
Lat. sulcus
Squamous suture
Sup. temporal gyrus
Parasplenial gyrus
Parietal b.
Sup. temporal sulcus
Angular gyrus
Precuneus
Lunate sulcus
Parieto-occipital sulcus
Sup. occipital gyrus
Cuneus
Occipital polar gyrus
Sup. ramus of calcarine sulcus
Lambdoid suture
Occipital b.
Middle nasal meatus
Levator labii superioris m.
Vitreous humor
Med. rectus m.
Sup. oblique m.
Lat. rectus m.
Temporal m.
Nu. accumbens
Ant. commissure
Third ventricle
Putamen
Globus pallidus, ext. segment
Globus pallidus, int. segment
Middle cerebral a.
Claustrum
Int. capsule
Thalamus
Caudate nu., tail
Lat. ventricle
Fornix, crus
Corpus callosum
Inf. sagittal sinus
Sup. sagittal sinus

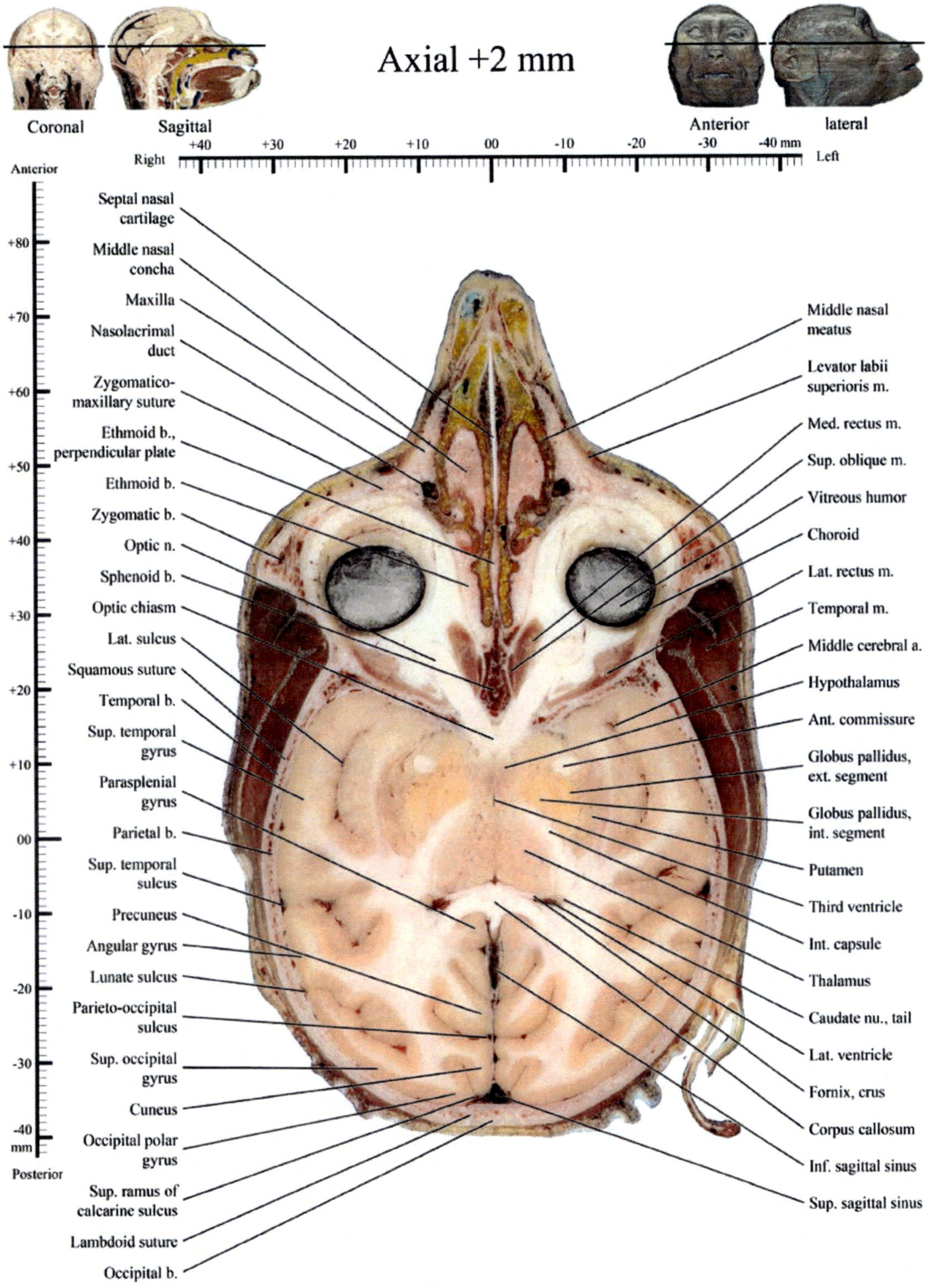

Axial +2 mm
Coronal
Sagittal
Anterior
lateral
Right
Left
+40
+30
+20
+10
00
-10
-20
-30
-40 mm
Anterior
+80
+70
+60
+50
+40
+30
+20
+10
00
-10
-20
-30
-40 mm
Posterior
Septal nasal cartilage
Middle nasal concha
Maxilla
Nasolacrimal duct
Zygomatico-maxillary suture
Ethmoid b., perpendicular plate
Ethmoid b.
Zygomatic b.
Optic n.
Sphenoid b.
Optic chiasm
Lat. sulcus
Squamous suture
Temporal b.
Sup. temporal gyrus
Parasplenial gyrus
Parietal b.
Sup. temporal sulcus
Precuneus
Angular gyrus
Lunate sulcus
Parieto-occipital sulcus
Sup. occipital gyrus
Cuneus
Occipital polar gyrus
Sup. ramus of calcarine sulcus
Lambdoid suture
Occipital b.
Middle nasal meatus
Levator labii superioris m.
Med. rectus m.
Sup. oblique m.
Vitreous humor
Choroid
Lat. rectus m.
Temporal m.
Middle cerebral a.
Hypothalamus
Ant. commissure
Globus pallidus, ext. segment
Globus pallidus, int. segment
Putamen
Third ventricle
Int. capsule
Thalamus
Caudate nu., tail
Lat. ventricle
Fornix, crus
Corpus callosum
Inf. sagittal sinus
Sup. sagittal sinus

Axial +1 mm

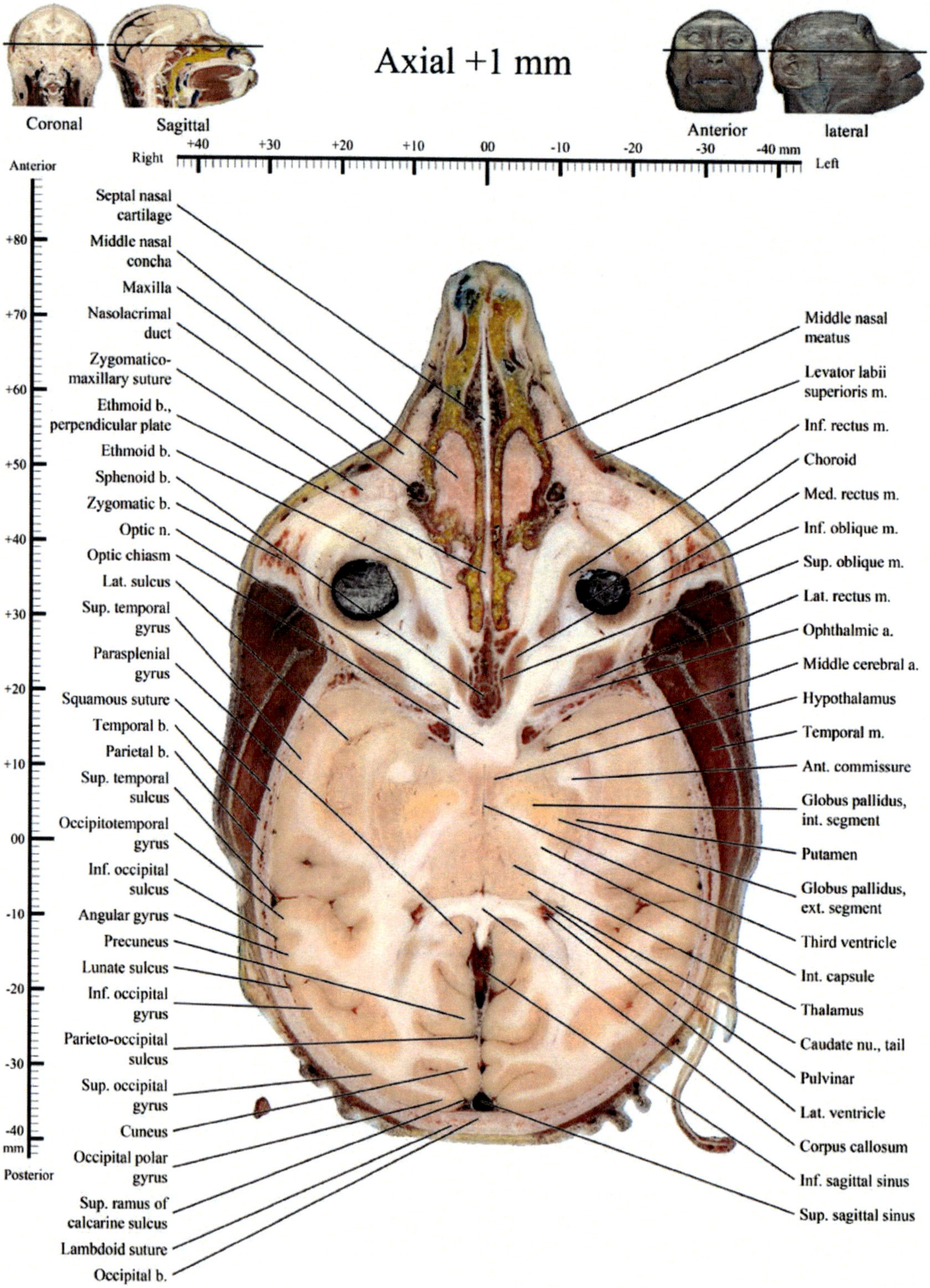

Coronal
Sagittal
Anterior
lateral
Right
+40
+30
+20
+10
00
-10
-20
-30
-40 mm
Left
Anterior
+80
+70
+60
+50
+40
+30
+20
+10
00
-10
-20
-30
-40
mm
Posterior
Septal nasal cartilage
Middle nasal concha
Maxilla
Nasolacrimal duct
Zygomatico-maxillary suture
Ethmoid b., perpendicular plate
Ethmoid b.
Sphenoid b.
Zygomatic b.
Optic n.
Optic chiasm
Lat. sulcus
Sup. temporal gyrus
Parasplenial gyrus
Squamous suture
Temporal b.
Parietal b.
Sup. temporal sulcus
Occipitotemporal gyrus
Inf. occipital sulcus
Angular gyrus
Precuneus
Lunate sulcus
Inf. occipital gyrus
Parieto-occipital sulcus
Sup. occipital gyrus
Cuneus
Occipital polar gyrus
Sup. ramus of calcarine sulcus
Lambdoid suture
Occipital b.
Middle nasal meatus
Levator labii superioris m.
Inf. rectus m.
Choroid
Med. rectus m.
Inf. oblique m.
Sup. oblique m.
Lat. rectus m.
Ophthalmic a.
Middle cerebral a.
Hypothalamus
Temporal m.
Ant. commissure
Globus pallidus, int. segment
Putamen
Globus pallidus, ext. segment
Third ventricle
Int. capsule
Thalamus
Caudate nu., tail
Pulvinar
Lat. ventricle
Corpus callosum
Inf. sagittal sinus
Sup. sagittal sinus

Axial 0 mm

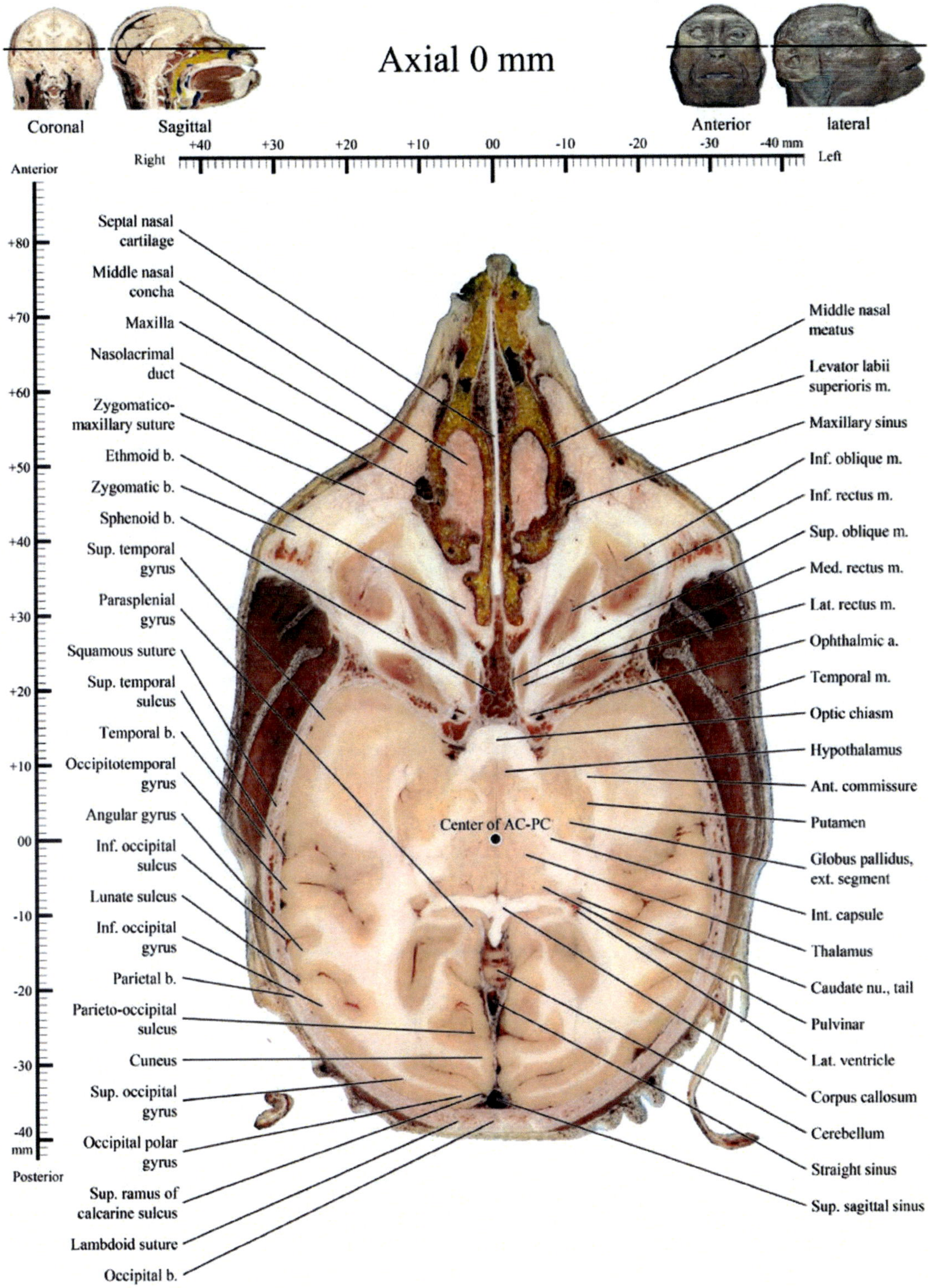
Coronal
Sagittal
Anterior
lateral
Right
+40
+30
+20
+10
00
-10
-20
-30
-40 mm
Left
Anterior
+80
+70
+60
+50
+40
+30
+20
+10
00
-10
-20
-30
-40
mm
Posterior
Septal nasal cartilage
Middle nasal concha
Maxilla
Nasolacrimal duct
Zygomatico-maxillary suture
Ethmoid b.
Zygomatic b.
Sphenoid b.
Sup. temporal gyrus
Parasplenial gyrus
Squamous suture
Sup. temporal sulcus
Temporal b.
Occipitotemporal gyrus
Angular gyrus
Inf. occipital sulcus
Lunate sulcus
Inf. occipital gyrus
Parietal b.
Parieto-occipital sulcus
Cuneus
Sup. occipital gyrus
Occipital polar gyrus
Sup. ramus of calcarine sulcus
Lambdoid suture
Occipital b.
Center of AC-PC
Middle nasal meatus
Levator labii superioris m.
Maxillary sinus
Inf. oblique m.
Inf. rectus m.
Sup. oblique m.
Med. rectus m.
Lat. rectus m.
Ophthalmic a.
Temporal m.
Optic chiasm
Hypothalamus
Ant. commissure
Putamen
Globus pallidus, ext. segment
Int. capsule
Thalamus
Caudate nu., tail
Pulvinar
Lat. ventricle
Corpus callosum
Cerebellum
Straight sinus
Sup. sagittal sinus

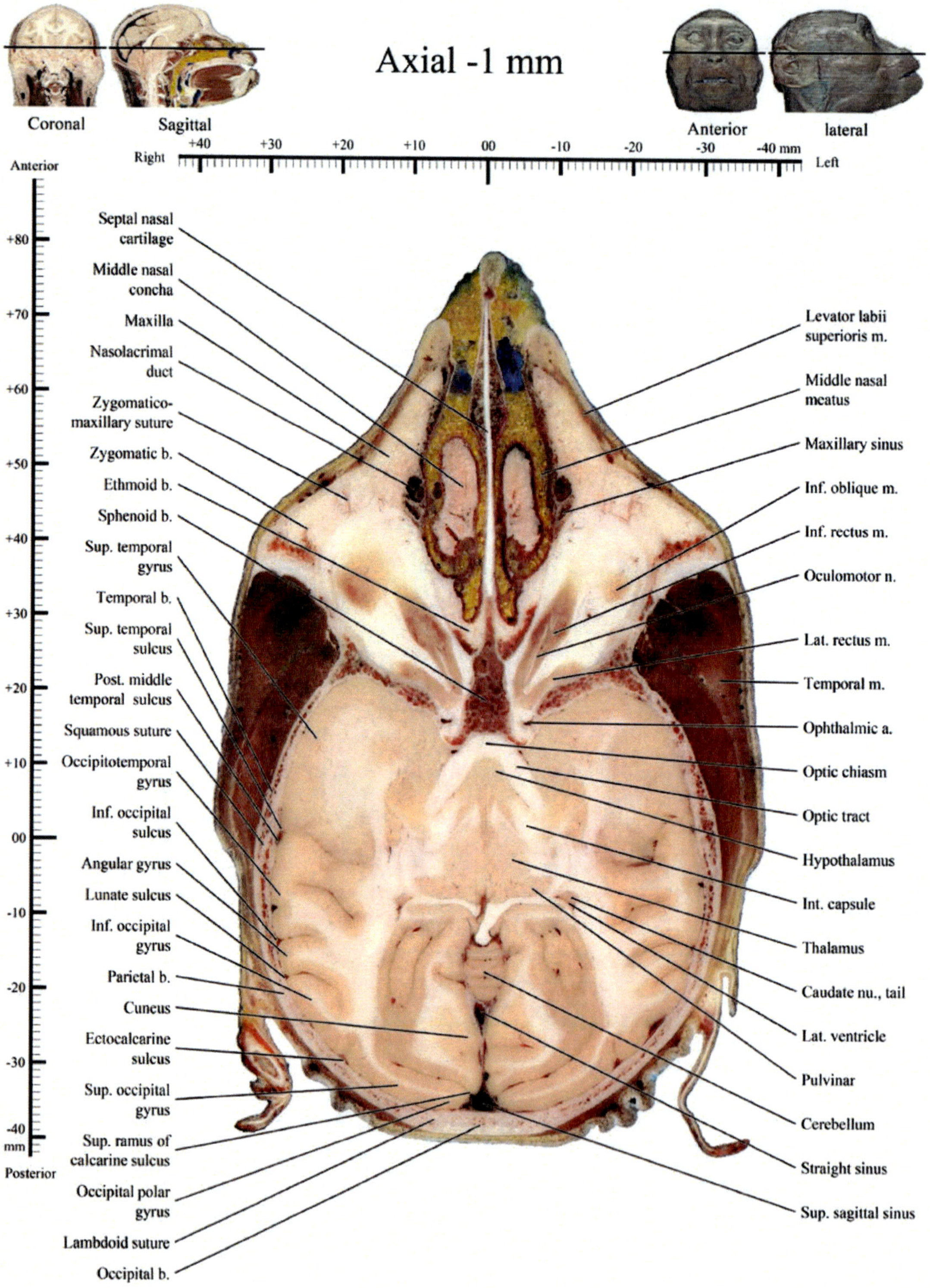
Coronal
Sagittal
Axial -1 mm
Anterior
lateral
Right
+40
+30
+20
+10
00
-10
-20
-30
-40 mm
Left
Anterior
+80
+70
+60
+50
+40
+30
+20
+10
00
-10
-20
-30
-40 mm
Posterior
Septal nasal cartilage
Middle nasal concha
Maxilla
Nasolacrimal duct
Zygomatico-maxillary suture
Zygomatic b.
Ethmoid b.
Sphenoid b.
Sup. temporal gyrus
Temporal b.
Sup. temporal sulcus
Post. middle temporal sulcus
Squamous suture
Occipitotemporal gyrus
Inf. occipital sulcus
Angular gyrus
Lunate sulcus
Inf. occipital gyrus
Parietal b.
Cuneus
Ectocalcarine sulcus
Sup. occipital gyrus
Sup. ramus of calcarine sulcus
Occipital polar gyrus
Lambdoid suture
Occipital b.
Levator labii superioris m.
Middle nasal meatus
Maxillary sinus
Inf. oblique m.
Inf. rectus m.
Oculomotor n.
Lat. rectus m.
Temporal m.
Ophthalmic a.
Optic chiasm
Optic tract
Hypothalamus
Int. capsule
Thalamus
Caudate nu., tail
Lat. ventricle
Pulvinar
Cerebellum
Straight sinus
Sup. sagittal sinus

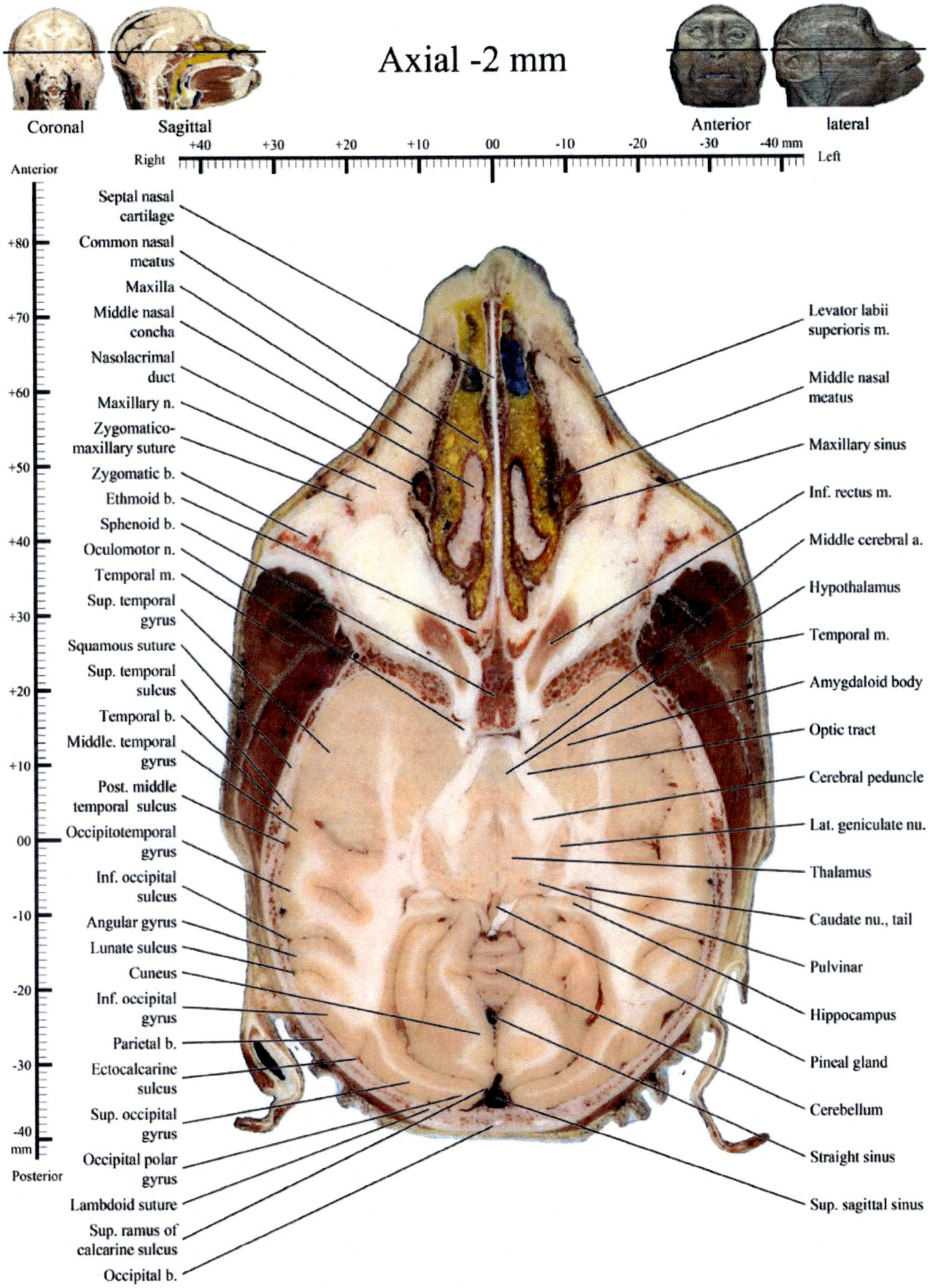
Axial -2 mm
Coronal
Sagittal
Anterior
lateral
Right
Left
+40 +30 +20 +10 00 -10 -20 -30 -40 mm
Anterior
Posterior
+80 +70 +60 +50 +40 +30 +20 +10 00 -10 -20 -30 -40 mm
Septal nasal cartilage
Common nasal meatus
Maxilla
Middle nasal concha
Nasolacrimal duct
Maxillary n.
Zygomatico-maxillary suture
Zygomatic b.
Ethmoid b.
Sphenoid b.
Oculomotor n.
Temporal m.
Sup. temporal gyrus
Squamous suture
Sup. temporal sulcus
Temporal b.
Middle. temporal gyrus
Post. middle temporal sulcus
Occipitotemporal gyrus
Inf. occipital sulcus
Angular gyrus
Lunate sulcus
Cuneus
Inf. occipital gyrus
Parietal b.
Ectocalcarine sulcus
Sup. occipital gyrus
Occipital polar gyrus
Lambdoid suture
Sup. ramus of calcarine sulcus
Occipital b.
Levator labii superioris m.
Middle nasal meatus
Maxillary sinus
Inf. rectus m.
Middle cerebral a.
Hypothalamus
Temporal m.
Amygdaloid body
Optic tract
Cerebral peduncle
Lat. geniculate nu.
Thalamus
Caudate nu., tail
Pulvinar
Hippocampus
Pineal gland
Cerebellum
Straight sinus
Sup. sagittal sinus

Axial -3 mm

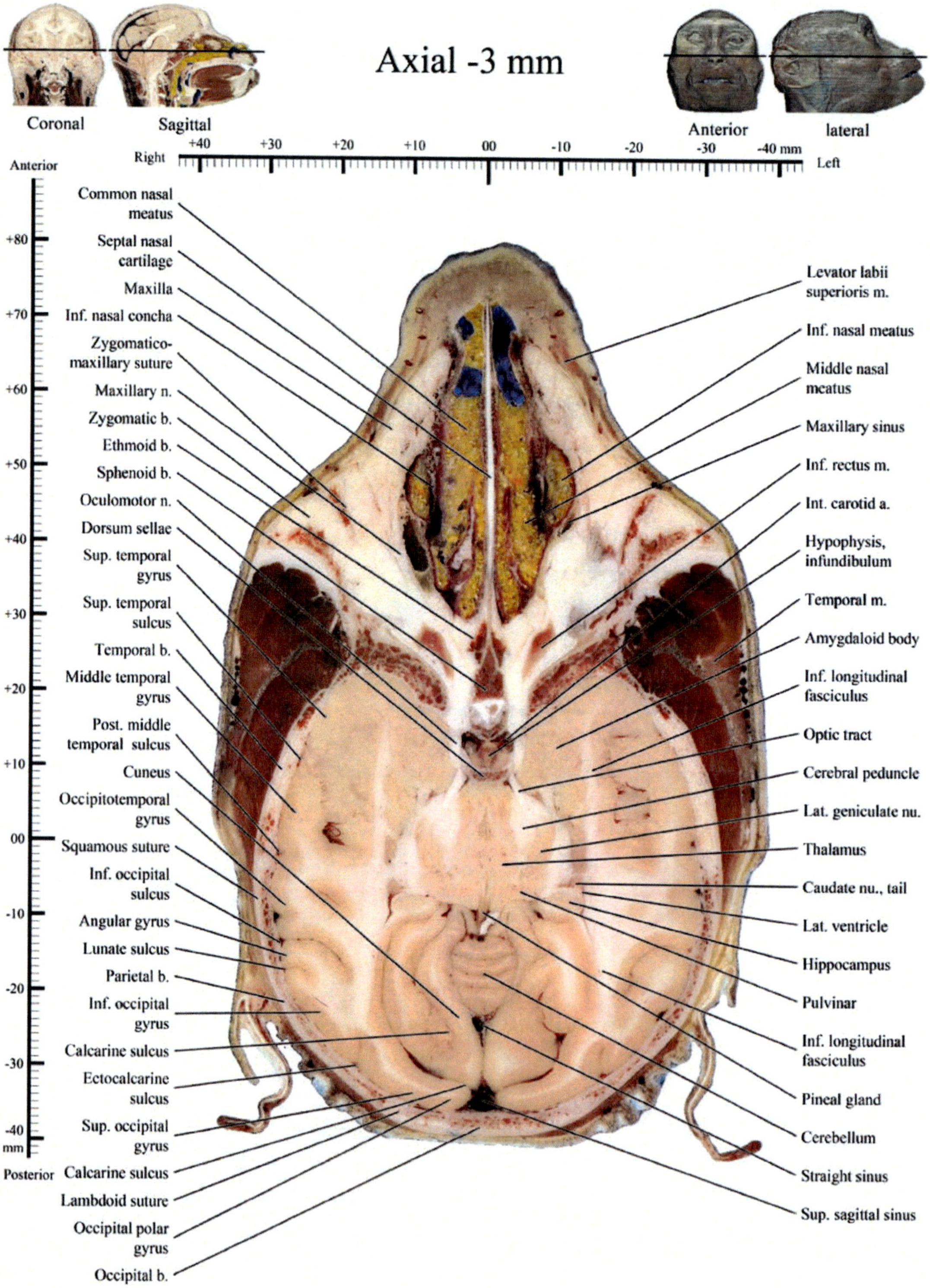
Coronal
Sagittal
Anterior
lateral
Right
+40
+30
+20
+10
00
-10
-20
-30
-40 mm
Left
Anterior
+80
+70
+60
+50
+40
+30
+20
+10
00
-10
-20
-30
-40
mm
Posterior
Common nasal meatus
Septal nasal cartilage
Maxilla
Inf. nasal concha
Zygomatico-maxillary suture
Maxillary n.
Zygomatic b.
Ethmoid b.
Sphenoid b.
Oculomotor n.
Dorsum sellae
Sup. temporal gyrus
Sup. temporal sulcus
Temporal b.
Middle temporal gyrus
Post. middle temporal sulcus
Cuneus
Occipitotemporal gyrus
Squamous suture
Inf. occipital sulcus
Angular gyrus
Lunate sulcus
Parietal b.
Inf. occipital gyrus
Calcarine sulcus
Ectocalcarine sulcus
Sup. occipital gyrus
Calcarine sulcus
Lambdoid suture
Occipital polar gyrus
Occipital b.
Levator labii superioris m.
Inf. nasal meatus
Middle nasal meatus
Maxillary sinus
Inf. rectus m.
Int. carotid a.
Hypophysis, infundibulum
Temporal m.
Amygdaloid body
Inf. longitudinal fasciculus
Optic tract
Cerebral peduncle
Lat. geniculate nu.
Thalamus
Caudate nu., tail
Lat. ventricle
Hippocampus
Pulvinar
Inf. longitudinal fasciculus
Pineal gland
Cerebellum
Straight sinus
Sup. sagittal sinus

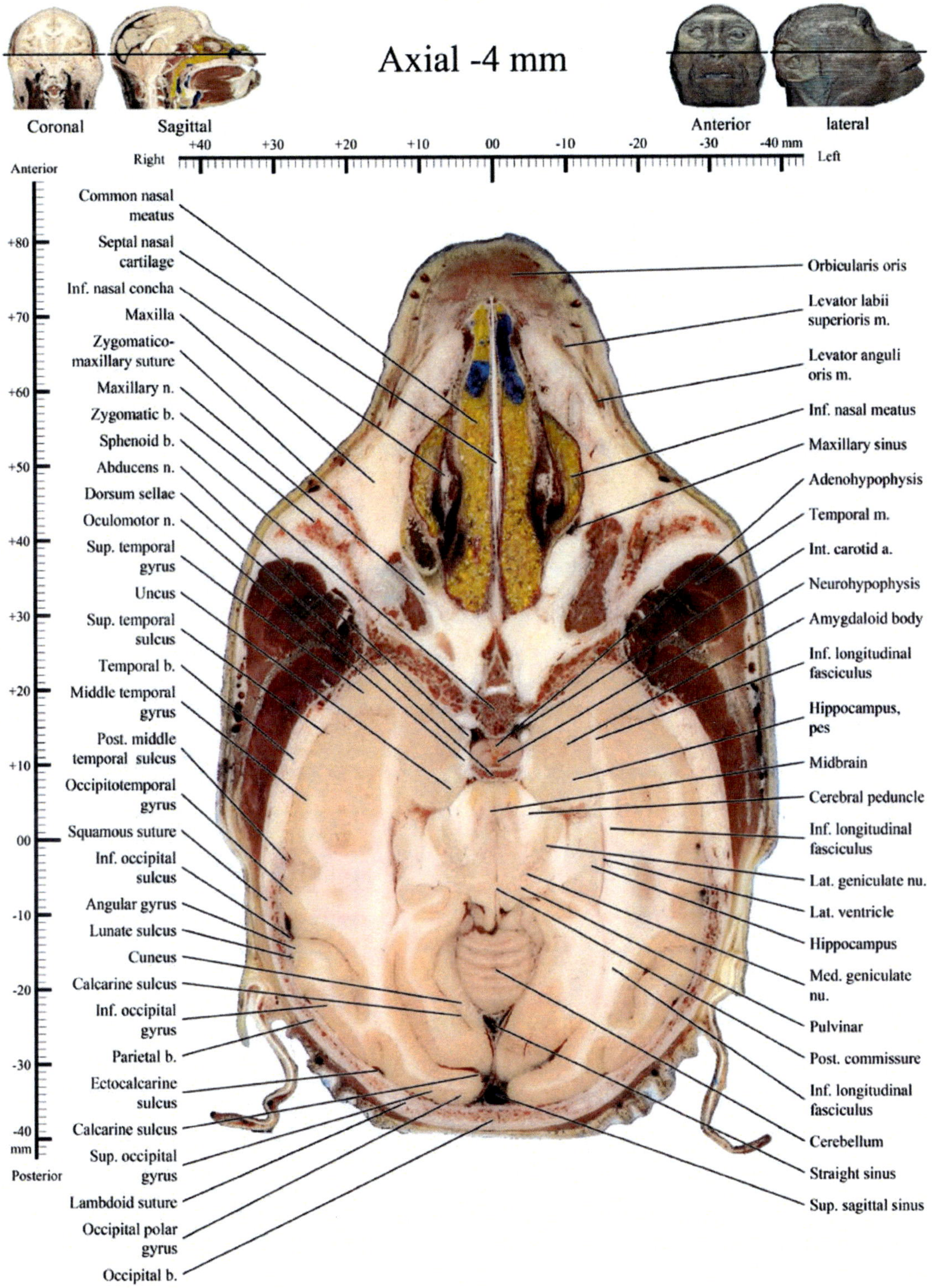
Axial -4 mm
Coronal
Sagittal
Anterior
lateral
Right
Left
+40 +30 +20 +10 00 -10 -20 -30 -40 mm
Anterior
+80 +70 +60 +50 +40 +30 +20 +10 00 -10 -20 -30 -40 mm
Posterior
Common nasal meatus
Septal nasal cartilage
Inf. nasal concha
Maxilla
Zygomatico-maxillary suture
Maxillary n.
Zygomatic b.
Sphenoid b.
Abducens n.
Dorsum sellae
Oculomotor n.
Sup. temporal gyrus
Uncus
Sup. temporal sulcus
Temporal b.
Middle temporal gyrus
Post. middle temporal sulcus
Occipitotemporal gyrus
Squamous suture
Inf. occipital sulcus
Angular gyrus
Lunate sulcus
Cuneus
Calcarine sulcus
Inf. occipital gyrus
Parietal b.
Ectocalcarine sulcus
Calcarine sulcus
Sup. occipital gyrus
Lambdoid suture
Occipital polar gyrus
Occipital b.
Orbicularis oris
Levator labii superioris m.
Levator anguli oris m.
Inf. nasal meatus
Maxillary sinus
Adenohypophysis
Temporal m.
Int. carotid a.
Neurohypophysis
Amygdaloid body
Inf. longitudinal fasciculus
Hippocampus, pes
Midbrain
Cerebral peduncle
Inf. longitudinal fasciculus
Lat. geniculate nu.
Lat. ventricle
Hippocampus
Med. geniculate nu.
Pulvinar
Post. commissure
Inf. longitudinal fasciculus
Cerebellum
Straight sinus
Sup. sagittal sinus

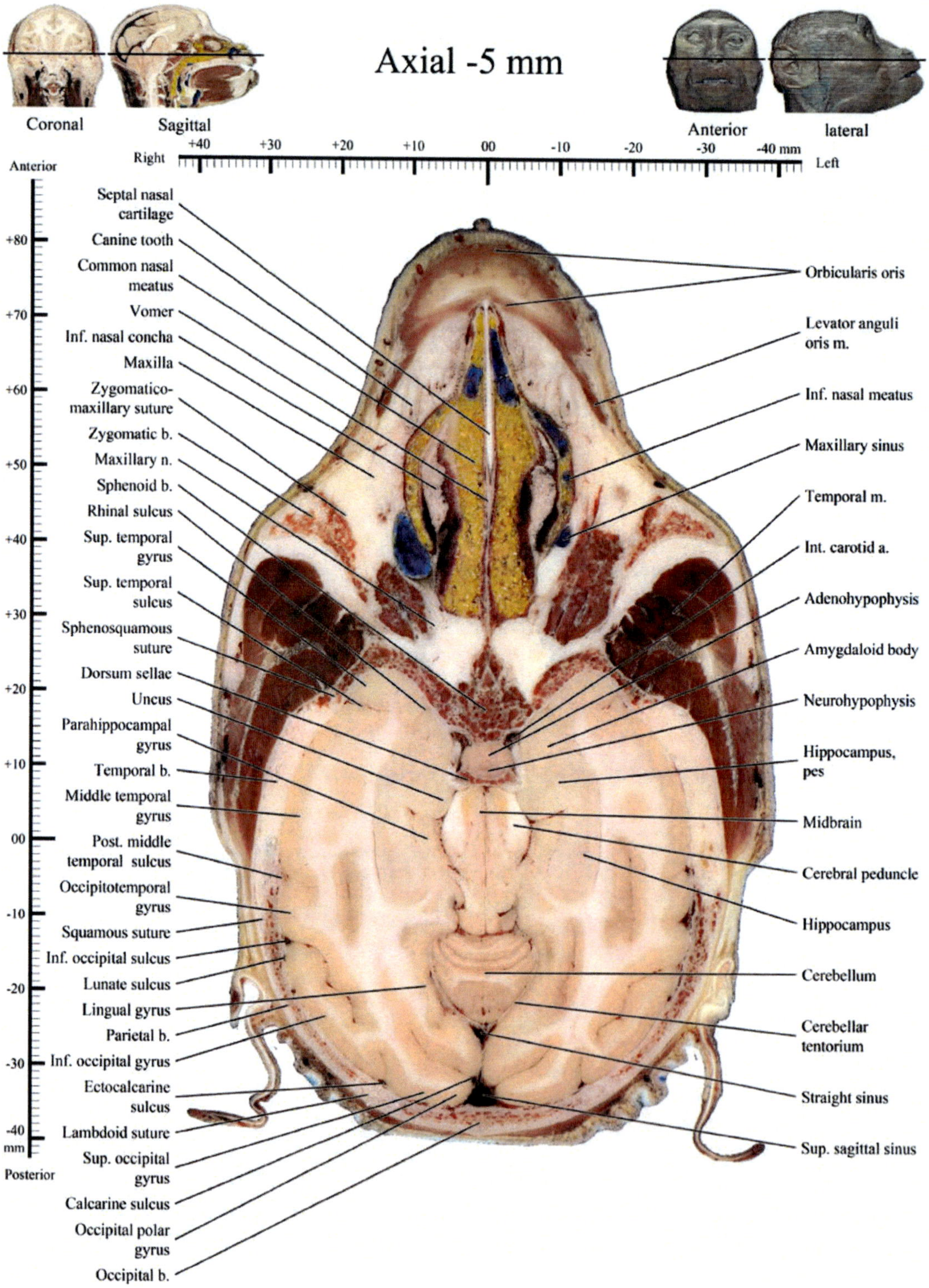
Axial -5 mm
Coronal
Sagittal
Anterior
lateral
Right
Left
+40
+30
+20
+10
00
-10
-20
-30
-40 mm
Anterior
Posterior
+80
+70
+60
+50
+40
+30
+20
+10
00
-10
-20
-30
-40 mm
Septal nasal cartilage
Canine tooth
Common nasal meatus
Vomer
Inf. nasal concha
Maxilla
Zygomatico-maxillary suture
Zygomatic b.
Maxillary n.
Sphenoid b.
Rhinal sulcus
Sup. temporal gyrus
Sup. temporal sulcus
Sphenosquamous suture
Dorsum sellae
Uncus
Parahippocampal gyrus
Temporal b.
Middle temporal gyrus
Post. middle temporal sulcus
Occipitotemporal gyrus
Squamous suture
Inf. occipital sulcus
Lunate sulcus
Lingual gyrus
Parietal b.
Inf. occipital gyrus
Ectocalcarine sulcus
Lambdoid suture
Sup. occipital gyrus
Calcarine sulcus
Occipital polar gyrus
Occipital b.
Orbicularis oris
Levator anguli oris m.
Inf. nasal meatus
Maxillary sinus
Temporal m.
Int. carotid a.
Adenohypophysis
Amygdaloid body
Neurohypophysis
Hippocampus, pes
Midbrain
Cerebral peduncle
Hippocampus
Cerebellum
Cerebellar tentorium
Straight sinus
Sup. sagittal sinus

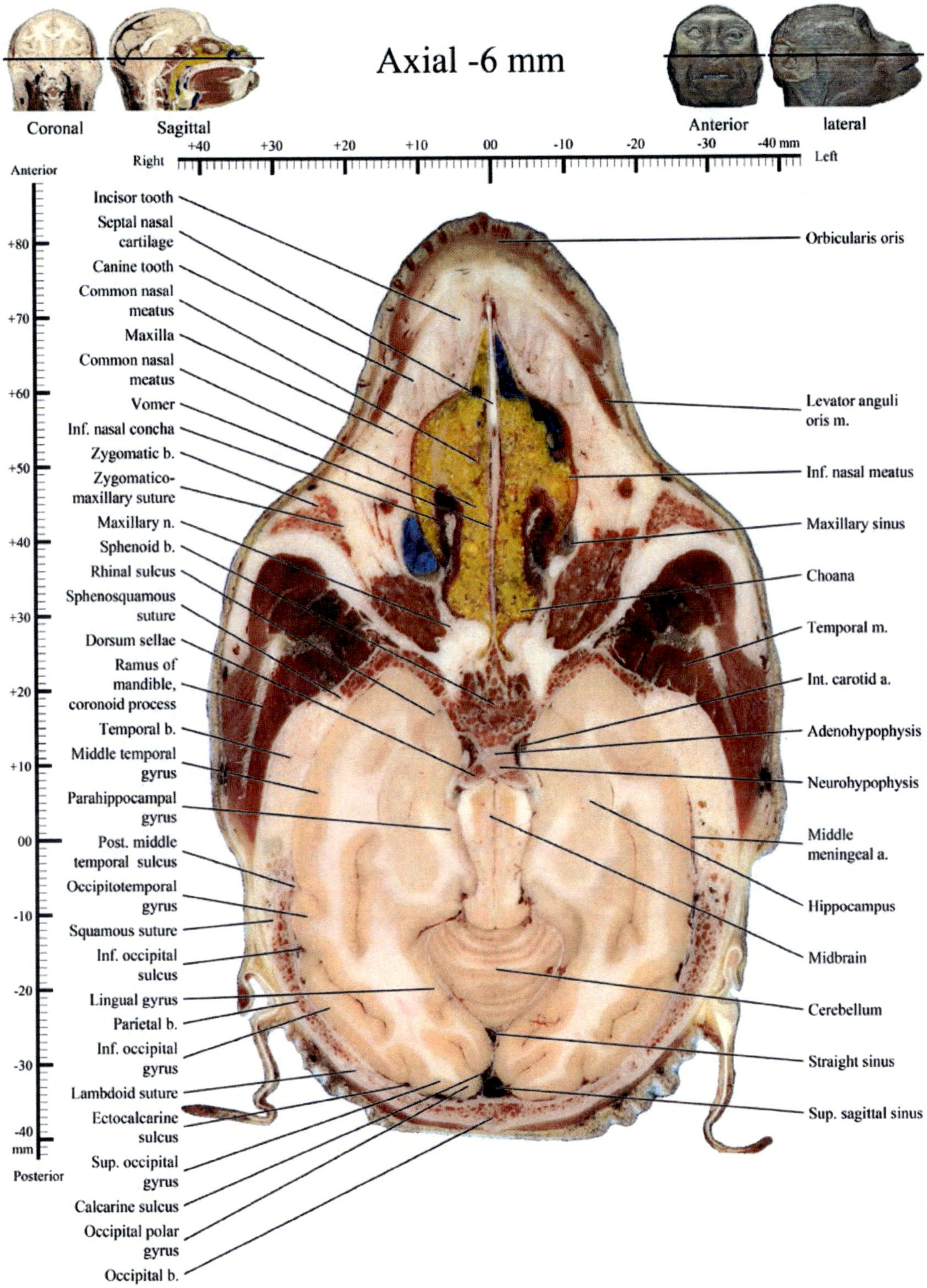
Axial -6 mm
Coronal
Sagittal
Anterior
lateral
Right
Left
+40 +30 +20 +10 00 -10 -20 -30 -40 mm
Anterior
Posterior
+80 +70 +60 +50 +40 +30 +20 +10 00 -10 -20 -30 -40 mm
Incisor tooth
Septal nasal cartilage
Canine tooth
Common nasal meatus
Maxilla
Common nasal meatus
Vomer
Inf. nasal concha
Zygomatic b.
Zygomatico-maxillary suture
Maxillary n.
Sphenoid b.
Rhinal sulcus
Sphenosquamous suture
Dorsum sellae
Ramus of mandible, coronoid process
Temporal b.
Middle temporal gyrus
Parahippocampal gyrus
Post. middle temporal sulcus
Occipitotemporal gyrus
Squamous suture
Inf. occipital sulcus
Lingual gyrus
Parietal b.
Inf. occipital gyrus
Lambdoid suture
Ectocalcarine sulcus
Sup. occipital gyrus
Calcarine sulcus
Occipital polar gyrus
Occipital b.
Orbicularis oris
Levator anguli oris m.
Inf. nasal meatus
Maxillary sinus
Choana
Temporal m.
Int. carotid a.
Adenohypophysis
Neurohypophysis
Middle meningeal a.
Hippocampus
Midbrain
Cerebellum
Straight sinus
Sup. sagittal sinus

Axial -7 mm

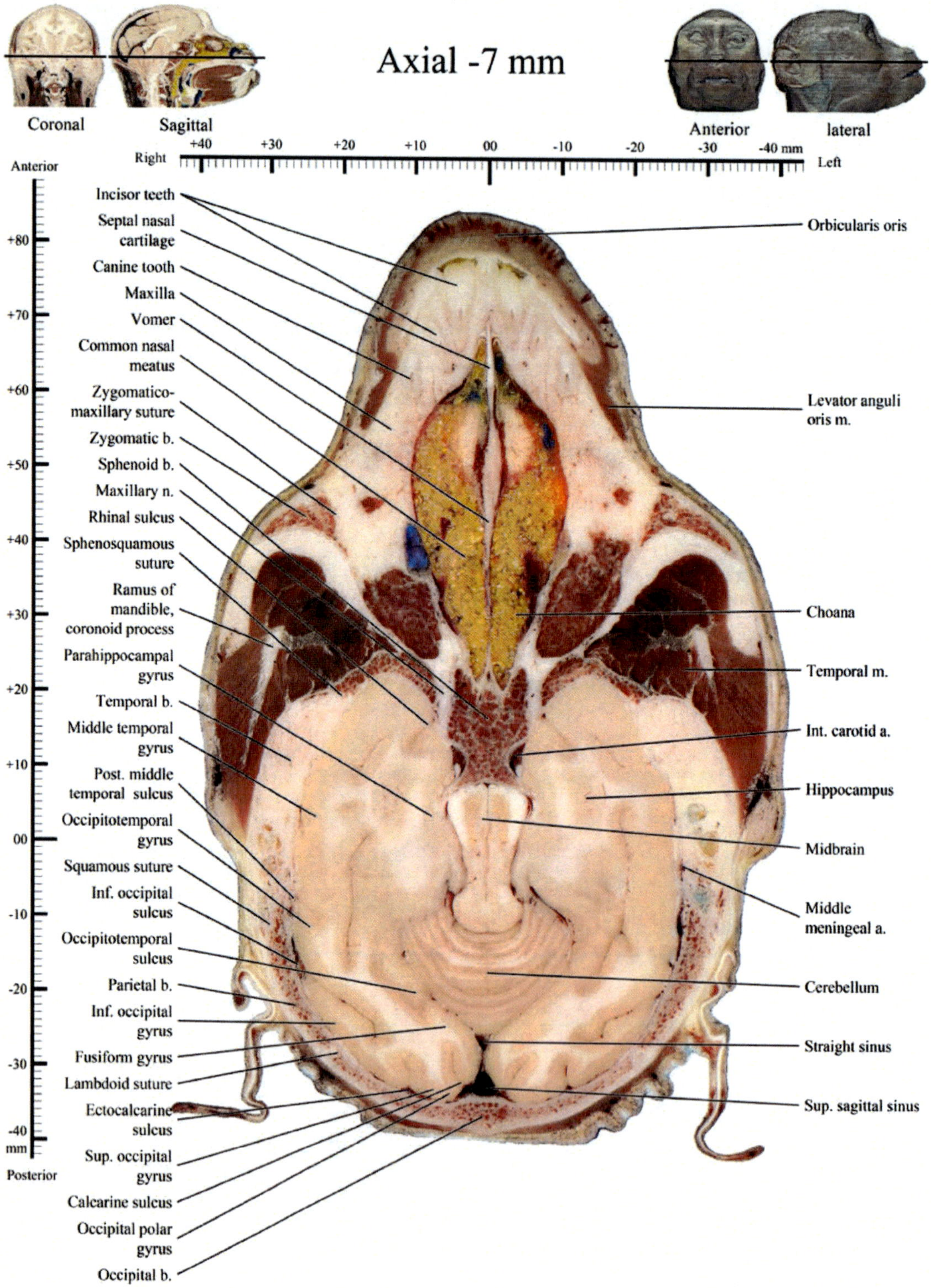
Coronal
Sagittal
Anterior
lateral
Right
Left
+40
+30
+20
+10
00
-10
-20
-30
-40 mm
Anterior
Posterior
+80
+70
+60
+50
+40
+30
+20
+10
00
-10
-20
-30
-40
mm
Incisor teeth
Septal nasal cartilage
Canine tooth
Maxilla
Vomer
Common nasal meatus
Zygomatico-maxillary suture
Zygomatic b.
Sphenoid b.
Maxillary n.
Rhinal sulcus
Sphenosquamous suture
Ramus of mandible, coronoid process
Parahippocampal gyrus
Temporal b.
Middle temporal gyrus
Post. middle temporal sulcus
Occipitotemporal gyrus
Squamous suture
Inf. occipital sulcus
Occipitotemporal sulcus
Parietal b.
Inf. occipital gyrus
Fusiform gyrus
Lambdoid suture
Ectocalcarine sulcus
Sup. occipital gyrus
Calcarine sulcus
Occipital polar gyrus
Occipital b.
Orbicularis oris
Levator anguli oris m.
Choana
Temporal m.
Int. carotid a.
Hippocampus
Midbrain
Middle meningeal a.
Cerebellum
Straight sinus
Sup. sagittal sinus

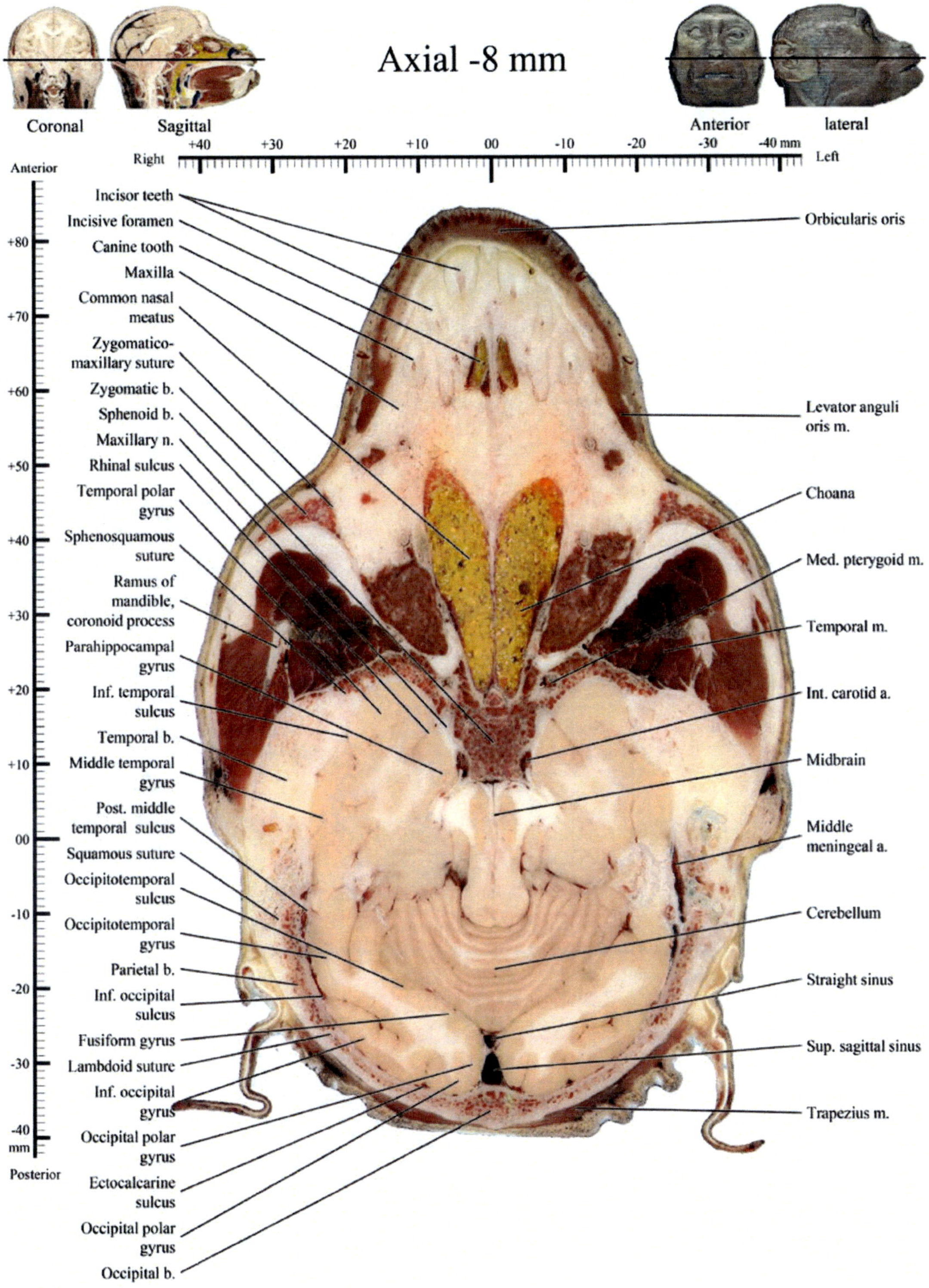
Axial -8 mm
Coronal
Sagittal
Anterior
lateral
Right
Left
+40 +30 +20 +10 00 -10 -20 -30 -40 mm
Anterior
+80
+70
+60
+50
+40
+30
+20
+10
00
-10
-20
-30
-40 mm
Posterior
Incisor teeth
Incisive foramen
Canine tooth
Maxilla
Common nasal meatus
Zygomatico-maxillary suture
Zygomatic b.
Sphenoid b.
Maxillary n.
Rhinal sulcus
Temporal polar gyrus
Sphenosquamous suture
Ramus of mandible, coronoid process
Parahippocampal gyrus
Inf. temporal sulcus
Temporal b.
Middle temporal gyrus
Post. middle temporal sulcus
Squamous suture
Occipitotemporal sulcus
Occipitotemporal gyrus
Parietal b.
Inf. occipital sulcus
Fusiform gyrus
Lambdoid suture
Inf. occipital gyrus
Occipital polar gyrus
Ectocalcarine sulcus
Occipital polar gyrus
Occipital b.
Orbicularis oris
Levator anguli oris m.
Choana
Med. pterygoid m.
Temporal m.
Int. carotid a.
Midbrain
Middle meningeal a.
Cerebellum
Straight sinus
Sup. sagittal sinus
Trapezius m.

Axial -9 mm

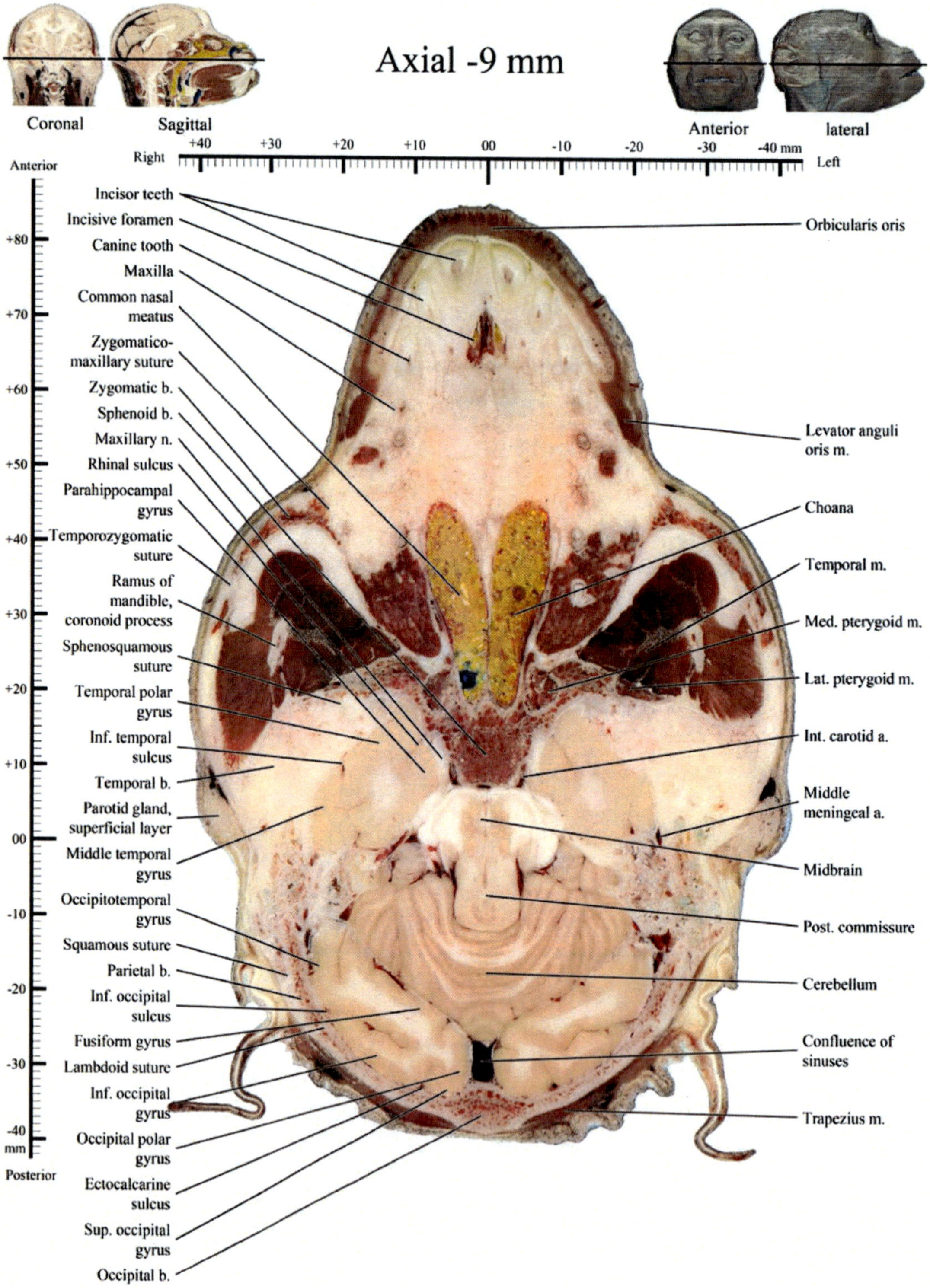
Coronal
Sagittal
Anterior
lateral
Right
+40
+30
+20
+10
00
-10
-20
-30
-40 mm
Left
Anterior
+80
+70
+60
+50
+40
+30
+20
+10
00
-10
-20
-30
-40 mm
Posterior
Incisor teeth
Incisive foramen
Canine tooth
Maxilla
Common nasal meatus
Zygomatico-maxillary suture
Zygomatic b.
Sphenoid b.
Maxillary n.
Rhinal sulcus
Parahippocampal gyrus
Temporozygomatic suture
Ramus of mandible, coronoid process
Sphenosquamous suture
Temporal polar gyrus
Inf. temporal sulcus
Temporal b.
Parotid gland, superficial layer
Middle temporal gyrus
Occipitotemporal gyrus
Squamous suture
Parietal b.
Inf. occipital sulcus
Fusiform gyrus
Lambdoid suture
Inf. occipital gyrus
Occipital polar gyrus
Ectocalcarine sulcus
Sup. occipital gyrus
Occipital b.
Orbicularis oris
Levator anguli oris m.
Choana
Temporal m.
Med. pterygoid m.
Lat. pterygoid m.
Int. carotid a.
Middle meningeal a.
Midbrain
Post. commissure
Cerebellum
Confluence of sinuses
Trapezius m.

Axial -10 mm

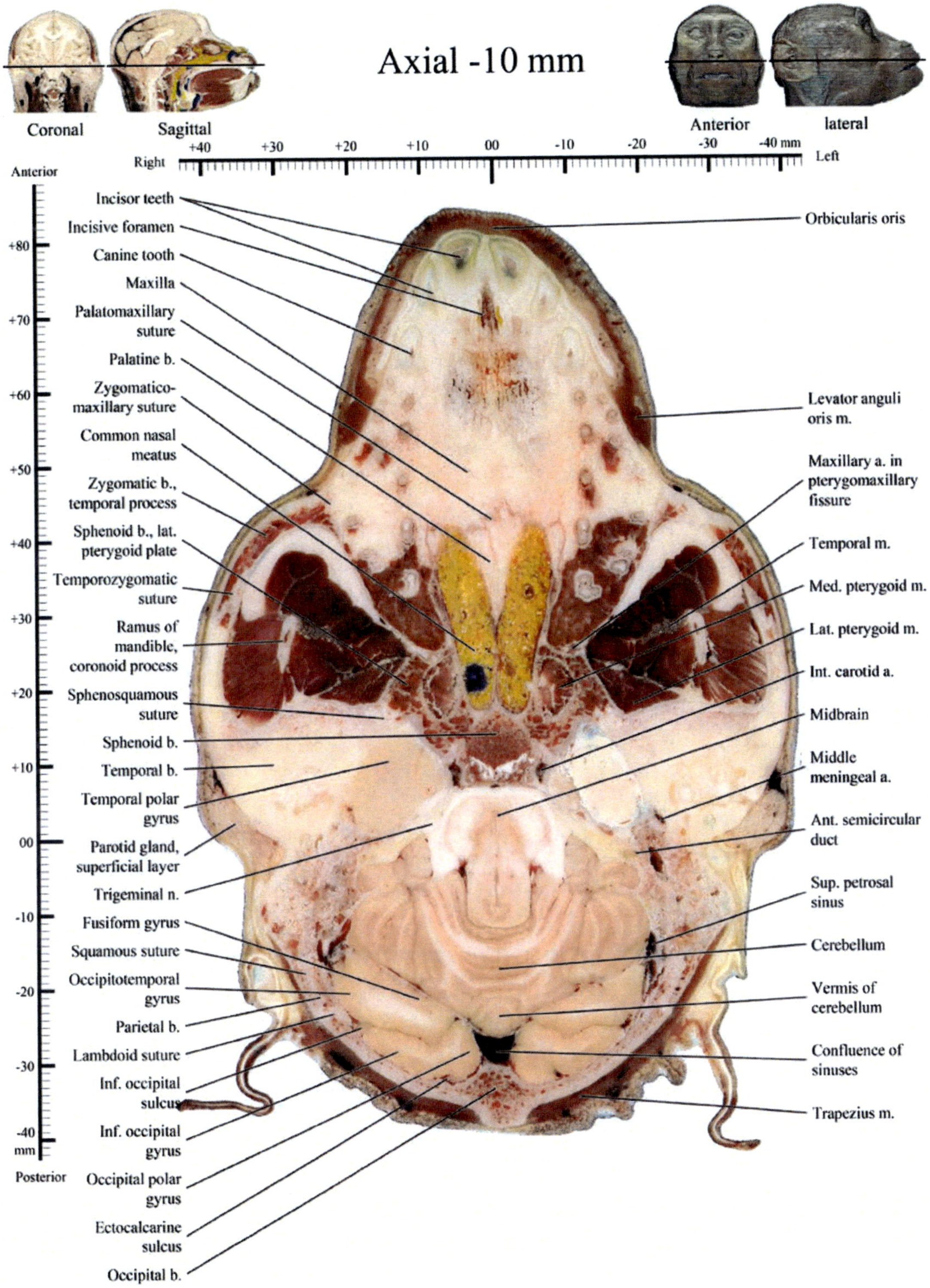
Coronal
Sagittal
Anterior
lateral
Right
Left
Anterior
Posterior
+40
+30
+20
+10
00
-10
-20
-30
-40 mm
+80
+70
+60
+50
+40
+30
+20
+10
00
-10
-20
-30
-40
mm
Incisor teeth
Incisive foramen
Canine tooth
Maxilla
Palatomaxillary suture
Palatine b.
Zygomatico-maxillary suture
Common nasal meatus
Zygomatic b., temporal process
Sphenoid b., lat. pterygoid plate
Temporozygomatic suture
Ramus of mandible, coronoid process
Sphenosquamous suture
Sphenoid b.
Temporal b.
Temporal polar gyrus
Parotid gland, superficial layer
Trigeminal n.
Fusiform gyrus
Squamous suture
Occipitotemporal gyrus
Parietal b.
Lambdoid suture
Inf. occipital sulcus
Inf. occipital gyrus
Occipital polar gyrus
Ectocalcarine sulcus
Occipital b.
Orbicularis oris
Levator anguli oris m.
Maxillary a. in pterygomaxillary fissure
Temporal m.
Med. pterygoid m.
Lat. pterygoid m.
Int. carotid a.
Midbrain
Middle meningeal a.
Ant. semicircular duct
Sup. petrosal sinus
Cerebellum
Vermis of cerebellum
Confluence of sinuses
Trapezius m.

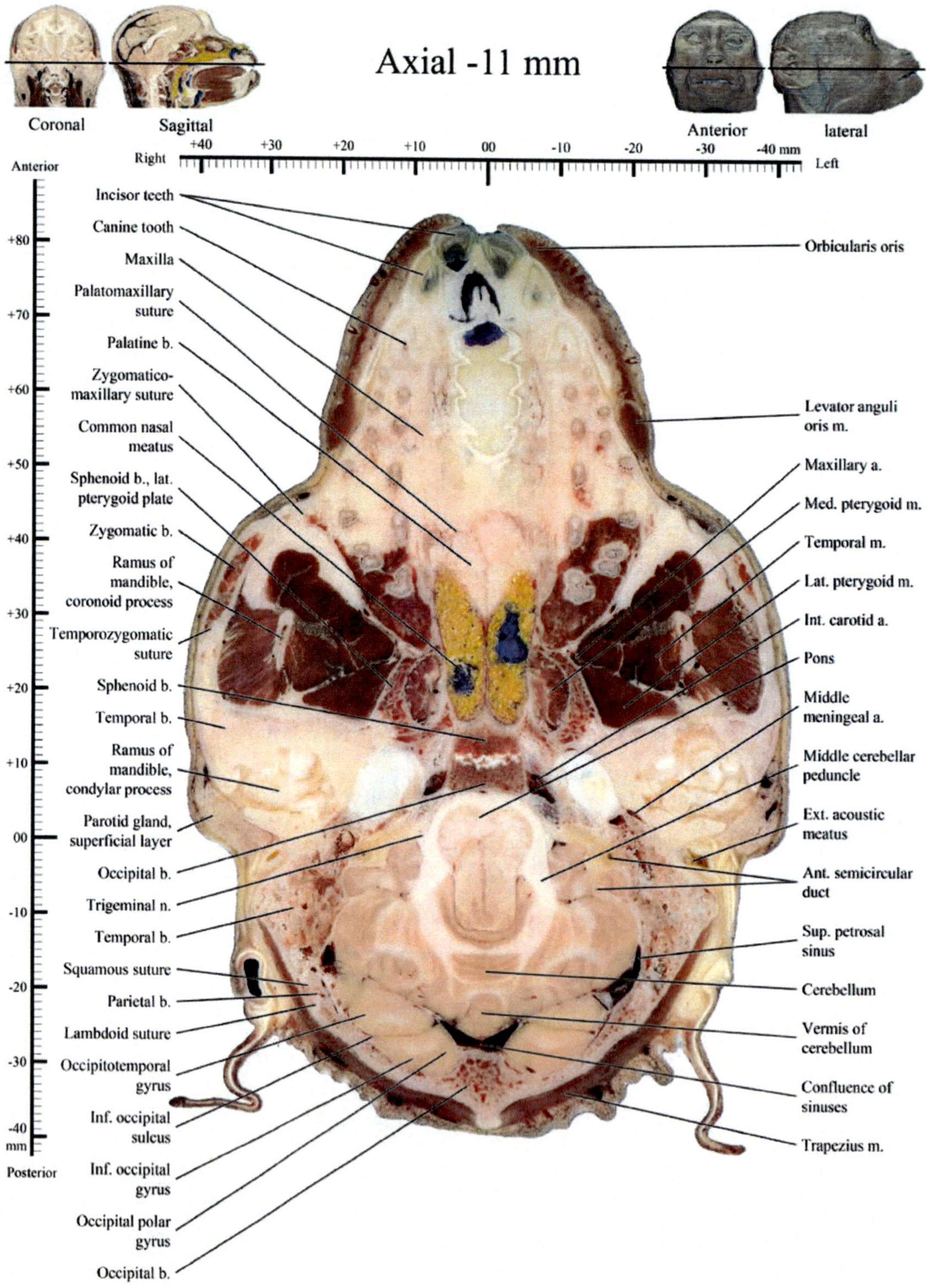
Axial -11 mm
Coronal
Sagittal
Anterior
lateral
Right
+40
+30
+20
+10
00
-10
-20
-30
-40 mm
Left
Anterior
+80
+70
+60
+50
+40
+30
+20
+10
00
-10
-20
-30
-40 mm
Posterior
Incisor teeth
Canine tooth
Maxilla
Palatomaxillary suture
Palatine b.
Zygomatico-maxillary suture
Common nasal meatus
Sphenoid b., lat. pterygoid plate
Zygomatic b.
Ramus of mandible, coronoid process
Temporozygomatic suture
Sphenoid b.
Temporal b.
Ramus of mandible, condylar process
Parotid gland, superficial layer
Occipital b.
Trigeminal n.
Temporal b.
Squamous suture
Parietal b.
Lambdoid suture
Occipitotemporal gyrus
Inf. occipital sulcus
Inf. occipital gyrus
Occipital polar gyrus
Occipital b.
Orbicularis oris
Levator anguli oris m.
Maxillary a.
Med. pterygoid m.
Temporal m.
Lat. pterygoid m.
Int. carotid a.
Pons
Middle meningeal a.
Middle cerebellar peduncle
Ext. acoustic meatus
Ant. semicircular duct
Sup. petrosal sinus
Cerebellum
Vermis of cerebellum
Confluence of sinuses
Trapezius m.

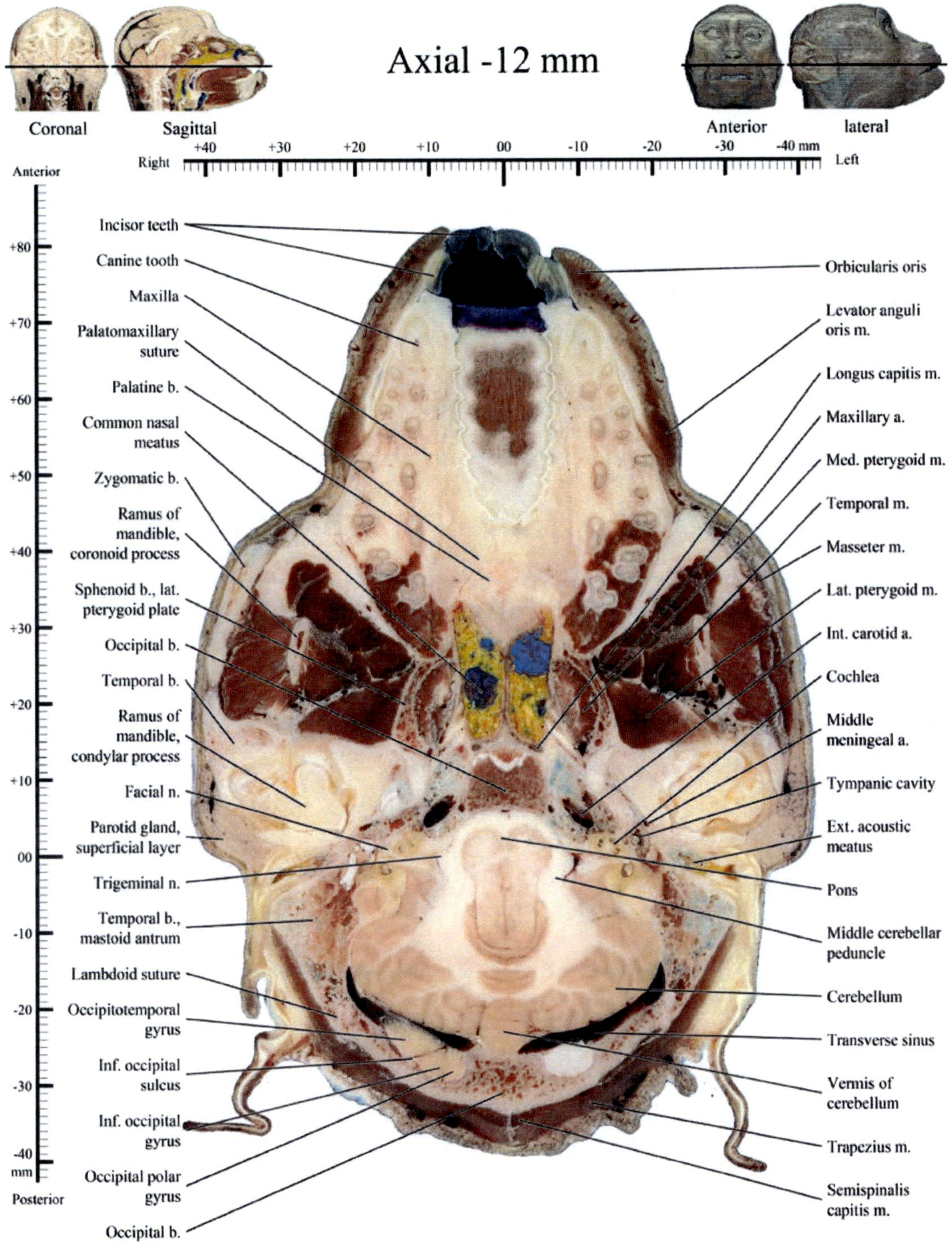

Axial -12 mm
Coronal
Sagittal
Anterior
lateral
Right
Left
+40
+30
+20
+10
00
-10
-20
-30
-40 mm
Anterior
Posterior
+80
+70
+60
+50
+40
+30
+20
+10
00
-10
-20
-30
-40
mm
Incisor teeth
Canine tooth
Maxilla
Palatomaxillary suture
Palatine b.
Common nasal meatus
Zygomatic b.
Ramus of mandible, coronoid process
Sphenoid b., lat. pterygoid plate
Occipital b.
Temporal b.
Ramus of mandible, condylar process
Facial n.
Parotid gland, superficial layer
Trigeminal n.
Temporal b., mastoid antrum
Lambdoid suture
Occipitotemporal gyrus
Inf. occipital sulcus
Inf. occipital gyrus
Occipital polar gyrus
Occipital b.
Orbicularis oris
Levator anguli oris m.
Longus capitis m.
Maxillary a.
Med. pterygoid m.
Temporal m.
Masseter m.
Lat. pterygoid m.
Int. carotid a.
Cochlea
Middle meningeal a.
Tympanic cavity
Ext. acoustic meatus
Pons
Middle cerebellar peduncle
Cerebellum
Transverse sinus
Vermis of cerebellum
Trapezius m.
Semispinalis capitis m.

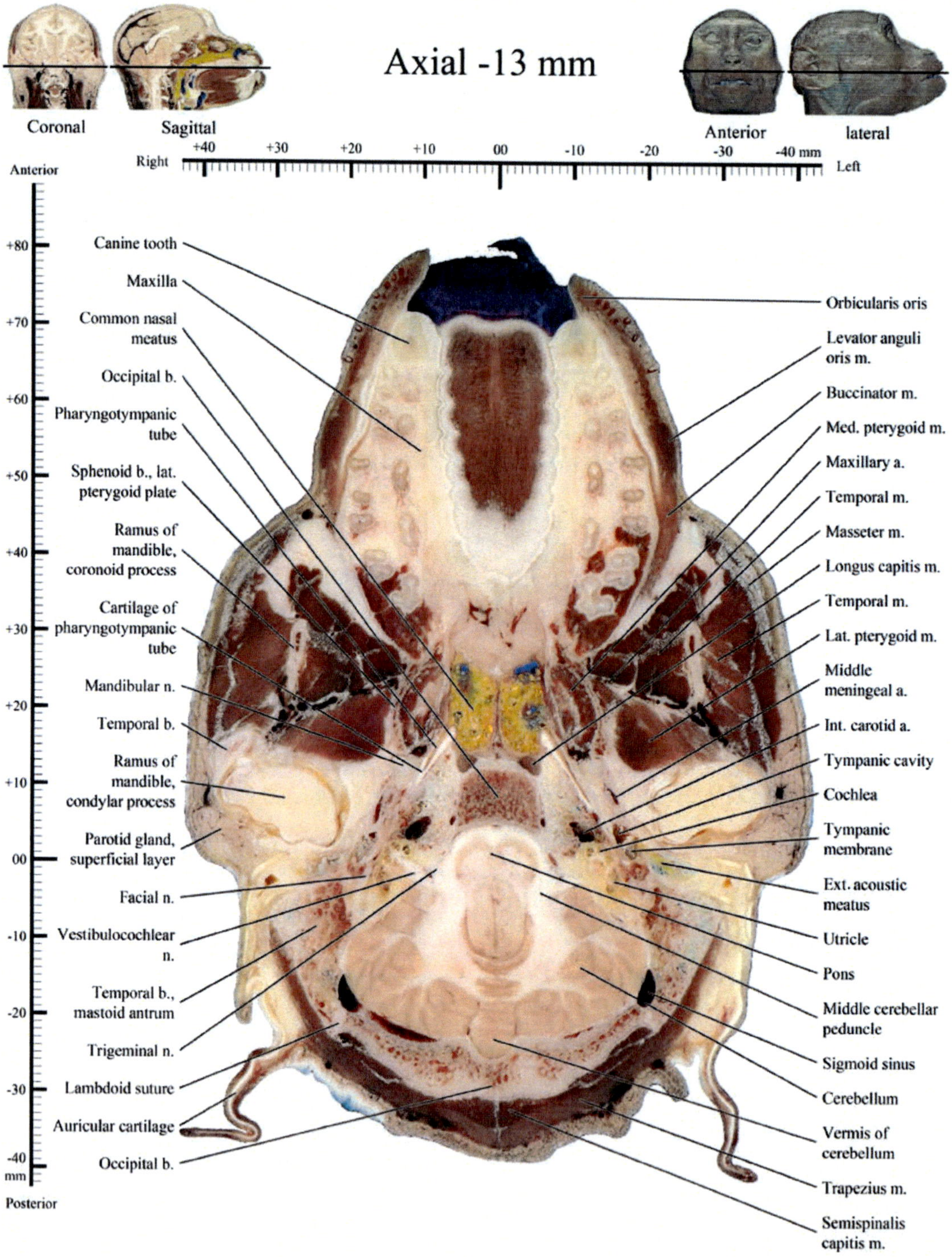
Axial -13 mm
Coronal
Sagittal
Anterior
lateral
Right
+40
+30
+20
+10
00
-10
-20
-30
-40 mm
Left
Anterior
Posterior
Canine tooth
Maxilla
Common nasal meatus
Occipital b.
Pharyngotympanic tube
Sphenoid b., lat. pterygoid plate
Ramus of mandible, coronoid process
Cartilage of pharyngotympanic tube
Mandibular n.
Temporal b.
Ramus of mandible, condylar process
Parotid gland, superficial layer
Facial n.
Vestibulocochlear n.
Temporal b., mastoid antrum
Trigeminal n.
Lambdoid suture
Auricular cartilage
Occipital b.
Orbicularis oris
Levator anguli oris m.
Buccinator m.
Med. pterygoid m.
Maxillary a.
Temporal m.
Masseter m.
Longus capitis m.
Temporal m.
Lat. pterygoid m.
Middle meningeal a.
Int. carotid a.
Tympanic cavity
Cochlea
Tympanic membrane
Ext. acoustic meatus
Utricle
Pons
Middle cerebellar peduncle
Sigmoid sinus
Cerebellum
Vermis of cerebellum
Trapezius m.
Semispinalis capitis m.

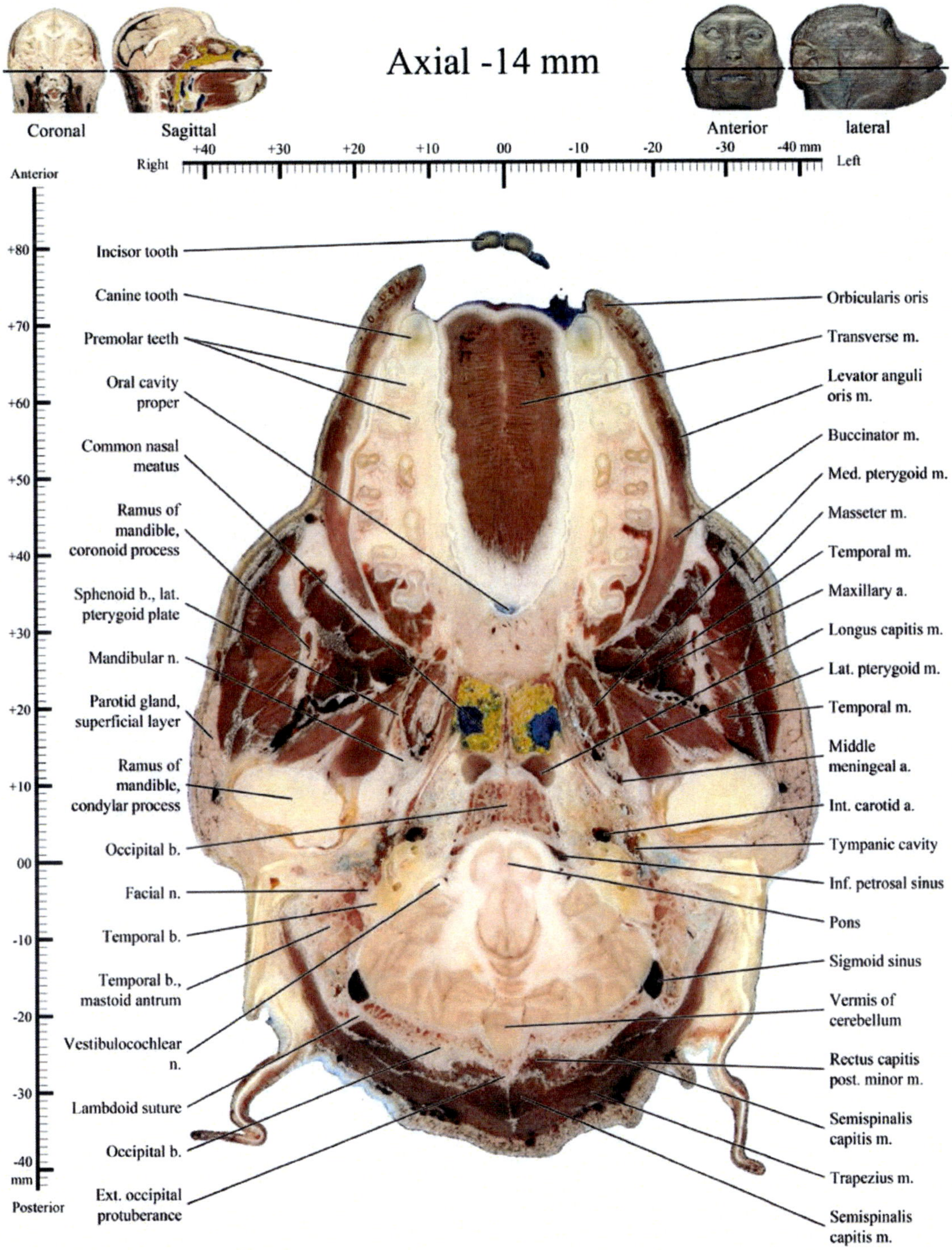
Coronal
Sagittal
Axial -14 mm
Anterior
lateral
Right
+40 +30 +20 +10 00 -10 -20 -30 -40 mm
Left
Anterior
+80
+70
+60
+50
+40
+30
+20
+10
00
-10
-20
-30
-40
mm
Posterior
Incisor tooth
Canine tooth
Premolar teeth
Oral cavity proper
Common nasal meatus
Ramus of mandible, coronoid process
Sphenoid b., lat. pterygoid plate
Mandibular n.
Parotid gland, superficial layer
Ramus of mandible, condylar process
Occipital b.
Facial n.
Temporal b.
Temporal b., mastoid antrum
Vestibulocochlear n.
Lambdoid suture
Occipital b.
Ext. occipital protuberance
Orbicularis oris
Transverse m.
Levator anguli oris m.
Buccinator m.
Med. pterygoid m.
Masseter m.
Temporal m.
Maxillary a.
Longus capitis m.
Lat. pterygoid m.
Temporal m.
Middle meningeal a.
Int. carotid a.
Tympanic cavity
Inf. petrosal sinus
Pons
Sigmoid sinus
Vermis of cerebellum
Rectus capitis post. minor m.
Semispinalis capitis m.
Trapezius m.
Semispinalis capitis m.

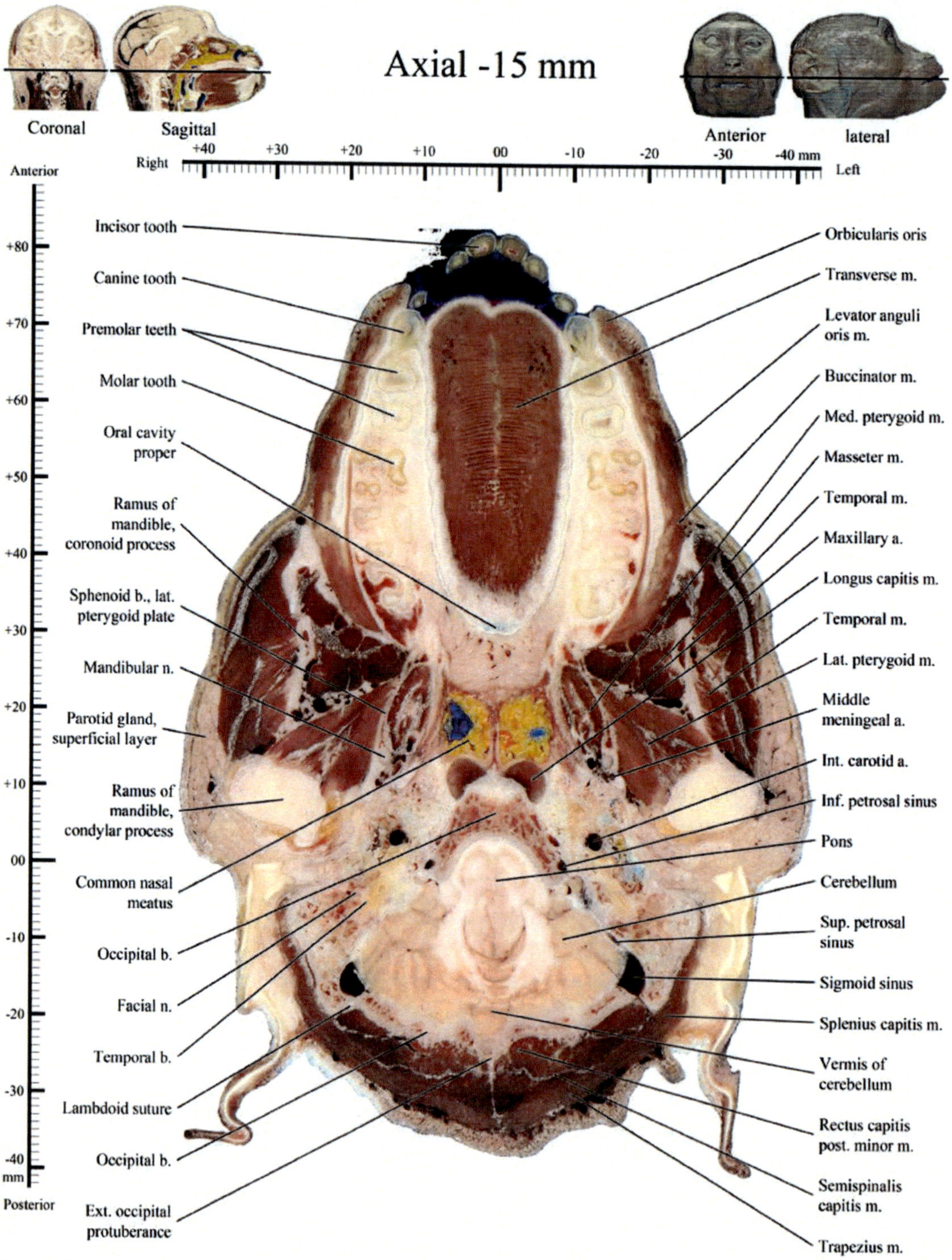
Axial -15 mm
Coronal
Sagittal
Anterior
lateral
Right
Left
Anterior
Posterior
Incisor tooth
Canine tooth
Premolar teeth
Molar tooth
Oral cavity proper
Ramus of mandible, coronoid process
Sphenoid b., lat. pterygoid plate
Mandibular n.
Parotid gland, superficial layer
Ramus of mandible, condylar process
Common nasal meatus
Occipital b.
Facial n.
Temporal b.
Lambdoid suture
Occipital b.
Ext. occipital protuberance
Orbicularis oris
Transverse m.
Levator anguli oris m.
Buccinator m.
Med. pterygoid m.
Masseter m.
Temporal m.
Maxillary a.
Longus capitis m.
Temporal m.
Lat. pterygoid m.
Middle meningeal a.
Int. carotid a.
Inf. petrosal sinus
Pons
Cerebellum
Sup. petrosal sinus
Sigmoid sinus
Splenius capitis m.
Vermis of cerebellum
Rectus capitis post. minor m.
Semispinalis capitis m.
Trapezius m.

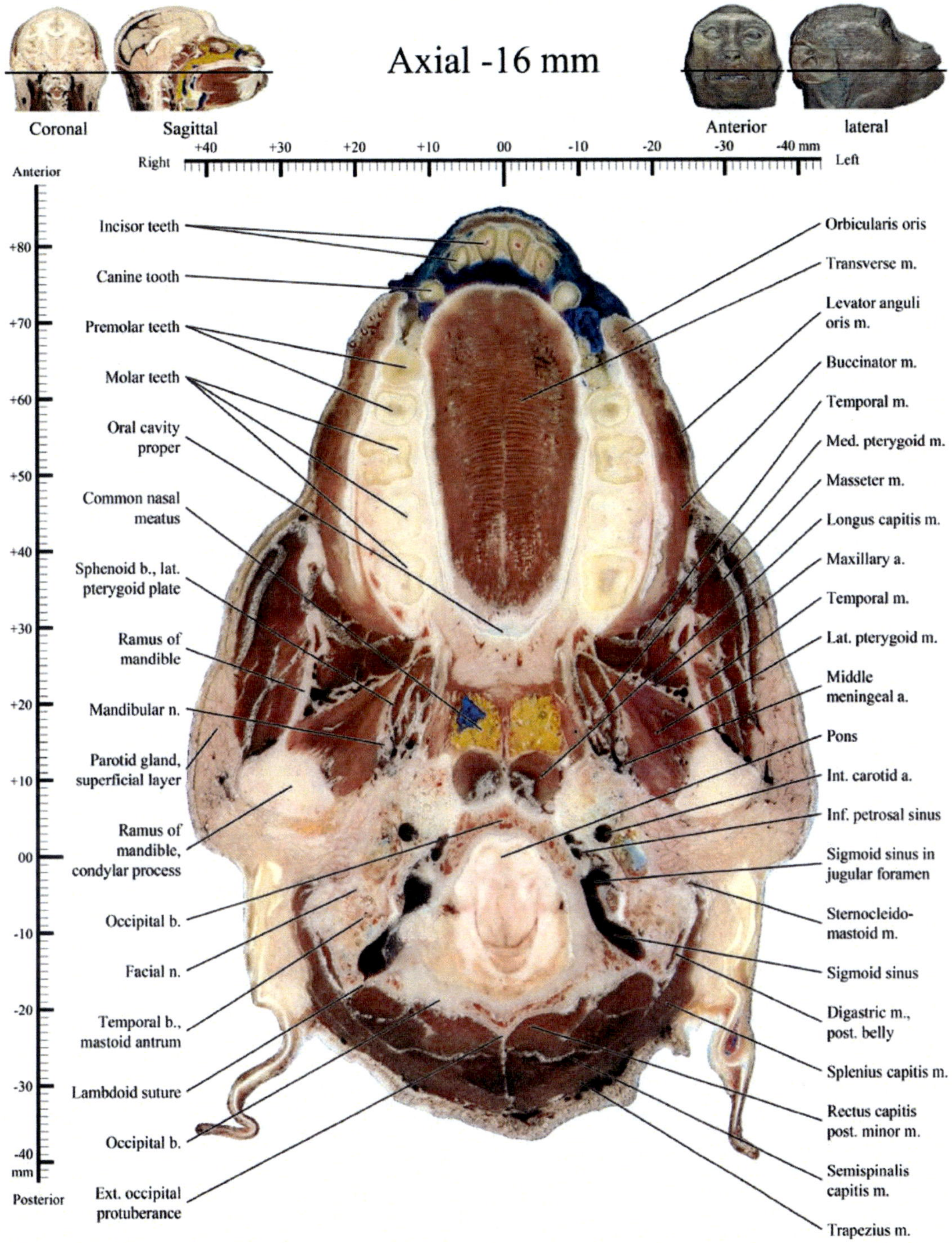

Axial -16 mm
Coronal
Sagittal
Anterior
lateral
Right
Left
+40 +30 +20 +10 00 -10 -20 -30 -40 mm
Anterior
Posterior
+80 +70 +60 +50 +40 +30 +20 +10 00 -10 -20 -30 -40 mm
Incisor teeth
Canine tooth
Premolar teeth
Molar teeth
Oral cavity proper
Common nasal meatus
Sphenoid b., lat. pterygoid plate
Ramus of mandible
Mandibular n.
Parotid gland, superficial layer
Ramus of mandible, condylar process
Occipital b.
Facial n.
Temporal b., mastoid antrum
Lambdoid suture
Occipital b.
Ext. occipital protuberance
Orbicularis oris
Transverse m.
Levator anguli oris m.
Buccinator m.
Temporal m.
Med. pterygoid m.
Masseter m.
Longus capitis m.
Maxillary a.
Temporal m.
Lat. pterygoid m.
Middle meningeal a.
Pons
Int. carotid a.
Inf. petrosal sinus
Sigmoid sinus in jugular foramen
Sternocleido-mastoid m.
Sigmoid sinus
Digastric m., post. belly
Splenius capitis m.
Rectus capitis post. minor m.
Semispinalis capitis m.
Trapezius m.

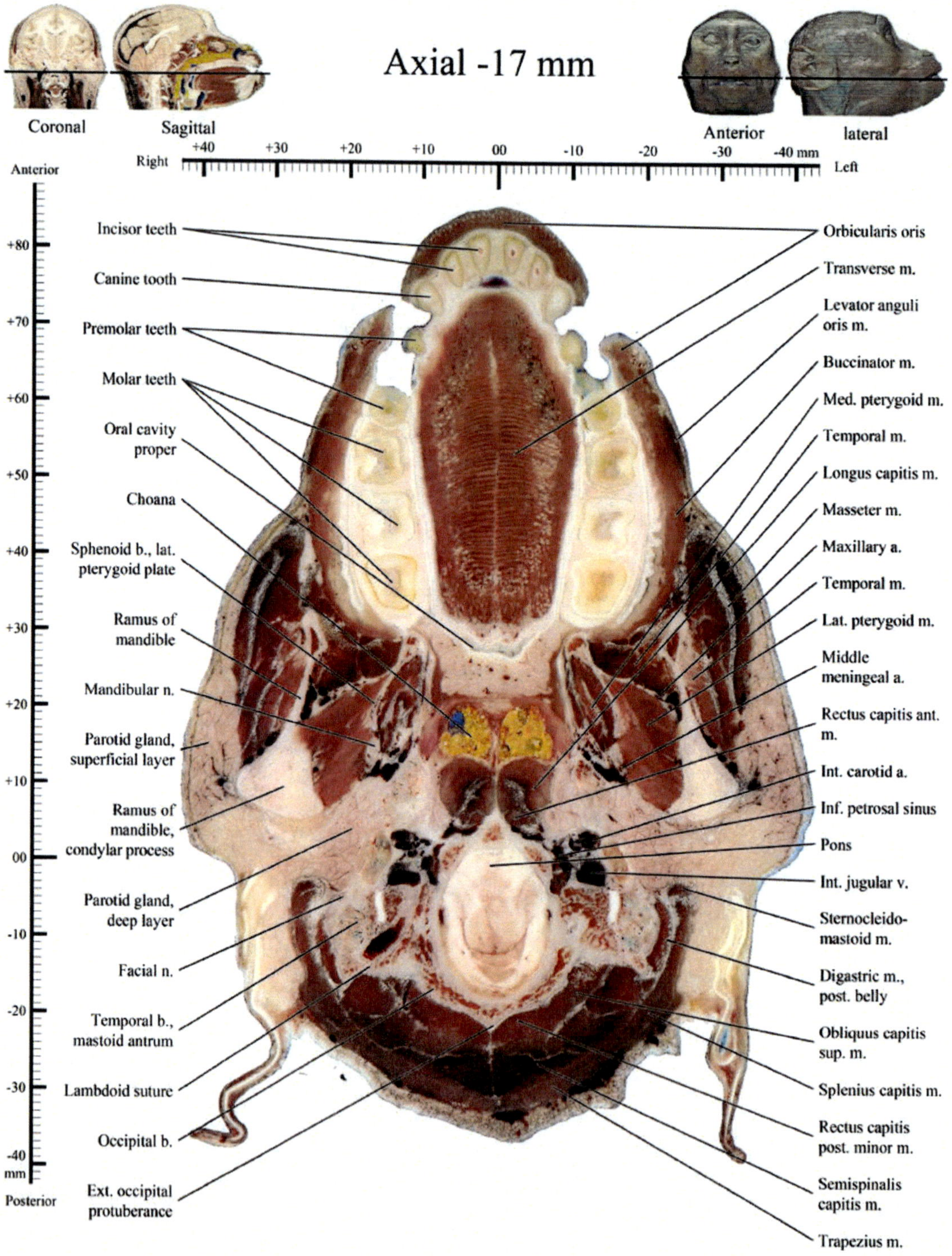
Axial -17 mm
Coronal
Sagittal
Anterior
lateral
Right
Left
+40
+30
+20
+10
00
-10
-20
-30
-40 mm
Anterior
+80
+70
+60
+50
+40
+30
+20
+10
00
-10
-20
-30
-40
mm
Posterior
Incisor teeth
Canine tooth
Premolar teeth
Molar teeth
Oral cavity proper
Choana
Sphenoid b., lat. pterygoid plate
Ramus of mandible
Mandibular n.
Parotid gland, superficial layer
Ramus of mandible, condylar process
Parotid gland, deep layer
Facial n.
Temporal b., mastoid antrum
Lambdoid suture
Occipital b.
Ext. occipital protuberance
Orbicularis oris
Transverse m.
Levator anguli oris m.
Buccinator m.
Med. pterygoid m.
Temporal m.
Longus capitis m.
Masseter m.
Maxillary a.
Temporal m.
Lat. pterygoid m.
Middle meningeal a.
Rectus capitis ant. m.
Int. carotid a.
Inf. petrosal sinus
Pons
Int. jugular v.
Sternocleido-mastoid m.
Digastric m., post. belly
Obliquus capitis sup. m.
Splenius capitis m.
Rectus capitis post. minor m.
Semispinalis capitis m.
Trapezius m.

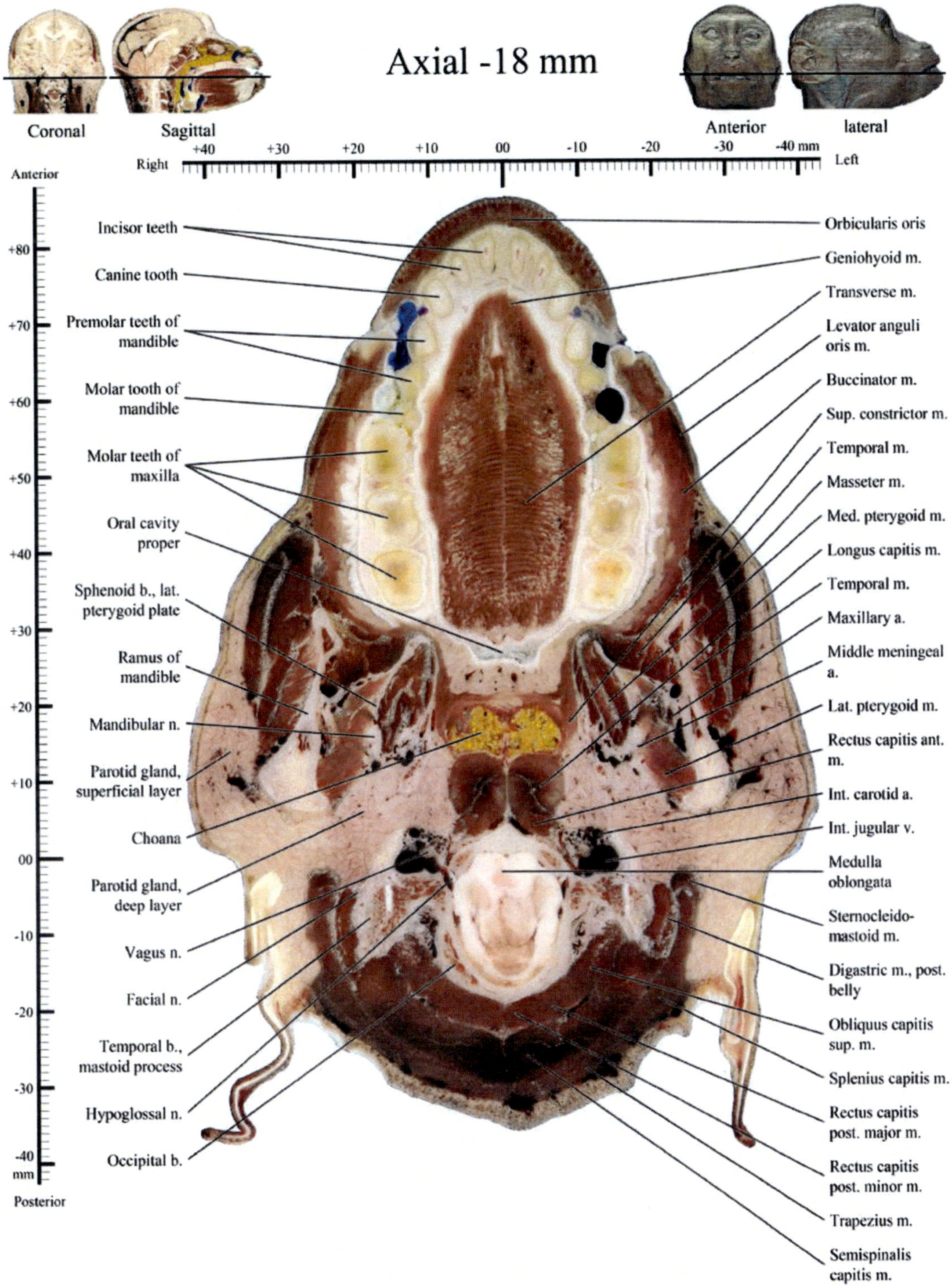
Coronal
Sagittal
Axial -18 mm
Anterior
lateral
Right
Left
+40
+30
+20
+10
00
-10
-20
-30
-40 mm
Anterior
Posterior
+80
+70
+60
+50
+40
+30
+20
+10
00
-10
-20
-30
-40 mm
Incisor teeth
Canine tooth
Premolar teeth of mandible
Molar tooth of mandible
Molar teeth of maxilla
Oral cavity proper
Sphenoid b., lat. pterygoid plate
Ramus of mandible
Mandibular n.
Parotid gland, superficial layer
Choana
Parotid gland, deep layer
Vagus n.
Facial n.
Temporal b., mastoid process
Hypoglossal n.
Occipital b.
Orbicularis oris
Geniohyoid m.
Transverse m.
Levator anguli oris m.
Buccinator m.
Sup. constrictor m.
Temporal m.
Masseter m.
Med. pterygoid m.
Longus capitis m.
Temporal m.
Maxillary a.
Middle meningeal a.
Lat. pterygoid m.
Rectus capitis ant. m.
Int. carotid a.
Int. jugular v.
Medulla oblongata
Sternocleido-mastoid m.
Digastric m., post. belly
Obliquus capitis sup. m.
Splenius capitis m.
Rectus capitis post. major m.
Rectus capitis post. minor m.
Trapezius m.
Semispinalis capitis m.

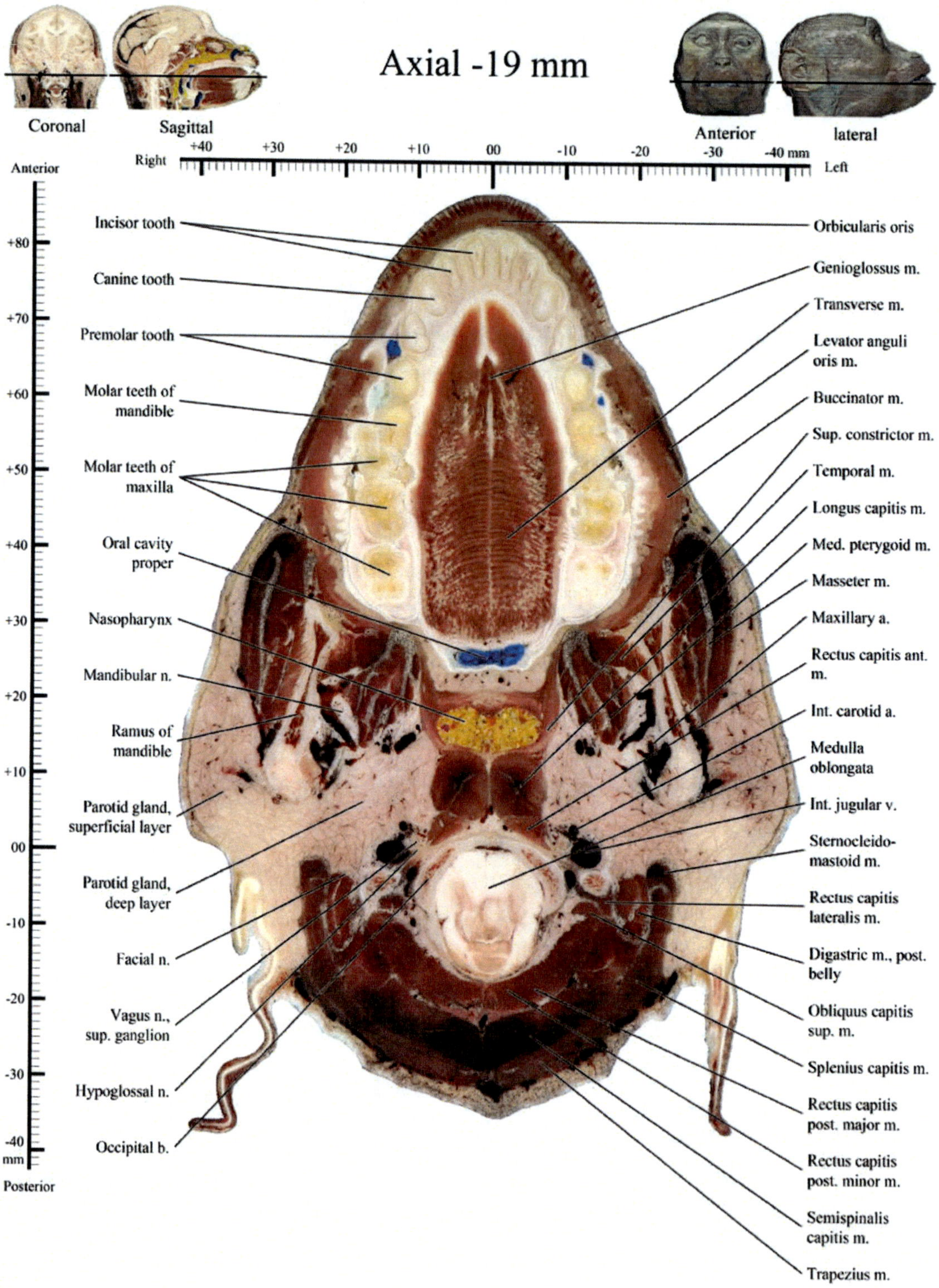
Axial -19 mm
Coronal
Sagittal
Anterior
lateral
Right
+40
+30
+20
+10
00
-10
-20
-30
-40 mm
Left
Anterior
+80
+70
+60
+50
+40
+30
+20
+10
00
-10
-20
-30
-40
mm
Posterior
Incisor tooth
Canine tooth
Premolar tooth
Molar teeth of mandible
Molar teeth of maxilla
Oral cavity proper
Nasopharynx
Mandibular n.
Ramus of mandible
Parotid gland, superficial layer
Parotid gland, deep layer
Facial n.
Vagus n., sup. ganglion
Hypoglossal n.
Occipital b.
Orbicularis oris
Genioglossus m.
Transverse m.
Levator anguli oris m.
Buccinator m.
Sup. constrictor m.
Temporal m.
Longus capitis m.
Med. pterygoid m.
Masseter m.
Maxillary a.
Rectus capitis ant. m.
Int. carotid a.
Medulla oblongata
Int. jugular v.
Sternocleido-mastoid m.
Rectus capitis lateralis m.
Digastric m., post. belly
Obliquus capitis sup. m.
Splenius capitis m.
Rectus capitis post. major m.
Rectus capitis post. minor m.
Semispinalis capitis m.
Trapezius m.

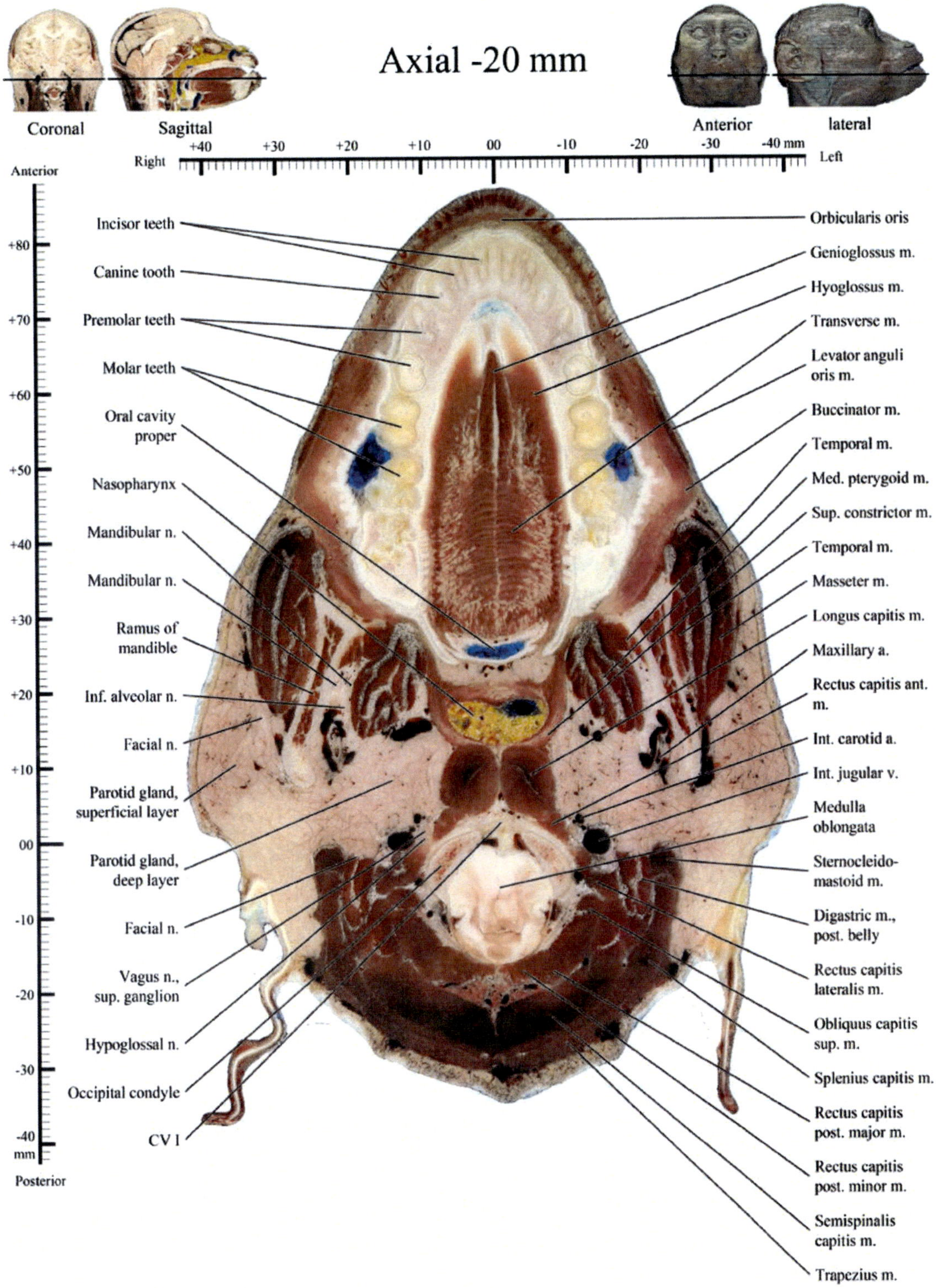

Axial -20 mm
Coronal
Sagittal
Anterior
lateral
Right
+40 +30 +20 +10 00 -10 -20 -30 -40 mm
Left
Anterior
+80 +70 +60 +50 +40 +30 +20 +10 00 -10 -20 -30 -40 mm
Posterior
Incisor teeth
Canine tooth
Premolar teeth
Molar teeth
Oral cavity proper
Nasopharynx
Mandibular n.
Mandibular n.
Ramus of mandible
Inf. alveolar n.
Facial n.
Parotid gland, superficial layer
Parotid gland, deep layer
Facial n.
Vagus n., sup. ganglion
Hypoglossal n.
Occipital condyle
CV I
Orbicularis oris
Genioglossus m.
Hyoglossus m.
Transverse m.
Levator anguli oris m.
Buccinator m.
Temporal m.
Med. pterygoid m.
Sup. constrictor m.
Temporal m.
Masseter m.
Longus capitis m.
Maxillary a.
Rectus capitis ant. m.
Int. carotid a.
Int. jugular v.
Medulla oblongata
Sternocleido-mastoid m.
Digastric m., post. belly
Rectus capitis lateralis m.
Obliquus capitis sup. m.
Splenius capitis m.
Rectus capitis post. major m.
Rectus capitis post. minor m.
Semispinalis capitis m.
Trapezius m.

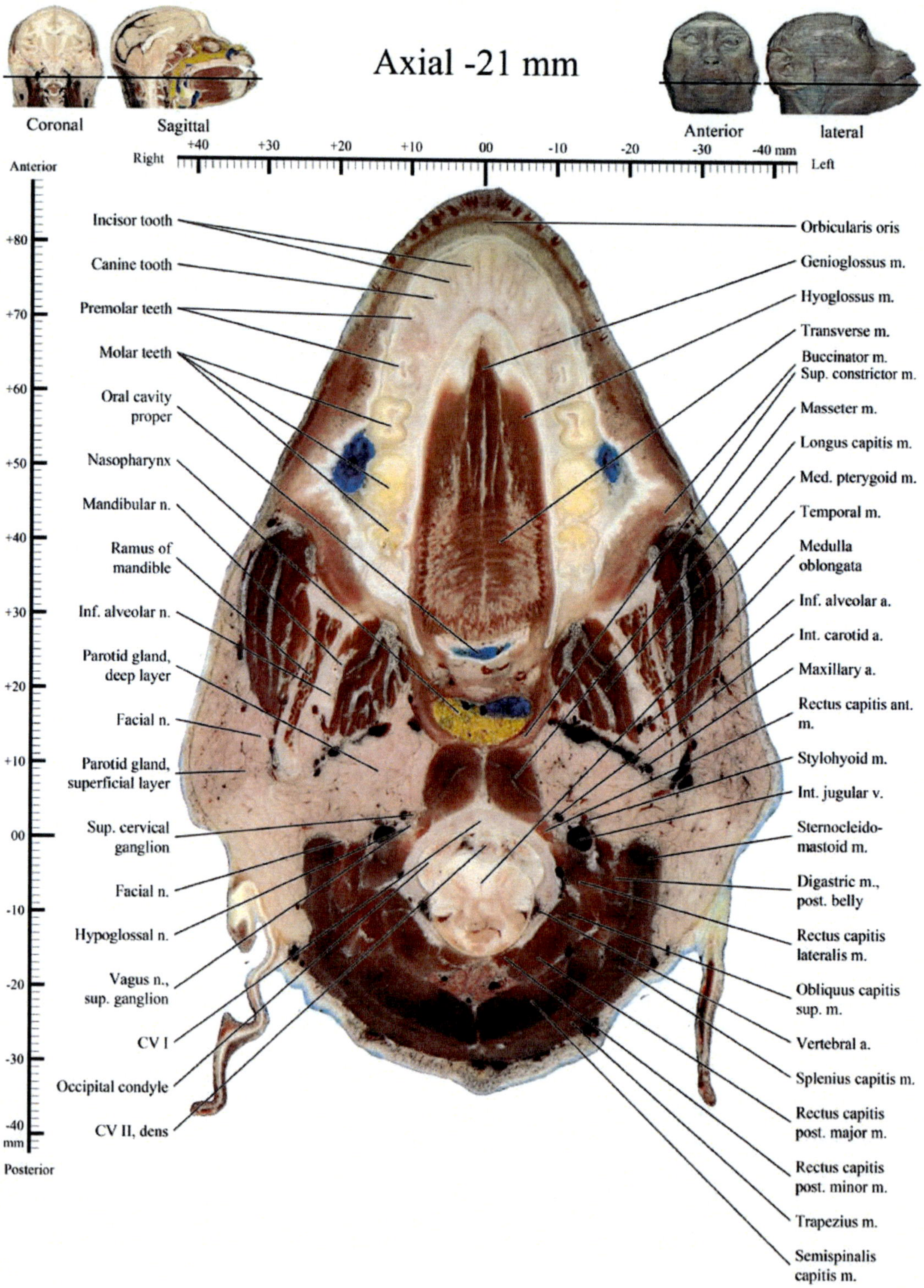
Axial -21 mm
Coronal
Sagittal
Anterior
lateral
Anterior
Right
Left
+40
+30
+20
+10
00
-10
-20
-30
-40 mm
+80
+70
+60
+50
+40
+30
+20
+10
00
-10
-20
-30
-40
mm
Posterior
Incisor tooth
Canine tooth
Premolar teeth
Molar teeth
Oral cavity proper
Nasopharynx
Mandibular n.
Ramus of mandible
Inf. alveolar n.
Parotid gland, deep layer
Facial n.
Parotid gland, superficial layer
Sup. cervical ganglion
Facial n.
Hypoglossal n.
Vagus n., sup. ganglion
CV I
Occipital condyle
CV II, dens
Orbicularis oris
Genioglossus m.
Hyoglossus m.
Transverse m.
Buccinator m.
Sup. constrictor m.
Masseter m.
Longus capitis m.
Med. pterygoid m.
Temporal m.
Medulla oblongata
Inf. alveolar a.
Int. carotid a.
Maxillary a.
Rectus capitis ant. m.
Stylohyoid m.
Int. jugular v.
Sternocleido-mastoid m.
Digastric m., post. belly
Rectus capitis lateralis m.
Obliquus capitis sup. m.
Vertebral a.
Splenius capitis m.
Rectus capitis post. major m.
Rectus capitis post. minor m.
Trapezius m.
Semispinalis capitis m.

Axial -22 mm

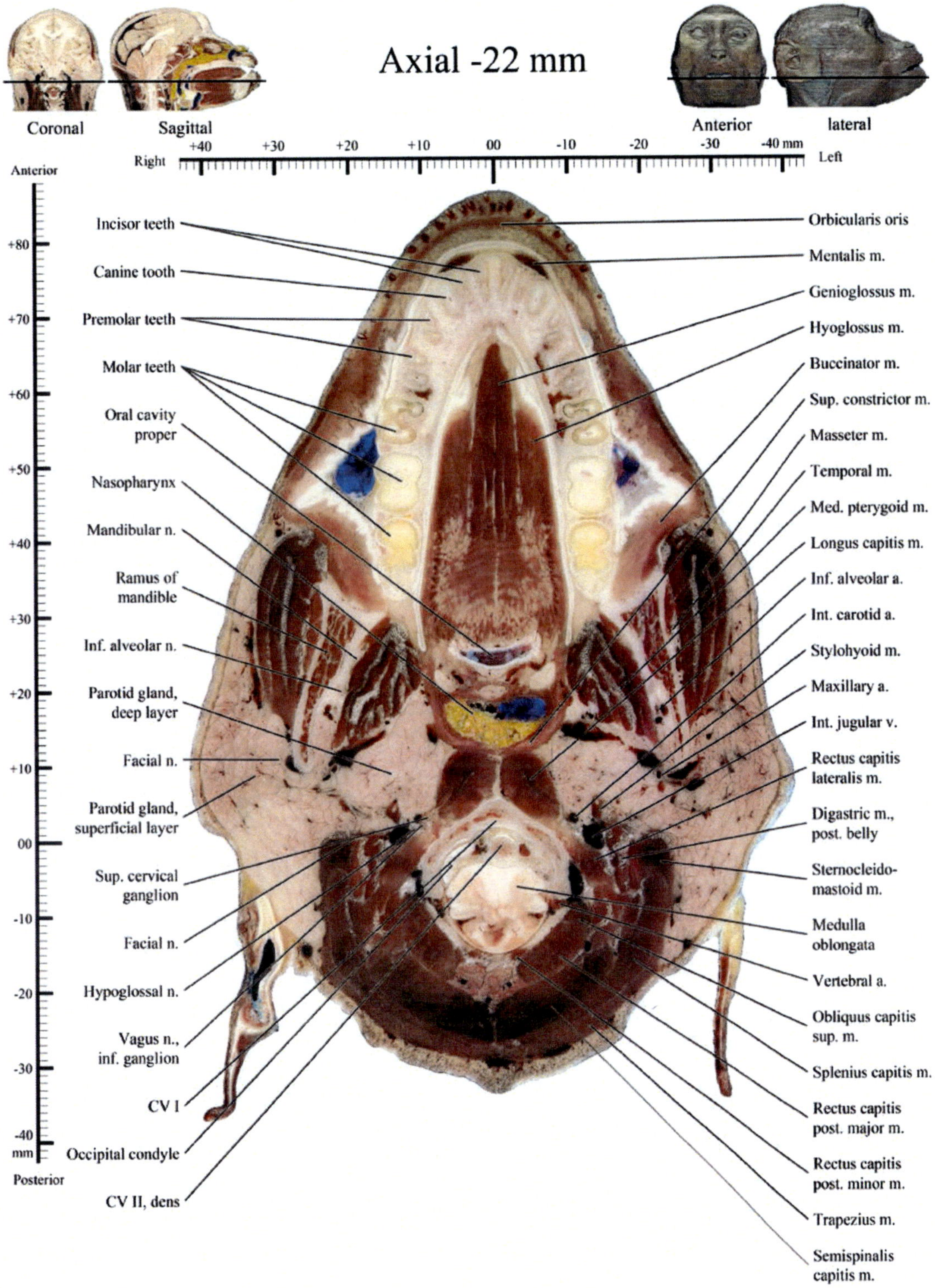

Coronal
Sagittal
Anterior
lateral
Right
Left
+40
+30
+20
+10
00
-10
-20
-30
-40 mm
Anterior
+80
+70
+60
+50
+40
+30
+20
+10
00
-10
-20
-30
-40 mm
Posterior
Incisor teeth
Canine tooth
Premolar teeth
Molar teeth
Oral cavity proper
Nasopharynx
Mandibular n.
Ramus of mandible
Inf. alveolar n.
Parotid gland, deep layer
Facial n.
Parotid gland, superficial layer
Sup. cervical ganglion
Facial n.
Hypoglossal n.
Vagus n., inf. ganglion
CV I
Occipital condyle
CV II, dens
Orbicularis oris
Mentalis m.
Genioglossus m.
Hyoglossus m.
Buccinator m.
Sup. constrictor m.
Masseter m.
Temporal m.
Med. pterygoid m.
Longus capitis m.
Inf. alveolar a.
Int. carotid a.
Stylohyoid m.
Maxillary a.
Int. jugular v.
Rectus capitis lateralis m.
Digastric m., post. belly
Sternocleido-mastoid m.
Medulla oblongata
Vertebral a.
Obliquus capitis sup. m.
Splenius capitis m.
Rectus capitis post. major m.
Rectus capitis post. minor m.
Trapezius m.
Semispinalis capitis m.

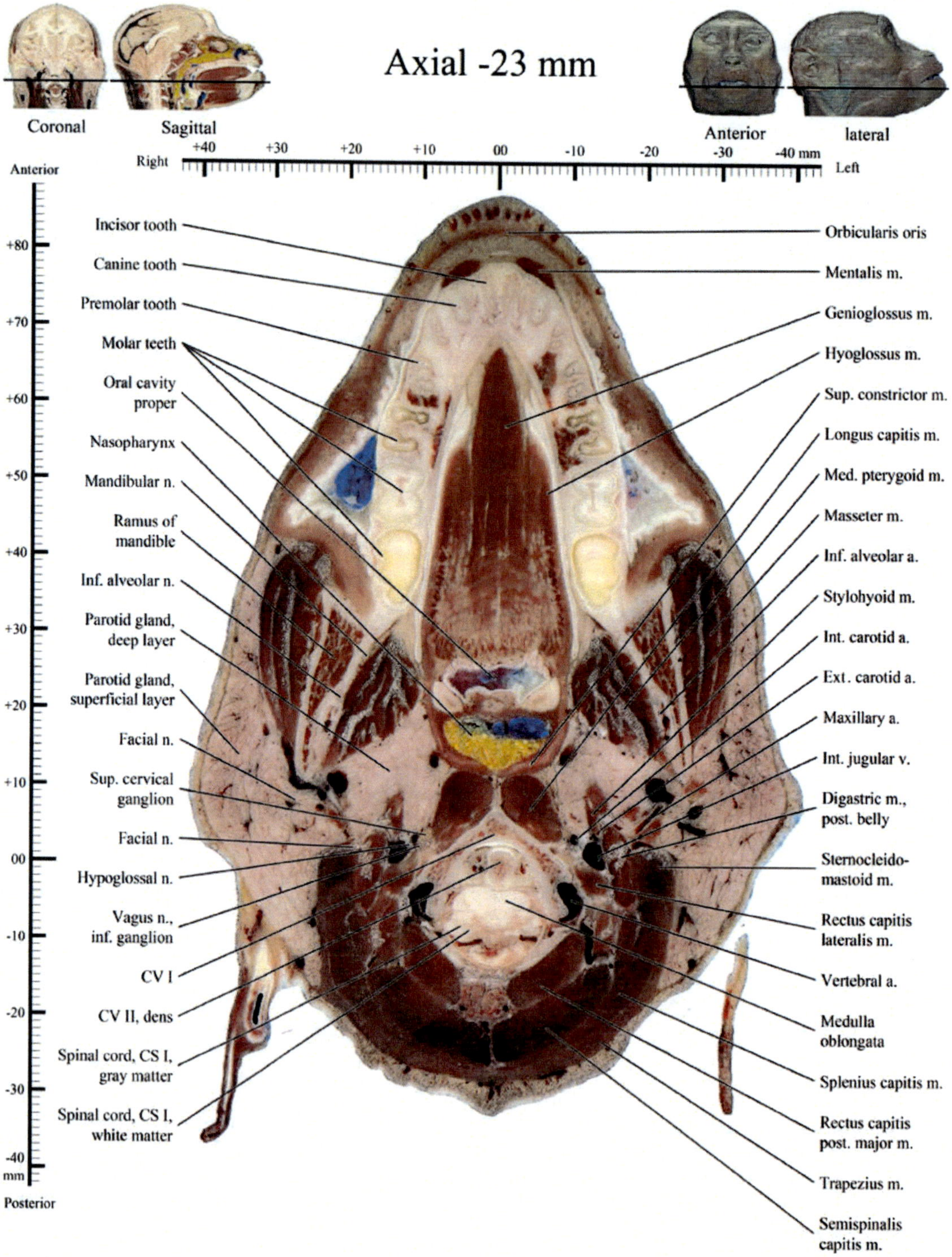
Coronal
Sagittal
Axial -23 mm
Anterior
lateral
Right
+40
+30
+20
+10
00
-10
-20
-30
-40 mm
Left
Anterior
+80
+70
+60
+50
+40
+30
+20
+10
00
-10
-20
-30
-40
mm
Posterior
Incisor tooth
Canine tooth
Premolar tooth
Molar teeth
Oral cavity proper
Nasopharynx
Mandibular n.
Ramus of mandible
Inf. alveolar n.
Parotid gland, deep layer
Parotid gland, superficial layer
Facial n.
Sup. cervical ganglion
Facial n.
Hypoglossal n.
Vagus n., inf. ganglion
CV I
CV II, dens
Spinal cord, CS I, gray matter
Spinal cord, CS I, white matter
Orbicularis oris
Mentalis m.
Genioglossus m.
Hyoglossus m.
Sup. constrictor m.
Longus capitis m.
Med. pterygoid m.
Masseter m.
Inf. alveolar a.
Stylohyoid m.
Int. carotid a.
Ext. carotid a.
Maxillary a.
Int. jugular v.
Digastric m., post. belly
Sternocleido-mastoid m.
Rectus capitis lateralis m.
Vertebral a.
Medulla oblongata
Splenius capitis m.
Rectus capitis post. major m.
Trapezius m.
Semispinalis capitis m.

Axial -24 mm

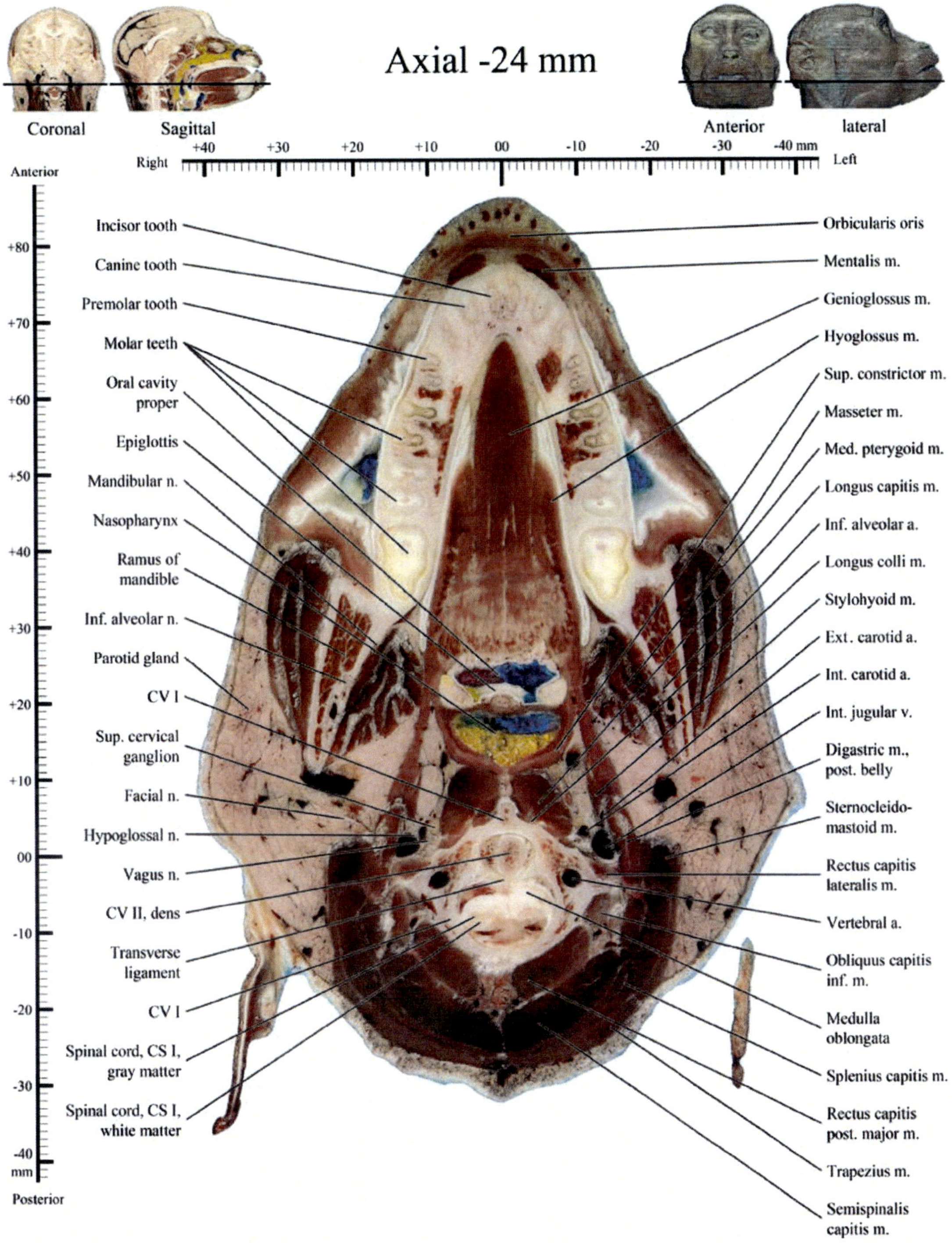
Coronal
Sagittal
Anterior
lateral
Right
Left
+40 +30 +20 +10 00 -10 -20 -30 -40 mm
Anterior
Posterior
+80 +70 +60 +50 +40 +30 +20 +10 00 -10 -20 -30 -40 mm
Incisor tooth
Canine tooth
Premolar tooth
Molar teeth
Oral cavity proper
Epiglottis
Mandibular n.
Nasopharynx
Ramus of mandible
Inf. alveolar n.
Parotid gland
CV I
Sup. cervical ganglion
Facial n.
Hypoglossal n.
Vagus n.
CV II, dens
Transverse ligament
CV I
Spinal cord, CS I, gray matter
Spinal cord, CS I, white matter
Orbicularis oris
Mentalis m.
Genioglossus m.
Hyoglossus m.
Sup. constrictor m.
Masseter m.
Med. pterygoid m.
Longus capitis m.
Inf. alveolar a.
Longus colli m.
Stylohyoid m.
Ext. carotid a.
Int. carotid a.
Int. jugular v.
Digastric m., post. belly
Sternocleido-mastoid m.
Rectus capitis lateralis m.
Vertebral a.
Obliquus capitis inf. m.
Medulla oblongata
Splenius capitis m.
Rectus capitis post. major m.
Trapezius m.
Semispinalis capitis m.

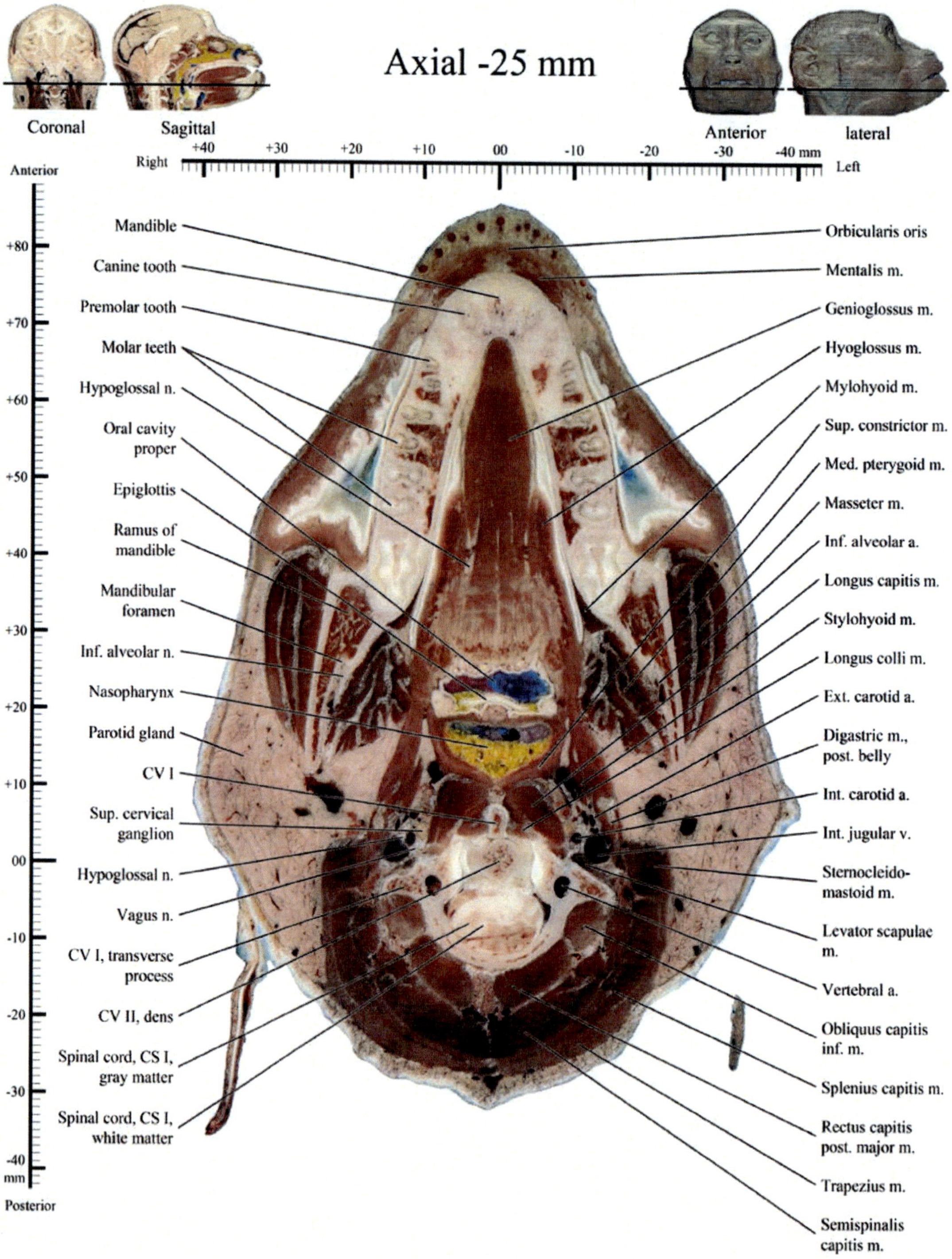
Axial -25 mm
Coronal
Sagittal
Anterior
lateral
Right
+40
+30
+20
+10
00
-10
-20
-30
-40 mm
Left
Anterior
+80
+70
+60
+50
+40
+30
+20
+10
00
-10
-20
-30
-40
mm
Posterior
Mandible
Canine tooth
Premolar tooth
Molar teeth
Hypoglossal n.
Oral cavity proper
Epiglottis
Ramus of mandible
Mandibular foramen
Inf. alveolar n.
Nasopharynx
Parotid gland
CV I
Sup. cervical ganglion
Hypoglossal n.
Vagus n.
CV I, transverse process
CV II, dens
Spinal cord, CS I, gray matter
Spinal cord, CS I, white matter
Orbicularis oris
Mentalis m.
Genioglossus m.
Hyoglossus m.
Mylohyoid m.
Sup. constrictor m.
Med. pterygoid m.
Masseter m.
Inf. alveolar a.
Longus capitis m.
Stylohyoid m.
Longus colli m.
Ext. carotid a.
Digastric m., post. belly
Int. carotid a.
Int. jugular v.
Sternocleido-mastoid m.
Levator scapulae m.
Vertebral a.
Obliquus capitis inf. m.
Splenius capitis m.
Rectus capitis post. major m.
Trapezius m.
Semispinalis capitis m.

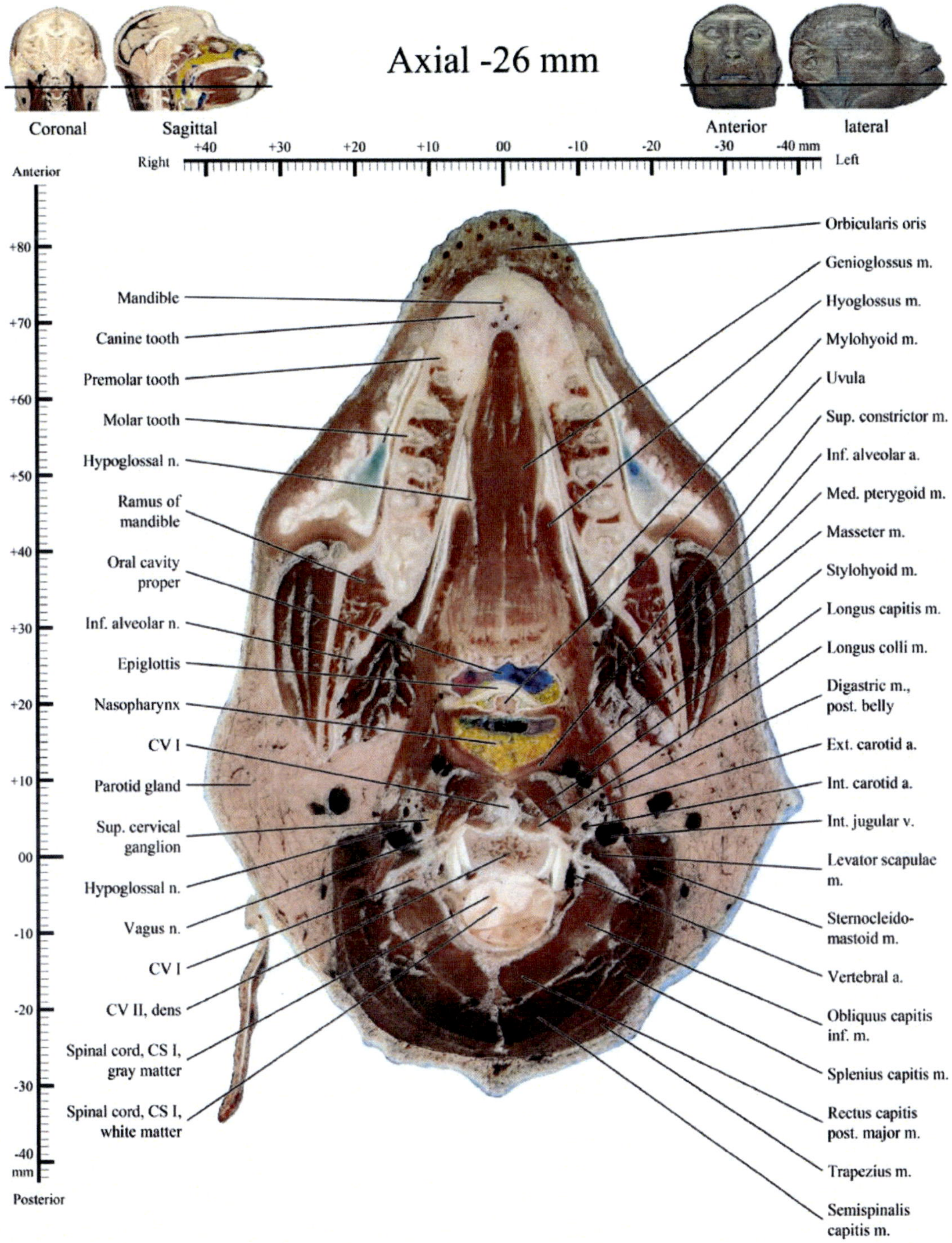

Coronal
Sagittal
Axial -26 mm
Anterior
lateral
+40 +30 +20 +10 00 -10 -20 -30 -40 mm
Right
Left
Anterior
+80
+70
+60
+50
+40
+30
+20
+10
00
-10
-20
-30
-40 mm
Posterior
Mandible
Canine tooth
Premolar tooth
Molar tooth
Hypoglossal n.
Ramus of mandible
Oral cavity proper
Inf. alveolar n.
Epiglottis
Nasopharynx
CV I
Parotid gland
Sup. cervical ganglion
Hypoglossal n.
Vagus n.
CV I
CV II, dens
Spinal cord, CS I, gray matter
Spinal cord, CS I, white matter
Orbicularis oris
Genioglossus m.
Hyoglossus m.
Mylohyoid m.
Uvula
Sup. constrictor m.
Inf. alveolar a.
Med. pterygoid m.
Masseter m.
Stylohyoid m.
Longus capitis m.
Longus colli m.
Digastric m., post. belly
Ext. carotid a.
Int. carotid a.
Int. jugular v.
Levator scapulae m.
Sternocleido-mastoid m.
Vertebral a.
Obliquus capitis inf. m.
Splenius capitis m.
Rectus capitis post. major m.
Trapezius m.
Semispinalis capitis m.

Axial -27 mm

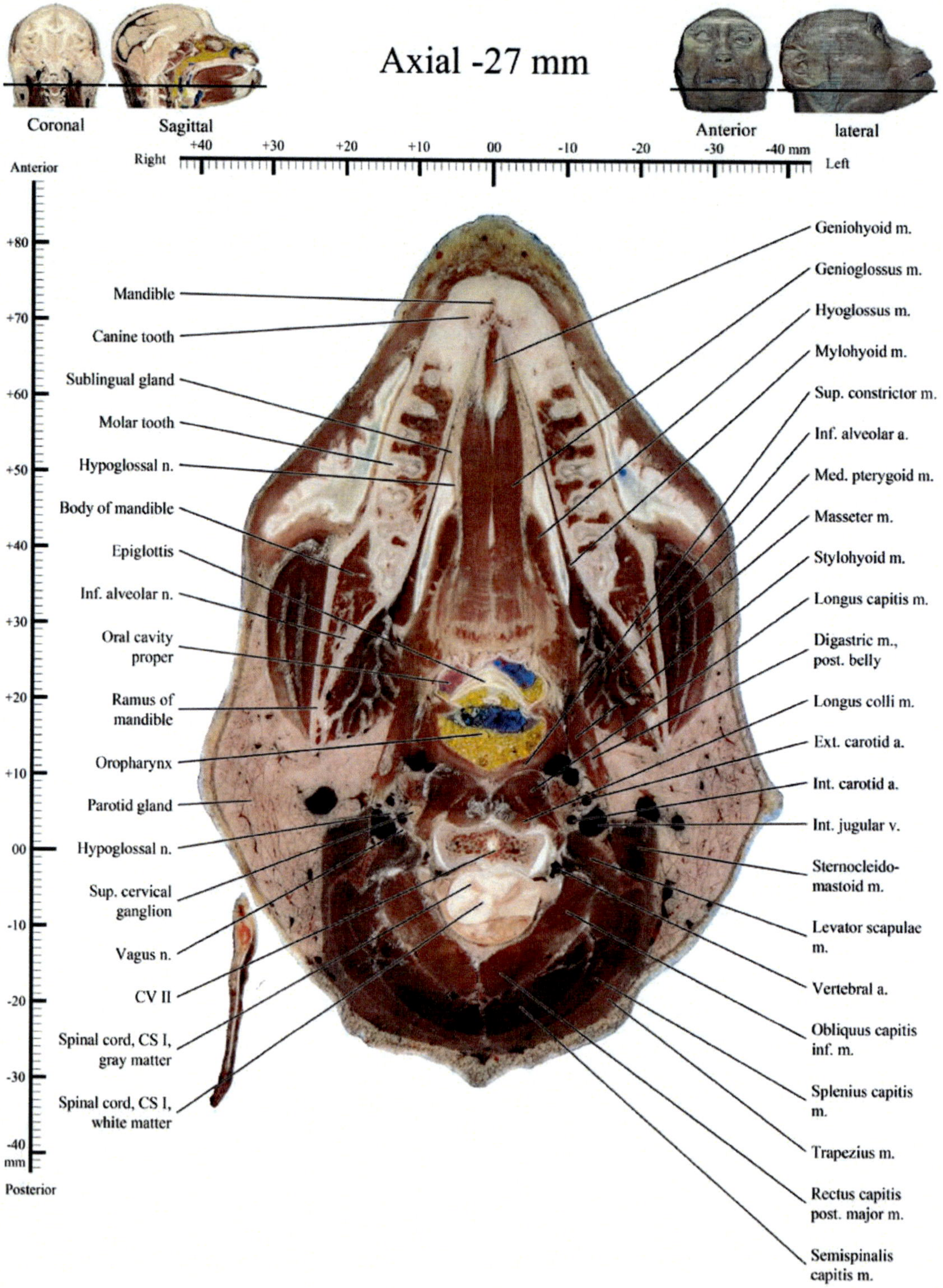
Coronal
Sagittal
Anterior
lateral
Right
+40 +30 +20 +10 00 -10 -20 -30 -40 mm
Left
Anterior
+80
+70
+60
+50
+40
+30
+20
+10
00
-10
-20
-30
-40 mm
Posterior
Mandible
Canine tooth
Sublingual gland
Molar tooth
Hypoglossal n.
Body of mandible
Epiglottis
Inf. alveolar n.
Oral cavity proper
Ramus of mandible
Oropharynx
Parotid gland
Hypoglossal n.
Sup. cervical ganglion
Vagus n.
CV II
Spinal cord, CS I, gray matter
Spinal cord, CS I, white matter
Geniohyoid m.
Genioglossus m.
Hyoglossus m.
Mylohyoid m.
Sup. constrictor m.
Inf. alveolar a.
Med. pterygoid m.
Masseter m.
Stylohyoid m.
Longus capitis m.
Digastric m., post. belly
Longus colli m.
Ext. carotid a.
Int. carotid a.
Int. jugular v.
Sternocleido-mastoid m.
Levator scapulae m.
Vertebral a.
Obliquus capitis inf. m.
Splenius capitis m.
Trapezius m.
Rectus capitis post. major m.
Semispinalis capitis m.

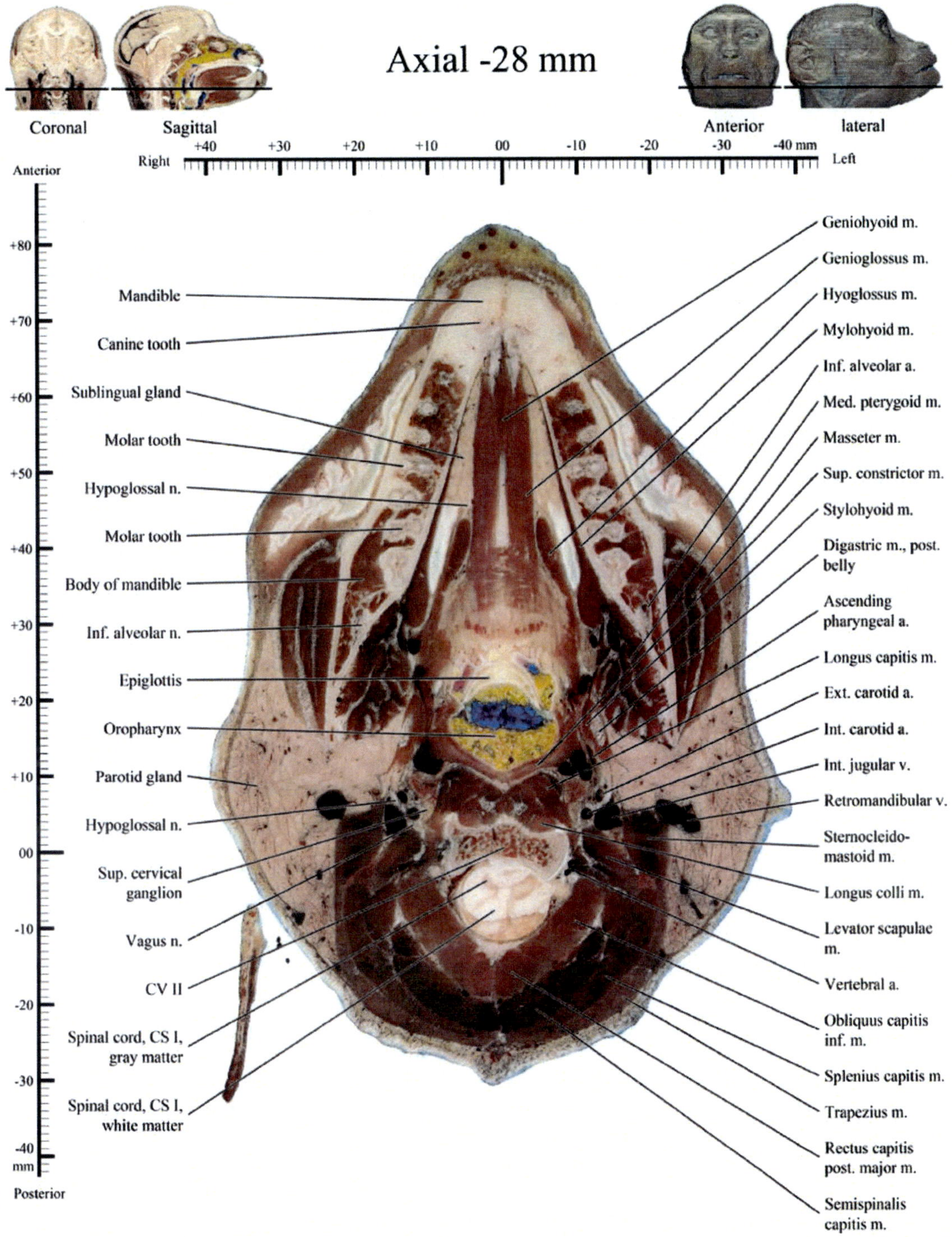
Axial -28 mm
Coronal
Sagittal
Anterior
lateral
Right
Left
+40
+30
+20
+10
00
-10
-20
-30
-40 mm
Anterior
Posterior
+80
+70
+60
+50
+40
+30
+20
+10
00
-10
-20
-30
-40 mm
Mandible
Canine tooth
Sublingual gland
Molar tooth
Hypoglossal n.
Molar tooth
Body of mandible
Inf. alveolar n.
Epiglottis
Oropharynx
Parotid gland
Hypoglossal n.
Sup. cervical ganglion
Vagus n.
CV II
Spinal cord, CS I, gray matter
Spinal cord, CS I, white matter
Geniohyoid m.
Genioglossus m.
Hyoglossus m.
Mylohyoid m.
Inf. alveolar a.
Med. pterygoid m.
Masseter m.
Sup. constrictor m.
Stylohyoid m.
Digastric m., post. belly
Ascending pharyngeal a.
Longus capitis m.
Ext. carotid a.
Int. carotid a.
Int. jugular v.
Retromandibular v.
Sternocleido-mastoid m.
Longus colli m.
Levator scapulae m.
Vertebral a.
Obliquus capitis inf. m.
Splenius capitis m.
Trapezius m.
Rectus capitis post. major m.
Semispinalis capitis m.

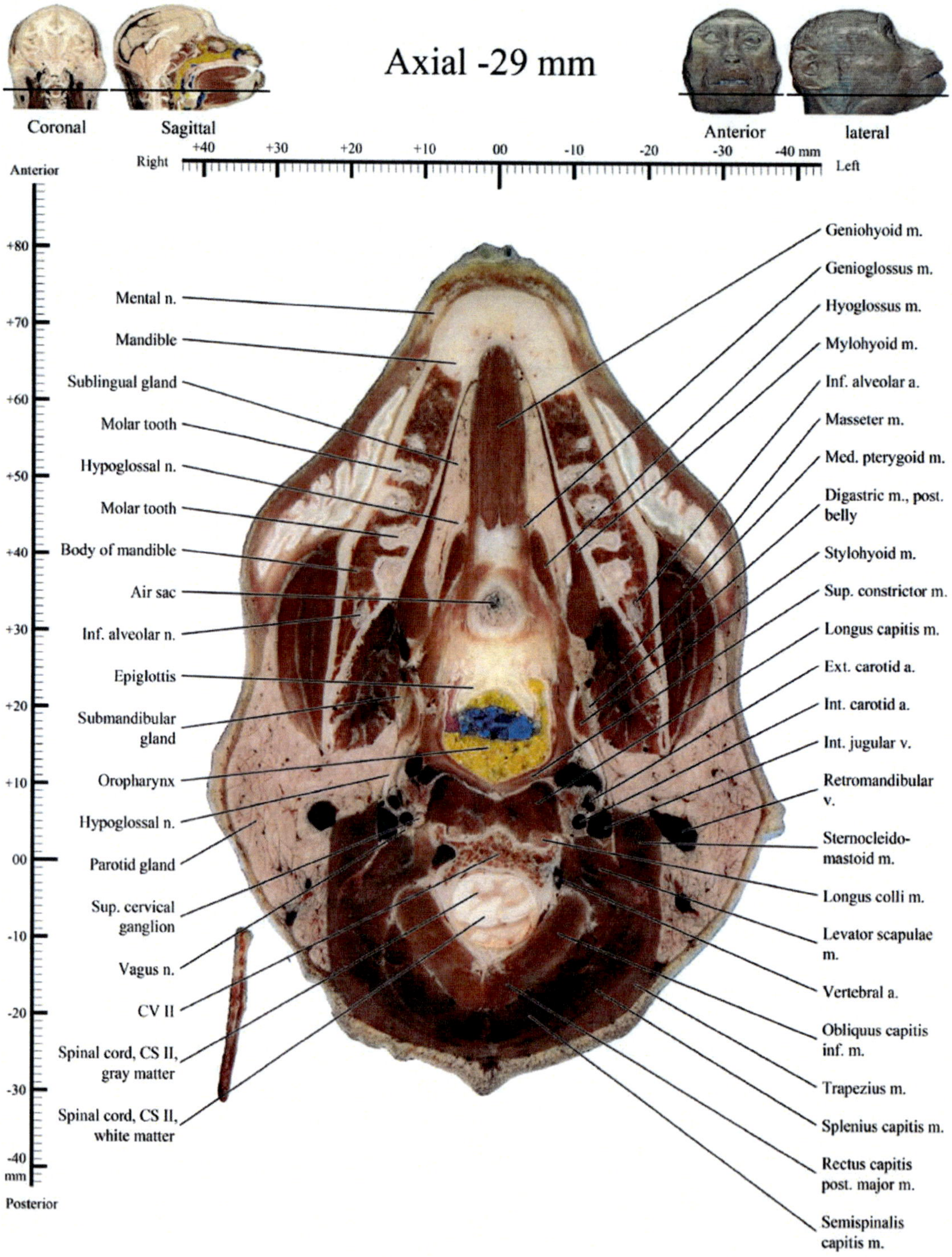
Axial -29 mm
Coronal
Sagittal
Anterior
lateral
Right
+40 +30 +20 +10 00 -10 -20 -30 -40 mm
Left
Anterior
+80
+70
+60
+50
+40
+30
+20
+10
00
-10
-20
-30
-40 mm
Posterior
Mental n.
Mandible
Sublingual gland
Molar tooth
Hypoglossal n.
Molar tooth
Body of mandible
Air sac
Inf. alveolar n.
Epiglottis
Submandibular gland
Oropharynx
Hypoglossal n.
Parotid gland
Sup. cervical ganglion
Vagus n.
CV II
Spinal cord, CS II, gray matter
Spinal cord, CS II, white matter
Geniohyoid m.
Genioglossus m.
Hyoglossus m.
Mylohyoid m.
Inf. alveolar a.
Masseter m.
Med. pterygoid m.
Digastric m., post. belly
Stylohyoid m.
Sup. constrictor m.
Longus capitis m.
Ext. carotid a.
Int. carotid a.
Int. jugular v.
Retromandibular v.
Sternocleido-mastoid m.
Longus colli m.
Levator scapulae m.
Vertebral a.
Obliquus capitis inf. m.
Trapezius m.
Splenius capitis m.
Rectus capitis post. major m.
Semispinalis capitis m.

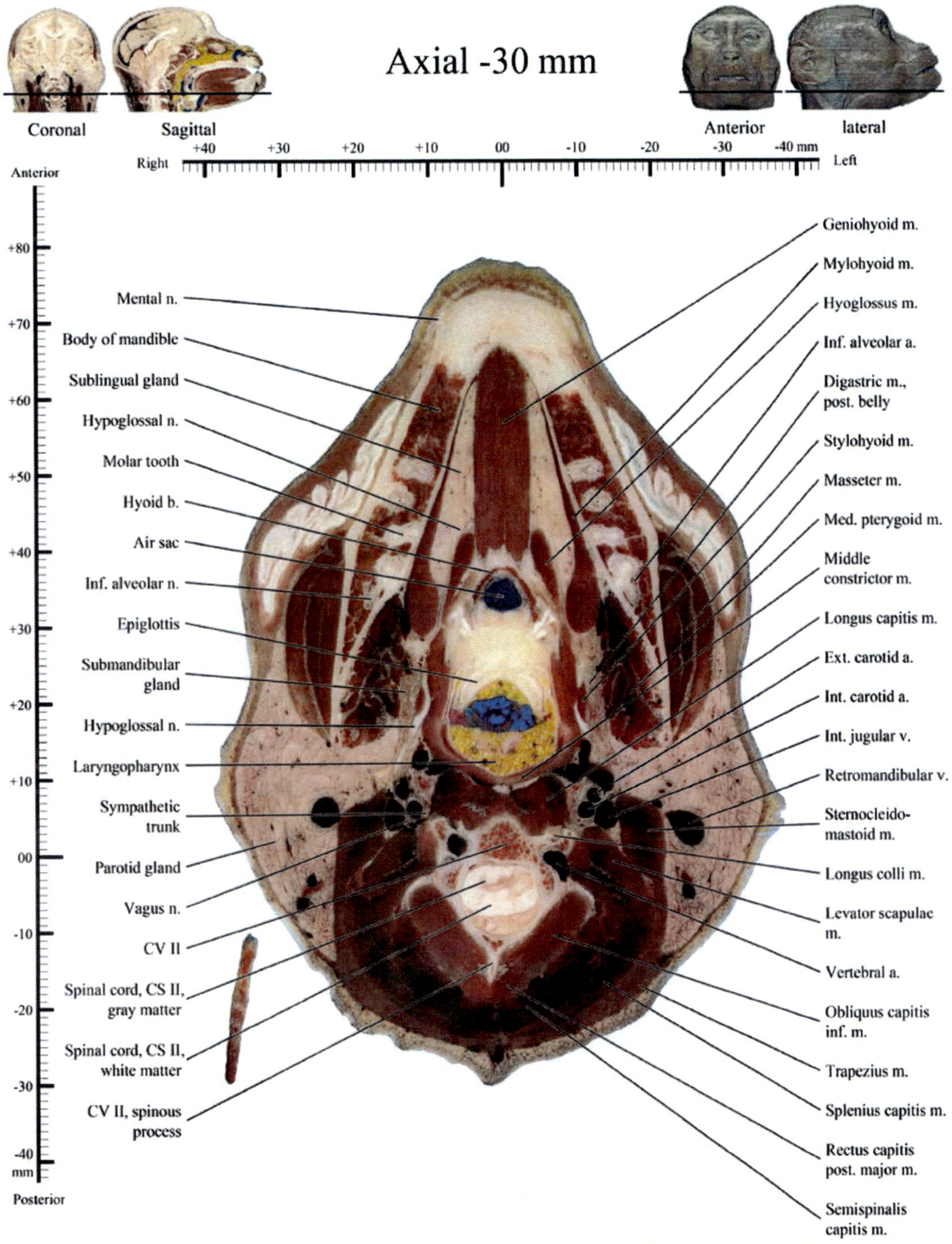
Axial -30 mm
Coronal
Sagittal
Anterior
lateral
Right
Left
+40
+30
+20
+10
00
-10
-20
-30
-40 mm
Anterior
+80
+70
+60
+50
+40
+30
+20
+10
00
-10
-20
-30
-40 mm
Posterior
Mental n.
Body of mandible
Sublingual gland
Hypoglossal n.
Molar tooth
Hyoid b.
Air sac
Inf. alveolar n.
Epiglottis
Submandibular gland
Hypoglossal n.
Laryngopharynx
Sympathetic trunk
Parotid gland
Vagus n.
CV II
Spinal cord, CS II, gray matter
Spinal cord, CS II, white matter
CV II, spinous process
Geniohyoid m.
Mylohyoid m.
Hyoglossus m.
Inf. alveolar a.
Digastric m., post. belly
Stylohyoid m.
Masseter m.
Med. pterygoid m.
Middle constrictor m.
Longus capitis m.
Ext. carotid a.
Int. carotid a.
Int. jugular v.
Retromandibular v.
Sternocleido-mastoid m.
Longus colli m.
Levator scapulae m.
Vertebral a.
Obliquus capitis inf. m.
Trapezius m.
Splenius capitis m.
Rectus capitis post. major m.
Semispinalis capitis m.

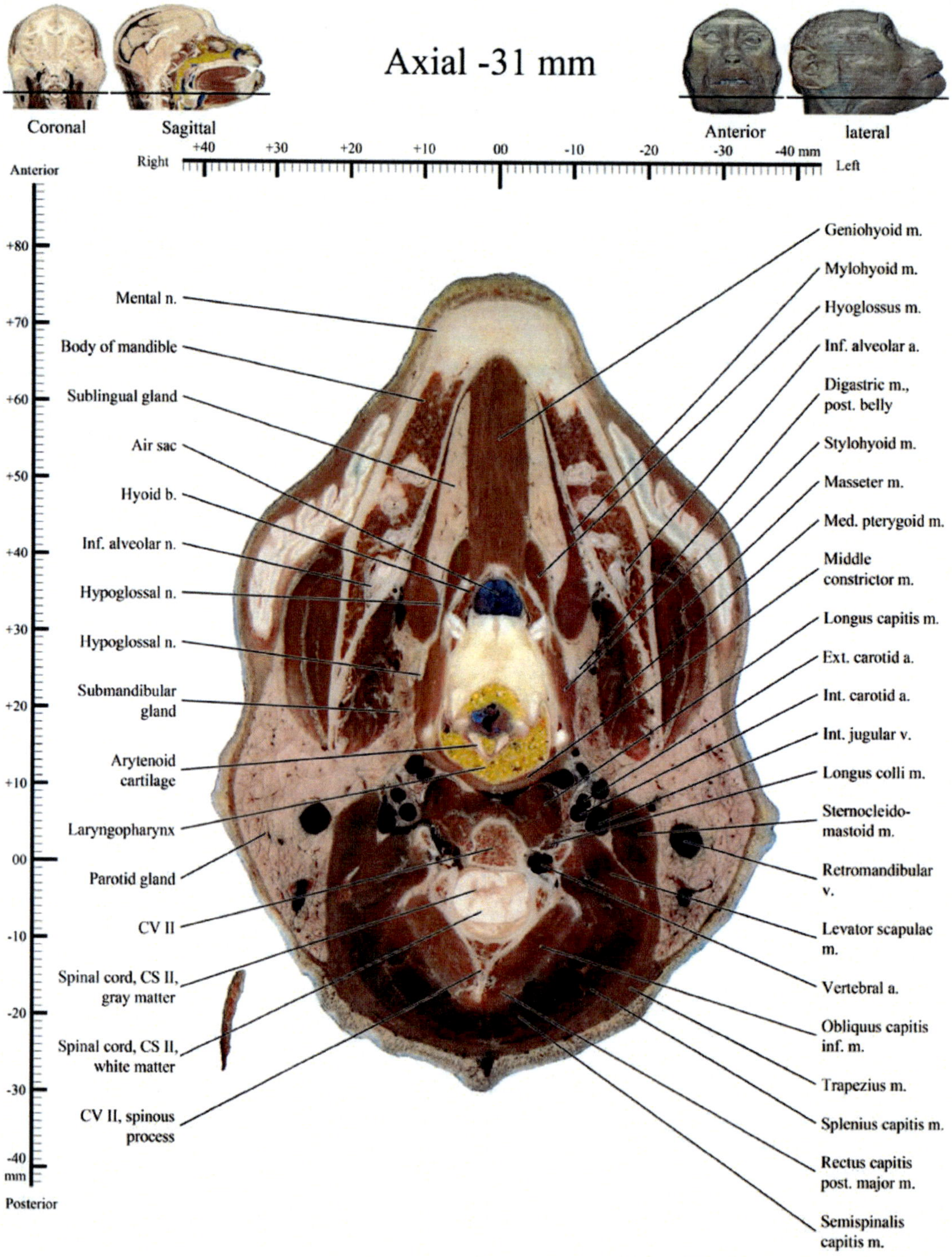
Axial -31 mm
Coronal
Sagittal
Anterior
lateral
Right
Left
+40
+30
+20
+10
00
-10
-20
-30
-40 mm
Anterior
Posterior
+80
+70
+60
+50
+40
+30
+20
+10
00
-10
-20
-30
-40 mm
Mental n.
Body of mandible
Sublingual gland
Air sac
Hyoid b.
Inf. alveolar n.
Hypoglossal n.
Hypoglossal n.
Submandibular gland
Arytenoid cartilage
Laryngopharynx
Parotid gland
CV II
Spinal cord, CS II, gray matter
Spinal cord, CS II, white matter
CV II, spinous process
Geniohyoid m.
Mylohyoid m.
Hyoglossus m.
Inf. alveolar a.
Digastric m., post. belly
Stylohyoid m.
Masseter m.
Med. pterygoid m.
Middle constrictor m.
Longus capitis m.
Ext. carotid a.
Int. carotid a.
Int. jugular v.
Longus colli m.
Sternocleido-mastoid m.
Retromandibular v.
Levator scapulae m.
Vertebral a.
Obliquus capitis inf. m.
Trapezius m.
Splenius capitis m.
Rectus capitis post. major m.
Semispinalis capitis m.

Axial -32 mm

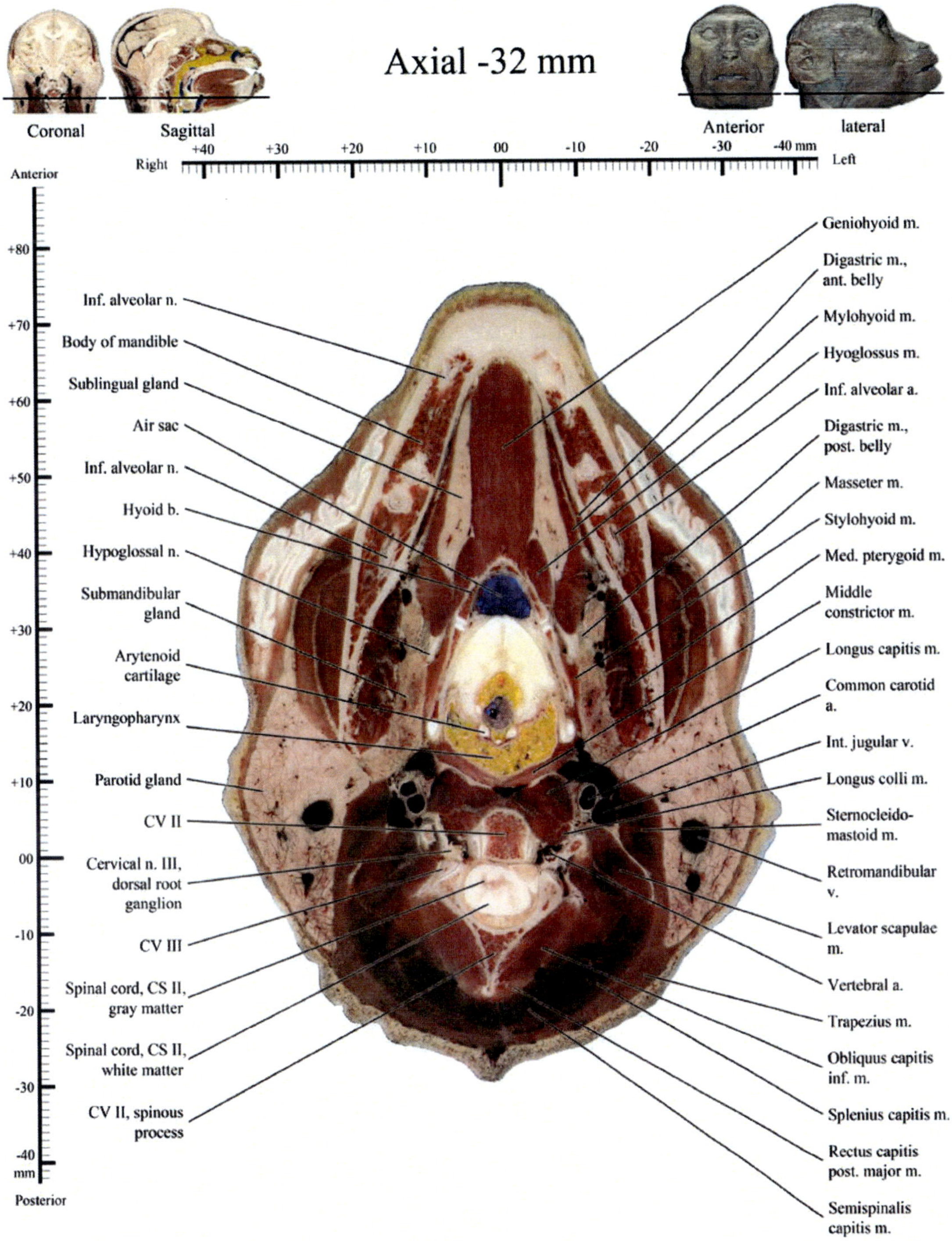

Coronal
Sagittal
Anterior
lateral
+40
+30
+20
+10
00
-10
-20
-30
-40 mm
Right
Left
Anterior
+80
+70
+60
+50
+40
+30
+20
+10
00
-10
-20
-30
-40
mm
Posterior
Inf. alveolar n.
Body of mandible
Sublingual gland
Air sac
Inf. alveolar n.
Hyoid b.
Hypoglossal n.
Submandibular gland
Arytenoid cartilage
Laryngopharynx
Parotid gland
CV II
Cervical n. III, dorsal root ganglion
CV III
Spinal cord, CS II, gray matter
Spinal cord, CS II, white matter
CV II, spinous process
Geniohyoid m.
Digastric m., ant. belly
Mylohyoid m.
Hyoglossus m.
Inf. alveolar a.
Digastric m., post. belly
Masseter m.
Stylohyoid m.
Med. pterygoid m.
Middle constrictor m.
Longus capitis m.
Common carotid a.
Int. jugular v.
Longus colli m.
Sternocleido-mastoid m.
Retromandibular v.
Levator scapulae m.
Vertebral a.
Trapezius m.
Obliquus capitis inf. m.
Splenius capitis m.
Rectus capitis post. major m.
Semispinalis capitis m.

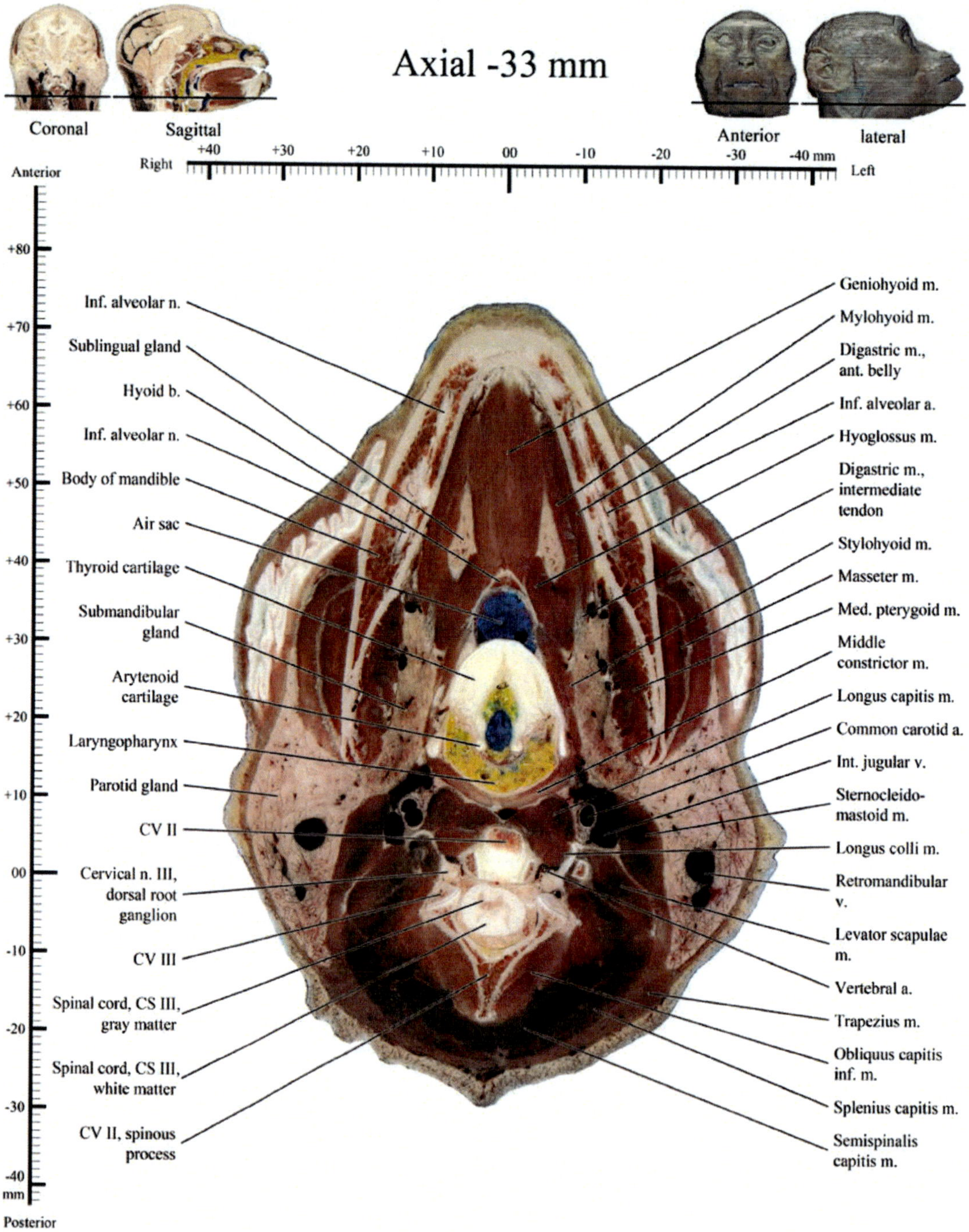
Axial -33 mm
Coronal
Sagittal
Anterior
lateral
Right
+40 +30 +20 +10 00 -10 -20 -30 -40 mm
Left
Anterior
+80 +70 +60 +50 +40 +30 +20 +10 00 -10 -20 -30 -40 mm
Posterior
Inf. alveolar n.
Sublingual gland
Hyoid b.
Inf. alveolar n.
Body of mandible
Air sac
Thyroid cartilage
Submandibular gland
Arytenoid cartilage
Laryngopharynx
Parotid gland
CV II
Cervical n. III, dorsal root ganglion
CV III
Spinal cord, CS III, gray matter
Spinal cord, CS III, white matter
CV II, spinous process
Geniohyoid m.
Mylohyoid m.
Digastric m., ant. belly
Inf. alveolar a.
Hyoglossus m.
Digastric m., intermediate tendon
Stylohyoid m.
Masseter m.
Med. pterygoid m.
Middle constrictor m.
Longus capitis m.
Common carotid a.
Int. jugular v.
Sternocleido-mastoid m.
Longus colli m.
Retromandibular v.
Levator scapulae m.
Vertebral a.
Trapezius m.
Obliquus capitis inf. m.
Splenius capitis m.
Semispinalis capitis m.

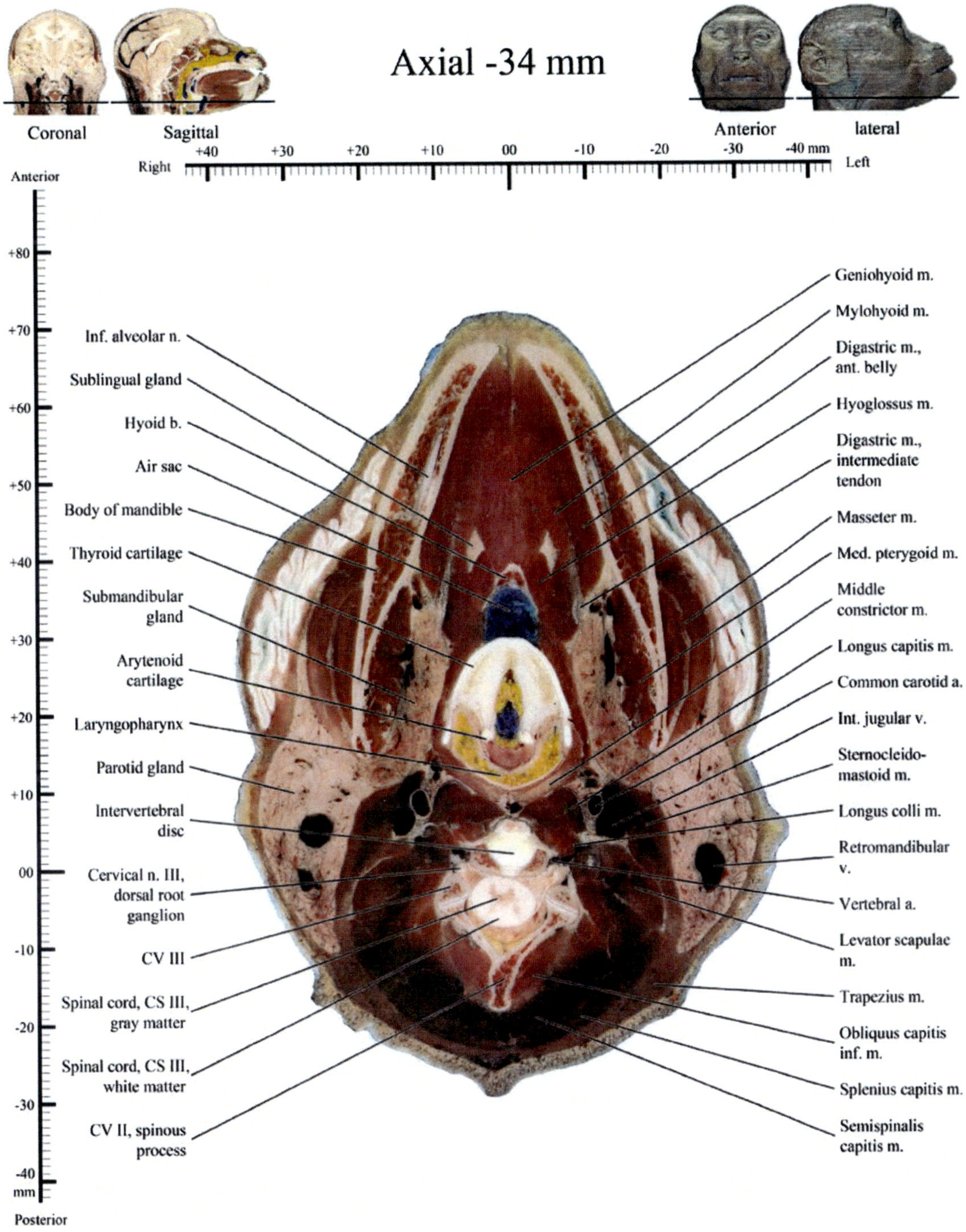

Axial -34 mm
Coronal
Sagittal
Anterior
lateral
Right
Left
+40
+30
+20
+10
00
-10
-20
-30
-40 mm
Anterior
+80
+70
+60
+50
+40
+30
+20
+10
00
-10
-20
-30
-40
mm
Posterior
Inf. alveolar n.
Sublingual gland
Hyoid b.
Air sac
Body of mandible
Thyroid cartilage
Submandibular gland
Arytenoid cartilage
Laryngopharynx
Parotid gland
Intervertebral disc
Cervical n. III, dorsal root ganglion
CV III
Spinal cord, CS III, gray matter
Spinal cord, CS III, white matter
CV II, spinous process
Geniohyoid m.
Mylohyoid m.
Digastric m., ant. belly
Hyoglossus m.
Digastric m., intermediate tendon
Masseter m.
Med. pterygoid m.
Middle constrictor m.
Longus capitis m.
Common carotid a.
Int. jugular v.
Sternocleido-mastoid m.
Longus colli m.
Retromandibular v.
Vertebral a.
Levator scapulae m.
Trapezius m.
Obliquus capitis inf. m.
Splenius capitis m.
Semispinalis capitis m.

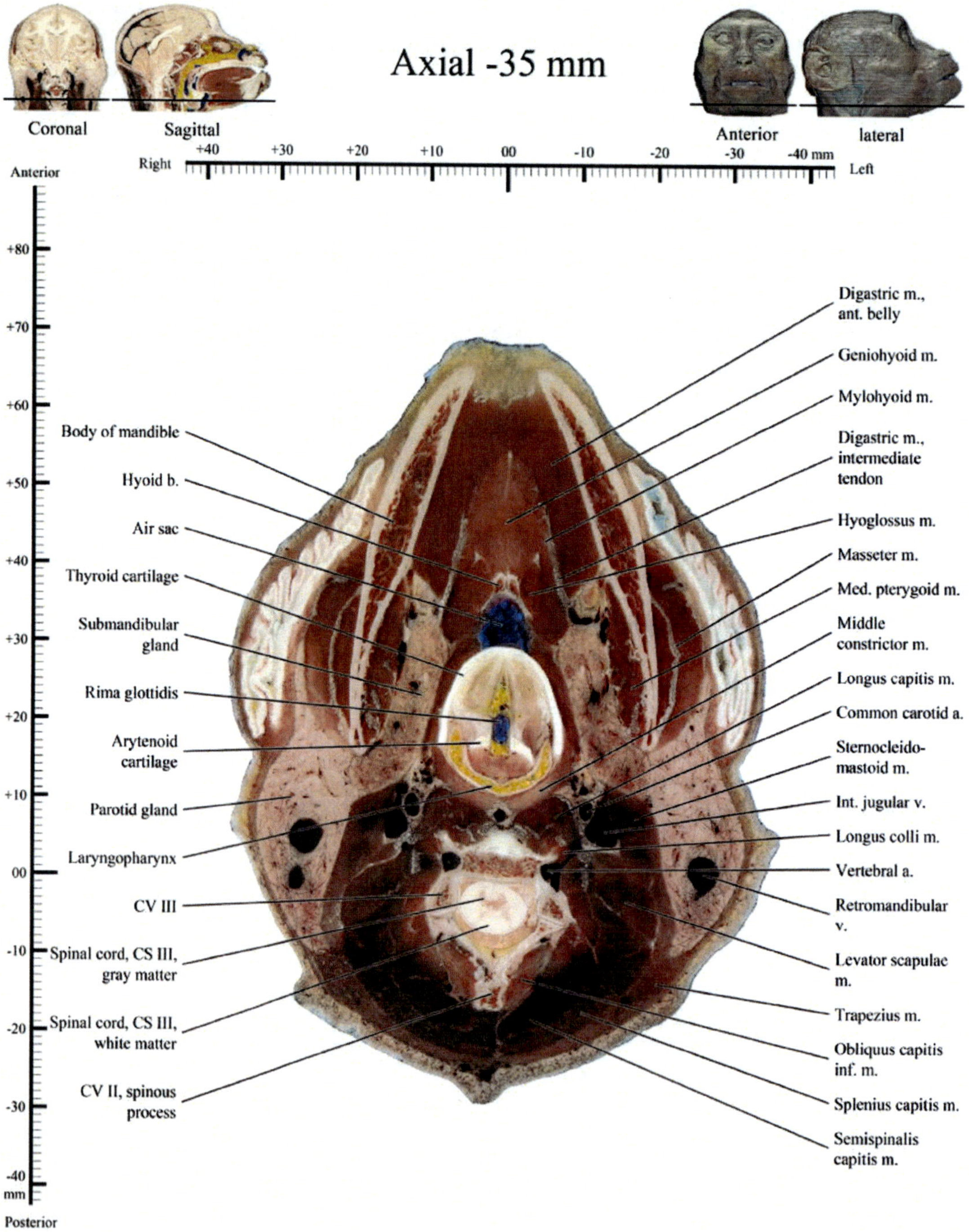
Axial -35 mm
Coronal
Sagittal
Anterior
lateral
Right
Left
+40
+30
+20
+10
00
-10
-20
-30
-40 mm
Anterior
+80
+70
+60
+50
+40
+30
+20
+10
00
-10
-20
-30
-40
mm
Posterior
Body of mandible
Hyoid b.
Air sac
Thyroid cartilage
Submandibular gland
Rima glottidis
Arytenoid cartilage
Parotid gland
Laryngopharynx
CV III
Spinal cord, CS III, gray matter
Spinal cord, CS III, white matter
CV II, spinous process
Digastric m., ant. belly
Geniohyoid m.
Mylohyoid m.
Digastric m., intermediate tendon
Hyoglossus m.
Masseter m.
Med. pterygoid m.
Middle constrictor m.
Longus capitis m.
Common carotid a.
Sternocleido-mastoid m.
Int. jugular v.
Longus colli m.
Vertebral a.
Retromandibular v.
Levator scapulae m.
Trapezius m.
Obliquus capitis inf. m.
Splenius capitis m.
Semispinalis capitis m.

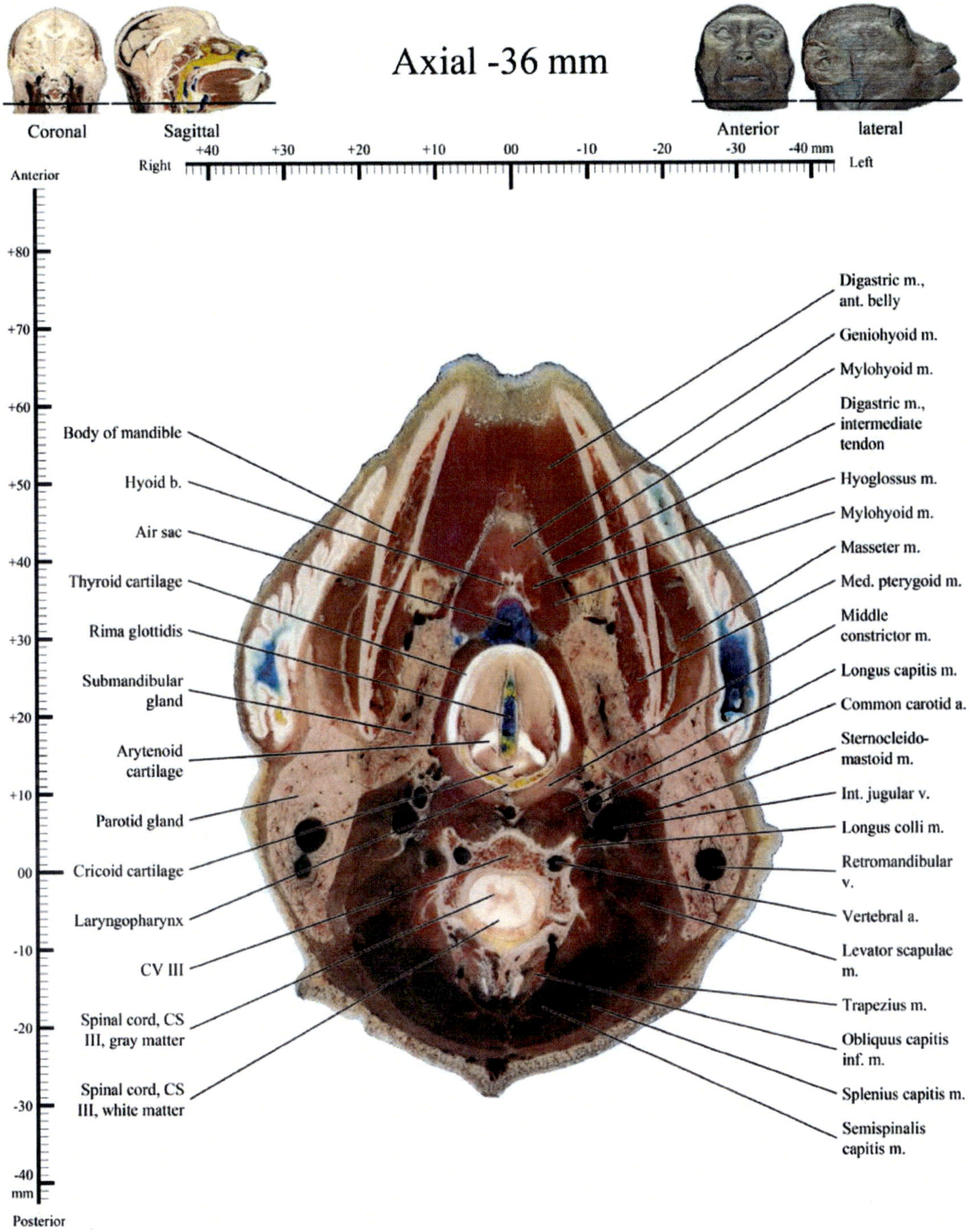
Axial -36 mm
Coronal
Sagittal
Anterior
lateral
Right
Left
+40
+30
+20
+10
00
-10
-20
-30
-40 mm
Anterior
+80
+70
+60
+50
+40
+30
+20
+10
00
-10
-20
-30
-40
mm
Posterior
Body of mandible
Hyoid b.
Air sac
Thyroid cartilage
Rima glottidis
Submandibular gland
Arytenoid cartilage
Parotid gland
Cricoid cartilage
Laryngopharynx
CV III
Spinal cord, CS III, gray matter
Spinal cord, CS III, white matter
Digastric m., ant. belly
Geniohyoid m.
Mylohyoid m.
Digastric m., intermediate tendon
Hyoglossus m.
Mylohyoid m.
Masseter m.
Med. pterygoid m.
Middle constrictor m.
Longus capitis m.
Common carotid a.
Sternocleido-mastoid m.
Int. jugular v.
Longus colli m.
Retromandibular v.
Vertebral a.
Levator scapulae m.
Trapezius m.
Obliquus capitis inf. m.
Splenius capitis m.
Semispinalis capitis m.

Axial -37 mm

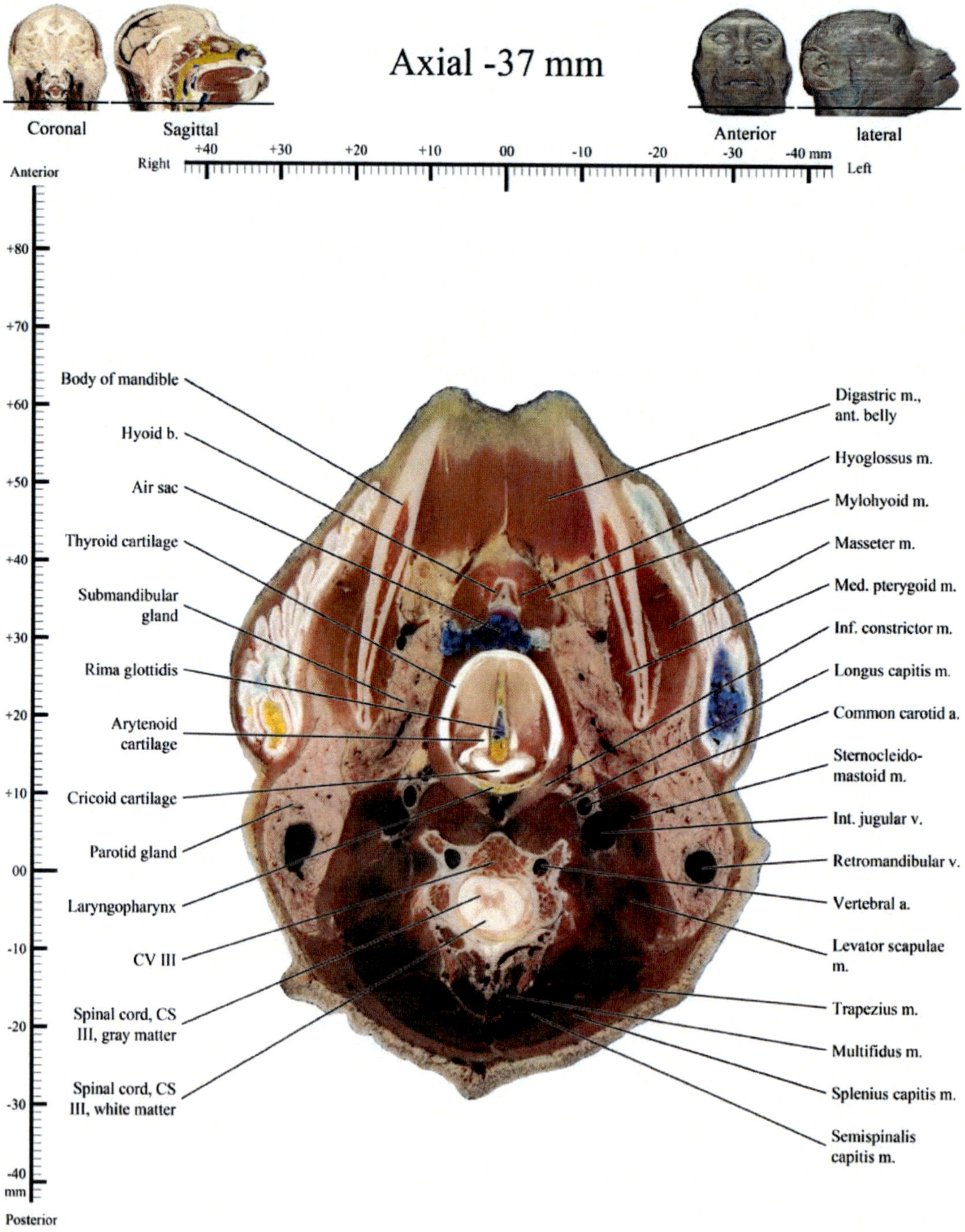
Coronal
Sagittal
Anterior
lateral
Right
+40
+30
+20
+10
00
-10
-20
-30
-40 mm
Left
Anterior
+80
+70
+60
+50
+40
+30
+20
+10
00
-10
-20
-30
-40
mm
Posterior
Body of mandible
Hyoid b.
Air sac
Thyroid cartilage
Submandibular gland
Rima glottidis
Arytenoid cartilage
Cricoid cartilage
Parotid gland
Laryngopharynx
CV III
Spinal cord, CS III, gray matter
Spinal cord, CS III, white matter
Digastric m., ant. belly
Hyoglossus m.
Mylohyoid m.
Masseter m.
Med. pterygoid m.
Inf. constrictor m.
Longus capitis m.
Common carotid a.
Sternocleido-mastoid m.
Int. jugular v.
Retromandibular v.
Vertebral a.
Levator scapulae m.
Trapezius m.
Multifidus m.
Splenius capitis m.
Semispinalis capitis m.

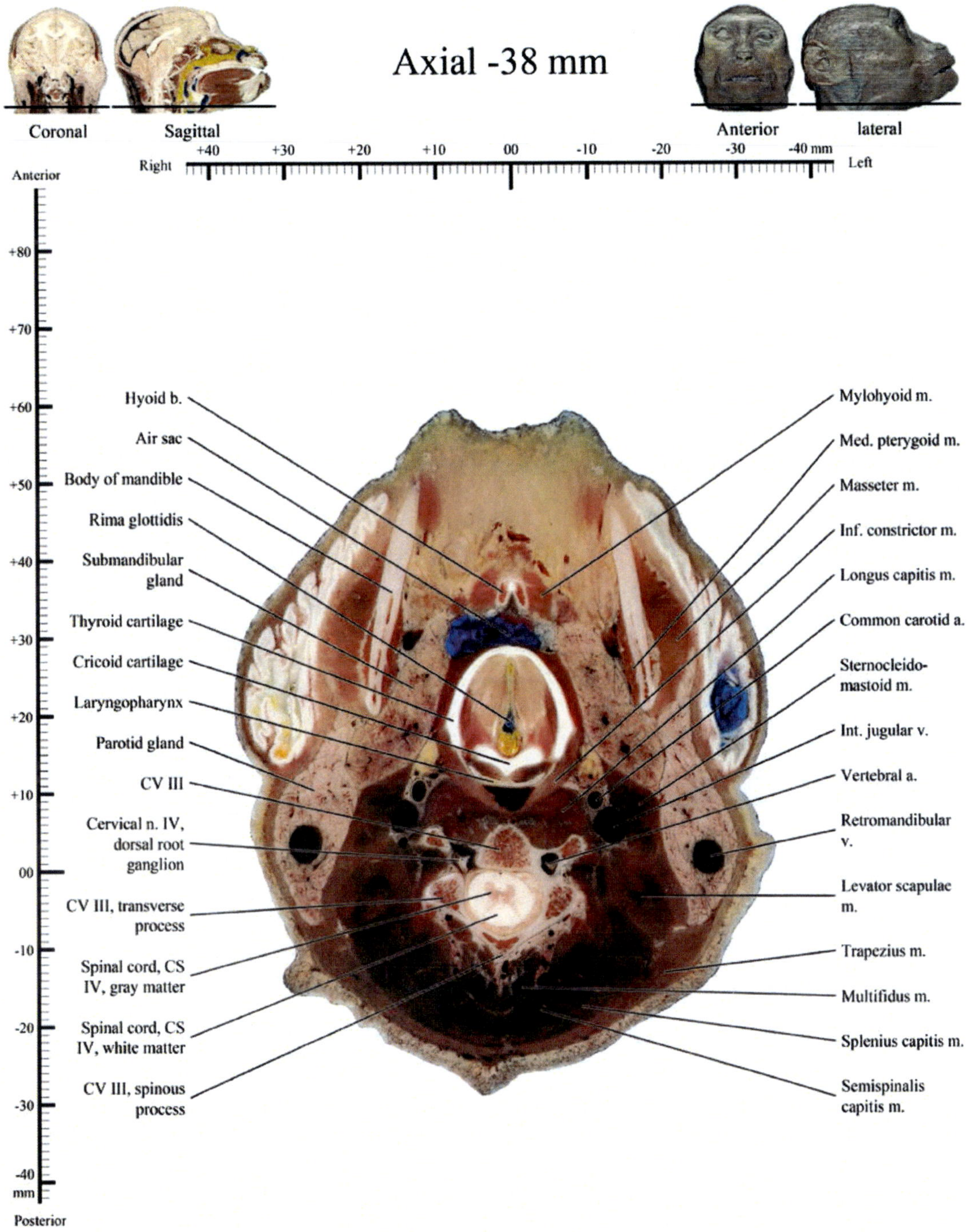
Axial -38 mm
Coronal
Sagittal
Anterior
lateral
Right
Left
+40
+30
+20
+10
00
-10
-20
-30
-40 mm
Anterior
+80
+70
+60
+50
+40
+30
+20
+10
00
-10
-20
-30
-40
mm
Posterior
Hyoid b.
Air sac
Body of mandible
Rima glottidis
Submandibular gland
Thyroid cartilage
Cricoid cartilage
Laryngopharynx
Parotid gland
CV III
Cervical n. IV, dorsal root ganglion
CV III, transverse process
Spinal cord, CS IV, gray matter
Spinal cord, CS IV, white matter
CV III, spinous process
Mylohyoid m.
Med. pterygoid m.
Masseter m.
Inf. constrictor m.
Longus capitis m.
Common carotid a.
Sternocleido-mastoid m.
Int. jugular v.
Vertebral a.
Retromandibular v.
Levator scapulae m.
Trapezius m.
Multifidus m.
Splenius capitis m.
Semispinalis capitis m.

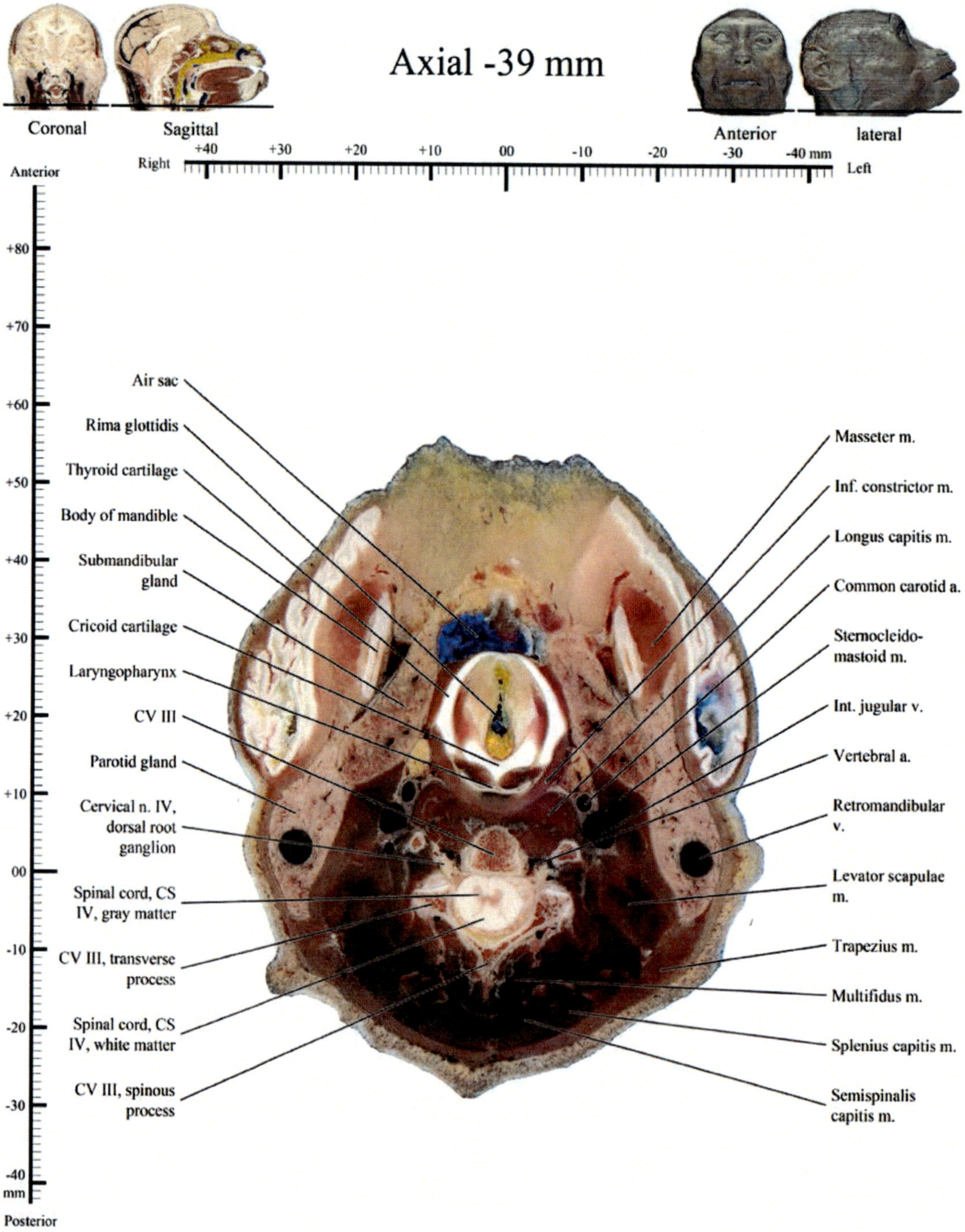
Coronal
Sagittal
Axial -39 mm
Anterior
lateral
+40 +30 +20 +10 00 -10 -20 -30 -40 mm
Right
Left
Anterior
+80
+70
+60
+50
+40
+30
+20
+10
00
-10
-20
-30
-40 mm
Posterior
Air sac
Rima glottidis
Thyroid cartilage
Body of mandible
Submandibular gland
Cricoid cartilage
Laryngopharynx
CV III
Parotid gland
Cervical n. IV, dorsal root ganglion
Spinal cord, CS IV, gray matter
CV III, transverse process
Spinal cord, CS IV, white matter
CV III, spinous process
Masseter m.
Inf. constrictor m.
Longus capitis m.
Common carotid a.
Sternocleido-mastoid m.
Int. jugular v.
Vertebral a.
Retromandibular v.
Levator scapulae m.
Trapezius m.
Multifidus m.
Splenius capitis m.
Semispinalis capitis m.

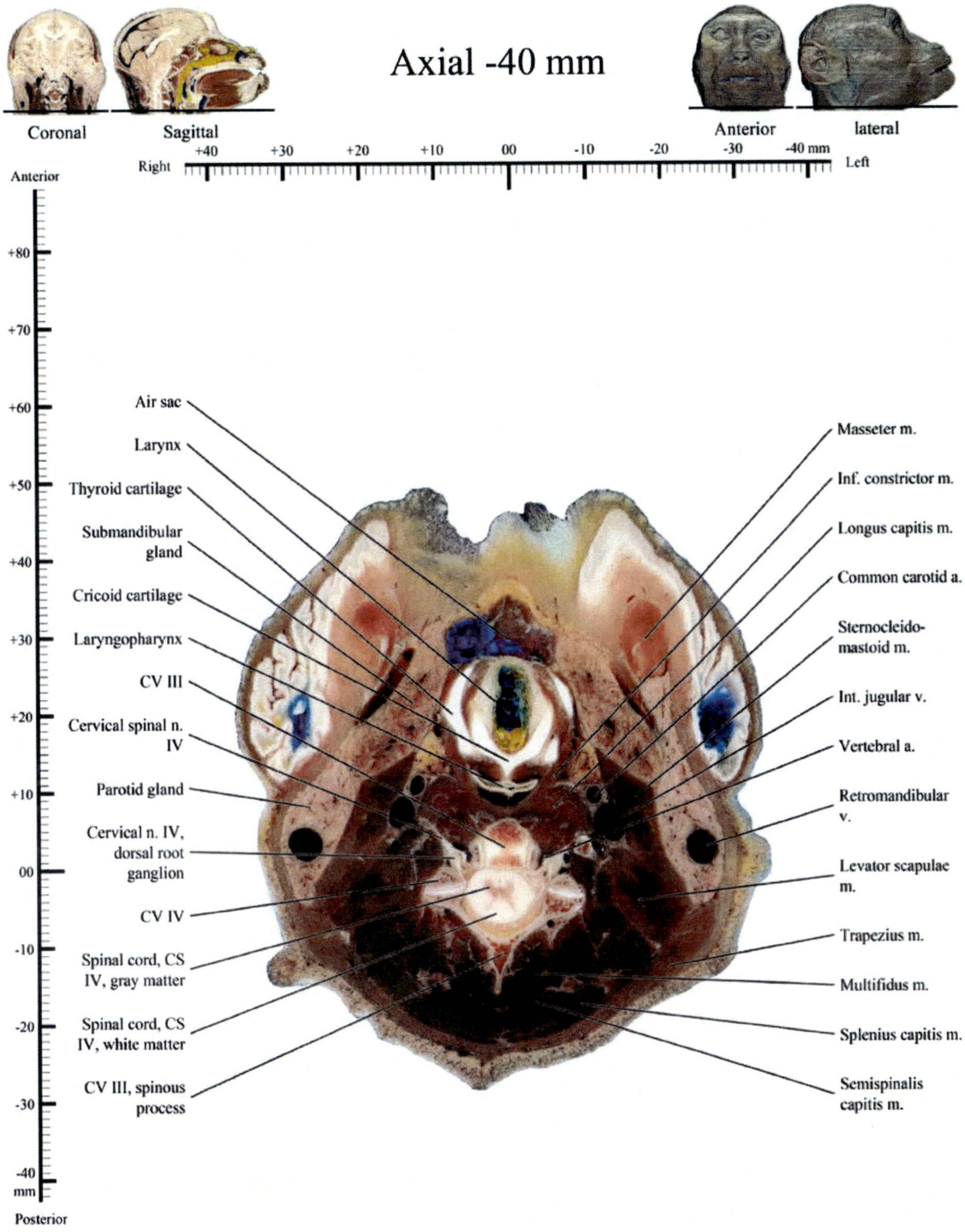
Coronal
Sagittal
Axial -40 mm
Anterior
lateral
+40 +30 +20 +10 00 -10 -20 -30 -40 mm
Right
Left
Anterior
+80
+70
+60
+50
+40
+30
+20
+10
00
-10
-20
-30
-40 mm
Posterior
Air sac
Larynx
Thyroid cartilage
Submandibular gland
Cricoid cartilage
Laryngopharynx
CV III
Cervical spinal n. IV
Parotid gland
Cervical n. IV, dorsal root ganglion
CV IV
Spinal cord, CS IV, gray matter
Spinal cord, CS IV, white matter
CV III, spinous process
Masseter m.
Inf. constrictor m.
Longus capitis m.
Common carotid a.
Sternocleido-mastoid m.
Int. jugular v.
Vertebral a.
Retromandibular v.
Levator scapulae m.
Trapezius m.
Multifidus m.
Splenius capitis m.
Semispinalis capitis m.

Axial -41 mm

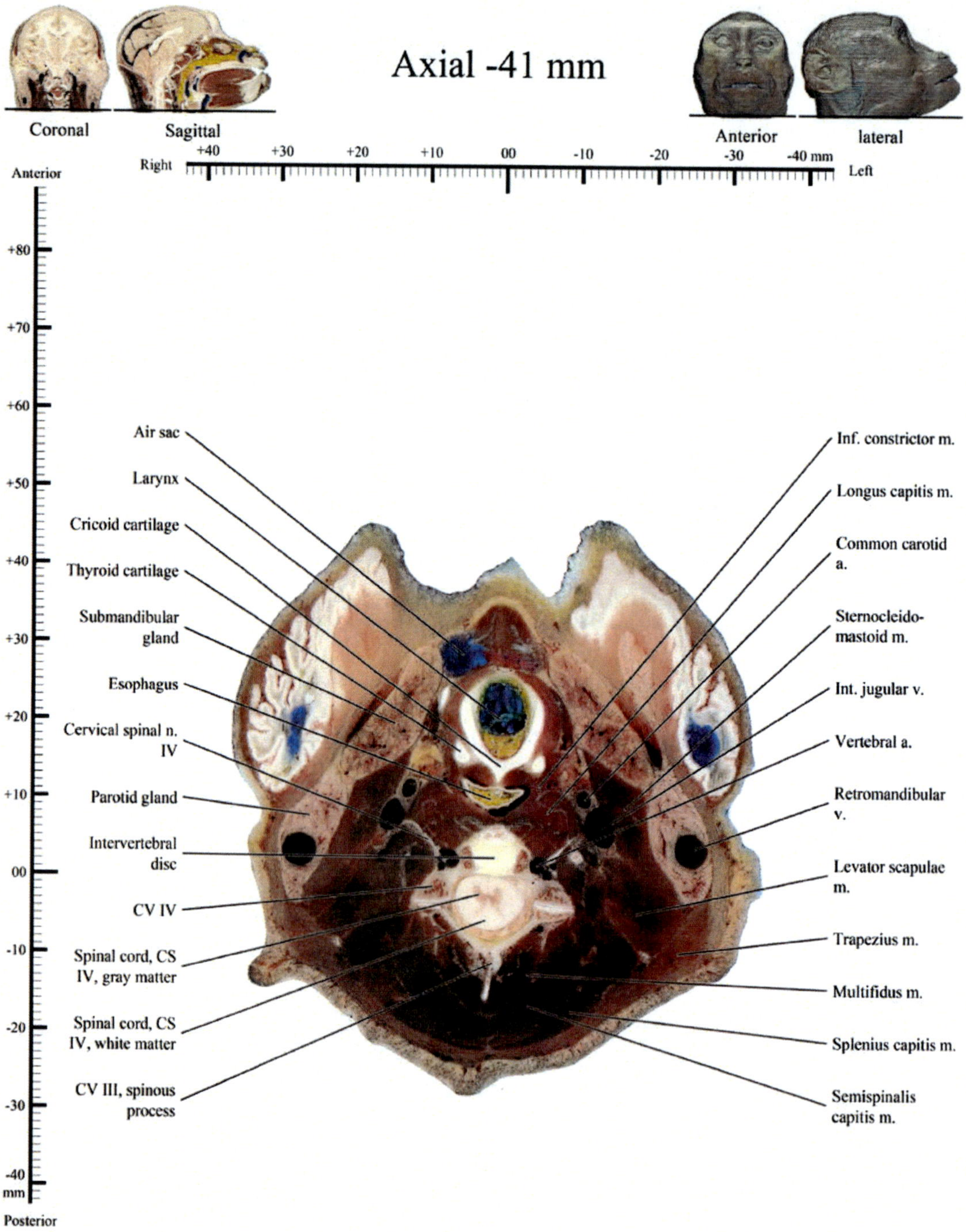
Coronal
Sagittal
Anterior
lateral
Right
+40
+30
+20
+10
00
-10
-20
-30
-40 mm
Left
Anterior
+80
+70
+60
+50
+40
+30
+20
+10
00
-10
-20
-30
-40
mm
Posterior
Air sac
Larynx
Cricoid cartilage
Thyroid cartilage
Submandibular gland
Esophagus
Cervical spinal n. IV
Parotid gland
Intervertebral disc
CV IV
Spinal cord, CS IV, gray matter
Spinal cord, CS IV, white matter
CV III, spinous process
Inf. constrictor m.
Longus capitis m.
Common carotid a.
Sternocleido-mastoid m.
Int. jugular v.
Vertebral a.
Retromandibular v.
Levator scapulae m.
Trapezius m.
Multifidus m.
Splenius capitis m.
Semispinalis capitis m.

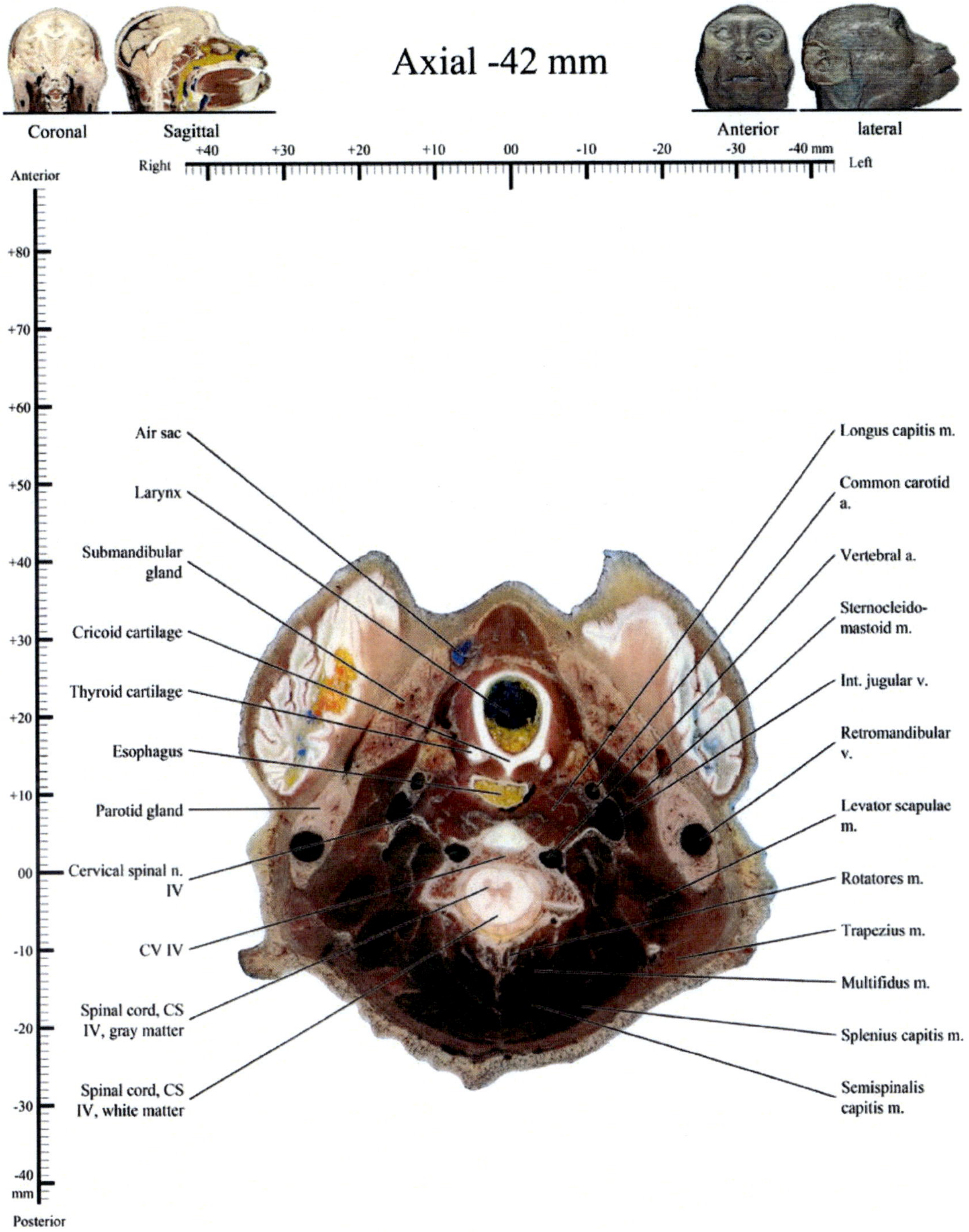

Coronal
Sagittal
Axial -42 mm
Anterior
lateral
Right
Left
+40
+30
+20
+10
00
-10
-20
-30
-40 mm
Anterior
+80
+70
+60
+50
+40
+30
+20
+10
00
-10
-20
-30
-40
mm
Posterior
Air sac
Larynx
Submandibular gland
Cricoid cartilage
Thyroid cartilage
Esophagus
Parotid gland
Cervical spinal n. IV
CV IV
Spinal cord, CS IV, gray matter
Spinal cord, CS IV, white matter
Longus capitis m.
Common carotid a.
Vertebral a.
Sternocleido-mastoid m.
Int. jugular v.
Retromandibular v.
Levator scapulae m.
Rotatores m.
Trapezius m.
Multifidus m.
Splenius capitis m.
Semispinalis capitis m.

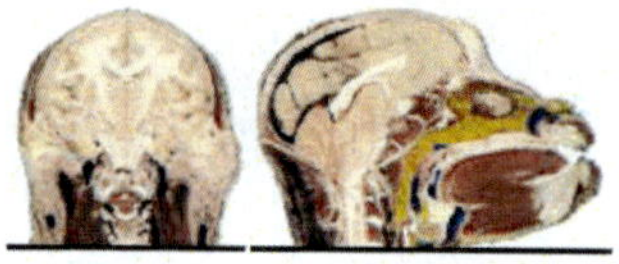
Coronal
Sagittal

Axial -43 mm

Anterior
lateral

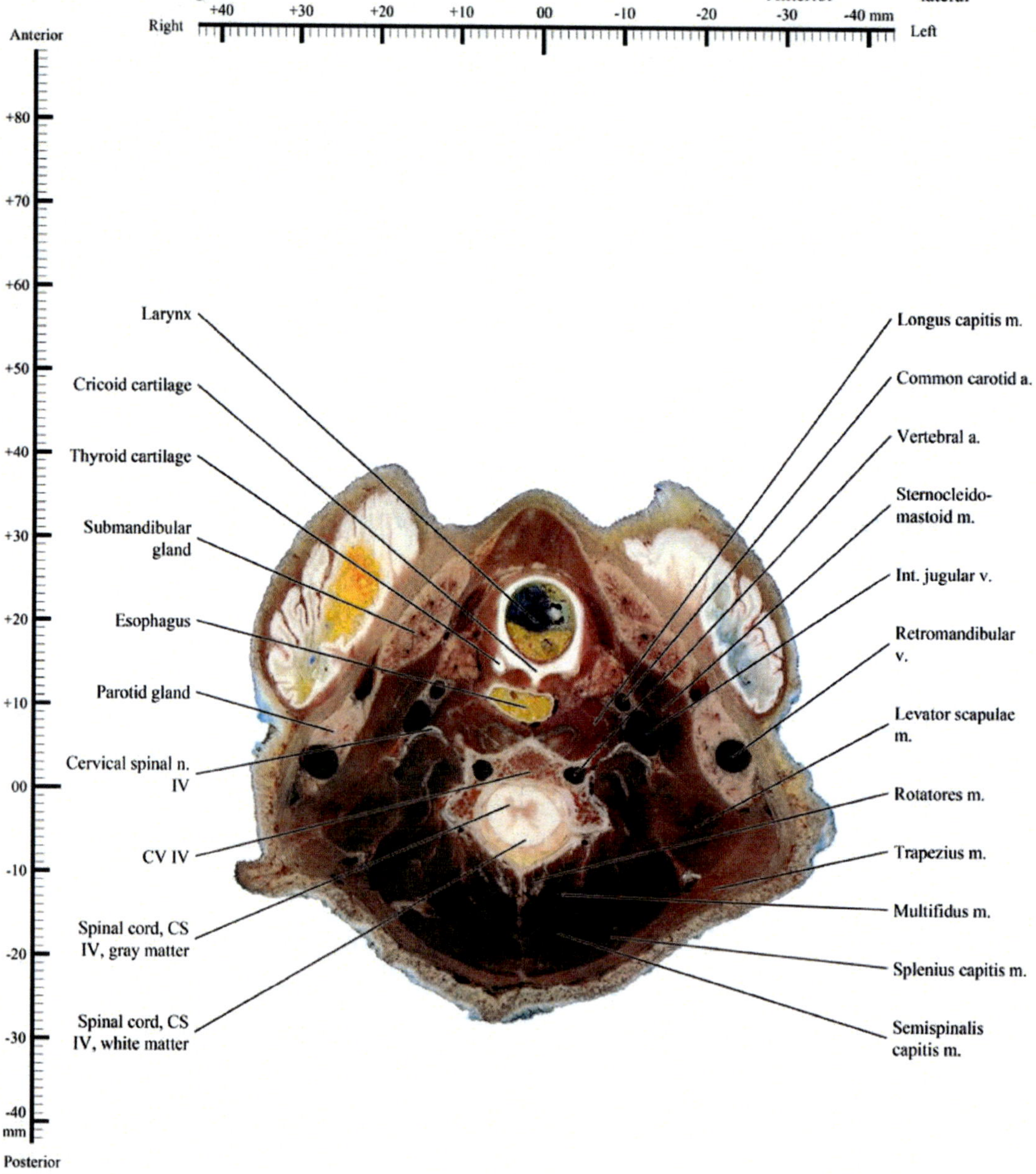
Right
+40
+30
+20
+10
00
-10
-20
-30
-40 mm
Left
Anterior
+80
+70
+60
+50
+40
+30
+20
+10
00
-10
-20
-30
-40
mm
Posterior
Larynx
Cricoid cartilage
Thyroid cartilage
Submandibular gland
Esophagus
Parotid gland
Cervical spinal n. IV
CV IV
Spinal cord, CS IV, gray matter
Spinal cord, CS IV, white matter
Longus capitis m.
Common carotid a.
Vertebral a.
Sternocleido-mastoid m.
Int. jugular v.
Retromandibular v.
Levator scapulae m.
Rotatores m.
Trapezius m.
Multifidus m.
Splenius capitis m.
Semispinalis capitis m.

3 Coronal

All images, 0.024 mm × 0.024 mm sized pixel

+63 mm ~ +42 mm, 7 mm intervals
+42 mm ~ +36 mm, 2 mm intervals
+36 mm ~ −34 mm, 1 mm intervals
−34 mm ~ −38 mm, 2 mm intervals

J. S. Park, *Cross-Sectional Atlas of Rhesus Monkey Head*, https://doi.org/10.1007/978-981-19-3254-0_3

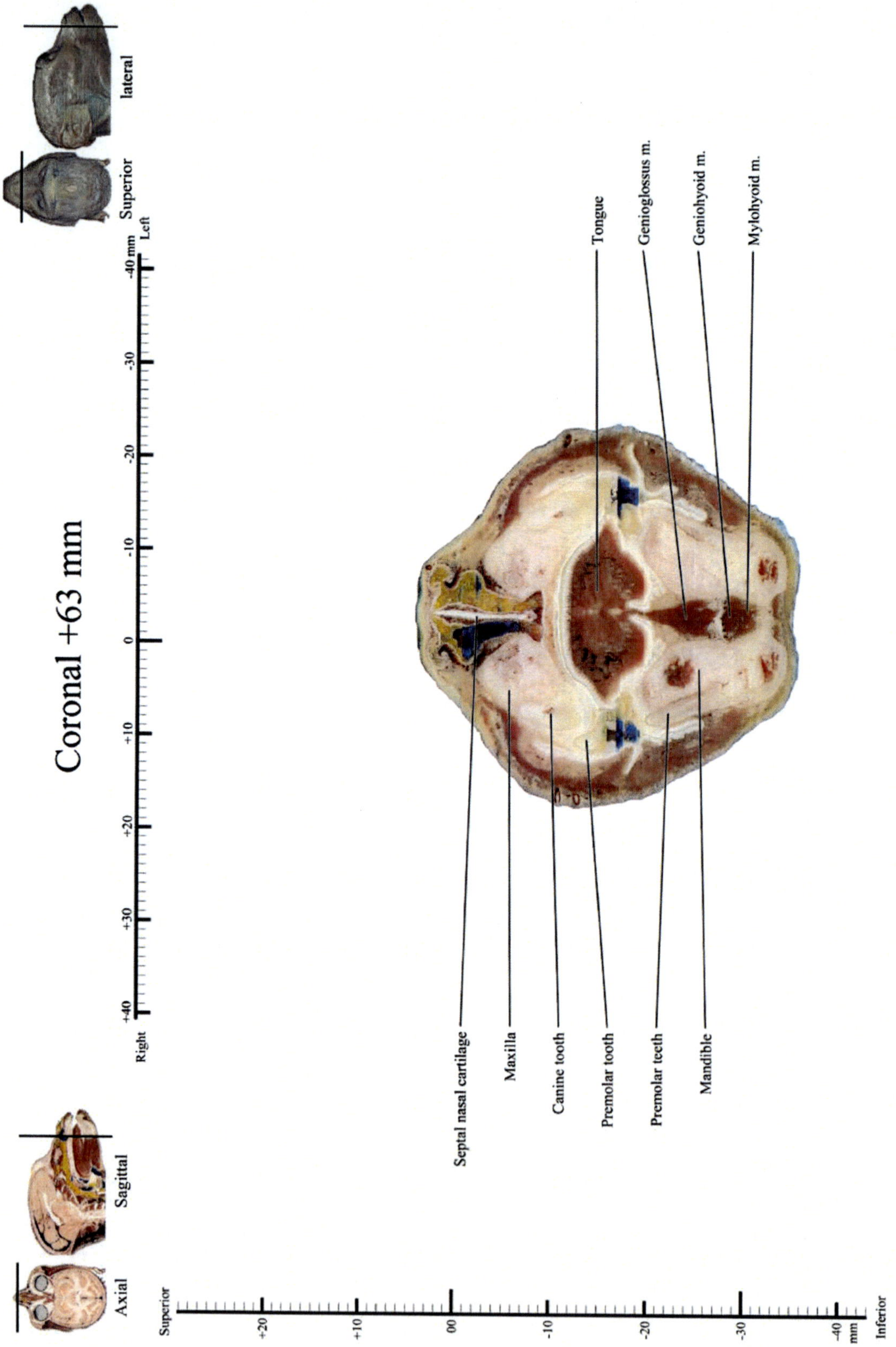
Coronal +63 mm
Superior
lateral
Axial
Sagittal
Right
Left
Superior
Inferior
mm
Tongue
Genioglossus m.
Geniohyoid m.
Mylohyoid m.
Septal nasal cartilage
Maxilla
Canine tooth
Premolar tooth
Premolar teeth
Mandible

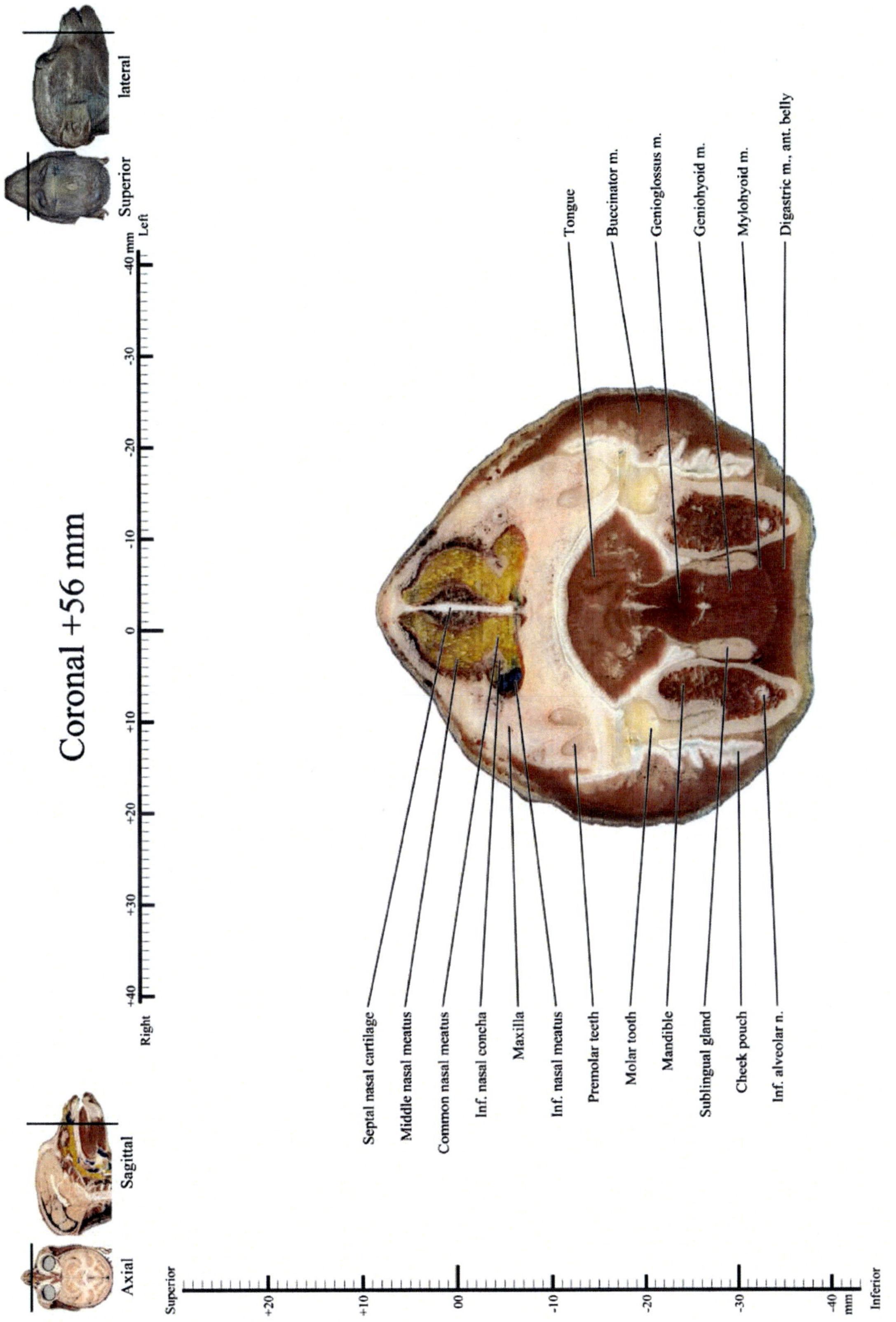
Coronal +56 mm
Axial
Sagittal
Superior
lateral
Right
Left
-40 mm
-30
-20
-10
0
+10
+20
+30
+40
Superior
Inferior
mm
Septal nasal cartilage
Middle nasal meatus
Common nasal meatus
Inf. nasal concha
Maxilla
Inf. nasal meatus
Premolar teeth
Molar tooth
Mandible
Sublingual gland
Cheek pouch
Inf. alveolar n.
Tongue
Buccinator m.
Genioglossus m.
Geniohyoid m.
Mylohyoid m.
Digastric m., ant. belly

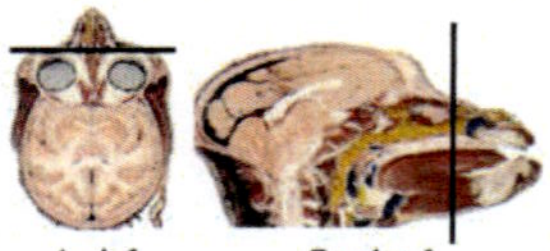

Coronal +49 mm

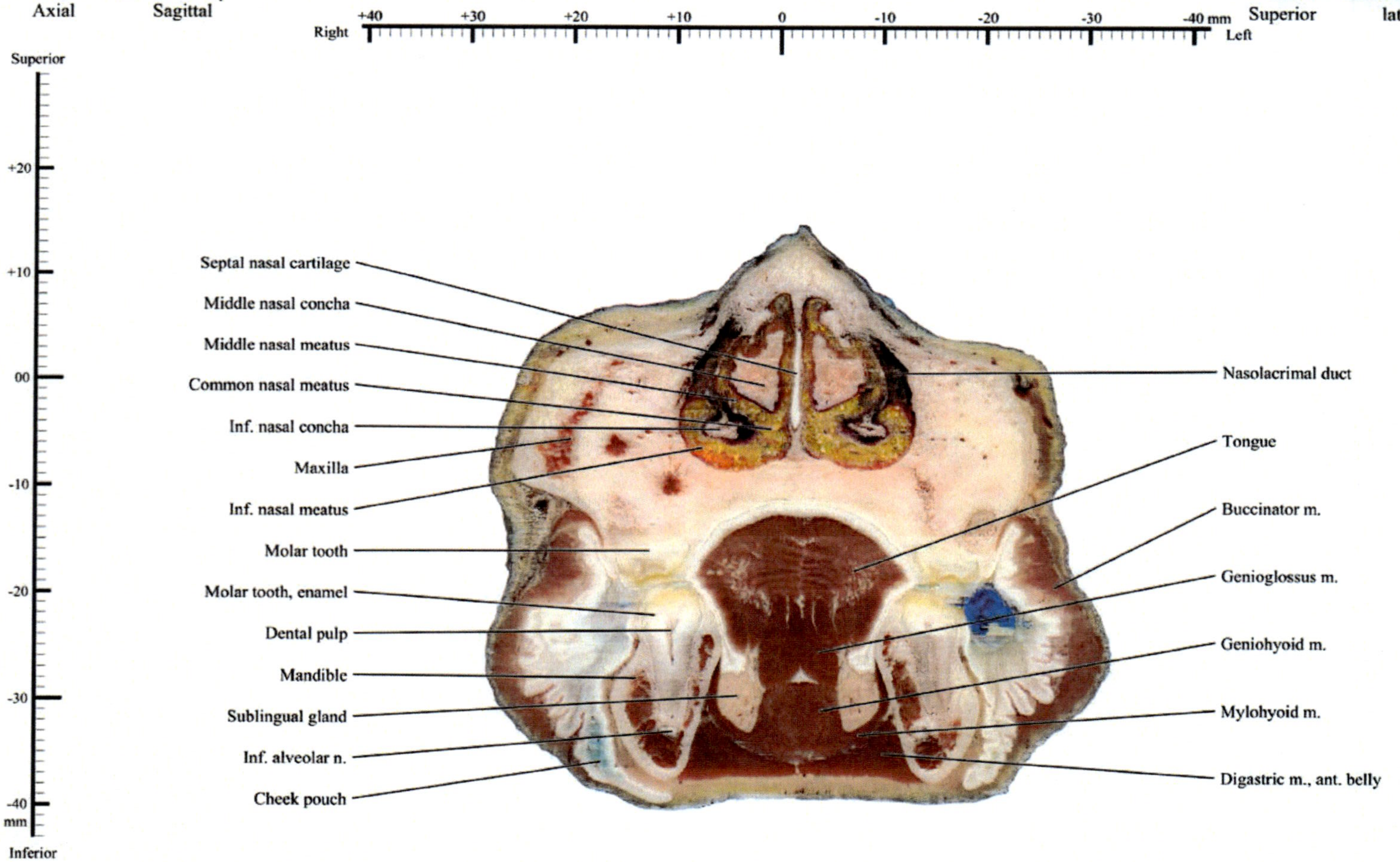

Coronal +42 mm

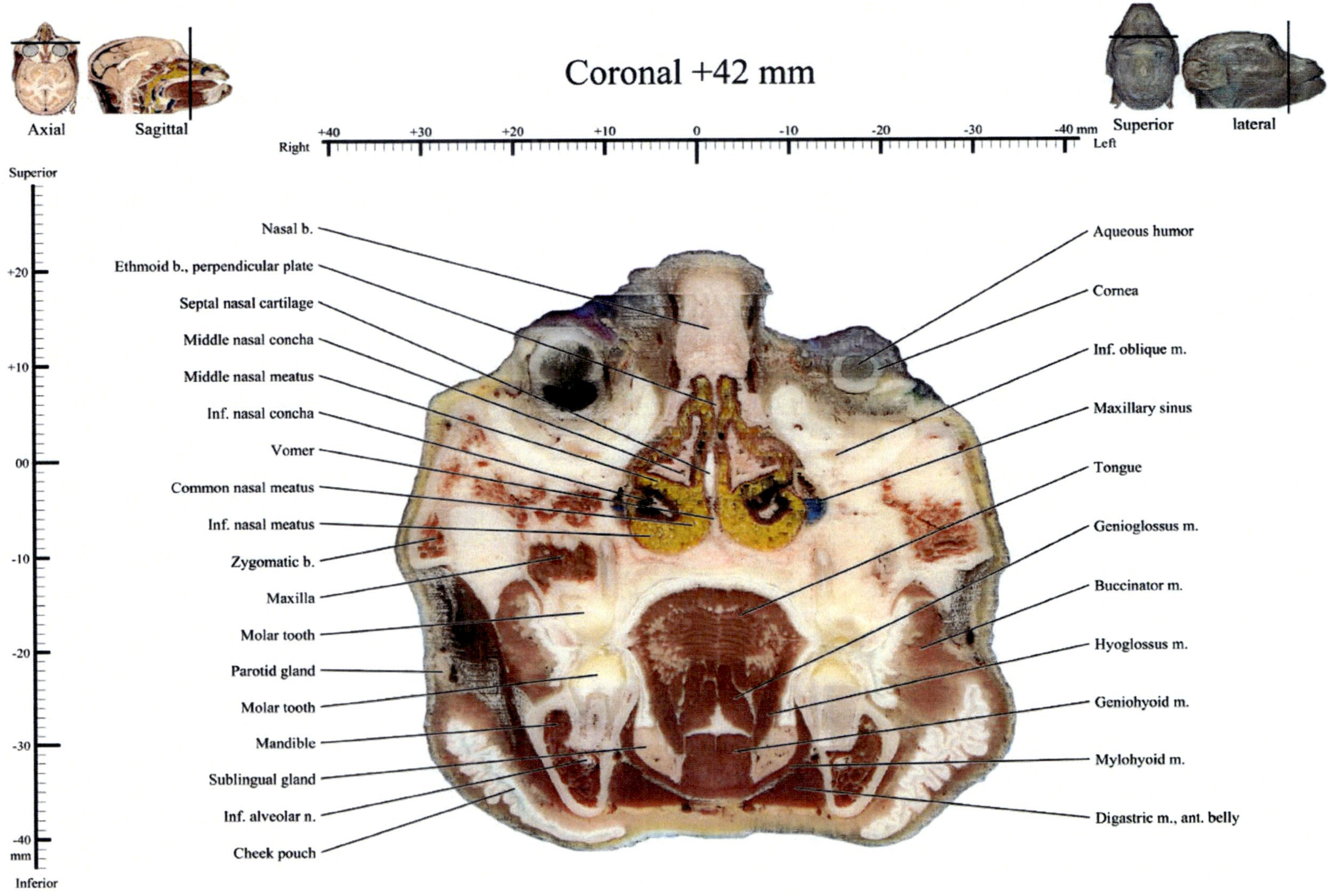

Coronal +40 mm

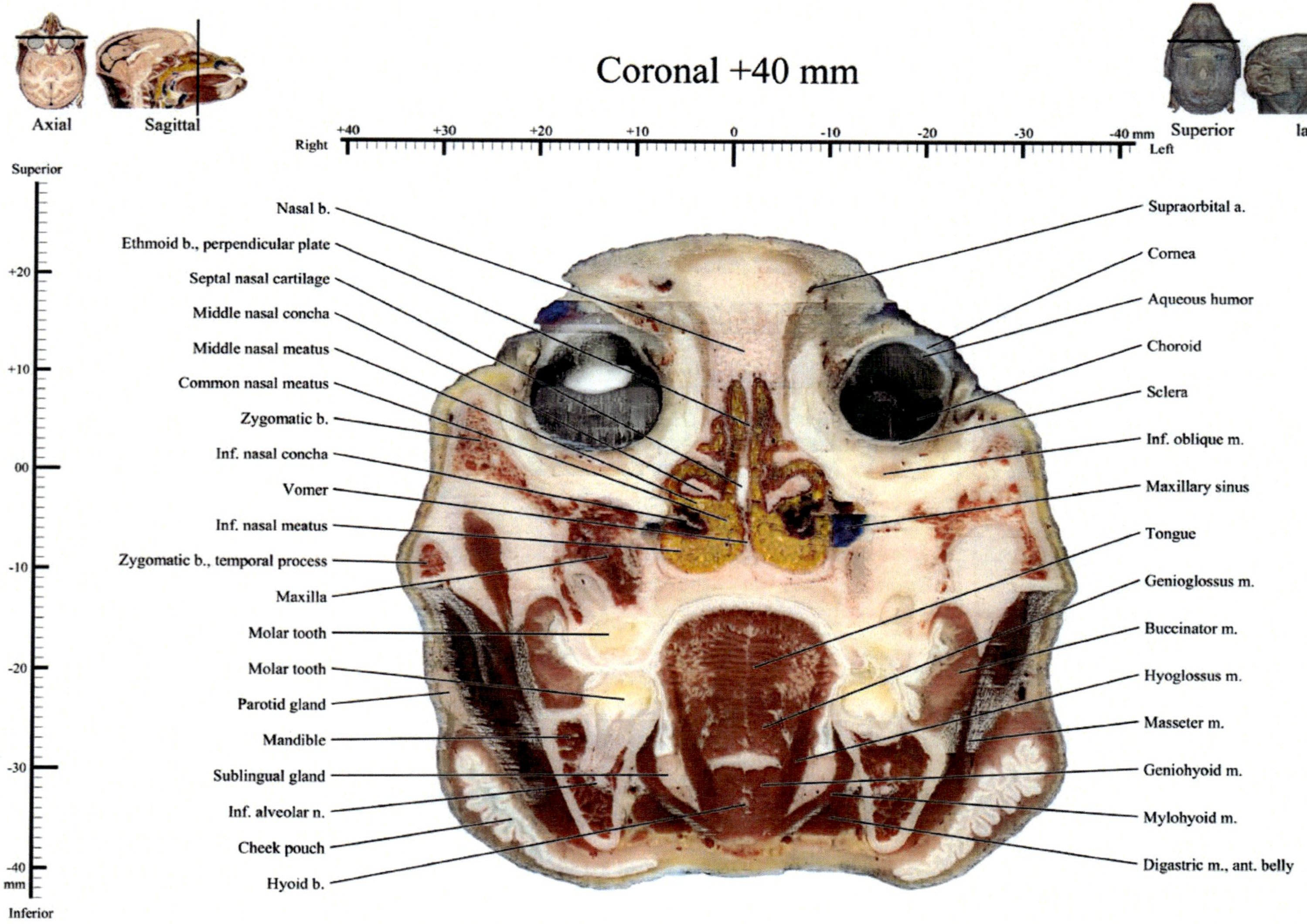

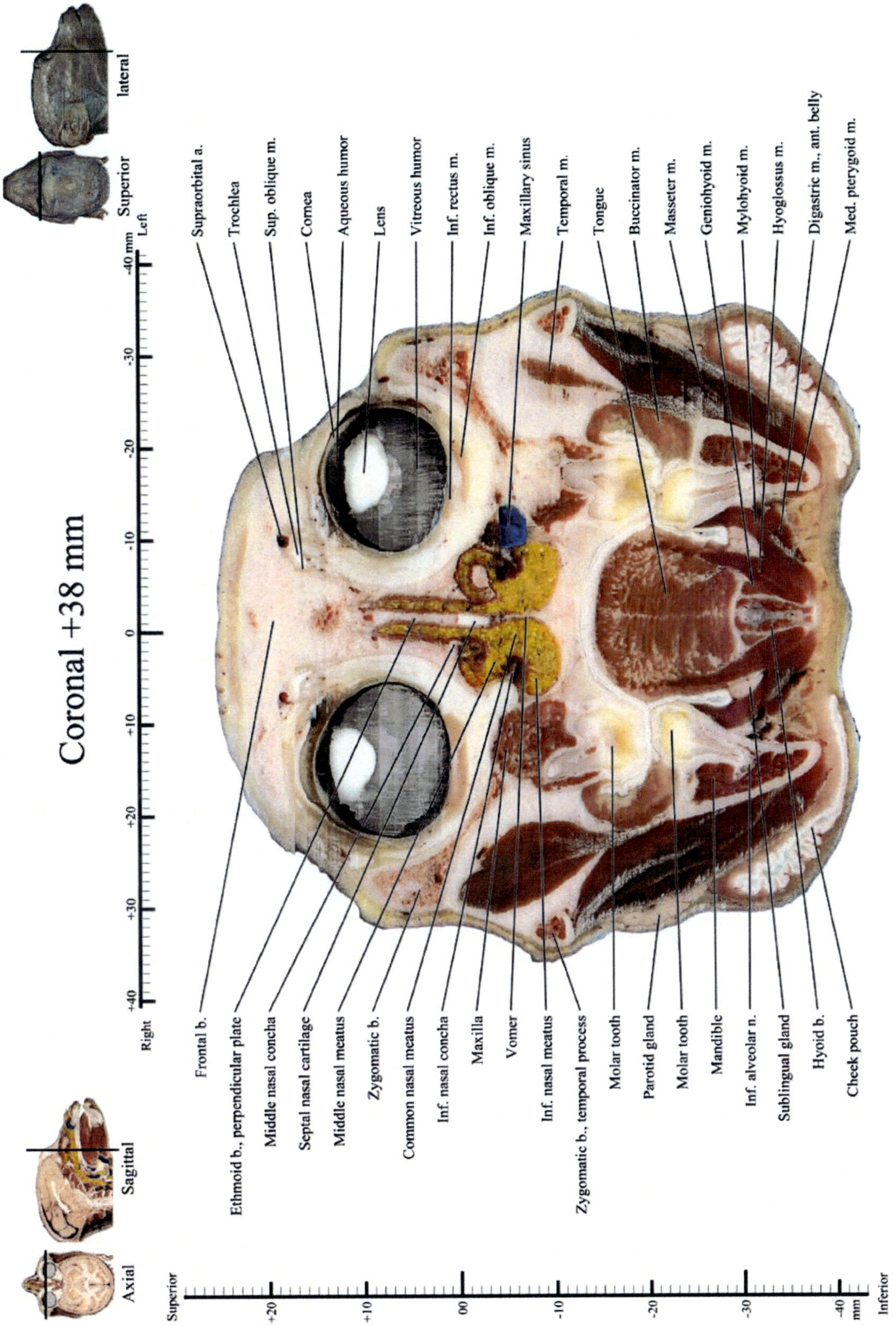
Coronal +38 mm
Right
Left
Superior
Inferior
lateral
Sagittal
Axial
mm
Supraorbital a.
Trochlea
Sup. oblique m.
Cornea
Aqueous humor
Lens
Vitreous humor
Inf. rectus m.
Inf. oblique m.
Maxillary sinus
Temporal m.
Tongue
Buccinator m.
Masseter m.
Geniohyoid m.
Mylohyoid m.
Hyoglossus m.
Digastric m., ant. belly
Med. pterygoid m.
Frontal b.
Ethmoid b., perpendicular plate
Middle nasal concha
Septal nasal cartilage
Middle nasal meatus
Zygomatic b.
Common nasal meatus
Inf. nasal concha
Maxilla
Vomer
Inf. nasal meatus
Zygomatic b., temporal process
Molar tooth
Parotid gland
Molar tooth
Mandible
Inf. alveolar n.
Sublingual gland
Hyoid b.
Cheek pouch

Coronal +36 mm

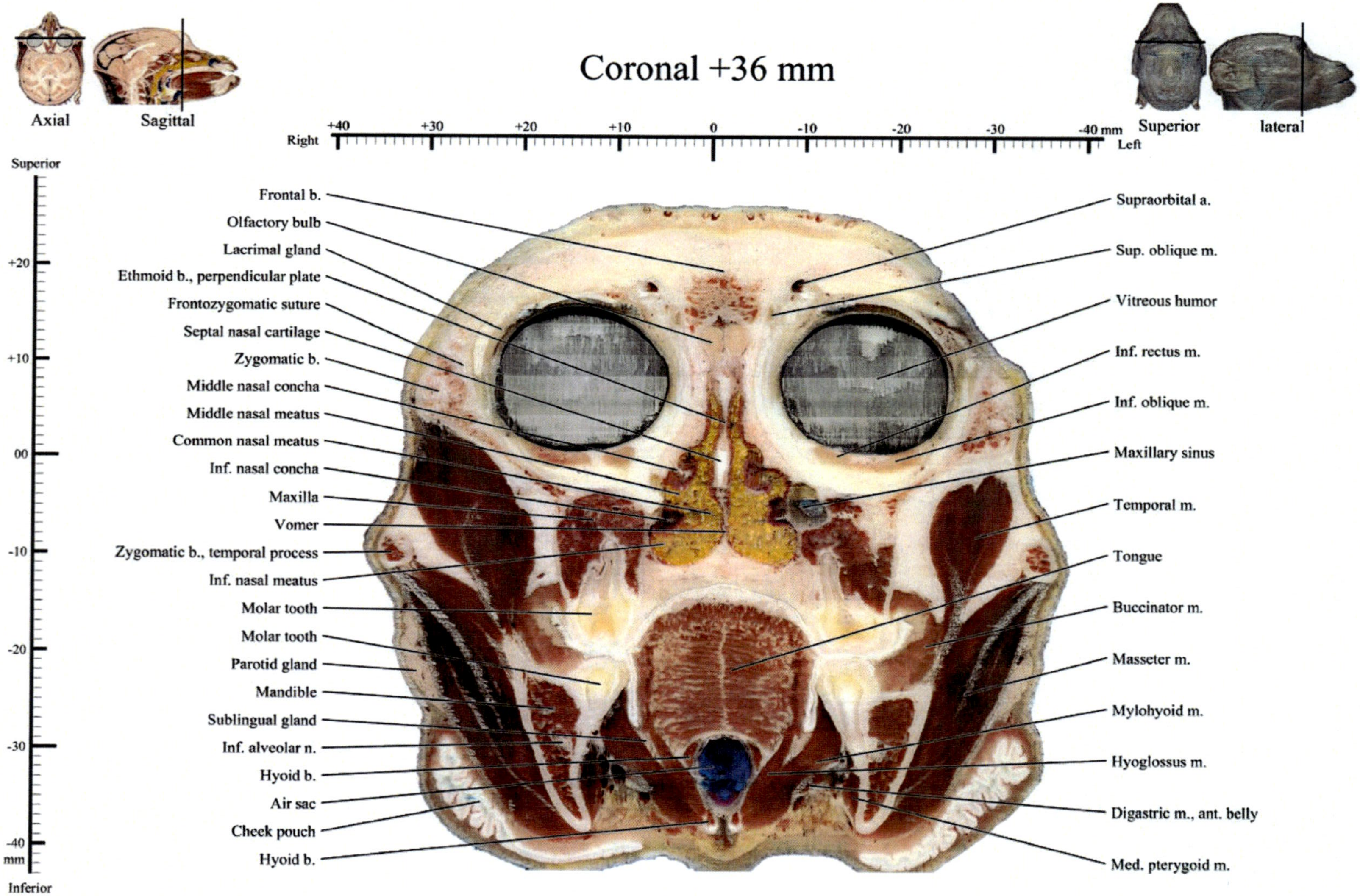

Coronal +35 mm

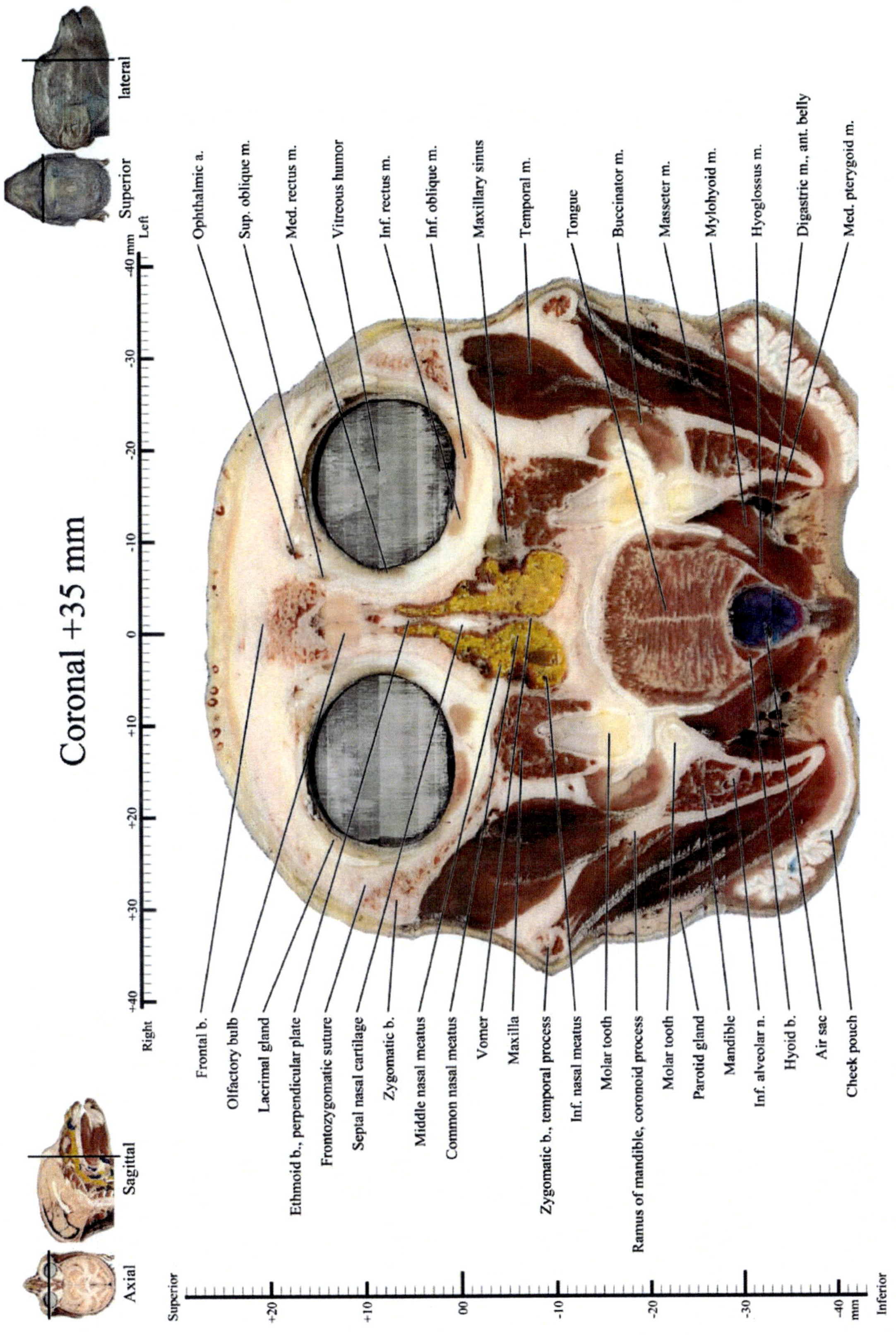

Coronal +34 mm

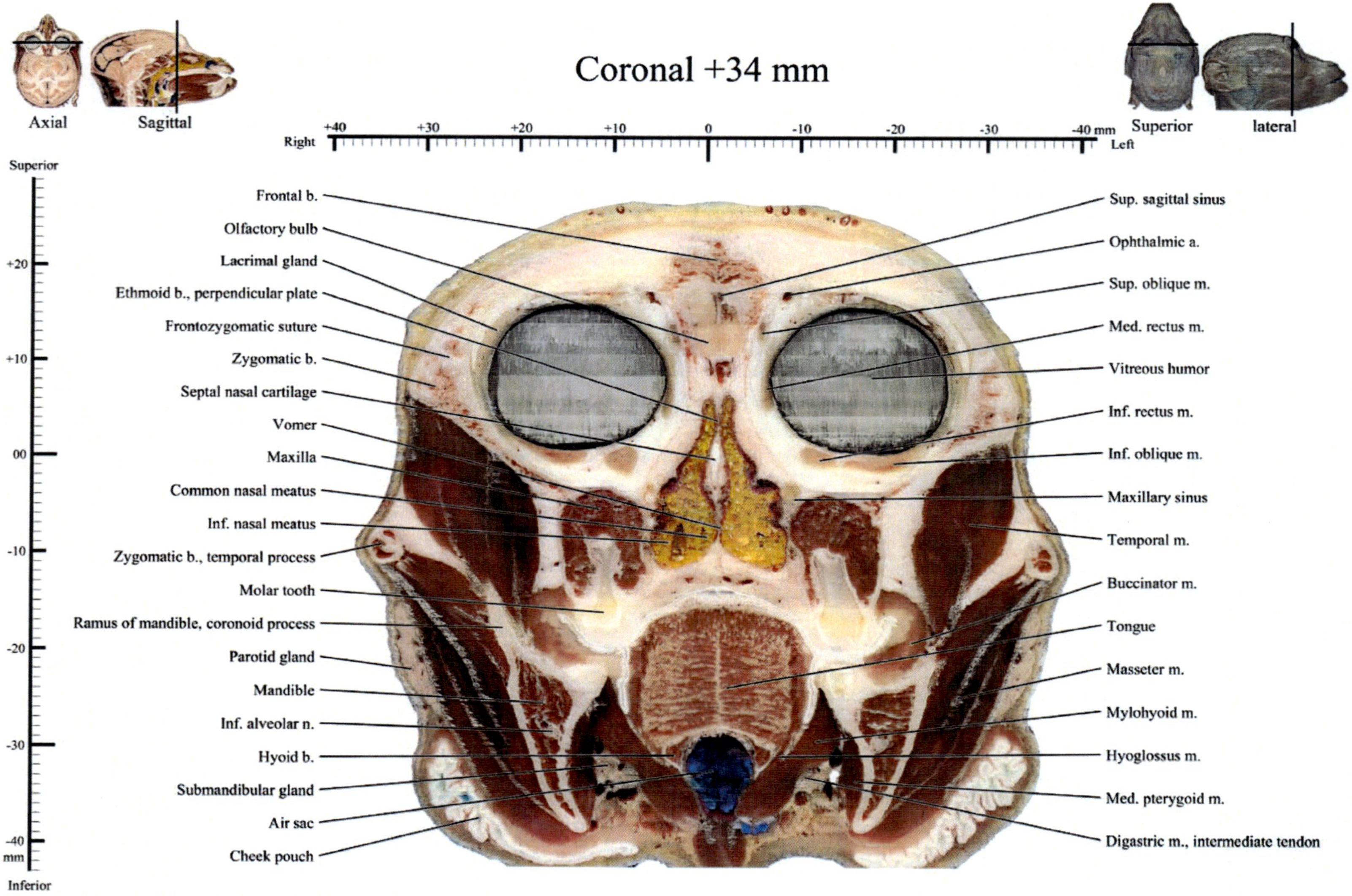

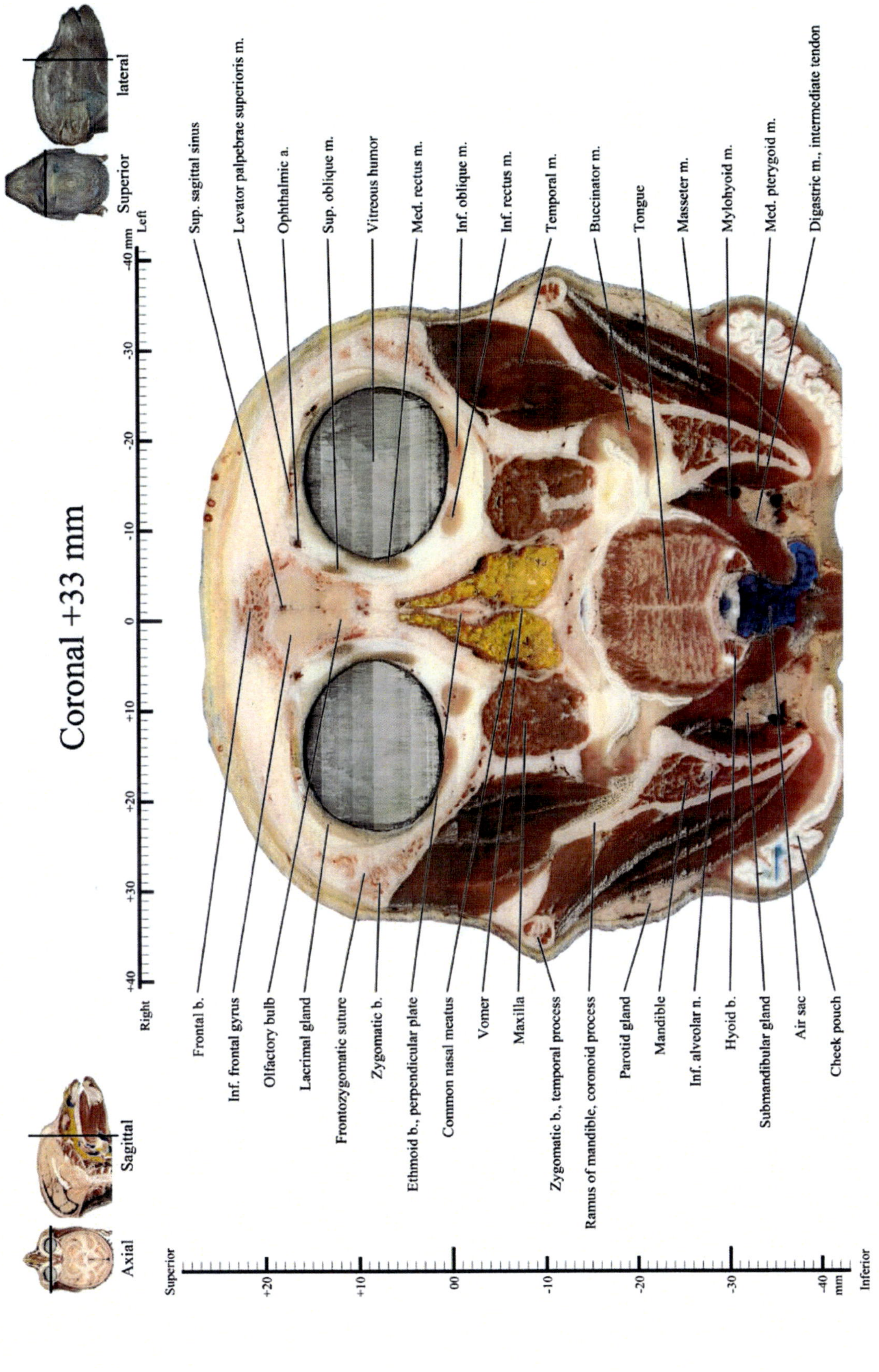
Coronal +33 mm
Superior
lateral
Axial
Sagittal
Right
Left
Superior
Inferior
mm
Sup. sagittal sinus
Levator palpebrae superioris m.
Ophthalmic a.
Sup. oblique m.
Vitreous humor
Med. rectus m.
Inf. oblique m.
Inf. rectus m.
Temporal m.
Buccinator m.
Tongue
Masseter m.
Mylohyoid m.
Med. pterygoid m.
Digastric m., intermediate tendon
Frontal b.
Inf. frontal gyrus
Olfactory bulb
Lacrimal gland
Frontozygomatic suture
Zygomatic b.
Ethmoid b., perpendicular plate
Common nasal meatus
Vomer
Maxilla
Zygomatic b., temporal process
Ramus of mandible, coronoid process
Parotid gland
Mandible
Inf. alveolar n.
Hyoid b.
Submandibular gland
Air sac
Cheek pouch

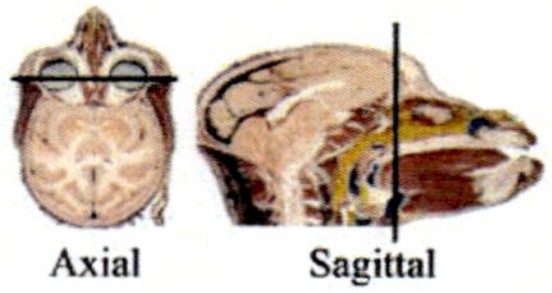

Coronal +32 mm

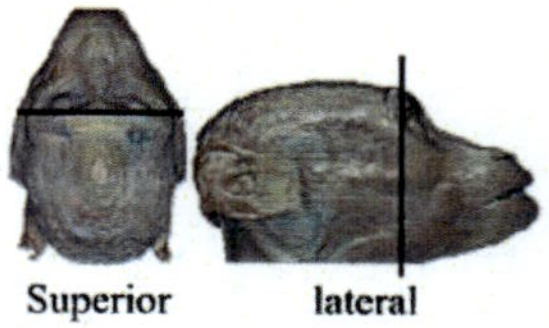

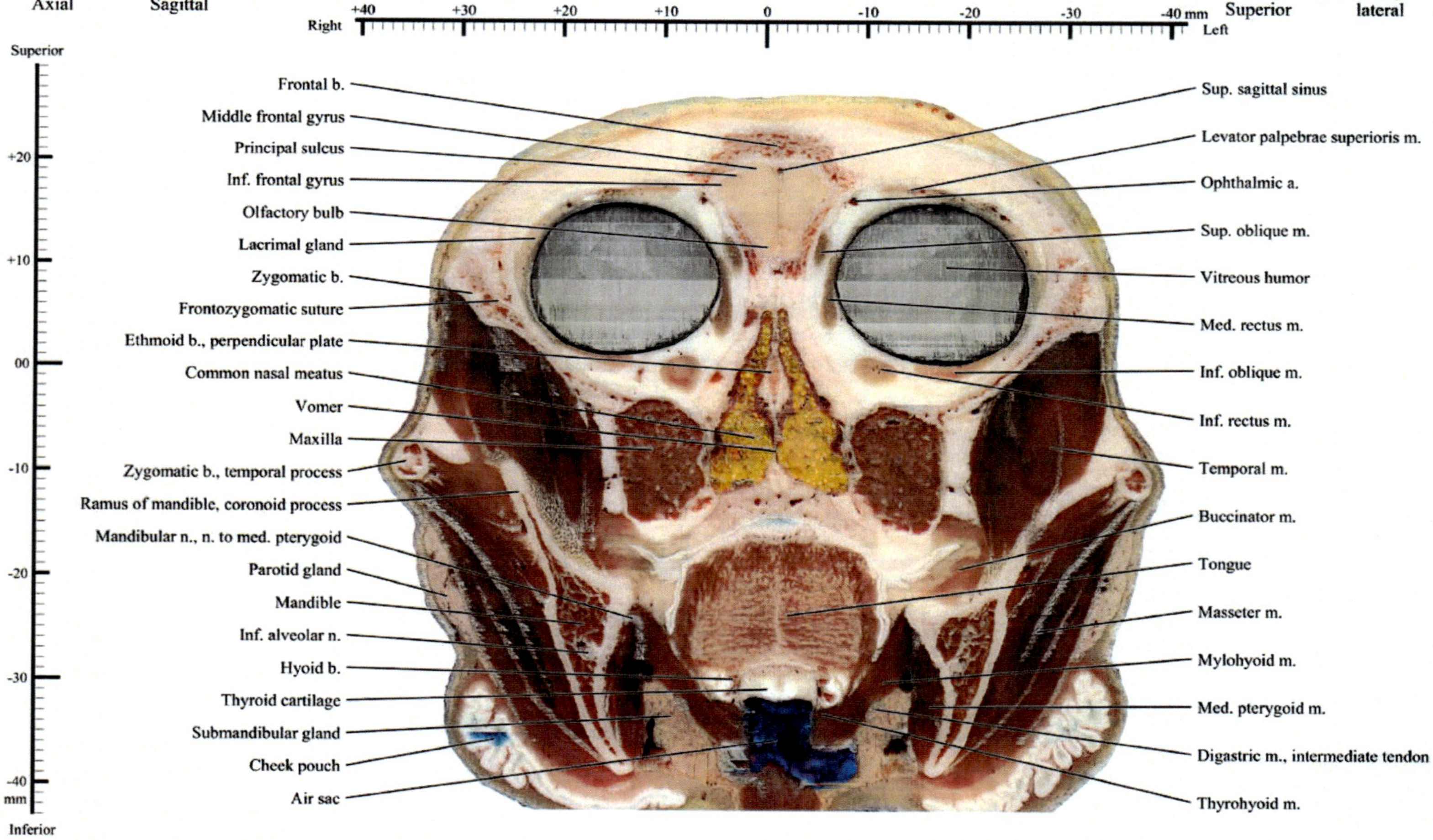

Coronal +31 mm

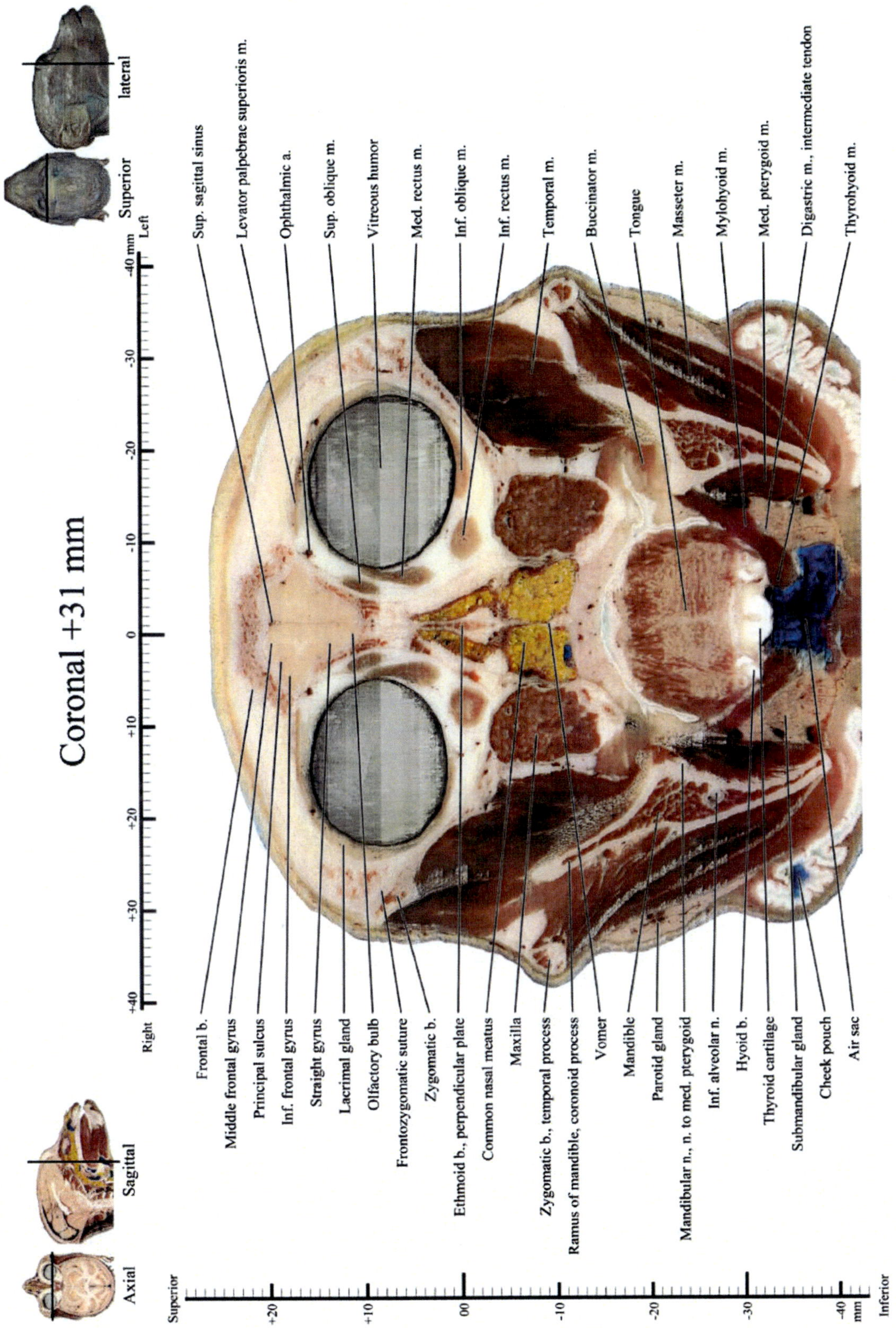

Coronal +30 mm

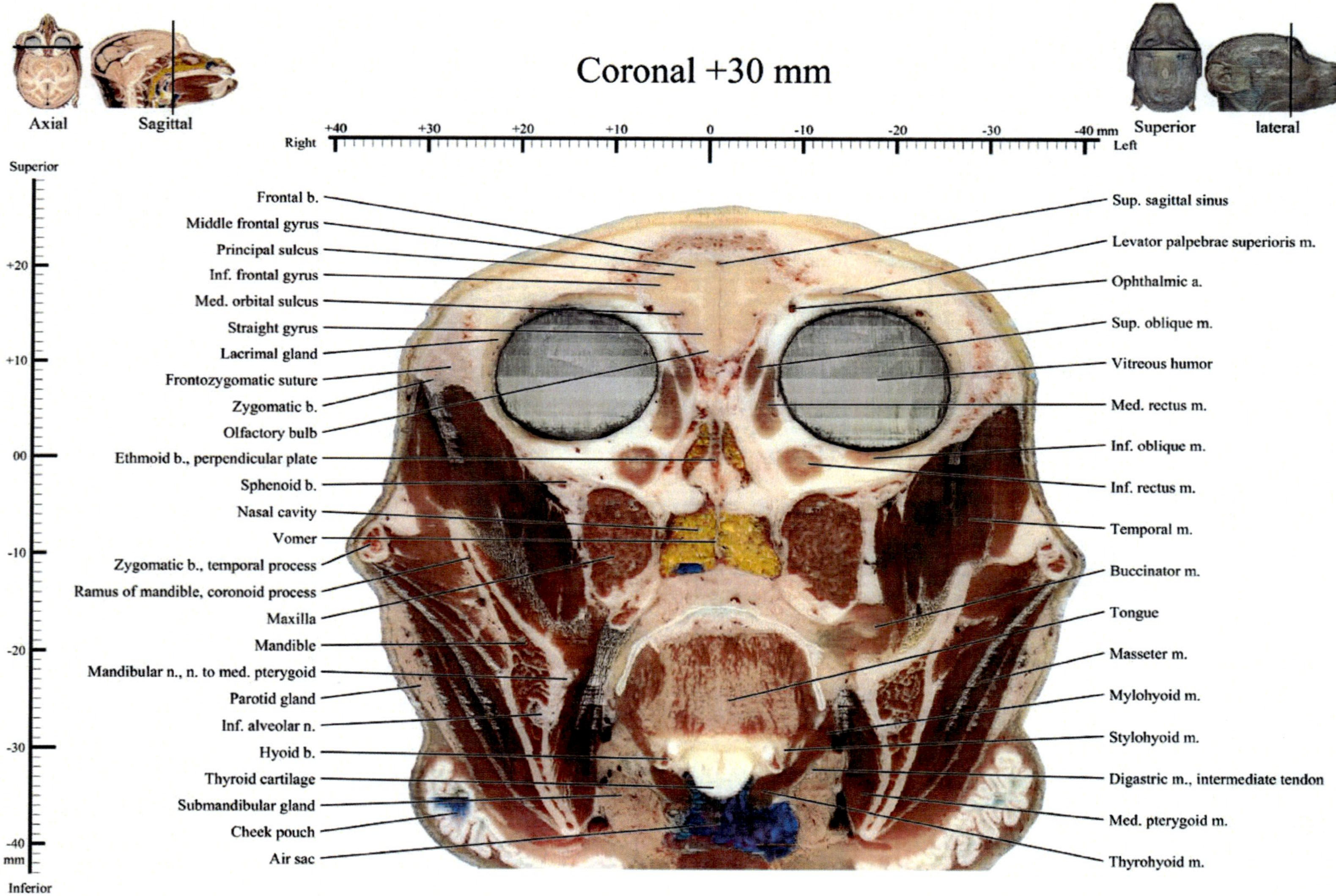

Coronal +29 mm

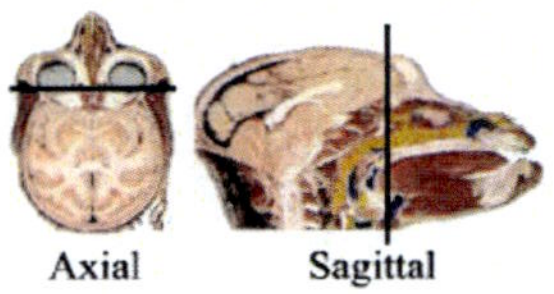

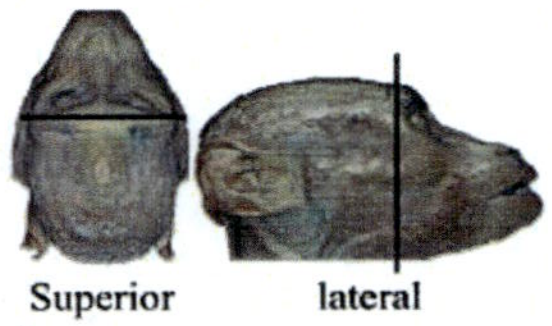

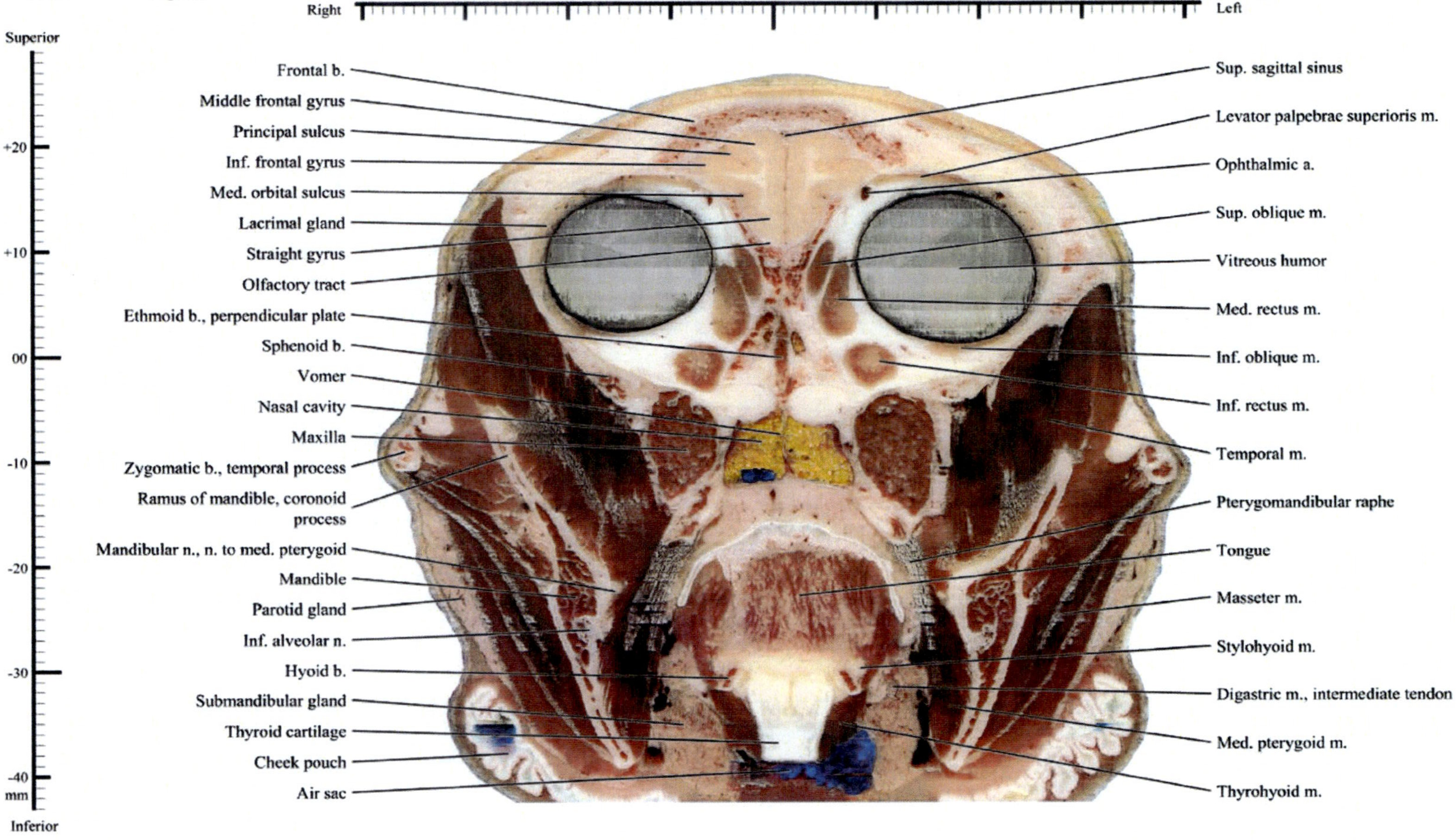

Coronal +28 mm

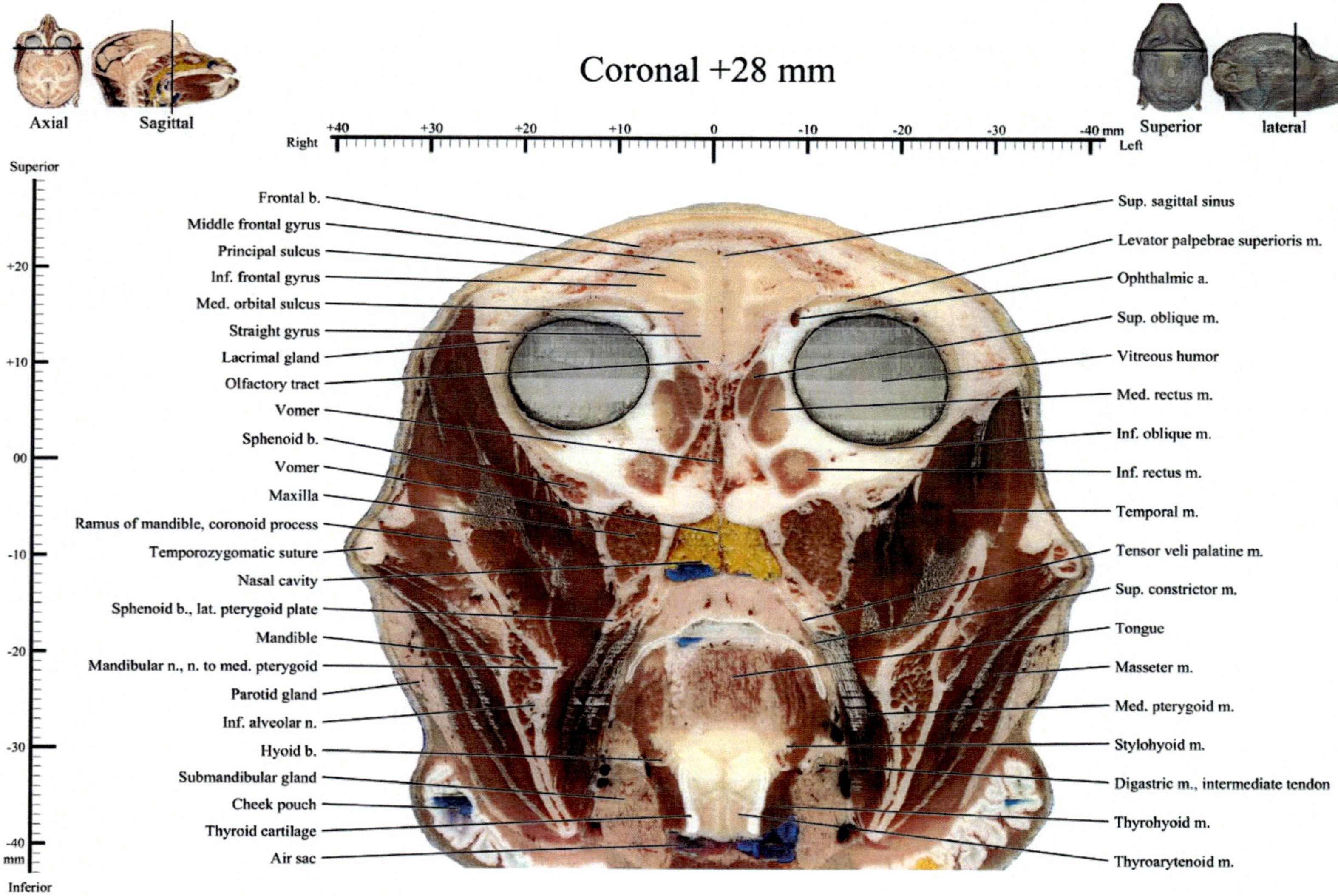

Coronal +27 mm

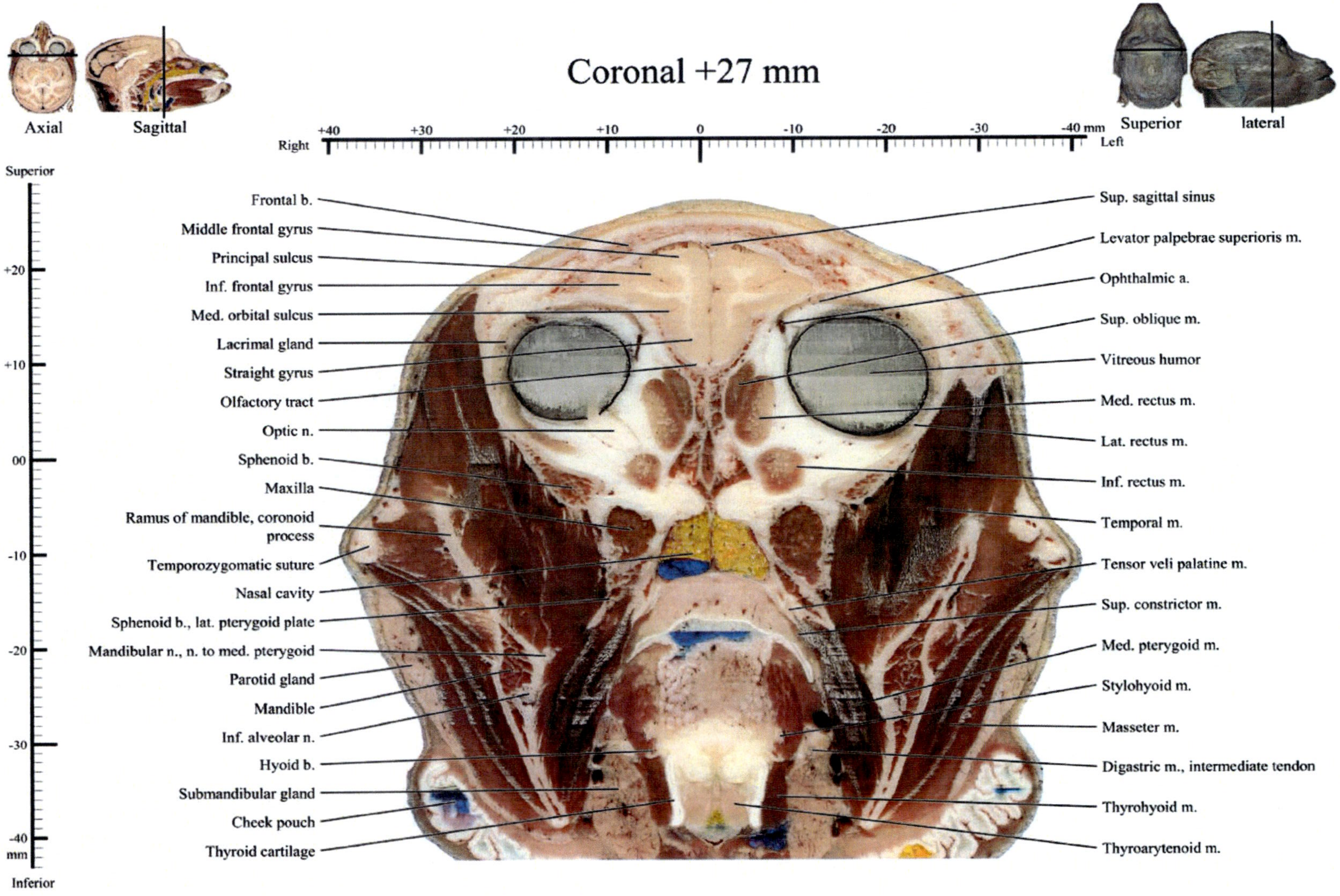

Coronal +26 mm

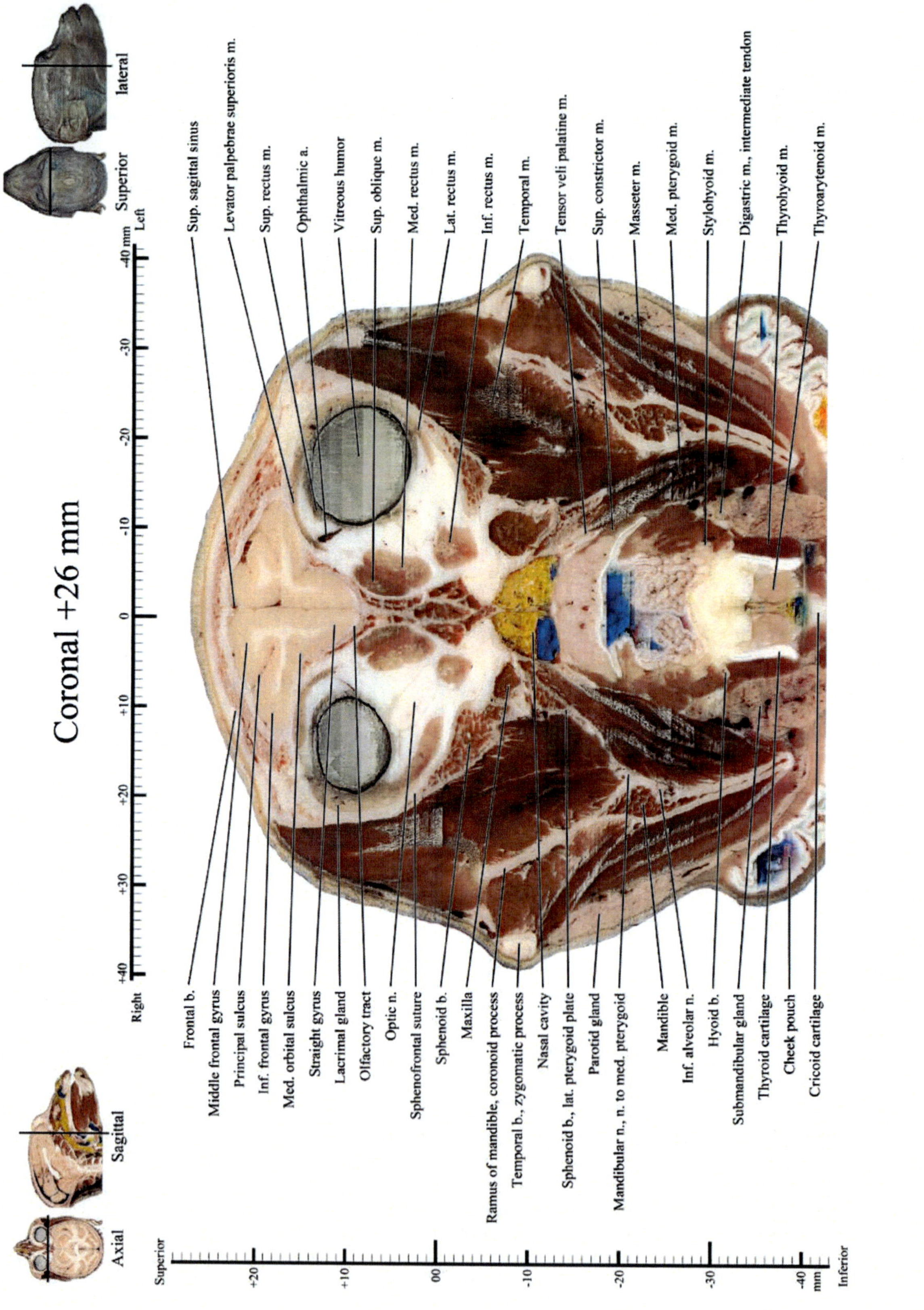

Coronal +25 mm

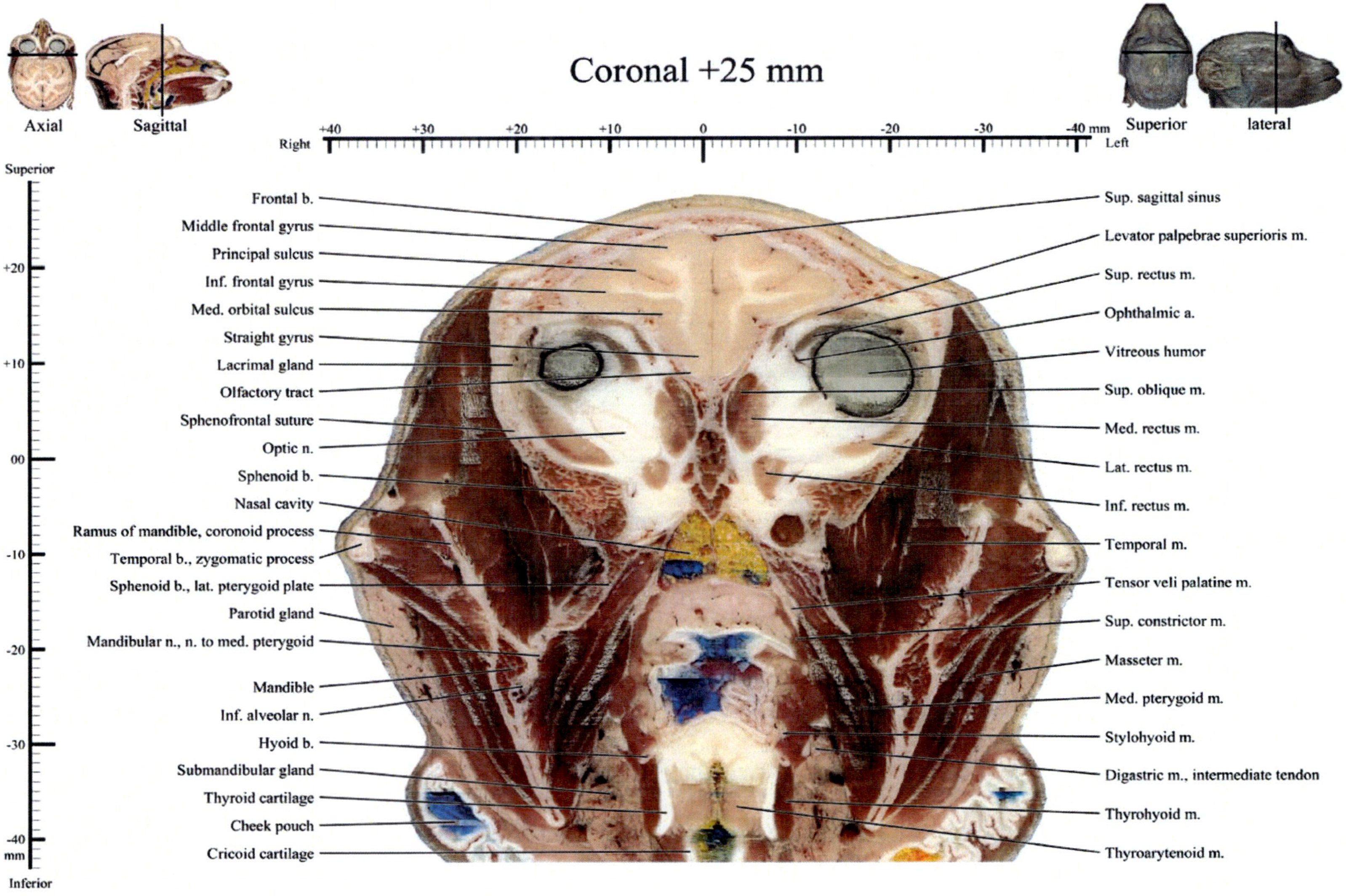

Coronal +24 mm

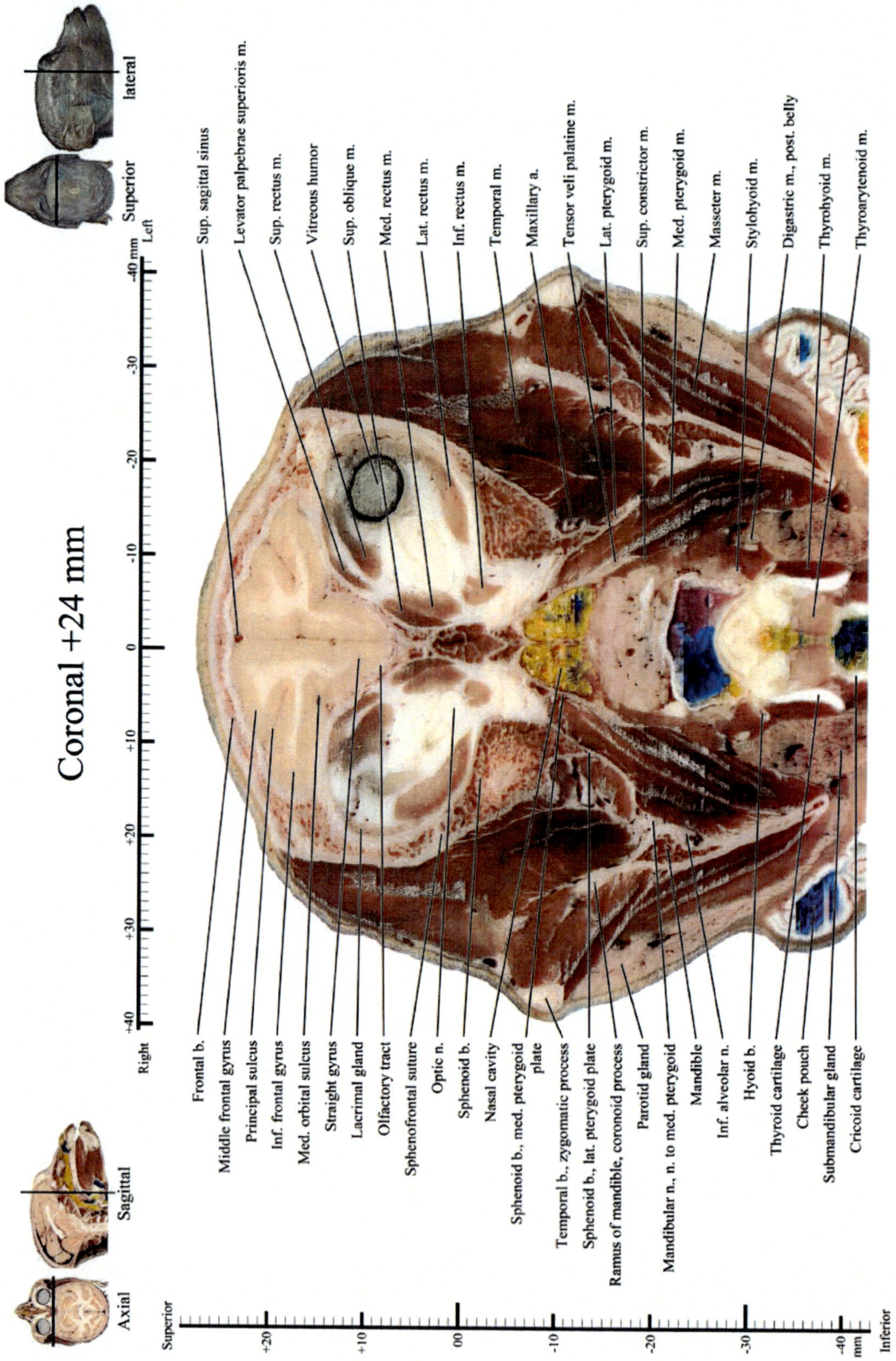

Coronal +23 mm

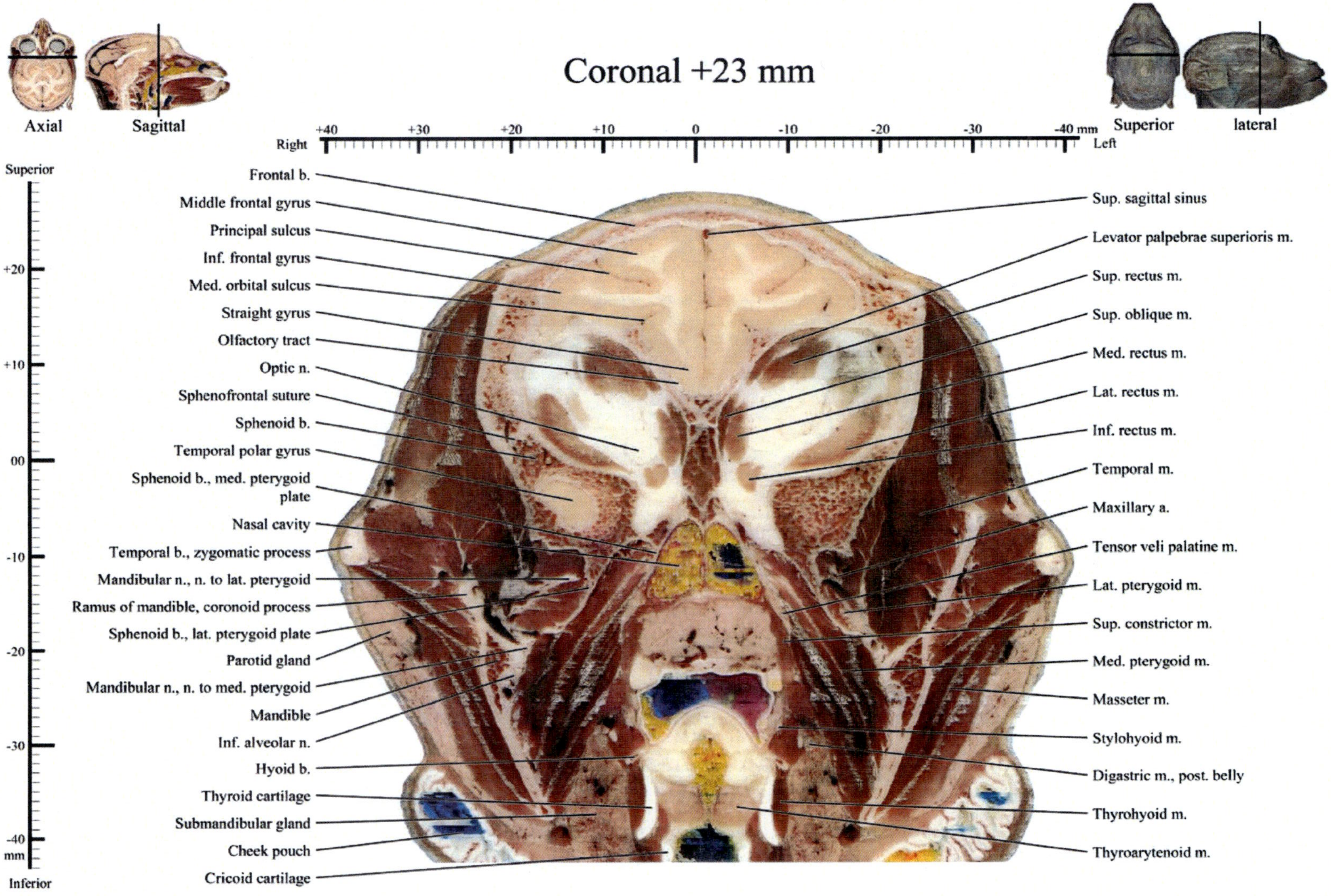

Coronal +22 mm

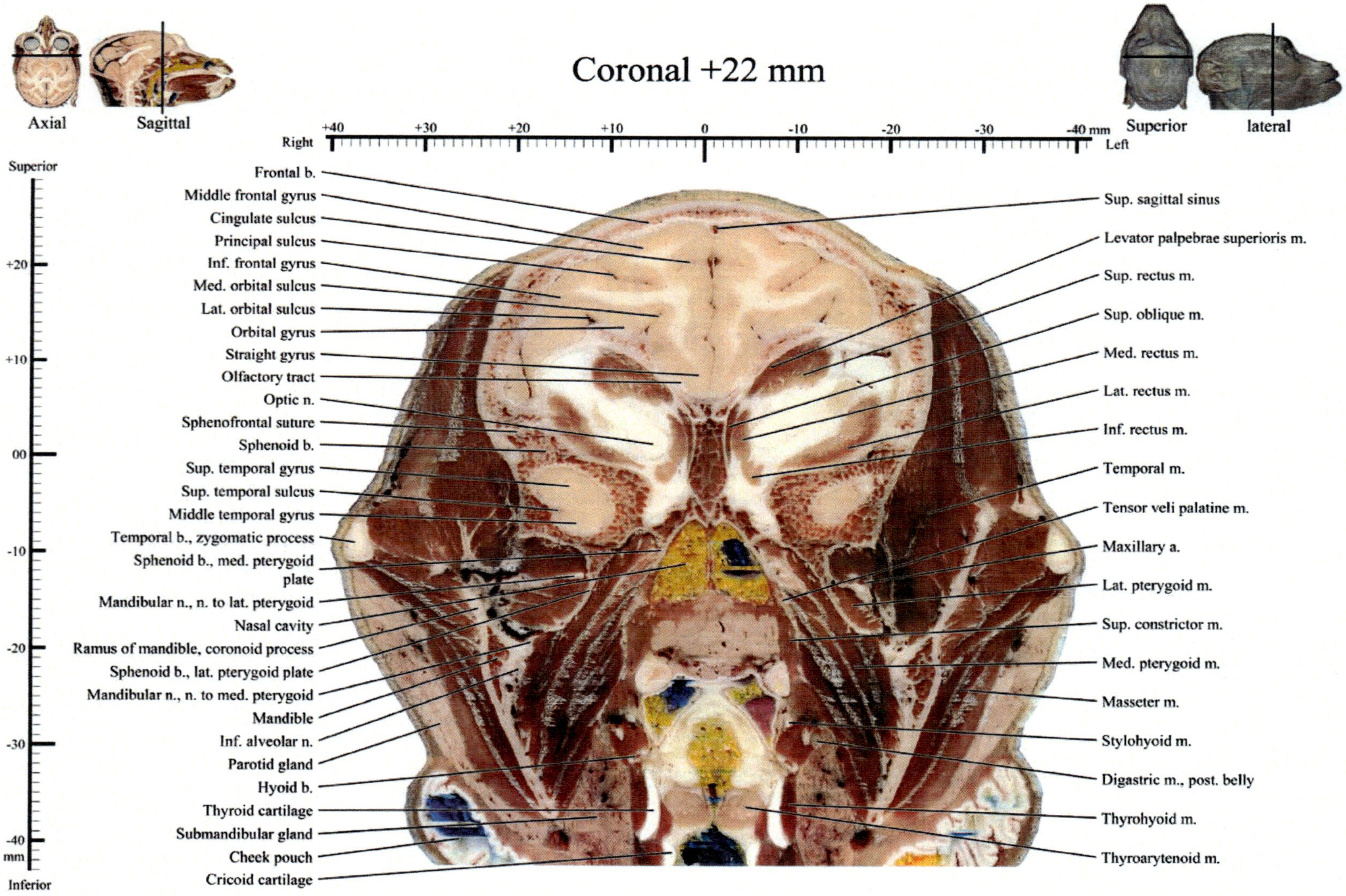

Coronal +21 mm

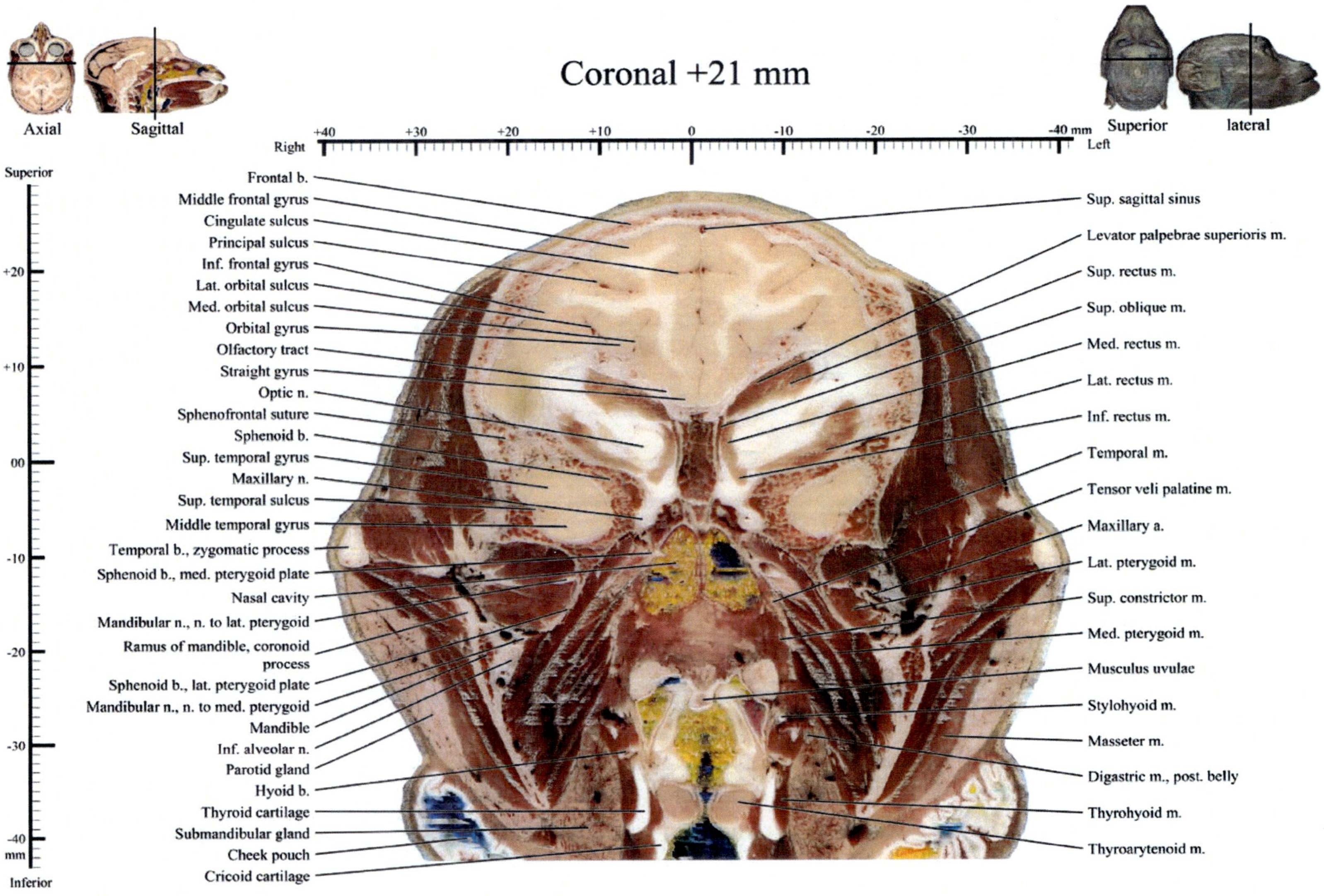

Coronal +20 mm

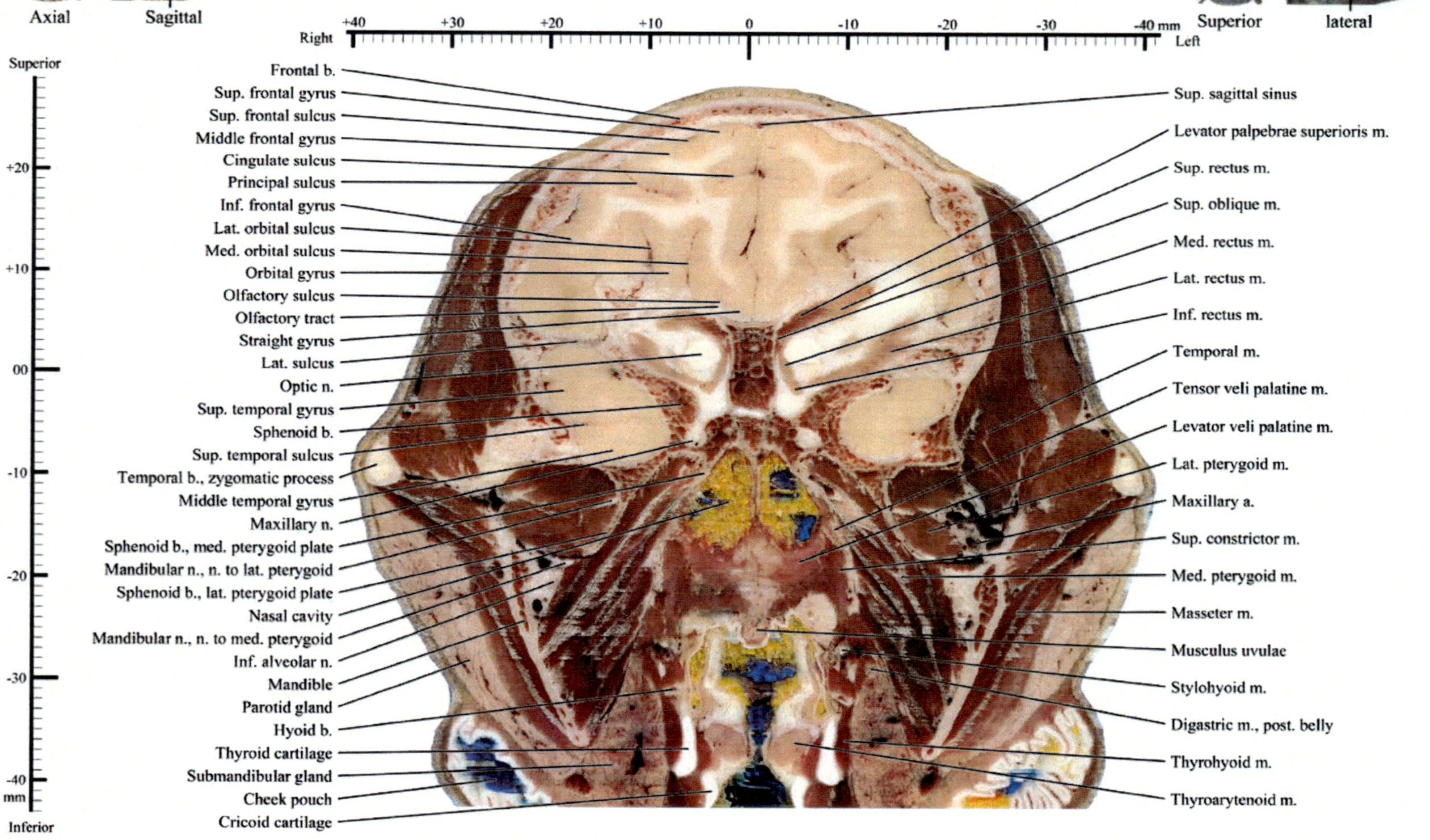

Coronal +19 mm

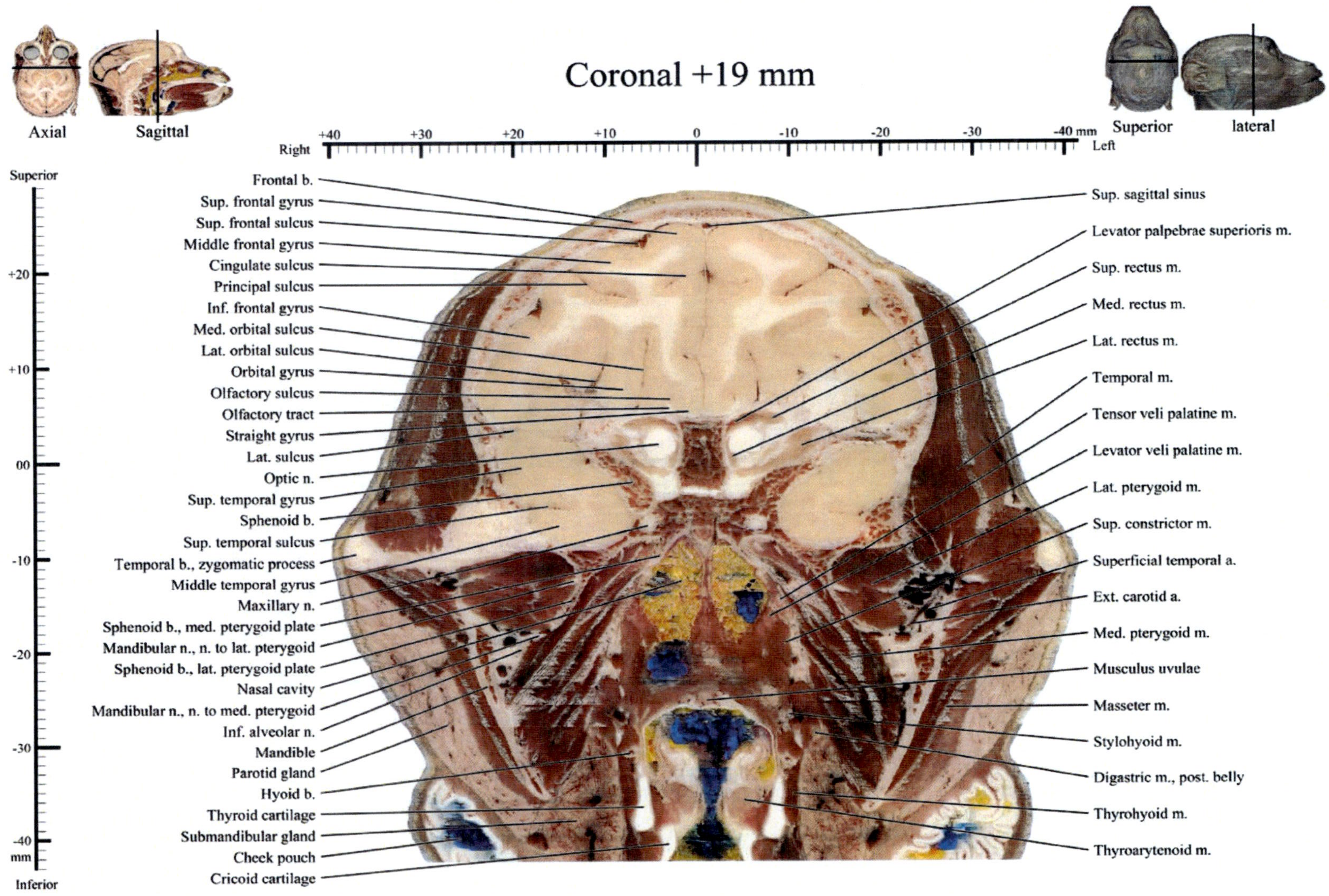

Coronal +18 mm

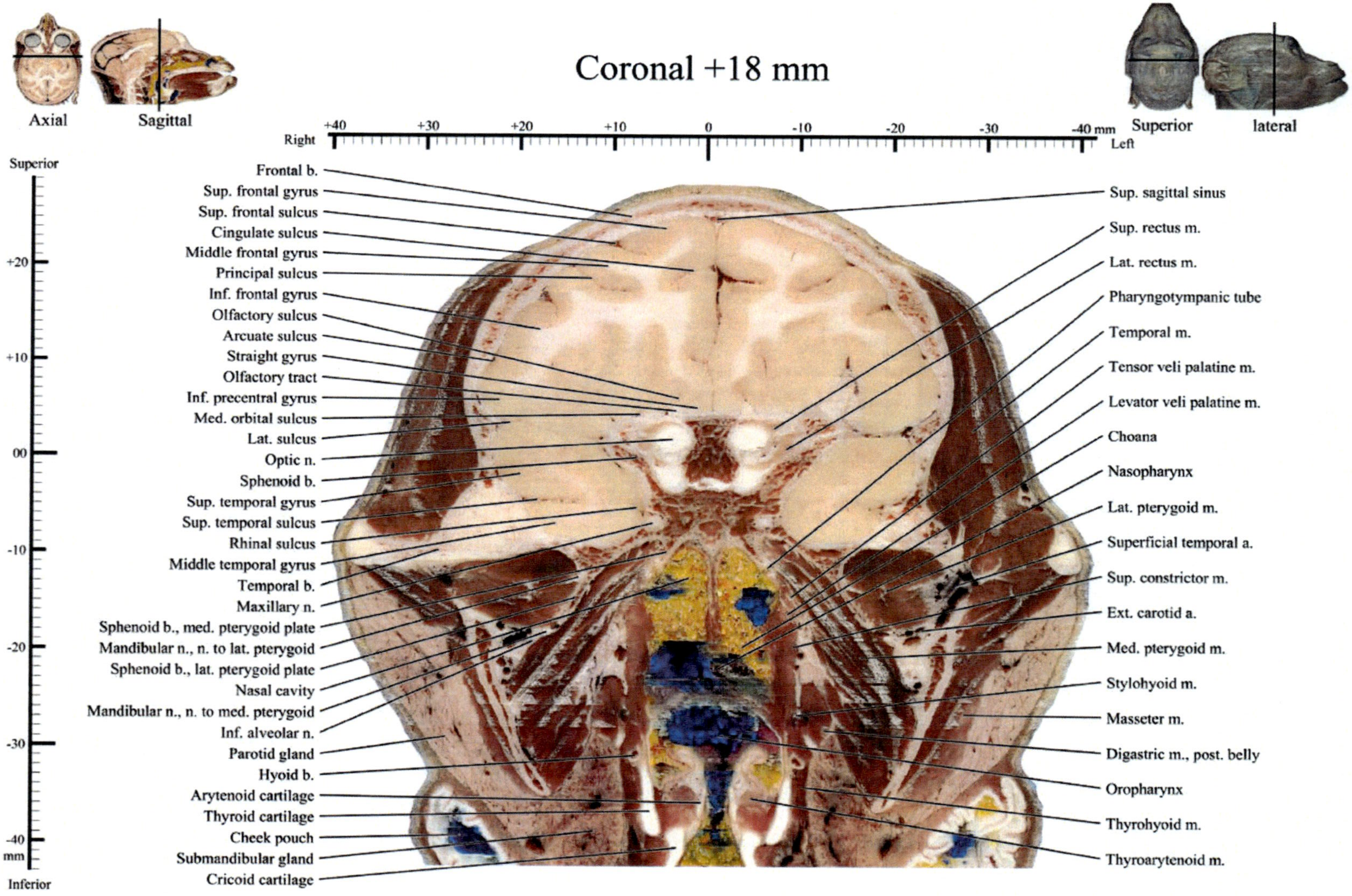

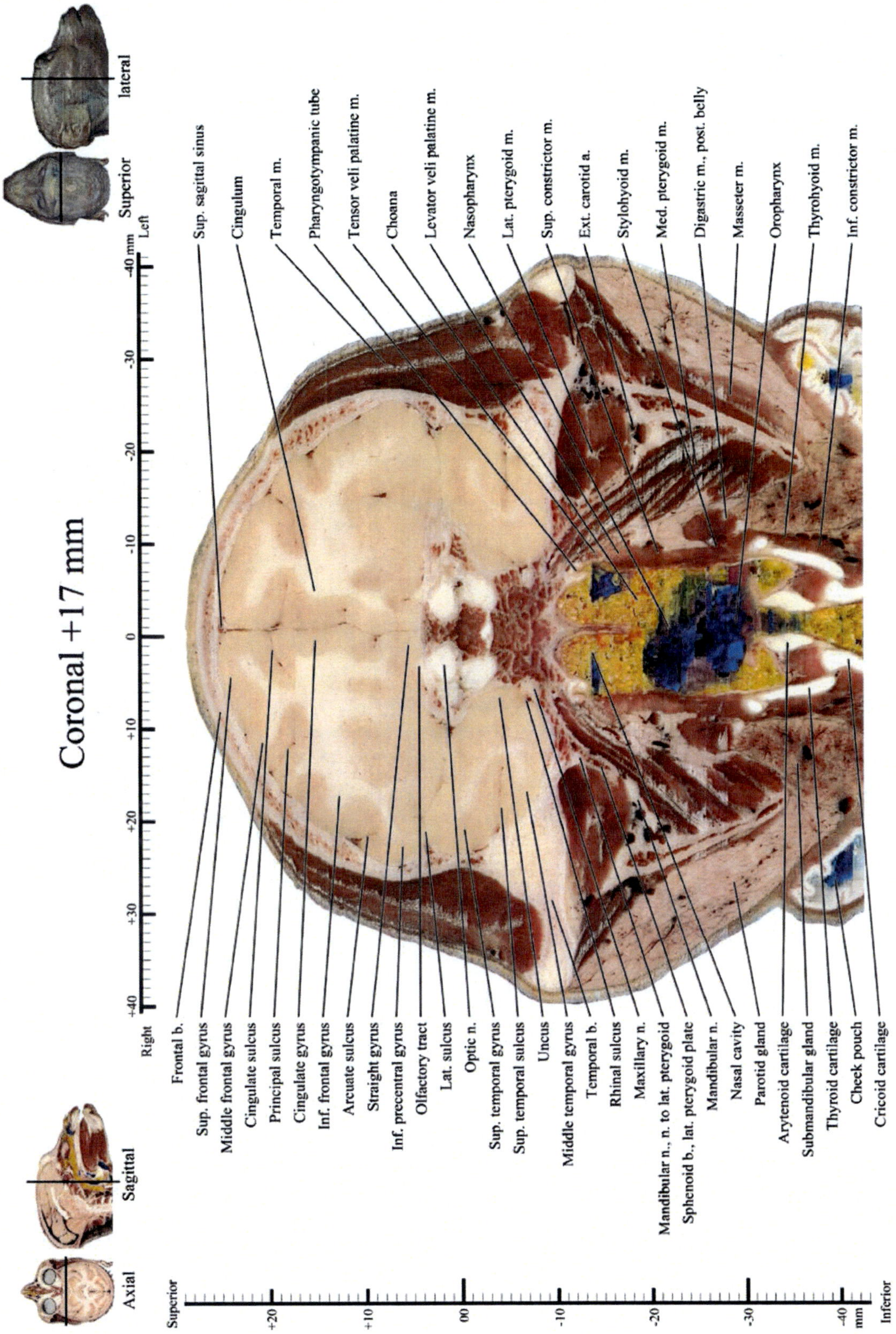
Coronal +17 mm
Superior
lateral
Axial
Sagittal
Right
Left
mm
Superior
Inferior
Sup. sagittal sinus
Cingulum
Temporal m.
Pharyngotympanic tube
Tensor veli palatine m.
Choana
Levator veli palatine m.
Nasopharynx
Lat. pterygoid m.
Sup. constrictor m.
Ext. carotid a.
Stylohyoid m.
Med. pterygoid m.
Digastric m., post. belly
Masseter m.
Oropharynx
Thyrohyoid m.
Inf. constrictor m.
Frontal b.
Sup. frontal gyrus
Middle frontal gyrus
Cingulate sulcus
Principal sulcus
Cingulate gyrus
Inf. frontal gyrus
Arcuate sulcus
Straight gyrus
Inf. precentral gyrus
Olfactory tract
Lat. sulcus
Optic n.
Sup. temporal gyrus
Sup. temporal sulcus
Uncus
Middle temporal gyrus
Temporal b.
Rhinal sulcus
Maxillary n.
Mandibular n., n. to lat. pterygoid
Sphenoid b., lat. pterygoid plate
Mandibular n.
Nasal cavity
Parotid gland
Arytenoid cartilage
Submandibular gland
Thyroid cartilage
Cheek pouch
Cricoid cartilage

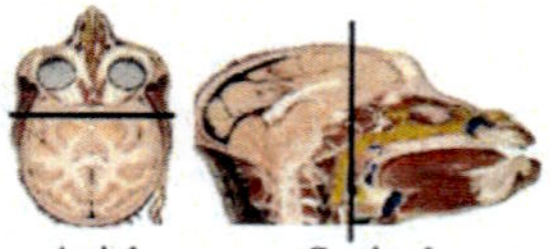

Coronal +16 mm

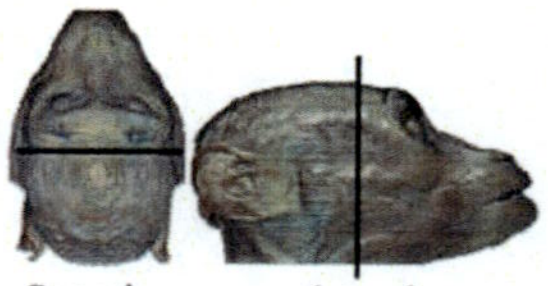

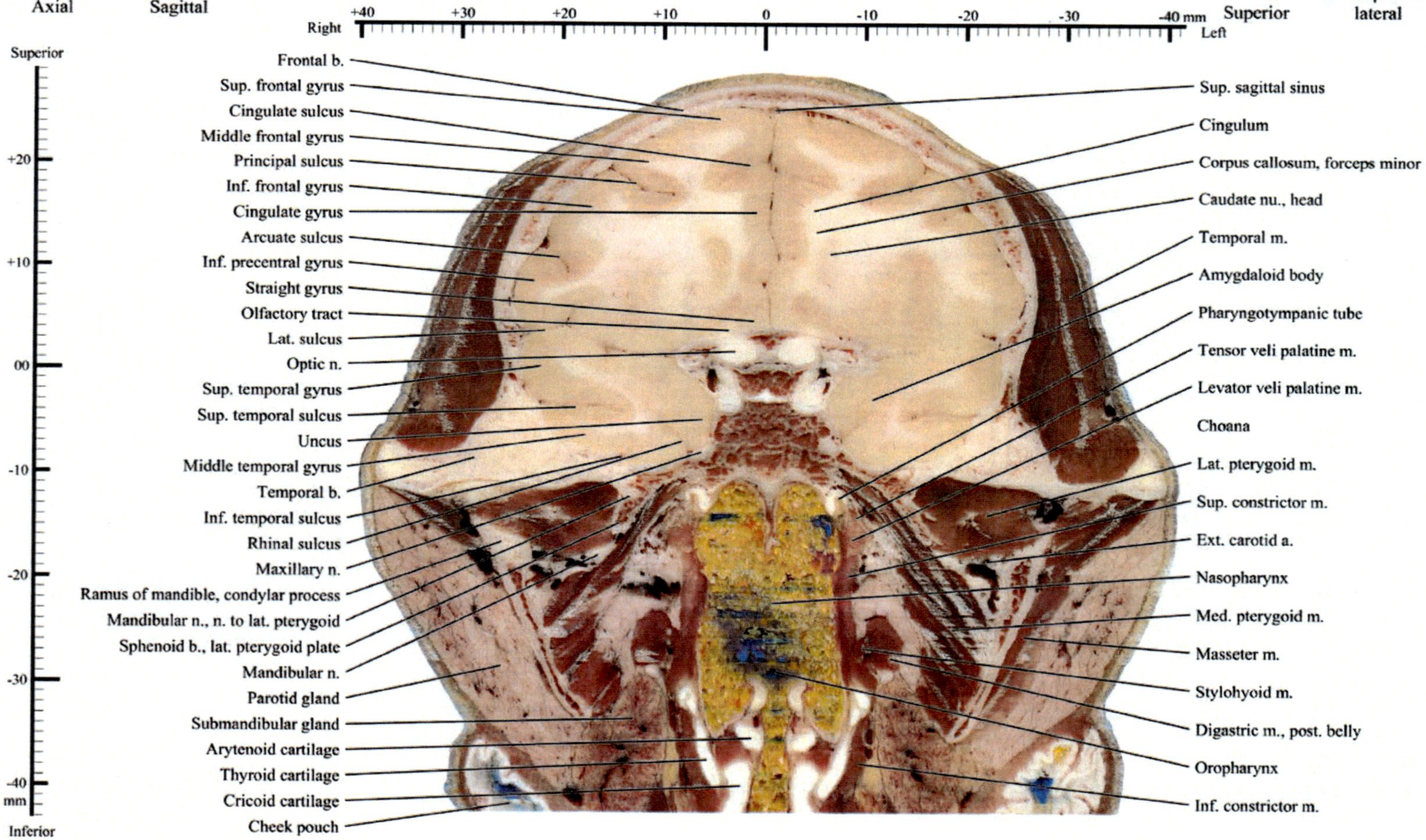

Coronal +15 mm

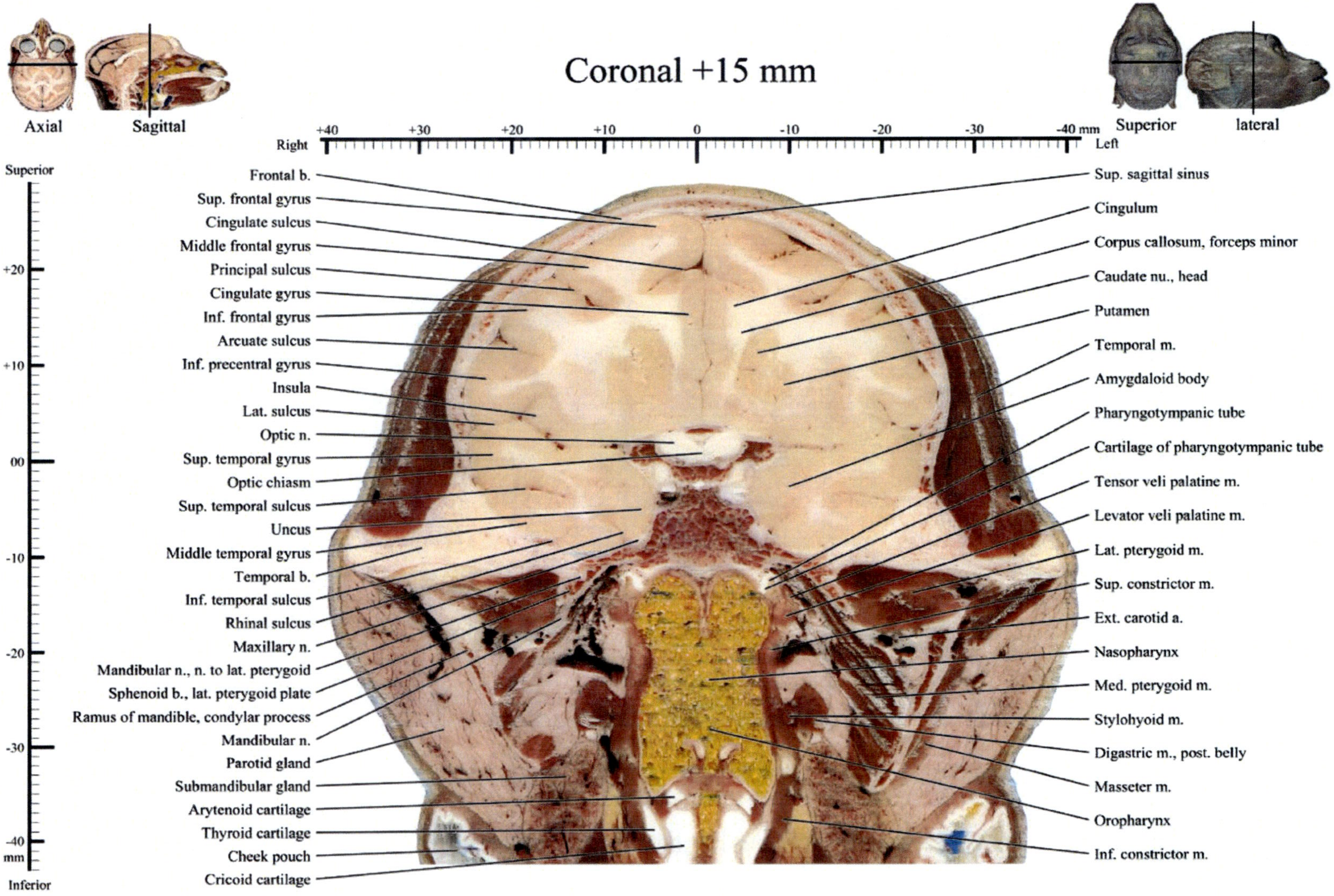

Coronal +14 mm

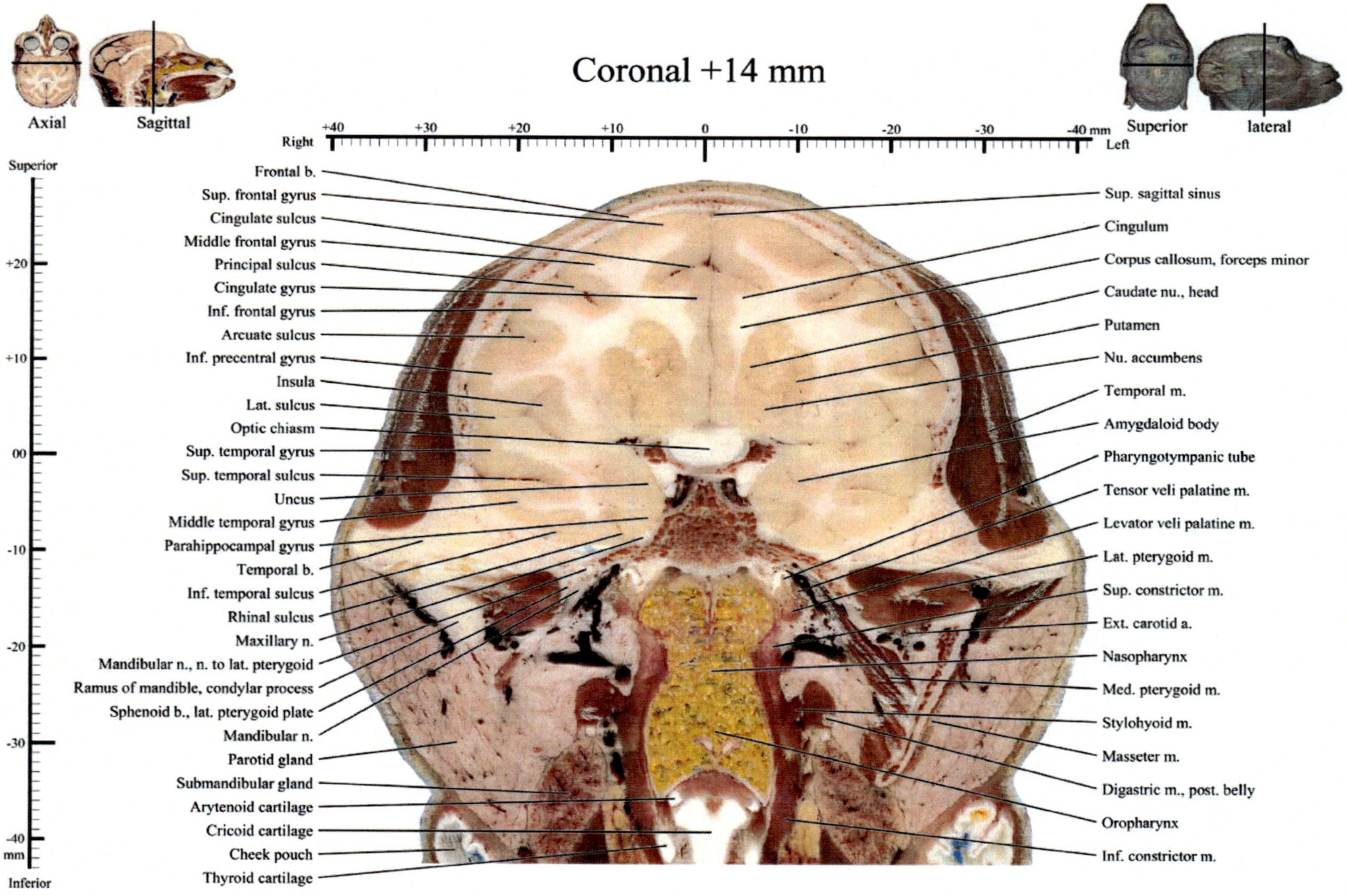

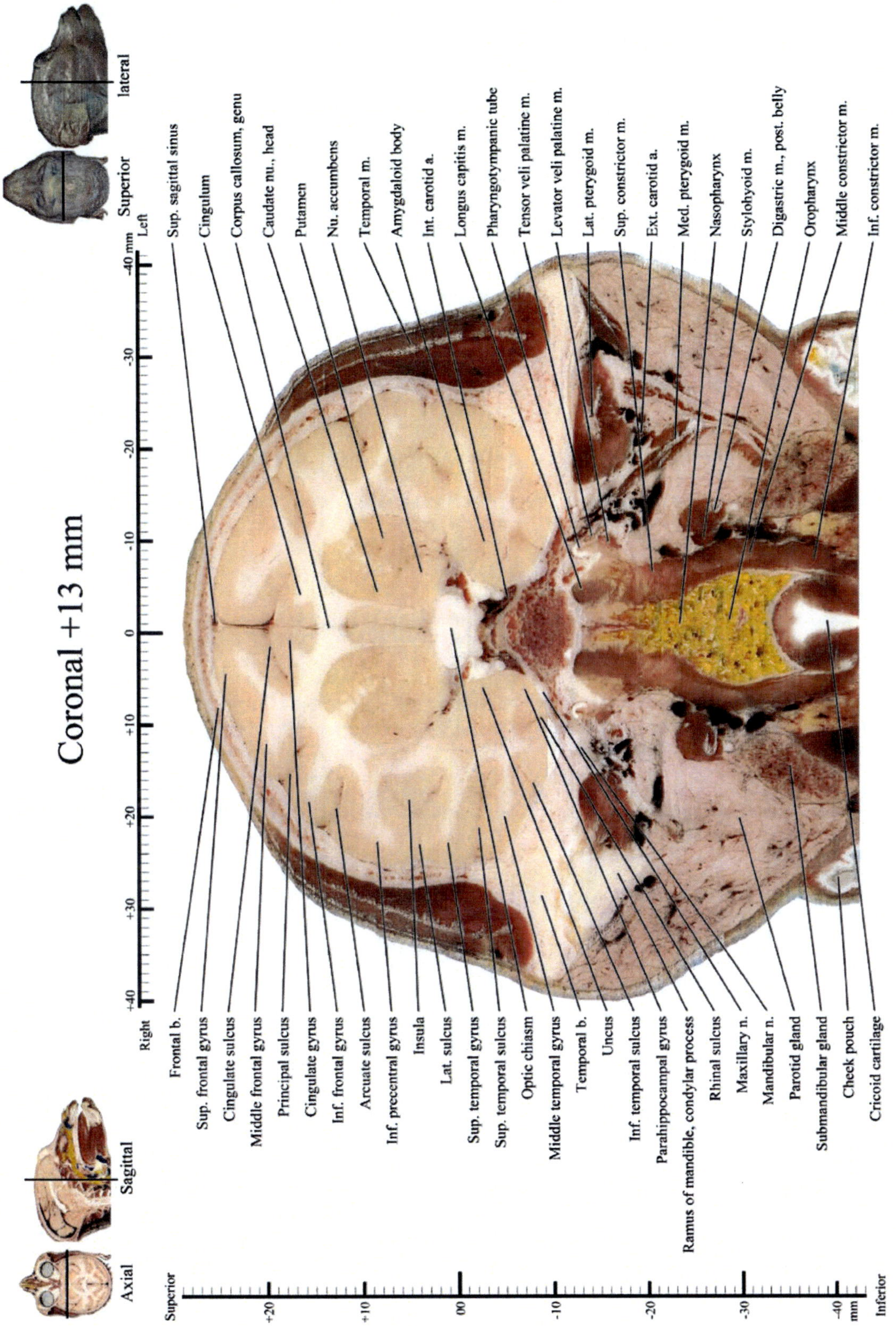
Coronal +13 mm
Superior
lateral
Axial
Sagittal
Right
Left
-40 mm
-30
-20
-10
0
+10
+20
+30
+40
Sup. sagittal sinus
Cingulum
Corpus callosum, genu
Caudate nu., head
Putamen
Nu. accumbens
Temporal m.
Amygdaloid body
Int. carotid a.
Longus capitis m.
Pharyngotympanic tube
Tensor veli palatine m.
Levator veli palatine m.
Lat. pterygoid m.
Sup. constrictor m.
Ext. carotid a.
Med. pterygoid m.
Nasopharynx
Stylohyoid m.
Digastric m., post. belly
Oropharynx
Middle constrictor m.
Inf. constrictor m.
Frontal b.
Sup. frontal gyrus
Cingulate sulcus
Middle frontal gyrus
Principal sulcus
Cingulate gyrus
Inf. frontal gyrus
Arcuate sulcus
Inf. precentral gyrus
Insula
Lat. sulcus
Sup. temporal gyrus
Sup. temporal sulcus
Optic chiasm
Middle temporal gyrus
Temporal b.
Uncus
Inf. temporal sulcus
Parahippocampal gyrus
Ramus of mandible, condylar process
Rhinal sulcus
Maxillary n.
Mandibular n.
Parotid gland
Submandibular gland
Cheek pouch
Cricoid cartilage
Superior
+20
+10
00
-10
-20
-30
-40
mm
Inferior

Coronal +12 mm

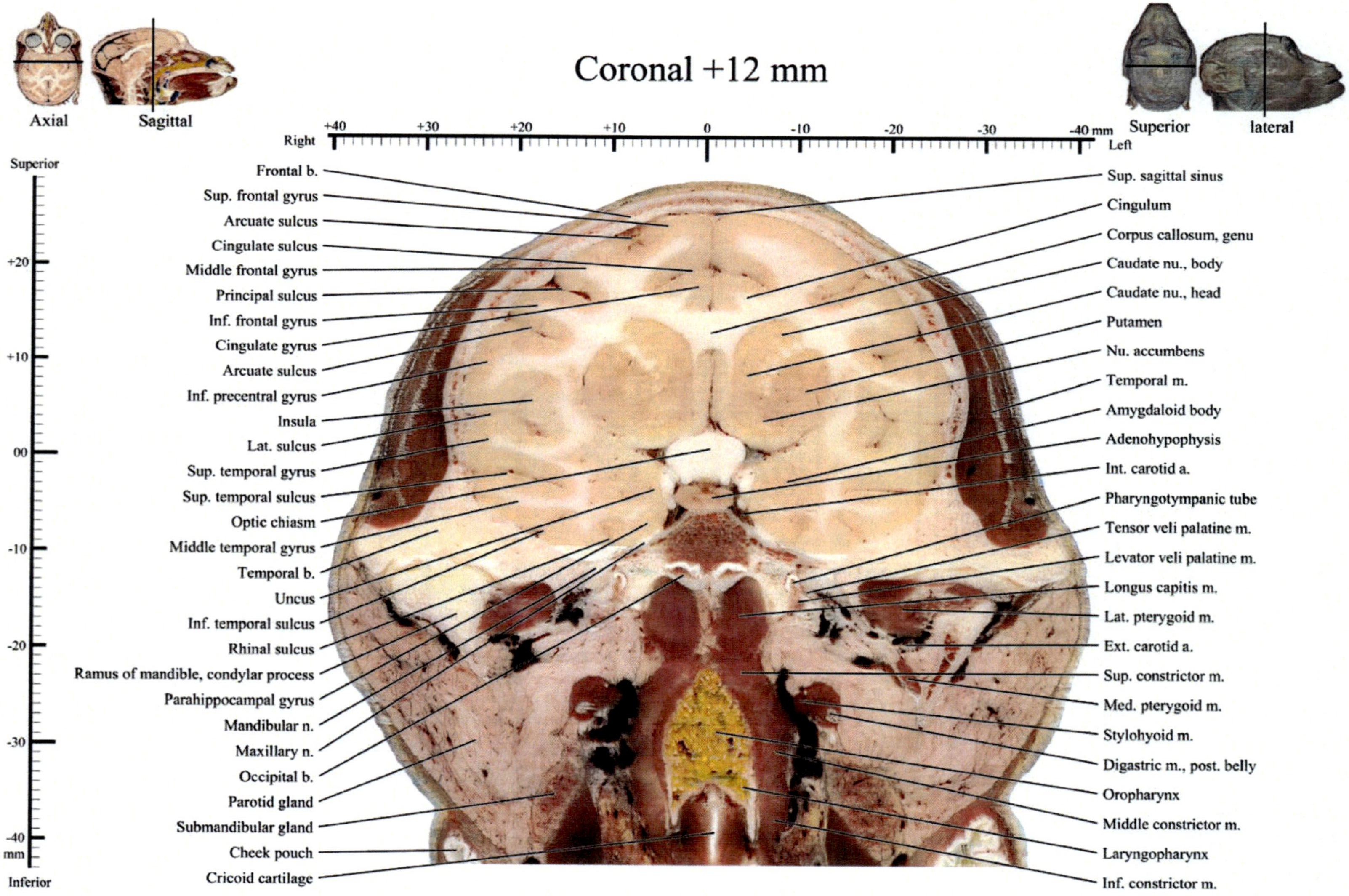

Coronal +11 mm

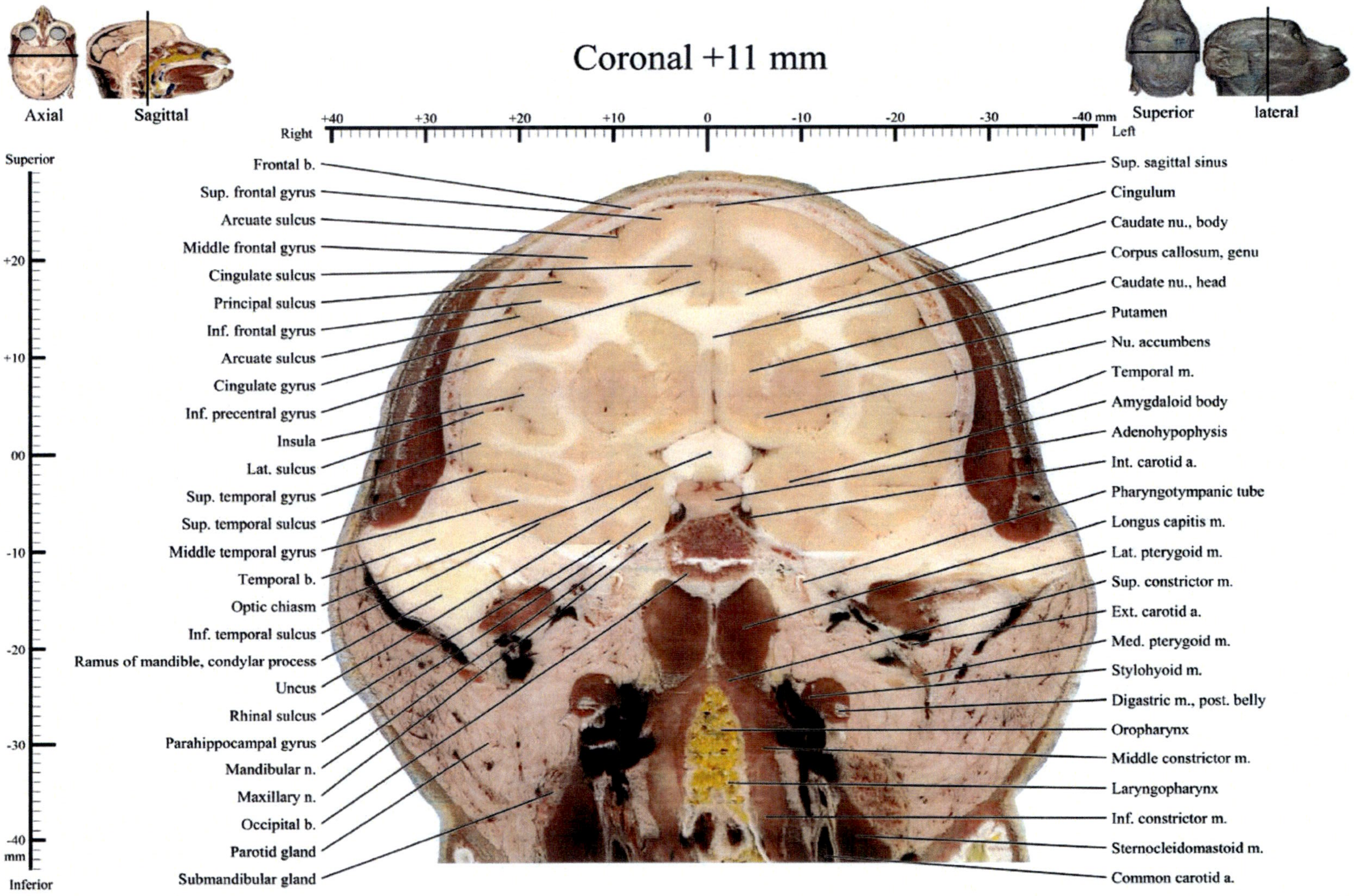

Coronal +10 mm

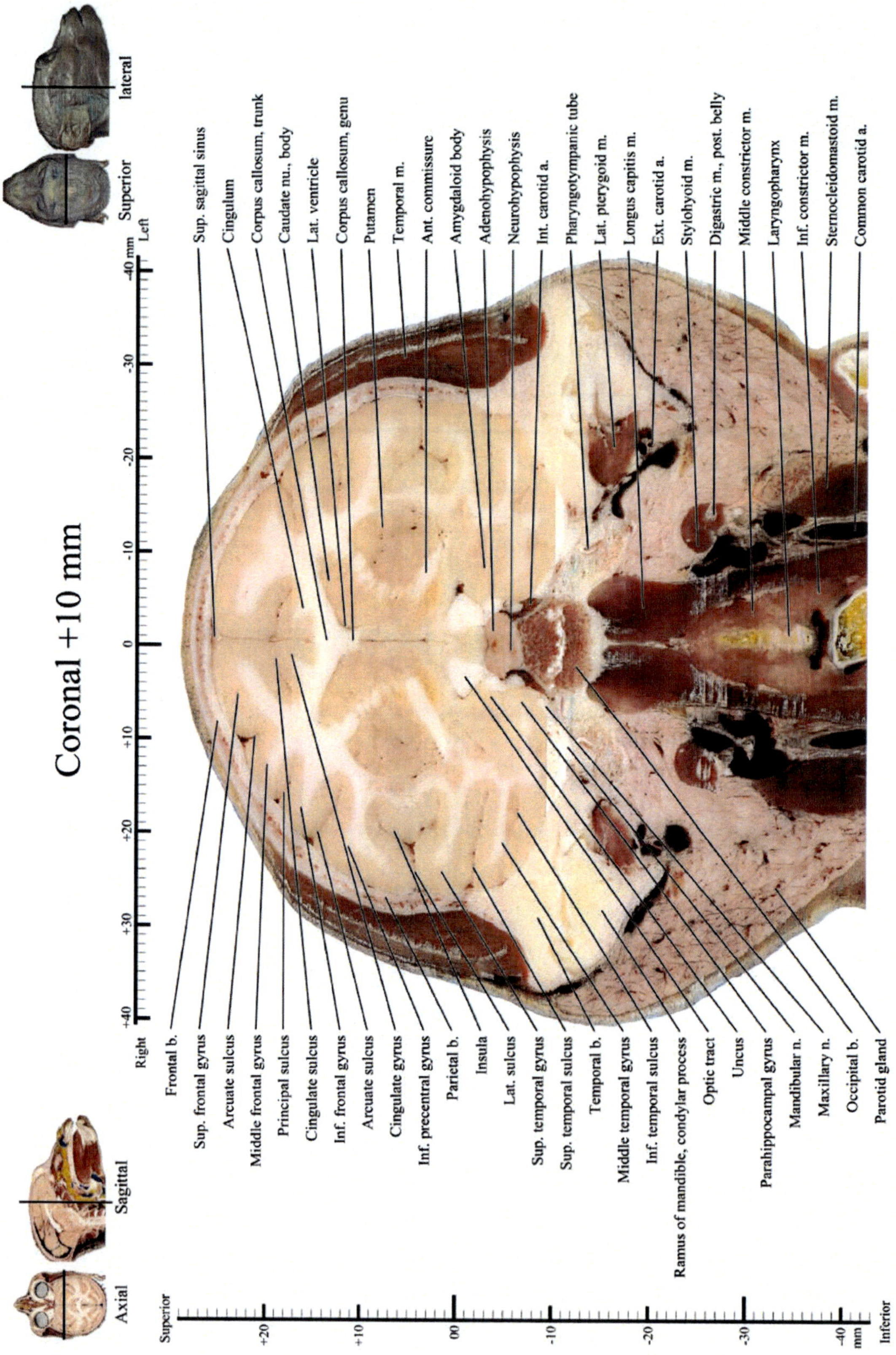

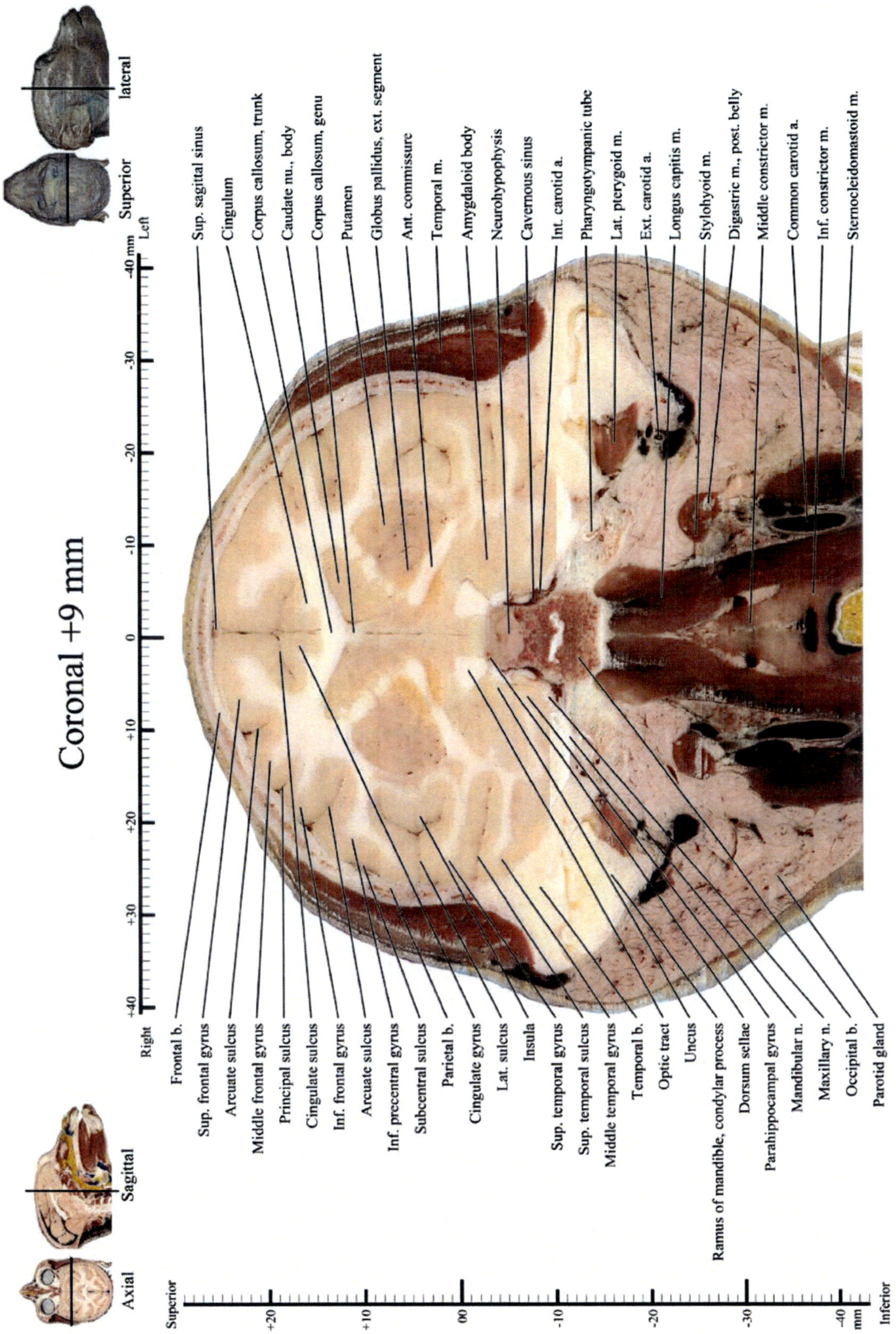
Coronal +9 mm
Axial
Sagittal
Superior
lateral
Right
Left
-40 mm
-30
-20
-10
0
+10
+20
+30
+40
Superior
+20
+10
00
-10
-20
-30
-40
mm
Inferior
Sup. sagittal sinus
Cingulum
Corpus callosum, trunk
Caudate nu., body
Corpus callosum, genu
Putamen
Globus pallidus, ext. segment
Ant. commissure
Temporal m.
Amygdaloid body
Neurohypophysis
Cavernous sinus
Int. carotid a.
Pharyngotympanic tube
Lat. pterygoid m.
Ext. carotid a.
Longus capitis m.
Stylohyoid m.
Digastric m., post. belly
Middle constrictor m.
Common carotid a.
Inf. constrictor m.
Sternocleidomastoid m.
Frontal b.
Sup. frontal gyrus
Arcuate sulcus
Middle frontal gyrus
Principal sulcus
Cingulate sulcus
Inf. frontal gyrus
Arcuate sulcus
Inf. precentral gyrus
Subcentral sulcus
Parietal b.
Cingulate gyrus
Lat. sulcus
Insula
Sup. temporal gyrus
Sup. temporal sulcus
Middle temporal gyrus
Temporal b.
Optic tract
Uncus
Ramus of mandible, condylar process
Dorsum sellae
Parahippocampal gyrus
Mandibular n.
Maxillary n.
Occipital b.
Parotid gland

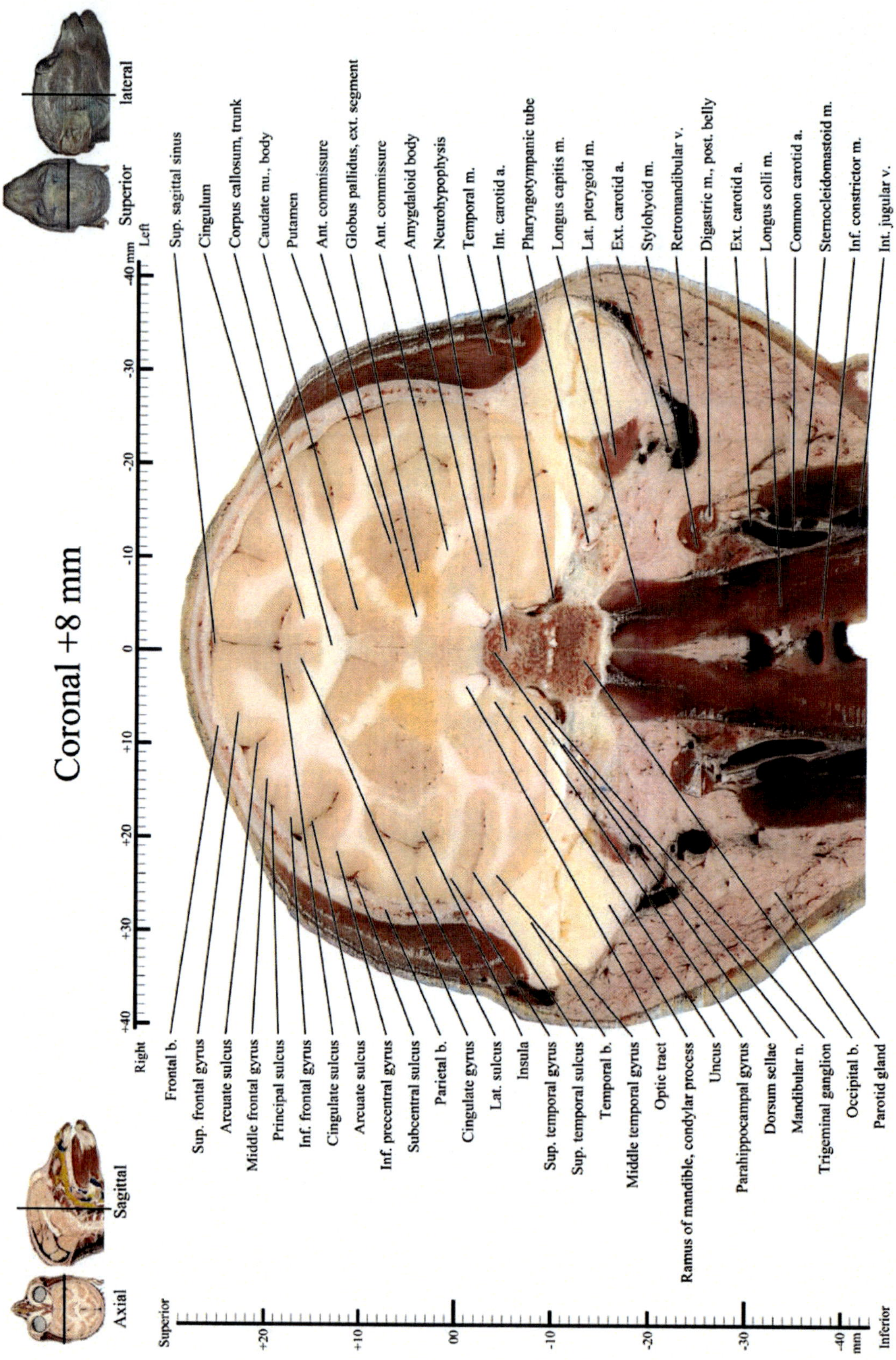
Coronal +8 mm
Superior
lateral
Axial
Sagittal
Right
Left
-40 mm
-30
-20
-10
0
+10
+20
+30
+40
Sup. sagittal sinus
Cingulum
Corpus callosum, trunk
Caudate nu., body
Putamen
Ant. commissure
Globus pallidus, ext. segment
Ant. commissure
Amygdaloid body
Neurohypophysis
Temporal m.
Int. carotid a.
Pharyngotympanic tube
Longus capitis m.
Lat. pterygoid m.
Ext. carotid a.
Stylohyoid m.
Retromandibular v.
Digastric m., post. belly
Ext. carotid a.
Longus colli m.
Common carotid a.
Sternocleidomastoid m.
Inf. constrictor m.
Int. jugular v.
Frontal b.
Sup. frontal gyrus
Arcuate sulcus
Middle frontal gyrus
Principal sulcus
Inf. frontal gyrus
Cingulate sulcus
Arcuate sulcus
Inf. precentral gyrus
Subcentral sulcus
Parietal b.
Cingulate gyrus
Lat. sulcus
Insula
Sup. temporal gyrus
Sup. temporal sulcus
Temporal b.
Middle temporal gyrus
Optic tract
Ramus of mandible, condylar process
Uncus
Parahippocampal gyrus
Dorsum sellae
Mandibular n.
Trigeminal ganglion
Occipital b.
Parotid gland
Superior
+20
+10
00
-10
-20
-30
-40
mm
Inferior

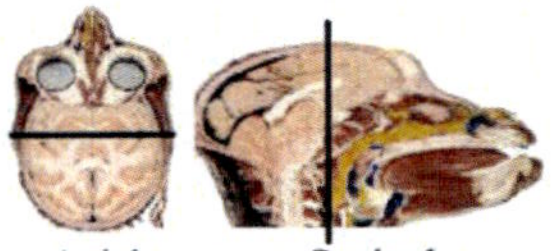
Axial
Sagittal

Coronal +7 mm

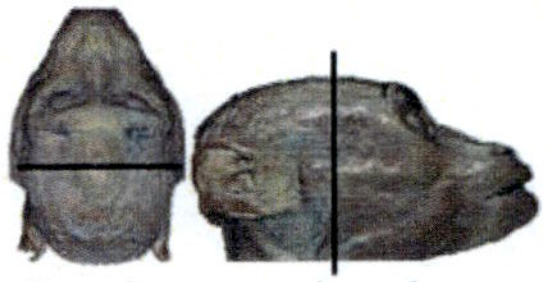
Superior
lateral

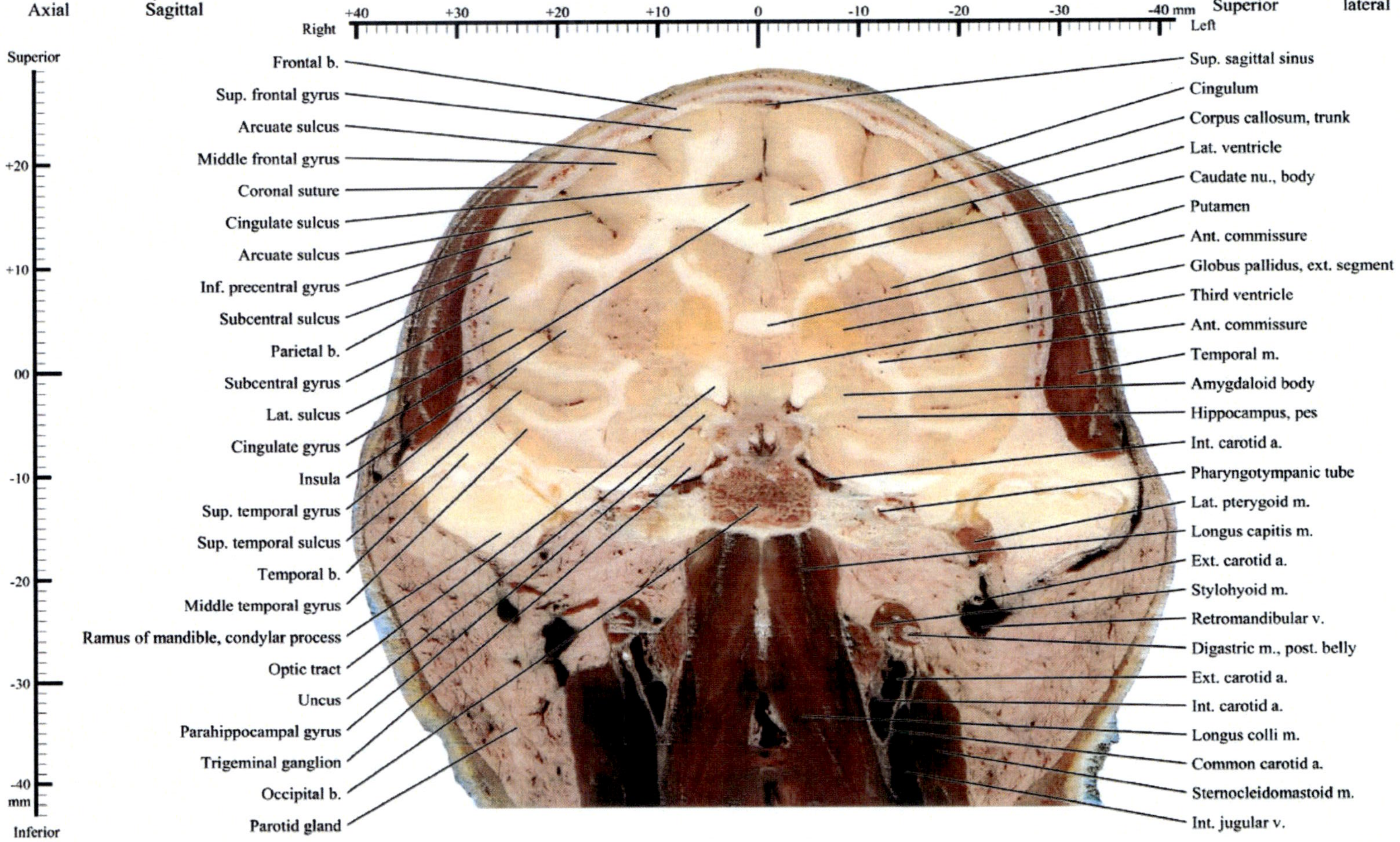
+40
+30
+20
+10
0
-10
-20
-30
-40 mm
Right
Left
Superior
+20
+10
00
-10
-20
-30
-40
mm
Inferior
Frontal b.
Sup. frontal gyrus
Arcuate sulcus
Middle frontal gyrus
Coronal suture
Cingulate sulcus
Arcuate sulcus
Inf. precentral gyrus
Subcentral sulcus
Parietal b.
Subcentral gyrus
Lat. sulcus
Cingulate gyrus
Insula
Sup. temporal gyrus
Sup. temporal sulcus
Temporal b.
Middle temporal gyrus
Ramus of mandible, condylar process
Optic tract
Uncus
Parahippocampal gyrus
Trigeminal ganglion
Occipital b.
Parotid gland
Sup. sagittal sinus
Cingulum
Corpus callosum, trunk
Lat. ventricle
Caudate nu., body
Putamen
Ant. commissure
Globus pallidus, ext. segment
Third ventricle
Ant. commissure
Temporal m.
Amygdaloid body
Hippocampus, pes
Int. carotid a.
Pharyngotympanic tube
Lat. pterygoid m.
Longus capitis m.
Ext. carotid a.
Stylohyoid m.
Retromandibular v.
Digastric m., post. belly
Ext. carotid a.
Int. carotid a.
Longus colli m.
Common carotid a.
Sternocleidomastoid m.
Int. jugular v.

Coronal +6 mm

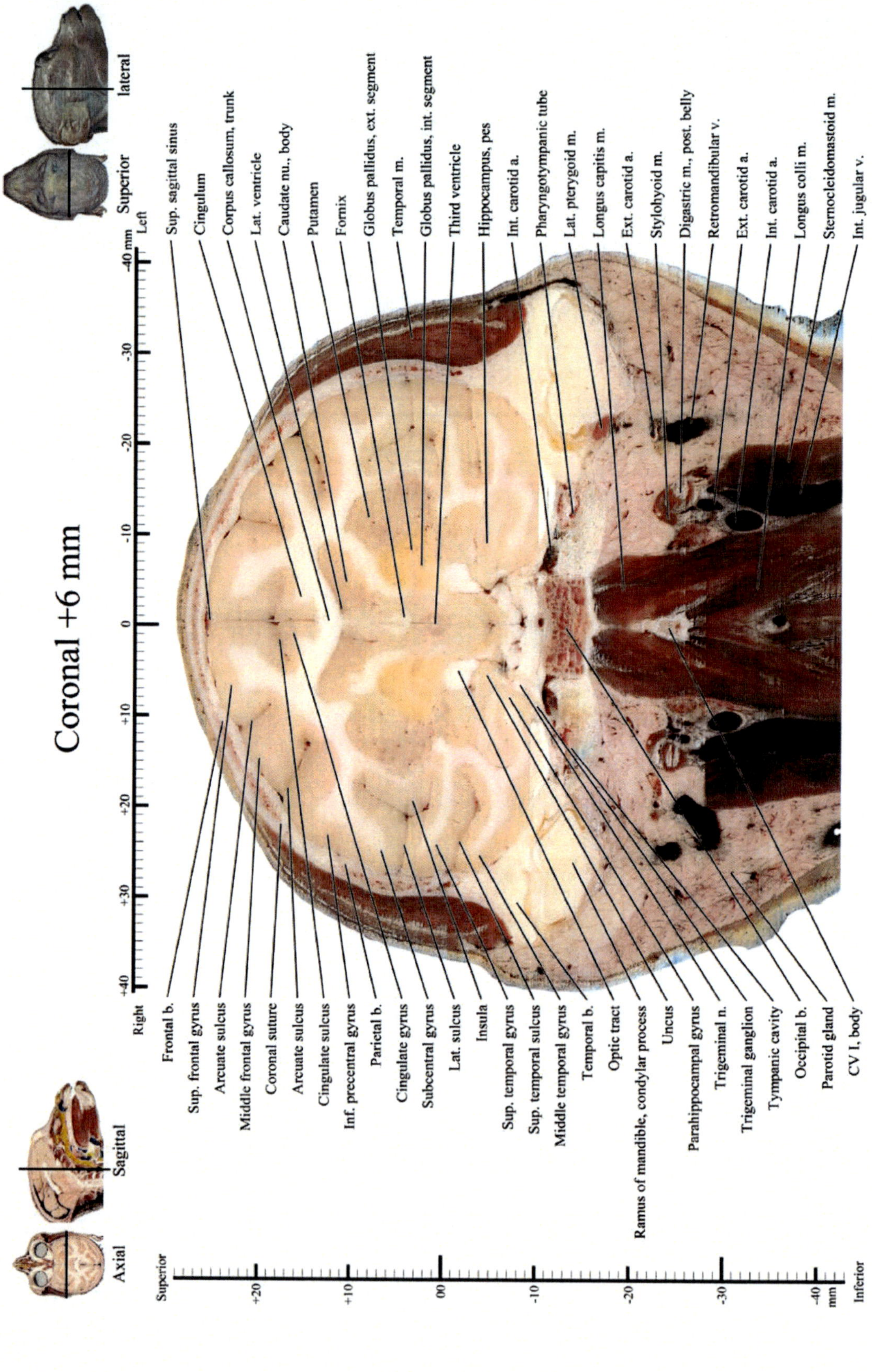
Superior
lateral
Axial
Sagittal
Left
Right
-40 mm
-30
-20
-10
0
+10
+20
+30
+40
Sup. sagittal sinus
Cingulum
Corpus callosum, trunk
Lat. ventricle
Caudate nu., body
Putamen
Fornix
Globus pallidus, ext. segment
Temporal m.
Globus pallidus, int. segment
Third ventricle
Hippocampus, pes
Int. carotid a.
Pharyngotympanic tube
Lat. pterygoid m.
Longus capitis m.
Ext. carotid a.
Stylohyoid m.
Digastric m., post. belly
Retromandibular v.
Ext. carotid a.
Int. carotid a.
Longus colli m.
Sternocleidomastoid m.
Int. jugular v.
Frontal b.
Sup. frontal gyrus
Arcuate sulcus
Middle frontal gyrus
Coronal suture
Arcuate sulcus
Cingulate sulcus
Inf. precentral gyrus
Parietal b.
Cingulate gyrus
Subcentral gyrus
Lat. sulcus
Insula
Sup. temporal gyrus
Sup. temporal sulcus
Middle temporal gyrus
Temporal b.
Optic tract
Ramus of mandible, condylar process
Uncus
Parahippocampal gyrus
Trigeminal n.
Trigeminal ganglion
Tympanic cavity
Occipital b.
Parotid gland
CV I, body
Superior
+20
+10
00
-10
-20
-30
-40
mm
Inferior

Coronal +5 mm

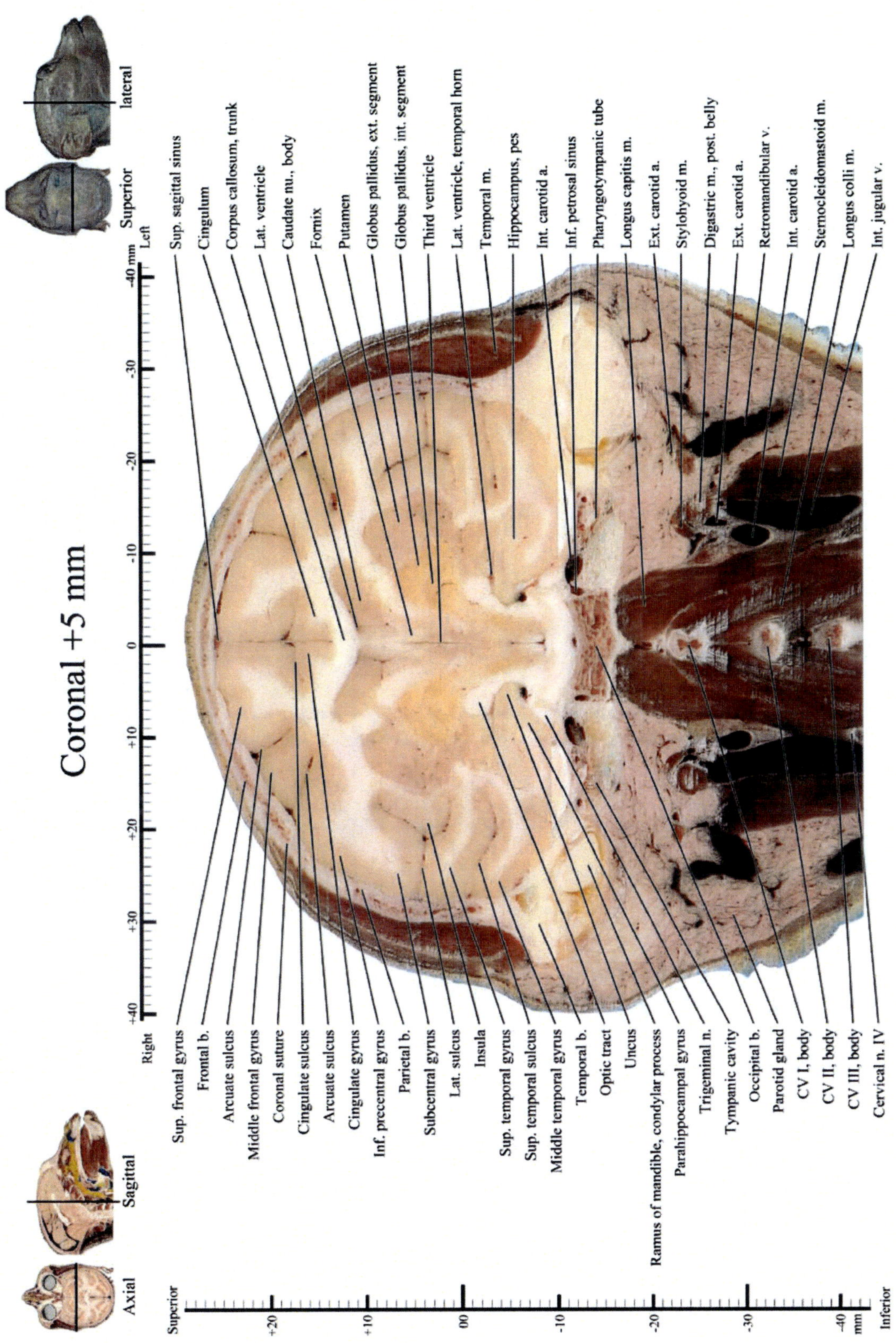
Superior
lateral
Axial
Sagittal
-40 mm
-30
-20
-10
0
+10
+20
+30
+40
Left
Right
Sup. sagittal sinus
Cingulum
Corpus callosum, trunk
Lat. ventricle
Caudate nu., body
Fornix
Putamen
Globus pallidus, ext. segment
Globus pallidus, int. segment
Third ventricle
Lat. ventricle, temporal horn
Temporal m.
Hippocampus, pes
Int. carotid a.
Inf. petrosal sinus
Pharyngotympanic tube
Longus capitis m.
Ext. carotid a.
Stylohyoid m.
Digastric m., post. belly
Ext. carotid a.
Retromandibular v.
Int. carotid a.
Sternocleidomastoid m.
Longus colli m.
Int. jugular v.
Sup. frontal gyrus
Frontal b.
Arcuate sulcus
Middle frontal gyrus
Coronal suture
Cingulate sulcus
Arcuate sulcus
Cingulate gyrus
Inf. precentral gyrus
Parietal b.
Subcentral gyrus
Lat. sulcus
Insula
Sup. temporal gyrus
Sup. temporal sulcus
Middle temporal gyrus
Temporal b.
Optic tract
Uncus
Ramus of mandible, condylar process
Parahippocampal gyrus
Trigeminal n.
Tympanic cavity
Occipital b.
Parotid gland
CV I, body
CV II, body
CV III, body
Cervical n. IV
Superior
+20
+10
00
-10
-20
-30
-40 mm
Inferior

Coronal +4 mm

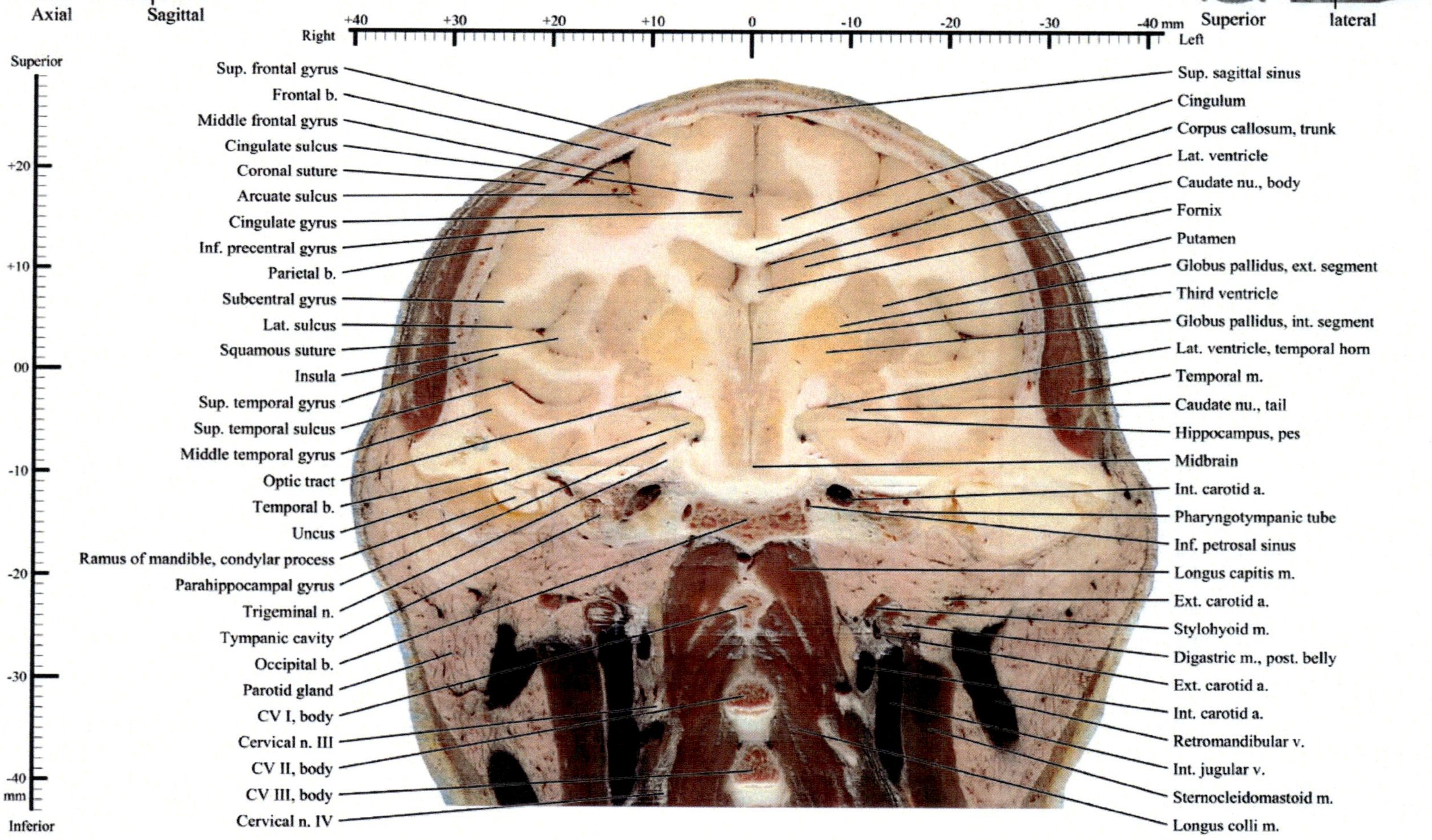
Axial
Sagittal
Superior
lateral
Right
Left
+40
+30
+20
+10
0
-10
-20
-30
-40 mm
Superior
+20
+10
00
-10
-20
-30
-40
mm
Inferior
Sup. frontal gyrus
Frontal b.
Middle frontal gyrus
Cingulate sulcus
Coronal suture
Arcuate sulcus
Cingulate gyrus
Inf. precentral gyrus
Parietal b.
Subcentral gyrus
Lat. sulcus
Squamous suture
Insula
Sup. temporal gyrus
Sup. temporal sulcus
Middle temporal gyrus
Optic tract
Temporal b.
Uncus
Ramus of mandible, condylar process
Parahippocampal gyrus
Trigeminal n.
Tympanic cavity
Occipital b.
Parotid gland
CV I, body
Cervical n. III
CV II, body
CV III, body
Cervical n. IV
Sup. sagittal sinus
Cingulum
Corpus callosum, trunk
Lat. ventricle
Caudate nu., body
Fornix
Putamen
Globus pallidus, ext. segment
Third ventricle
Globus pallidus, int. segment
Lat. ventricle, temporal horn
Temporal m.
Caudate nu., tail
Hippocampus, pes
Midbrain
Int. carotid a.
Pharyngotympanic tube
Inf. petrosal sinus
Longus capitis m.
Ext. carotid a.
Stylohyoid m.
Digastric m., post. belly
Ext. carotid a.
Int. carotid a.
Retromandibular v.
Int. jugular v.
Sternocleidomastoid m.
Longus colli m.

Coronal +3 mm

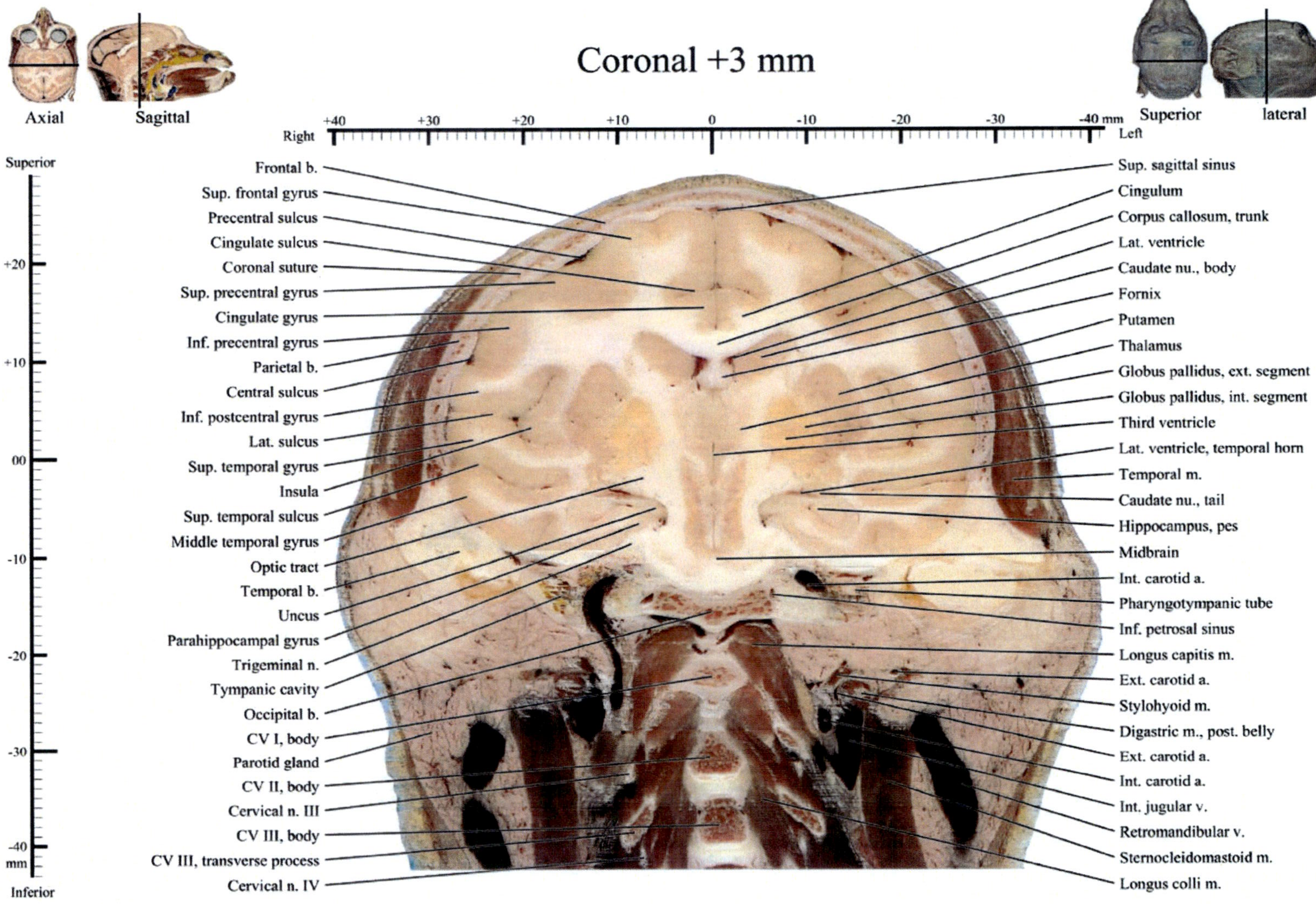
Axial
Sagittal
Superior
lateral
Right
Left
+40
+30
+20
+10
0
-10
-20
-30
-40 mm
Superior
+20
+10
00
-10
-20
-30
-40
mm
Inferior
Frontal b.
Sup. frontal gyrus
Precentral sulcus
Cingulate sulcus
Coronal suture
Sup. precentral gyrus
Cingulate gyrus
Inf. precentral gyrus
Parietal b.
Central sulcus
Inf. postcentral gyrus
Lat. sulcus
Sup. temporal gyrus
Insula
Sup. temporal sulcus
Middle temporal gyrus
Optic tract
Temporal b.
Uncus
Parahippocampal gyrus
Trigeminal n.
Tympanic cavity
Occipital b.
CV I, body
Parotid gland
CV II, body
Cervical n. III
CV III, body
CV III, transverse process
Cervical n. IV
Sup. sagittal sinus
Cingulum
Corpus callosum, trunk
Lat. ventricle
Caudate nu., body
Fornix
Putamen
Thalamus
Globus pallidus, ext. segment
Globus pallidus, int. segment
Third ventricle
Lat. ventricle, temporal horn
Temporal m.
Caudate nu., tail
Hippocampus, pes
Midbrain
Int. carotid a.
Pharyngotympanic tube
Inf. petrosal sinus
Longus capitis m.
Ext. carotid a.
Stylohyoid m.
Digastric m., post. belly
Ext. carotid a.
Int. carotid a.
Int. jugular v.
Retromandibular v.
Sternocleidomastoid m.
Longus colli m.

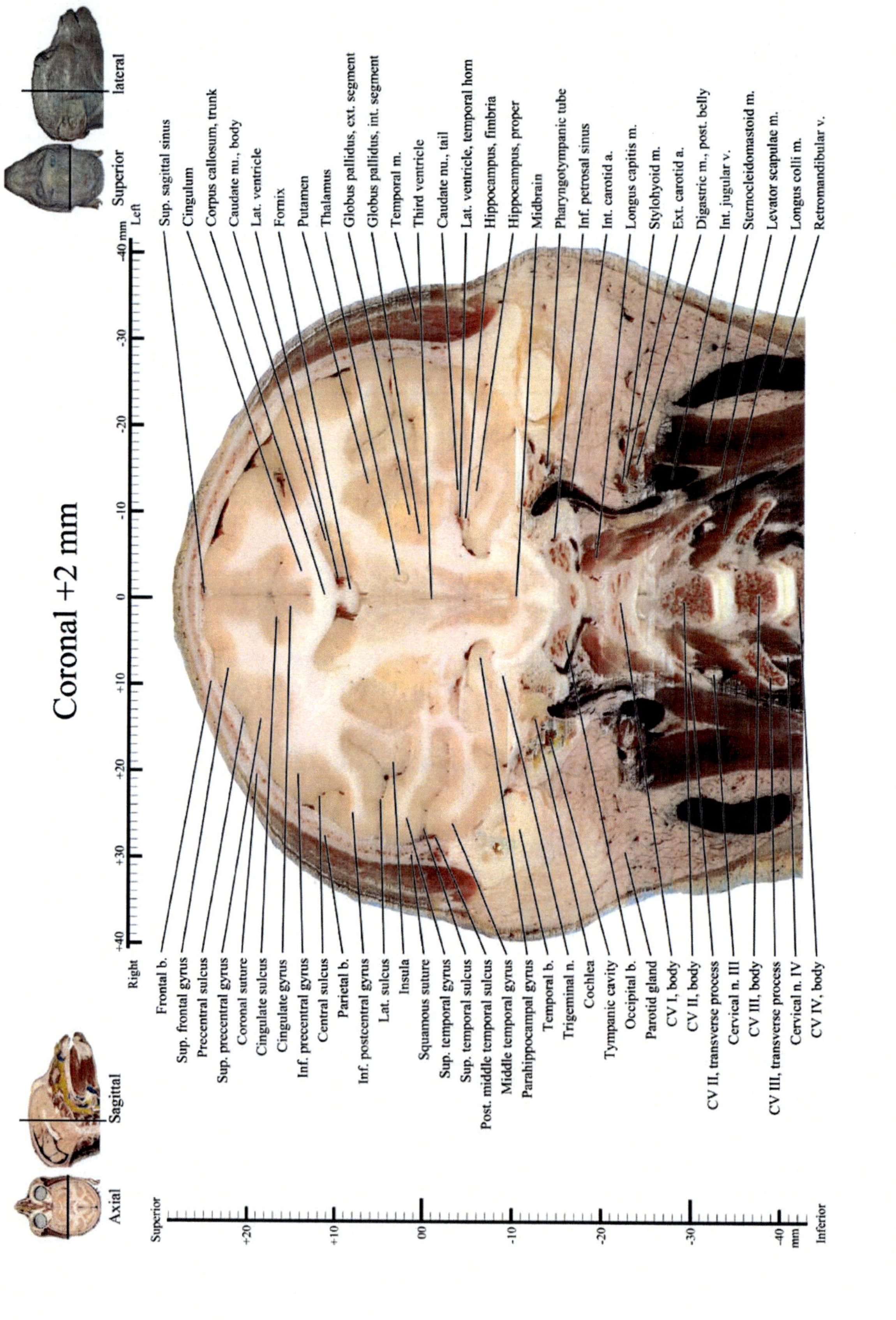
Coronal +2 mm
Superior
lateral
Axial
Sagittal
Left
Right
Superior
Inferior
mm
Sup. sagittal sinus
Cingulum
Corpus callosum, trunk
Caudate nu., body
Lat. ventricle
Fornix
Putamen
Thalamus
Globus pallidus, ext. segment
Globus pallidus, int. segment
Temporal m.
Third ventricle
Caudate nu., tail
Lat. ventricle, temporal horn
Hippocampus, fimbria
Hippocampus, proper
Midbrain
Pharyngotympanic tube
Inf. petrosal sinus
Int. carotid a.
Longus capitis m.
Stylohyoid m.
Ext. carotid a.
Digastric m., post. belly
Int. jugular v.
Sternocleidomastoid m.
Levator scapulae m.
Longus colli m.
Retromandibular v.
Frontal b.
Sup. frontal gyrus
Precentral sulcus
Sup. precentral gyrus
Coronal suture
Cingulate sulcus
Cingulate gyrus
Inf. precentral gyrus
Central sulcus
Parietal b.
Inf. postcentral gyrus
Lat. sulcus
Insula
Squamous suture
Sup. temporal gyrus
Sup. temporal sulcus
Post. middle temporal sulcus
Middle temporal gyrus
Parahippocampal gyrus
Temporal b.
Trigeminal n.
Cochlea
Tympanic cavity
Occipital b.
Parotid gland
CV I, body
CV II, body
CV II, transverse process
Cervical n. III
CV III, body
CV III, transverse process
Cervical n. IV
CV IV, body

Coronal +1 mm

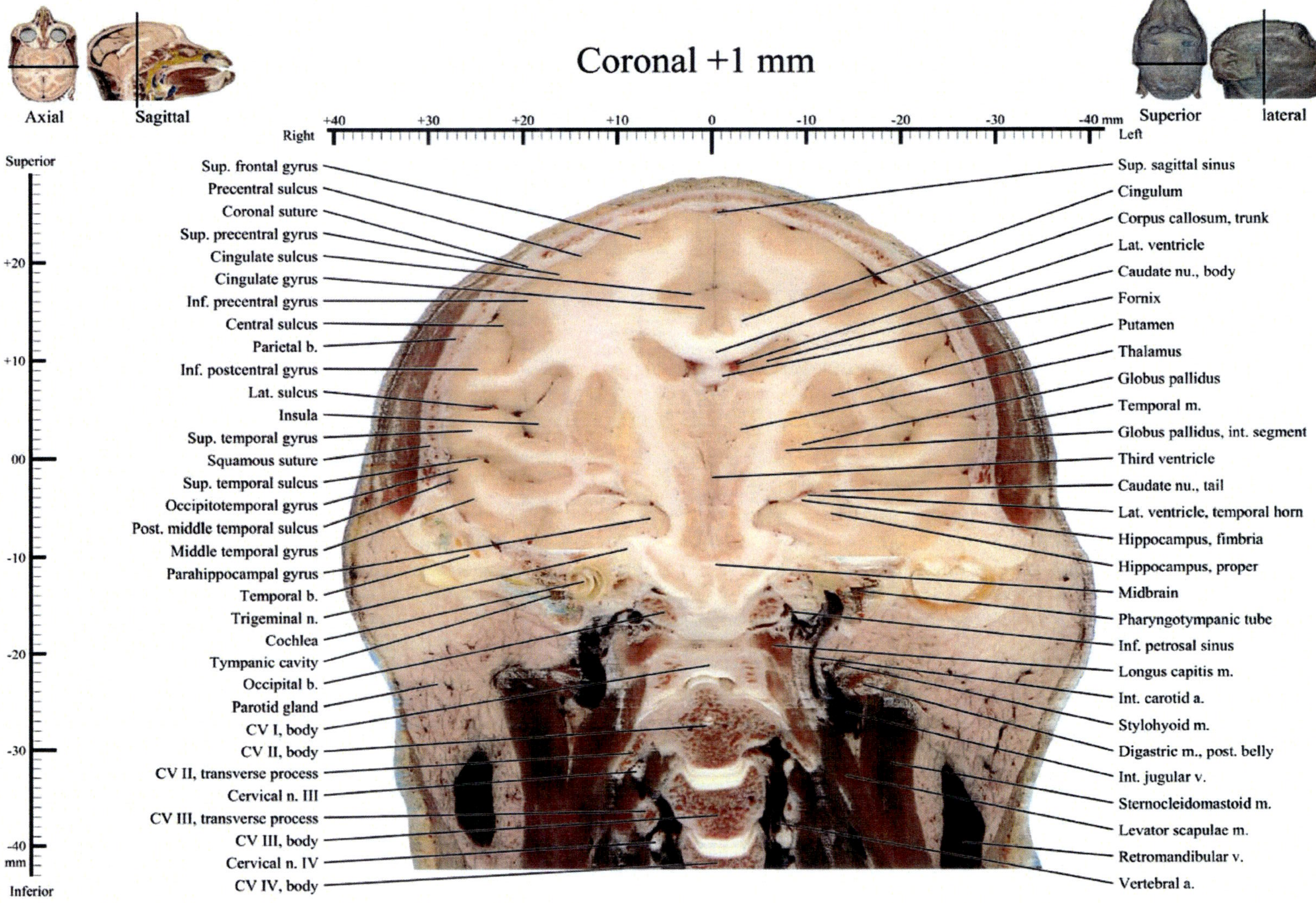
Axial
Sagittal
Superior
lateral
Right
Left
+40
+30
+20
+10
0
-10
-20
-30
-40 mm
Superior
+20
+10
00
-10
-20
-30
-40
mm
Inferior
Sup. frontal gyrus
Precentral sulcus
Coronal suture
Sup. precentral gyrus
Cingulate sulcus
Cingulate gyrus
Inf. precentral gyrus
Central sulcus
Parietal b.
Inf. postcentral gyrus
Lat. sulcus
Insula
Sup. temporal gyrus
Squamous suture
Sup. temporal sulcus
Occipitotemporal gyrus
Post. middle temporal sulcus
Middle temporal gyrus
Parahippocampal gyrus
Temporal b.
Trigeminal n.
Cochlea
Tympanic cavity
Occipital b.
Parotid gland
CV I, body
CV II, body
CV II, transverse process
Cervical n. III
CV III, transverse process
CV III, body
Cervical n. IV
CV IV, body
Sup. sagittal sinus
Cingulum
Corpus callosum, trunk
Lat. ventricle
Caudate nu., body
Fornix
Putamen
Thalamus
Globus pallidus
Temporal m.
Globus pallidus, int. segment
Third ventricle
Caudate nu., tail
Lat. ventricle, temporal horn
Hippocampus, fimbria
Hippocampus, proper
Midbrain
Pharyngotympanic tube
Inf. petrosal sinus
Longus capitis m.
Int. carotid a.
Stylohyoid m.
Digastric m., post. belly
Int. jugular v.
Sternocleidomastoid m.
Levator scapulae m.
Retromandibular v.
Vertebral a.

Coronal 0 mm

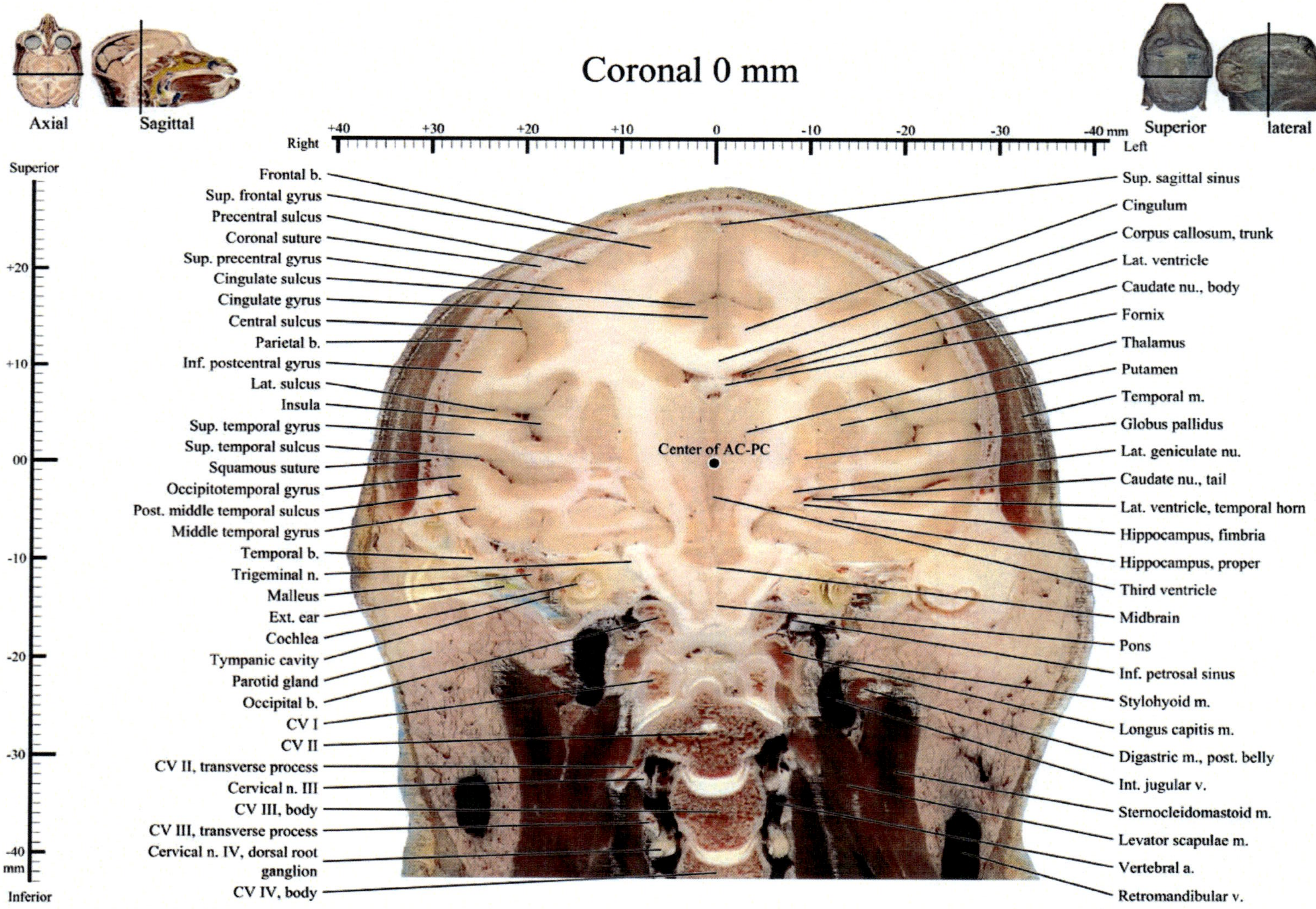
Axial
Sagittal
Superior
lateral
+40
+30
+20
+10
0
-10
-20
-30
-40 mm
Right
Left
Superior
+20
+10
00
-10
-20
-30
-40
mm
Inferior
Center of AC-PC
Frontal b.
Sup. frontal gyrus
Precentral sulcus
Coronal suture
Sup. precentral gyrus
Cingulate sulcus
Cingulate gyrus
Central sulcus
Parietal b.
Inf. postcentral gyrus
Lat. sulcus
Insula
Sup. temporal gyrus
Sup. temporal sulcus
Squamous suture
Occipitotemporal gyrus
Post. middle temporal sulcus
Middle temporal gyrus
Temporal b.
Trigeminal n.
Malleus
Ext. ear
Cochlea
Tympanic cavity
Parotid gland
Occipital b.
CV I
CV II
CV II, transverse process
Cervical n. III
CV III, body
CV III, transverse process
Cervical n. IV, dorsal root ganglion
CV IV, body
Sup. sagittal sinus
Cingulum
Corpus callosum, trunk
Lat. ventricle
Caudate nu., body
Fornix
Thalamus
Putamen
Temporal m.
Globus pallidus
Lat. geniculate nu.
Caudate nu., tail
Lat. ventricle, temporal horn
Hippocampus, fimbria
Hippocampus, proper
Third ventricle
Midbrain
Pons
Inf. petrosal sinus
Stylohyoid m.
Longus capitis m.
Digastric m., post. belly
Int. jugular v.
Sternocleidomastoid m.
Levator scapulae m.
Vertebral a.
Retromandibular v.

Coronal -0.5 mm

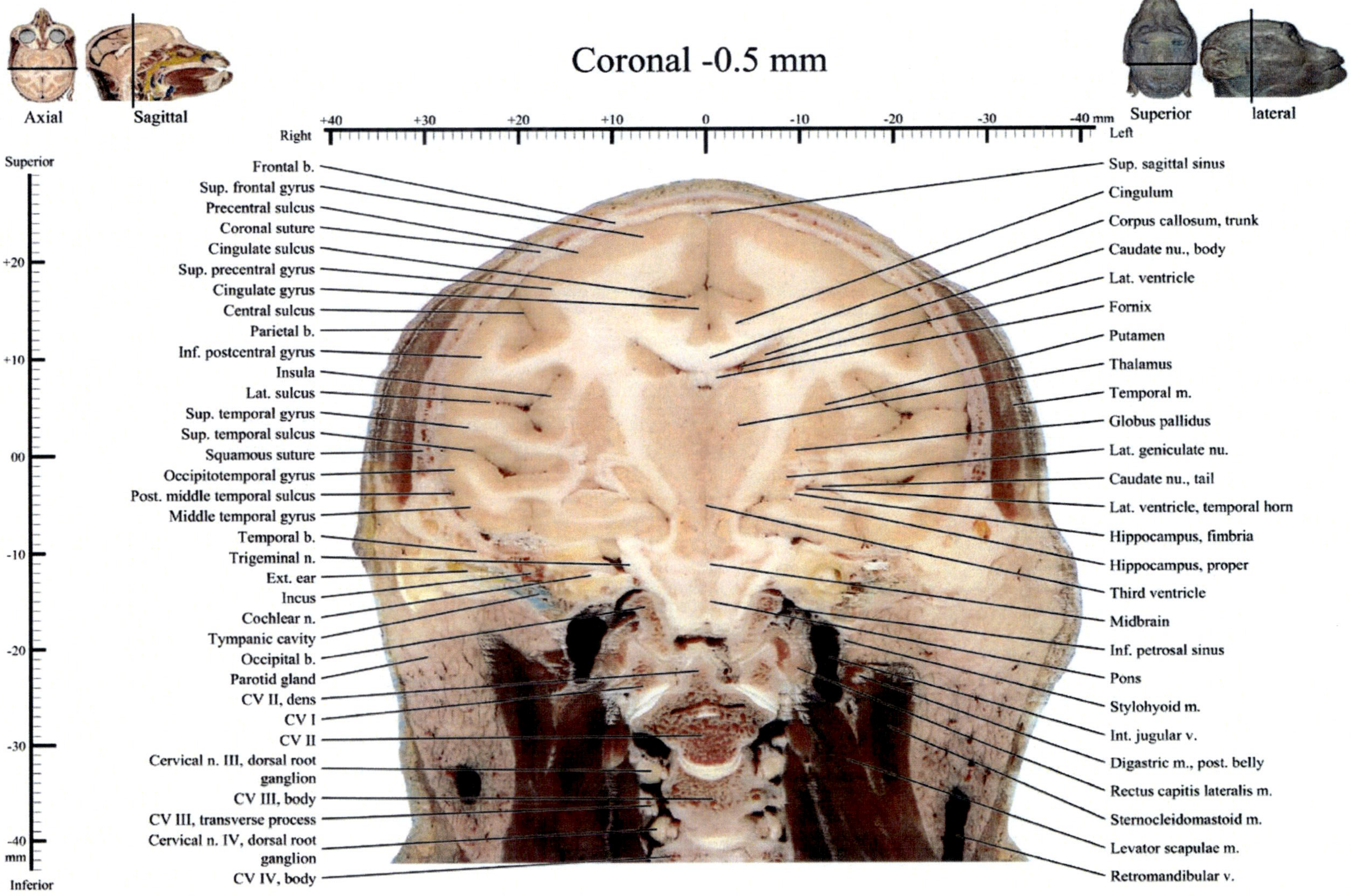

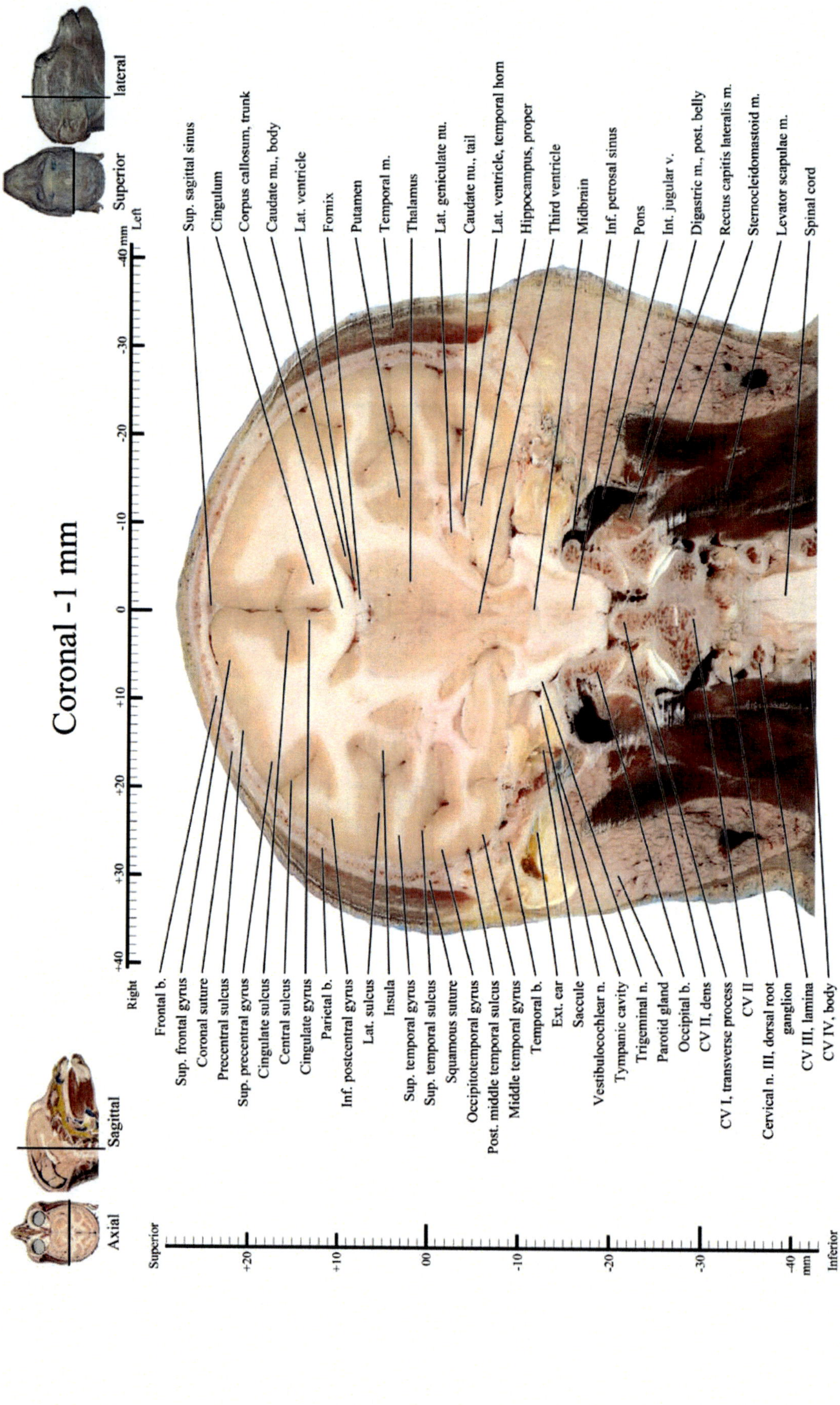
Coronal -1 mm
Axial
Sagittal
Superior
lateral
Superior
+20
+10
00
-10
-20
-30
-40
mm
Inferior
-40 mm
Left
-30
-20
-10
0
+10
+20
+30
+40
Right
Sup. sagittal sinus
Cingulum
Corpus callosum, trunk
Caudate nu., body
Lat. ventricle
Fornix
Putamen
Temporal m.
Thalamus
Lat. geniculate nu.
Caudate nu., tail
Lat. ventricle, temporal horn
Hippocampus, proper
Third ventricle
Midbrain
Inf. petrosal sinus
Pons
Int. jugular v.
Digastric m., post. belly
Rectus capitis lateralis m.
Sternocleidomastoid m.
Levator scapulae m.
Spinal cord
Frontal b.
Sup. frontal gyrus
Coronal suture
Precentral sulcus
Sup. precentral gyrus
Cingulate sulcus
Central sulcus
Cingulate gyrus
Parietal b.
Inf. postcentral gyrus
Lat. sulcus
Insula
Sup. temporal gyrus
Sup. temporal sulcus
Squamous suture
Occipitotemporal gyrus
Post. middle temporal sulcus
Middle temporal gyrus
Temporal b.
Ext. ear
Saccule
Vestibulocochlear n.
Tympanic cavity
Trigeminal n.
Parotid gland
Occipital b.
CV II, dens
CV I, transverse process
CV II
Cervical n. III, dorsal root ganglion
CV III, lamina
CV IV, body

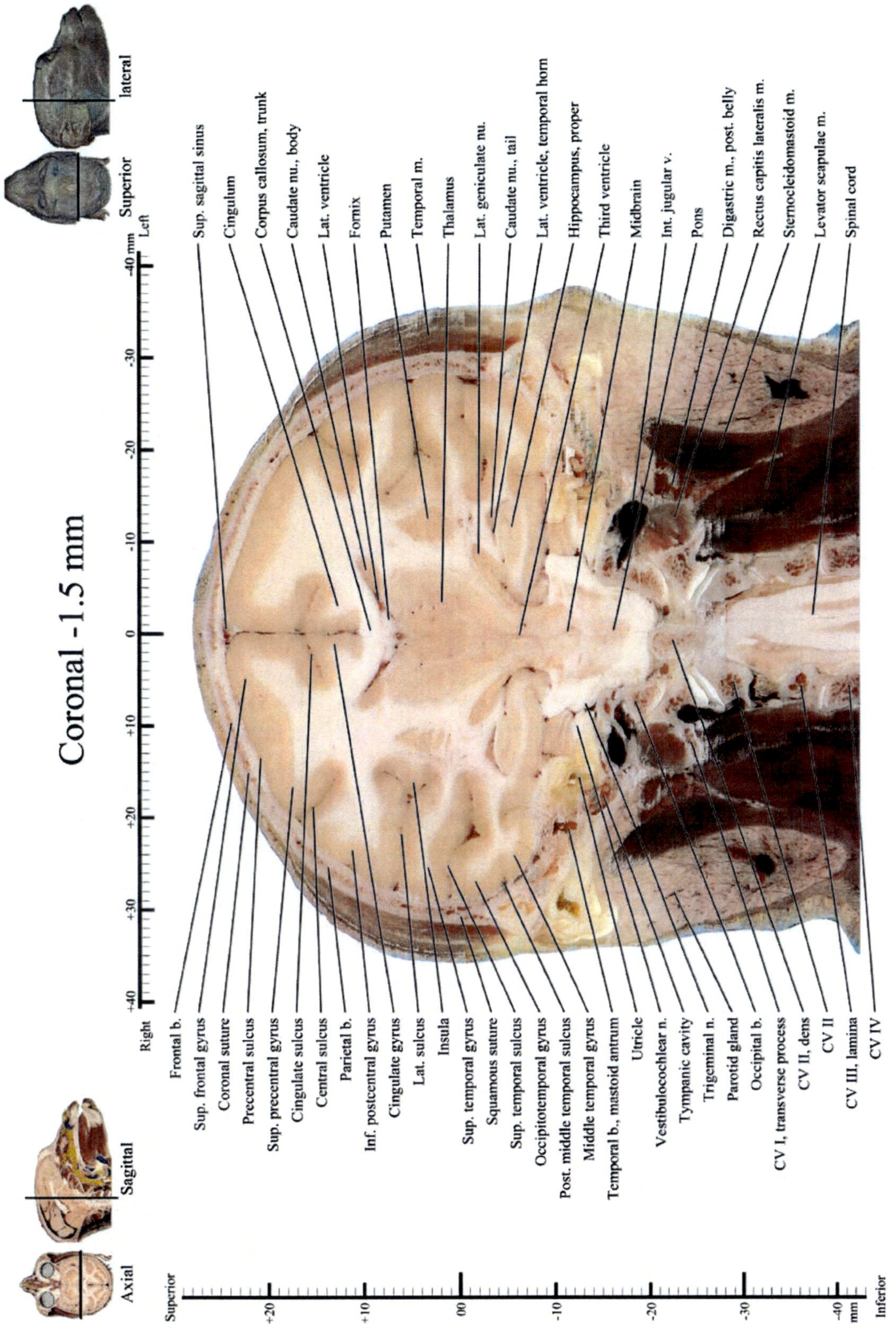

Coronal -1.5 mm
Axial
Sagittal
Superior
lateral
Superior
Inferior
Right
Left
mm
Frontal b.
Sup. frontal gyrus
Coronal suture
Precentral sulcus
Sup. precentral gyrus
Cingulate sulcus
Central sulcus
Parietal b.
Inf. postcentral gyrus
Cingulate gyrus
Lat. sulcus
Insula
Sup. temporal gyrus
Squamous suture
Sup. temporal sulcus
Occipitotemporal gyrus
Post. middle temporal sulcus
Middle temporal gyrus
Temporal b., mastoid antrum
Utricle
Vestibulocochlear n.
Tympanic cavity
Trigeminal n.
Parotid gland
Occipital b.
CV I, transverse process
CV II, dens
CV II
CV III, lamina
CV IV
Sup. sagittal sinus
Cingulum
Corpus callosum, trunk
Caudate nu., body
Lat. ventricle
Fornix
Putamen
Temporal m.
Thalamus
Lat. geniculate nu.
Caudate nu., tail
Lat. ventricle, temporal horn
Hippocampus, proper
Third ventricle
Midbrain
Int. jugular v.
Pons
Digastric m., post. belly
Rectus capitis lateralis m.
Sternocleidomastoid m.
Levator scapulae m.
Spinal cord

Coronal -2 mm

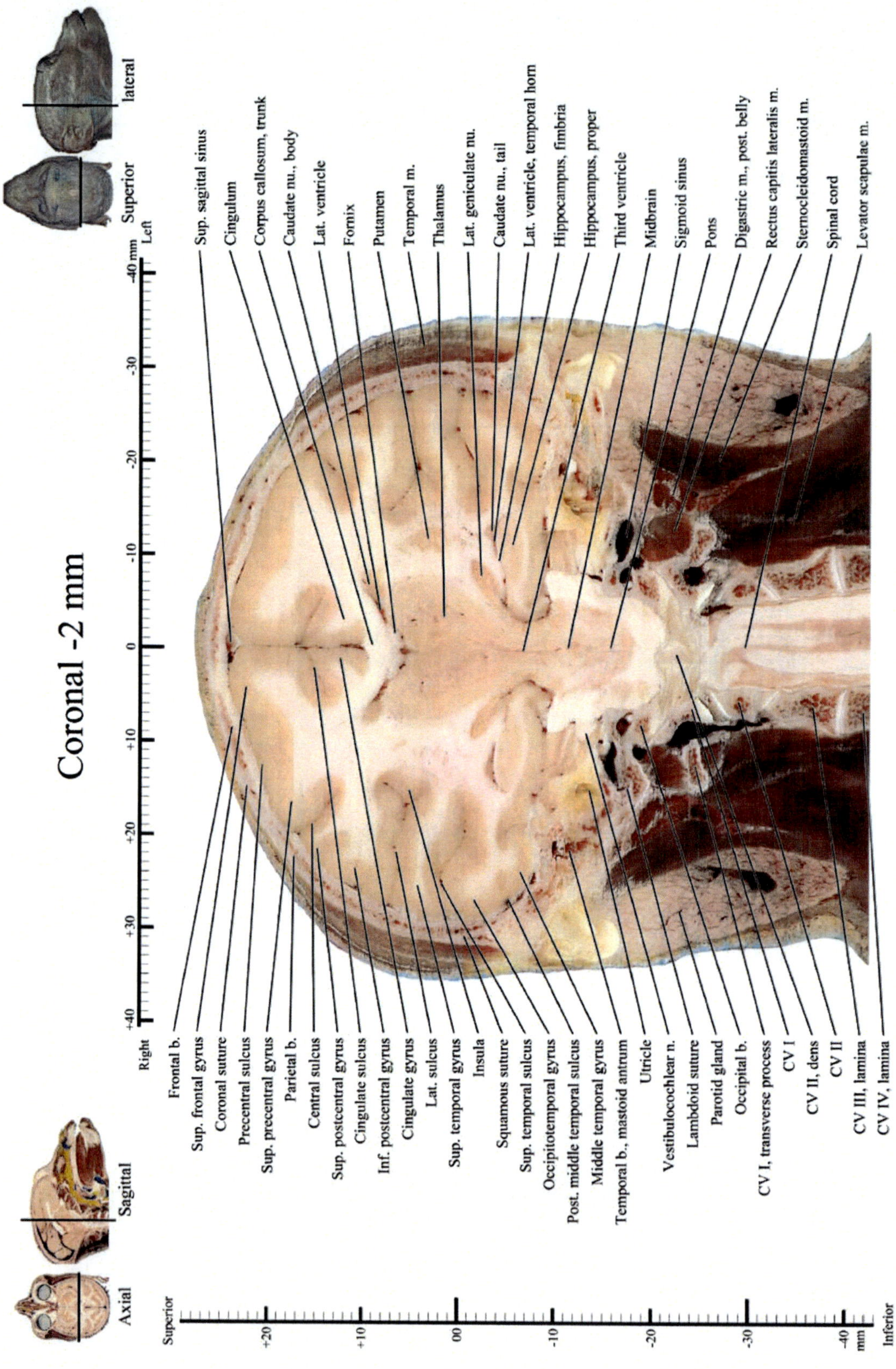
Axial
Sagittal
Superior
lateral
Superior
Inferior
Right
Left
-40 mm
+40
Sup. sagittal sinus
Cingulum
Corpus callosum, trunk
Caudate nu., body
Lat. ventricle
Fornix
Putamen
Temporal m.
Thalamus
Lat. geniculate nu.
Caudate nu., tail
Lat. ventricle, temporal horn
Hippocampus, fimbria
Hippocampus, proper
Third ventricle
Midbrain
Sigmoid sinus
Pons
Digastric m., post. belly
Rectus capitis lateralis m.
Sternocleidomastoid m.
Spinal cord
Levator scapulae m.
Frontal b.
Sup. frontal gyrus
Coronal suture
Precentral sulcus
Sup. precentral gyrus
Parietal b.
Central sulcus
Sup. postcentral gyrus
Cingulate sulcus
Inf. postcentral gyrus
Cingulate gyrus
Lat. sulcus
Sup. temporal gyrus
Insula
Squamous suture
Sup. temporal sulcus
Occipitotemporal gyrus
Post. middle temporal sulcus
Middle temporal gyrus
Temporal b., mastoid antrum
Utricle
Vestibulocochlear n.
Lambdoid suture
Parotid gland
Occipital b.
CV I, transverse process
CV I
CV II, dens
CV II
CV III, lamina
CV IV, lamina

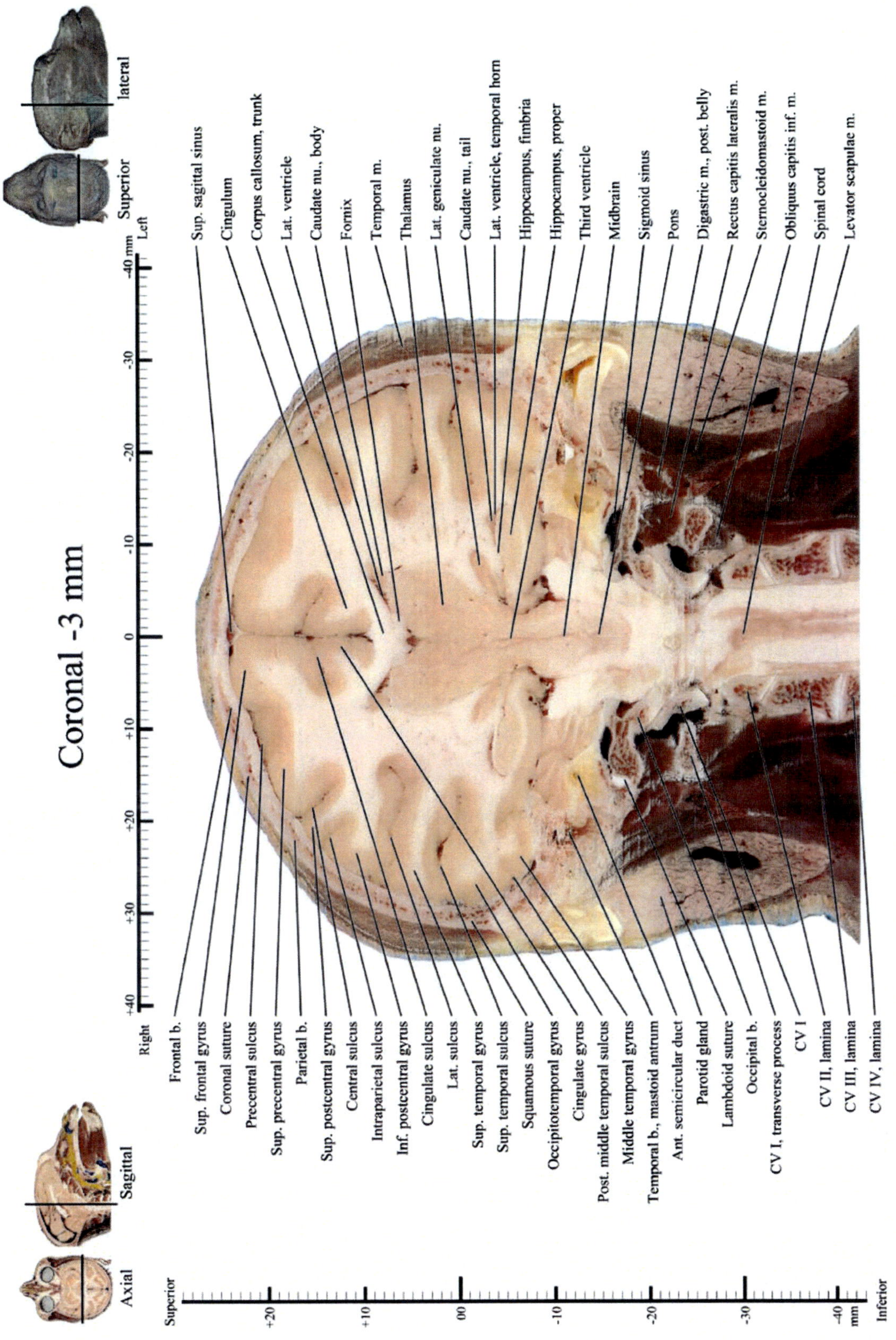
Coronal -3 mm
Superior
lateral
Axial
Sagittal
Right
Left
mm
Inferior
Sup. sagital sinus
Cingulum
Corpus callosum, trunk
Lat. ventricle
Caudate nu., body
Fornix
Temporal m.
Thalamus
Lat. geniculate nu.
Caudate nu., tail
Lat. ventricle, temporal horn
Hippocampus, fimbria
Hippocampus, proper
Third ventricle
Midbrain
Sigmoid sinus
Pons
Digastric m., post. belly
Rectus capitis lateralis m.
Sternocleidomastoid m.
Obliquus capitis inf. m.
Spinal cord
Levator scapulae m.
Frontal b.
Sup. frontal gyrus
Coronal suture
Precentral sulcus
Sup. precentral gyrus
Parietal b.
Sup. postcentral gyrus
Central sulcus
Intraparietal sulcus
Inf. postcentral gyrus
Cingulate sulcus
Lat. sulcus
Sup. temporal gyrus
Sup. temporal sulcus
Squamous suture
Occipitotemporal gyrus
Cingulate gyrus
Post. middle temporal sulcus
Middle temporal gyrus
Temporal b., mastoid antrum
Ant. semicircular duct
Parotid gland
Lambdoid suture
Occipital b.
CV I, transverse process
CV I
CV II, lamina
CV III, lamina
CV IV, lamina

Coronal -4 mm

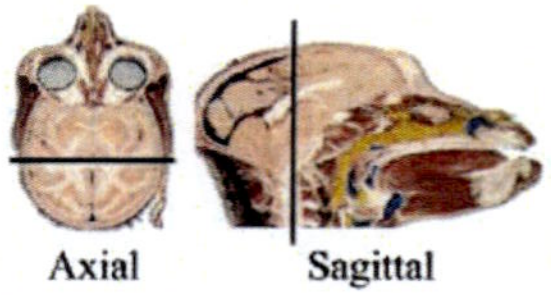
Axial
Sagittal

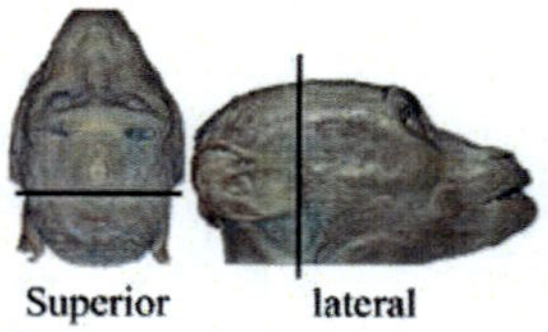
Superior
lateral

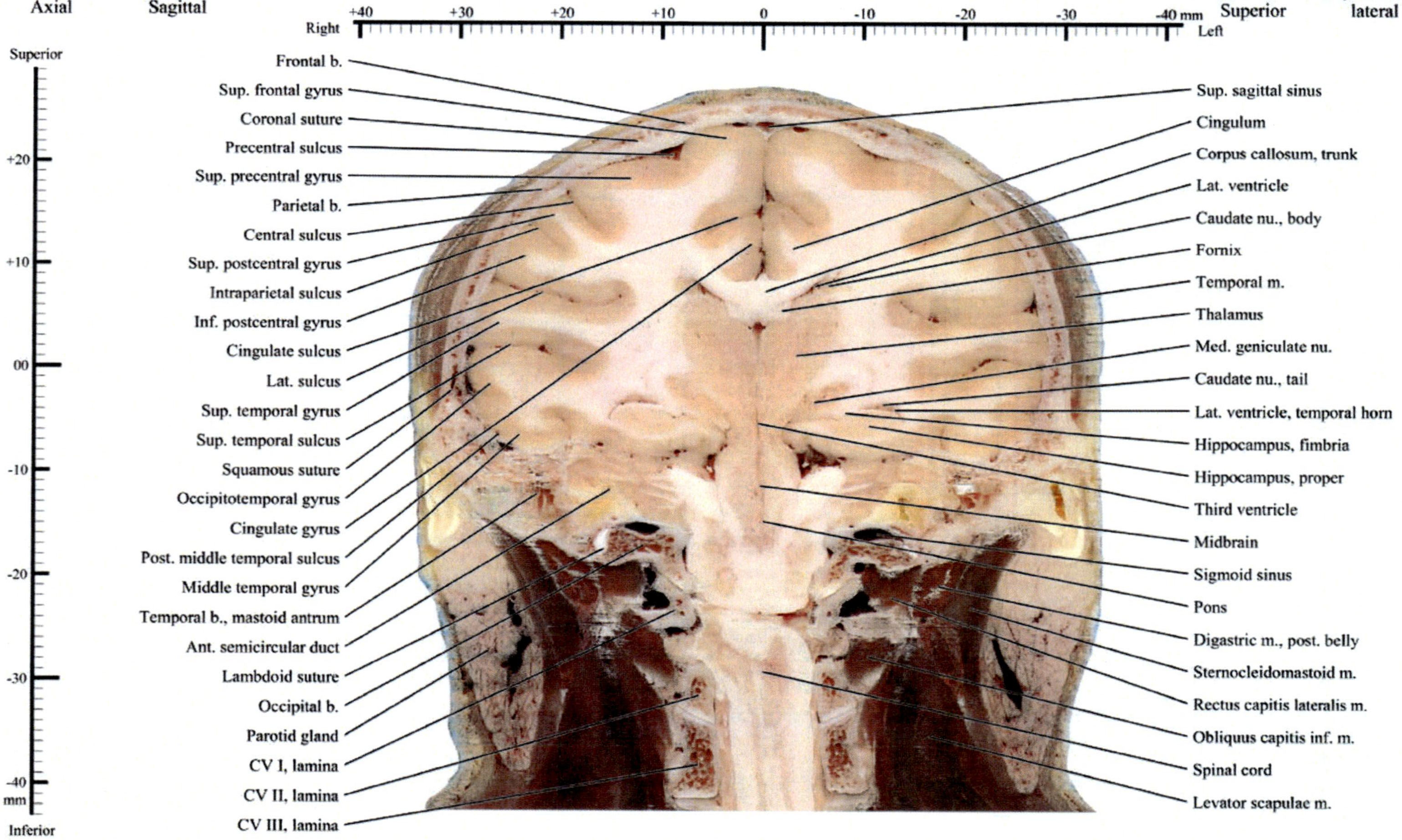
Right
Left
+40
+30
+20
+10
0
-10
-20
-30
-40 mm
Superior
Inferior
+20
+10
00
-10
-20
-30
-40
mm
Frontal b.
Sup. frontal gyrus
Coronal suture
Precentral sulcus
Sup. precentral gyrus
Parietal b.
Central sulcus
Sup. postcentral gyrus
Intraparietal sulcus
Inf. postcentral gyrus
Cingulate sulcus
Lat. sulcus
Sup. temporal gyrus
Sup. temporal sulcus
Squamous suture
Occipitotemporal gyrus
Cingulate gyrus
Post. middle temporal sulcus
Middle temporal gyrus
Temporal b., mastoid antrum
Ant. semicircular duct
Lambdoid suture
Occipital b.
Parotid gland
CV I, lamina
CV II, lamina
CV III, lamina
Sup. sagittal sinus
Cingulum
Corpus callosum, trunk
Lat. ventricle
Caudate nu., body
Fornix
Temporal m.
Thalamus
Med. geniculate nu.
Caudate nu., tail
Lat. ventricle, temporal horn
Hippocampus, fimbria
Hippocampus, proper
Third ventricle
Midbrain
Sigmoid sinus
Pons
Digastric m., post. belly
Sternocleidomastoid m.
Rectus capitis lateralis m.
Obliquus capitis inf. m.
Spinal cord
Levator scapulae m.

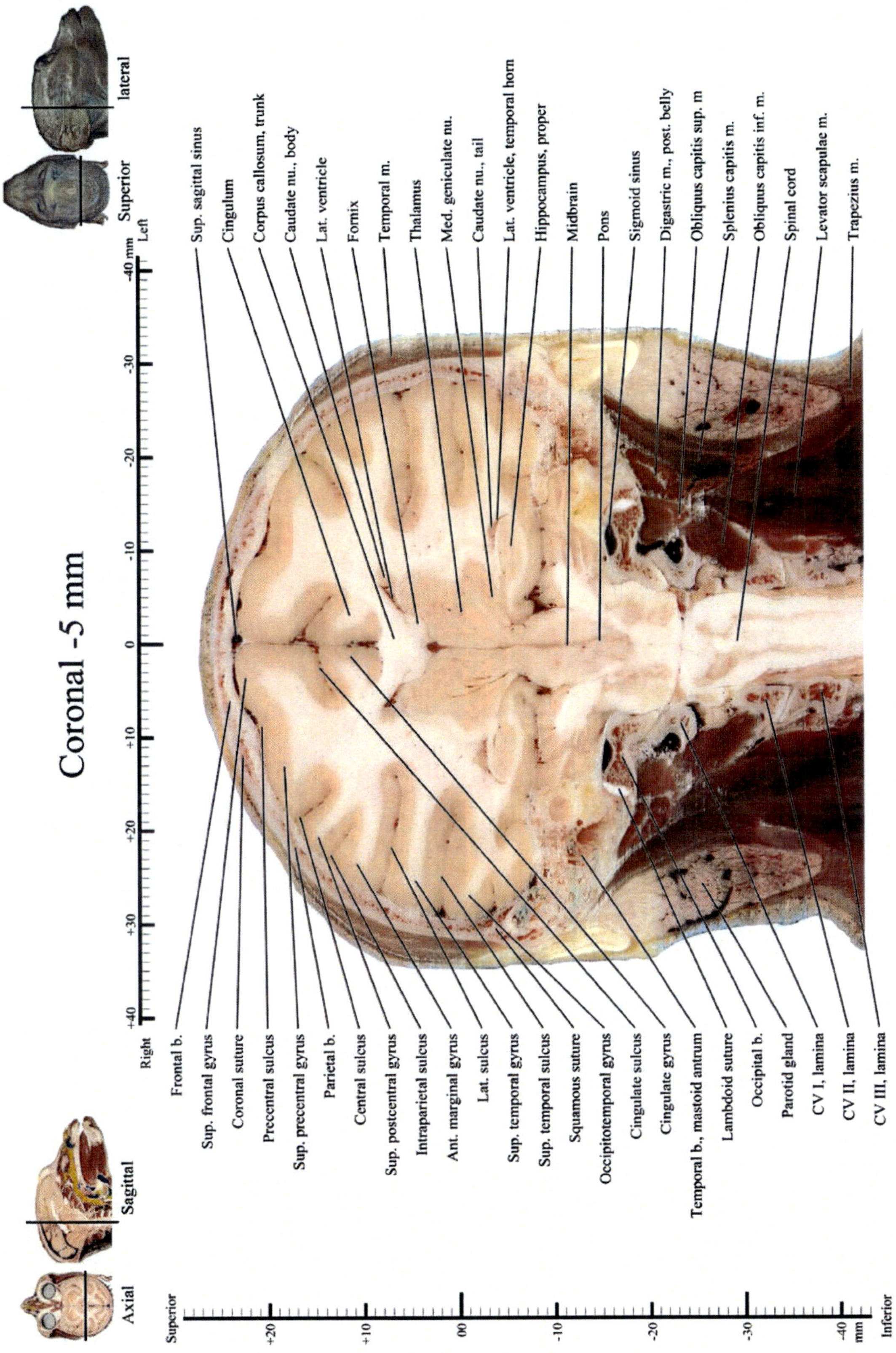
Coronal -5 mm
Axial
Sagittal
Superior
lateral
Right
Left
Superior
Inferior
mm
Sup. sagittal sinus
Cingulum
Corpus callosum, trunk
Caudate nu., body
Lat. ventricle
Fornix
Temporal m.
Thalamus
Med. geniculate nu.
Caudate nu., tail
Lat. ventricle, temporal horn
Hippocampus, proper
Midbrain
Pons
Sigmoid sinus
Digastric m., post. belly
Obliquus capitis sup. m
Splenius capitis m.
Obliquus capitis inf. m.
Spinal cord
Levator scapulae m.
Trapezius m.
Frontal b.
Sup. frontal gyrus
Coronal suture
Precentral sulcus
Sup. precentral gyrus
Parietal b.
Central sulcus
Sup. postcentral gyrus
Intraparietal sulcus
Ant. marginal gyrus
Lat. sulcus
Sup. temporal gyrus
Sup. temporal sulcus
Squamous suture
Occipitotemporal gyrus
Cingulate sulcus
Cingulate gyrus
Temporal b., mastoid antrum
Lambdoid suture
Occipital b.
Parotid gland
CV I, lamina
CV II, lamina
CV III, lamina

Coronal -6 mm

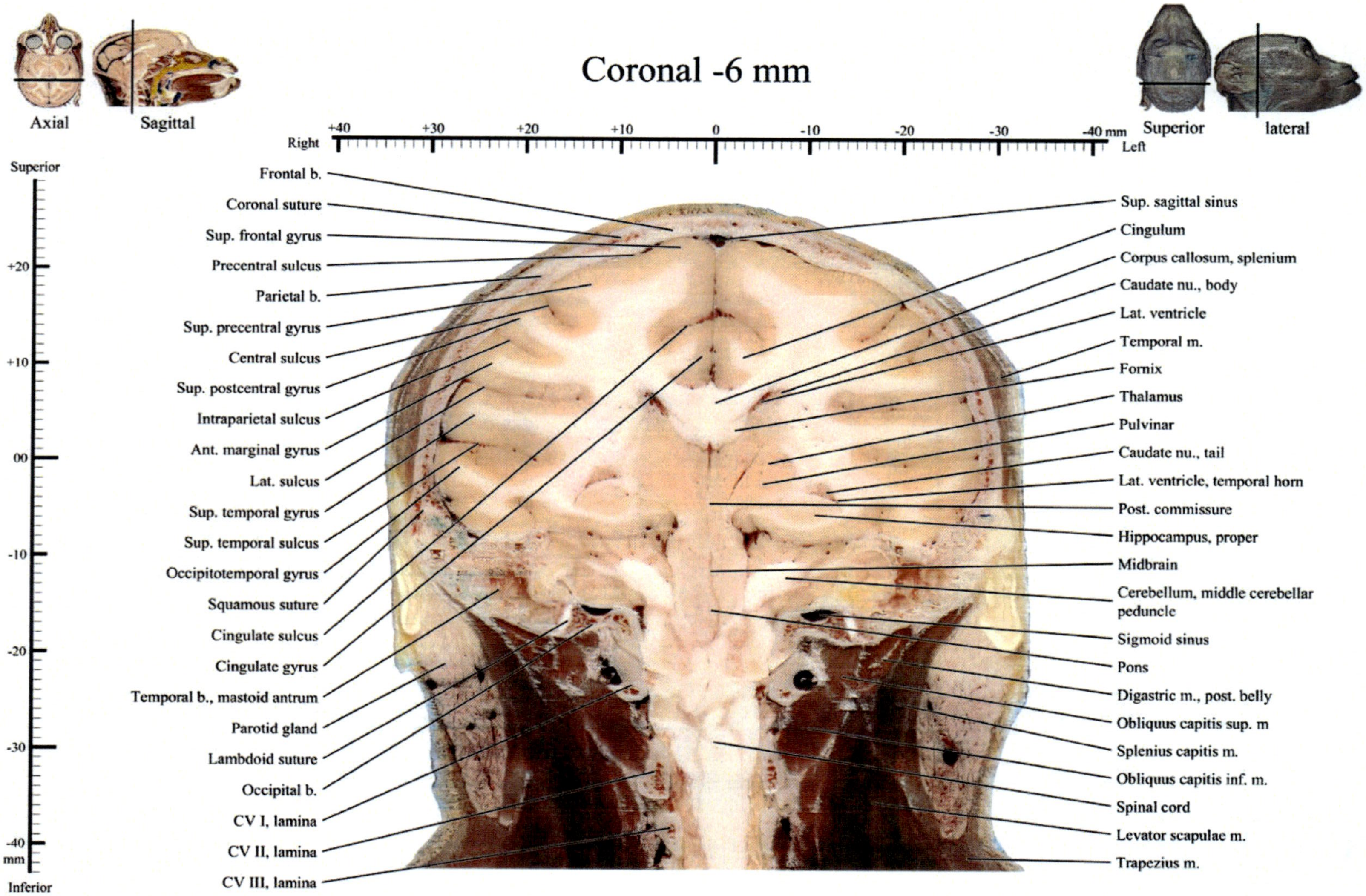
Axial
Sagittal
Superior
lateral
Right
Left
+40
+30
+20
+10
0
-10
-20
-30
-40 mm
Superior
+20
+10
00
-10
-20
-30
-40
mm
Inferior
Frontal b.
Coronal suture
Sup. frontal gyrus
Precentral sulcus
Parietal b.
Sup. precentral gyrus
Central sulcus
Sup. postcentral gyrus
Intraparietal sulcus
Ant. marginal gyrus
Lat. sulcus
Sup. temporal gyrus
Sup. temporal sulcus
Occipitotemporal gyrus
Squamous suture
Cingulate sulcus
Cingulate gyrus
Temporal b., mastoid antrum
Parotid gland
Lambdoid suture
Occipital b.
CV I, lamina
CV II, lamina
CV III, lamina
Sup. sagittal sinus
Cingulum
Corpus callosum, splenium
Caudate nu., body
Lat. ventricle
Temporal m.
Fornix
Thalamus
Pulvinar
Caudate nu., tail
Lat. ventricle, temporal horn
Post. commissure
Hippocampus, proper
Midbrain
Cerebellum, middle cerebellar peduncle
Sigmoid sinus
Pons
Digastric m., post. belly
Obliquus capitis sup. m
Splenius capitis m.
Obliquus capitis inf. m.
Spinal cord
Levator scapulae m.
Trapezius m.

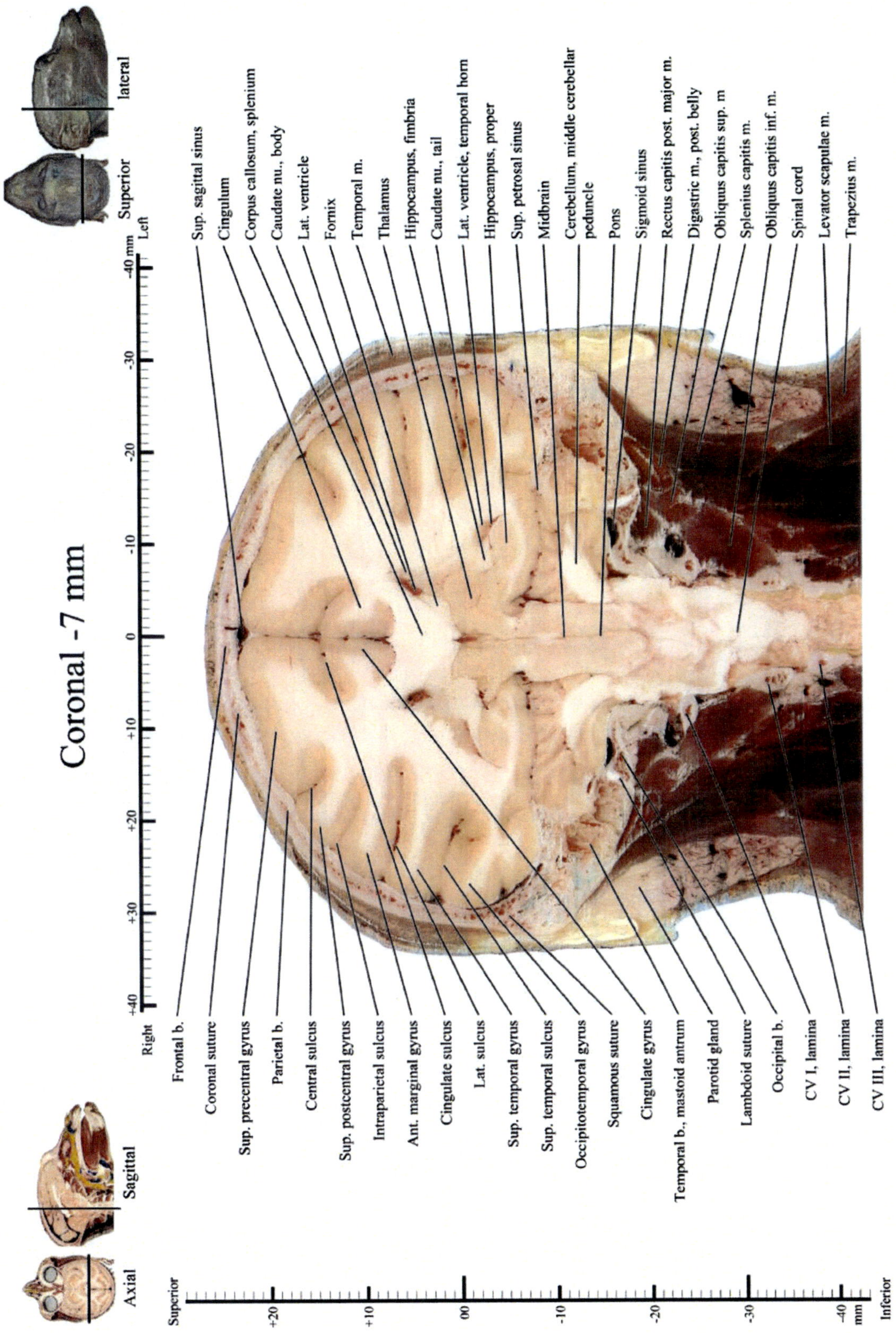
Coronal -7 mm
Superior
lateral
Sagittal
Axial
-40 mm
Left
-30
-20
-10
0
+10
+20
+30
+40
Right
Superior
+20
+10
00
-10
-20
-30
-40
mm
Inferior
Sup. sagittal sinus
Cingulum
Corpus callosum, splenium
Caudate nu., body
Lat. ventricle
Fornix
Temporal m.
Thalamus
Hippocampus, fimbria
Caudate nu., tail
Lat. ventricle, temporal horn
Hippocampus, proper
Sup. petrosal sinus
Midbrain
Cerebellum, middle cerebellar peduncle
Pons
Sigmoid sinus
Rectus capitis post. major m.
Digastric m., post. belly
Obliquus capitis sup. m
Splenius capitis m.
Obliquus capitis inf. m.
Spinal cord
Levator scapulae m.
Trapezius m.
Frontal b.
Coronal suture
Sup. precentral gyrus
Parietal b.
Central sulcus
Sup. postcentral gyrus
Intraparietal sulcus
Ant. marginal gyrus
Cingulate sulcus
Lat. sulcus
Sup. temporal gyrus
Sup. temporal sulcus
Occipitotemporal gyrus
Squamous suture
Cingulate gyrus
Temporal b., mastoid antrum
Parotid gland
Lambdoid suture
Occipital b.
CV I, lamina
CV II, lamina
CV III, lamina

Coronal -8 mm

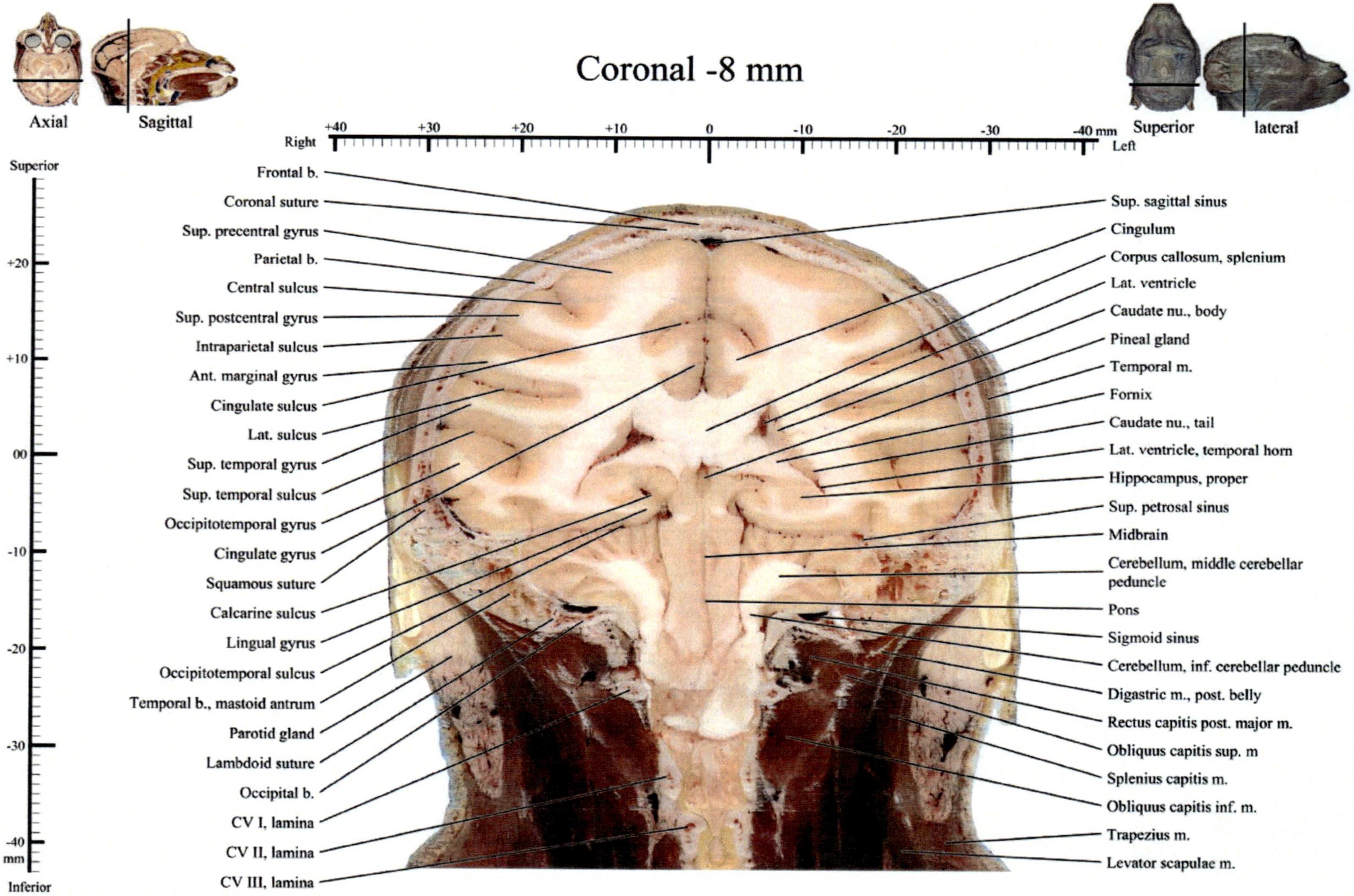
Axial
Sagittal
Superior
lateral
Right
Left
+40
+30
+20
+10
0
-10
-20
-30
-40 mm
Superior
Inferior
00
mm
Frontal b.
Coronal suture
Sup. precentral gyrus
Parietal b.
Central sulcus
Sup. postcentral gyrus
Intraparietal sulcus
Ant. marginal gyrus
Cingulate sulcus
Lat. sulcus
Sup. temporal gyrus
Sup. temporal sulcus
Occipitotemporal gyrus
Cingulate gyrus
Squamous suture
Calcarine sulcus
Lingual gyrus
Occipitotemporal sulcus
Temporal b., mastoid antrum
Parotid gland
Lambdoid suture
Occipital b.
CV I, lamina
CV II, lamina
CV III, lamina
Sup. sagittal sinus
Cingulum
Corpus callosum, splenium
Lat. ventricle
Caudate nu., body
Pineal gland
Temporal m.
Fornix
Caudate nu., tail
Lat. ventricle, temporal horn
Hippocampus, proper
Sup. petrosal sinus
Midbrain
Cerebellum, middle cerebellar peduncle
Pons
Sigmoid sinus
Cerebellum, inf. cerebellar peduncle
Digastric m., post. belly
Rectus capitis post. major m.
Obliquus capitis sup. m
Splenius capitis m.
Obliquus capitis inf. m.
Trapezius m.
Levator scapulae m.

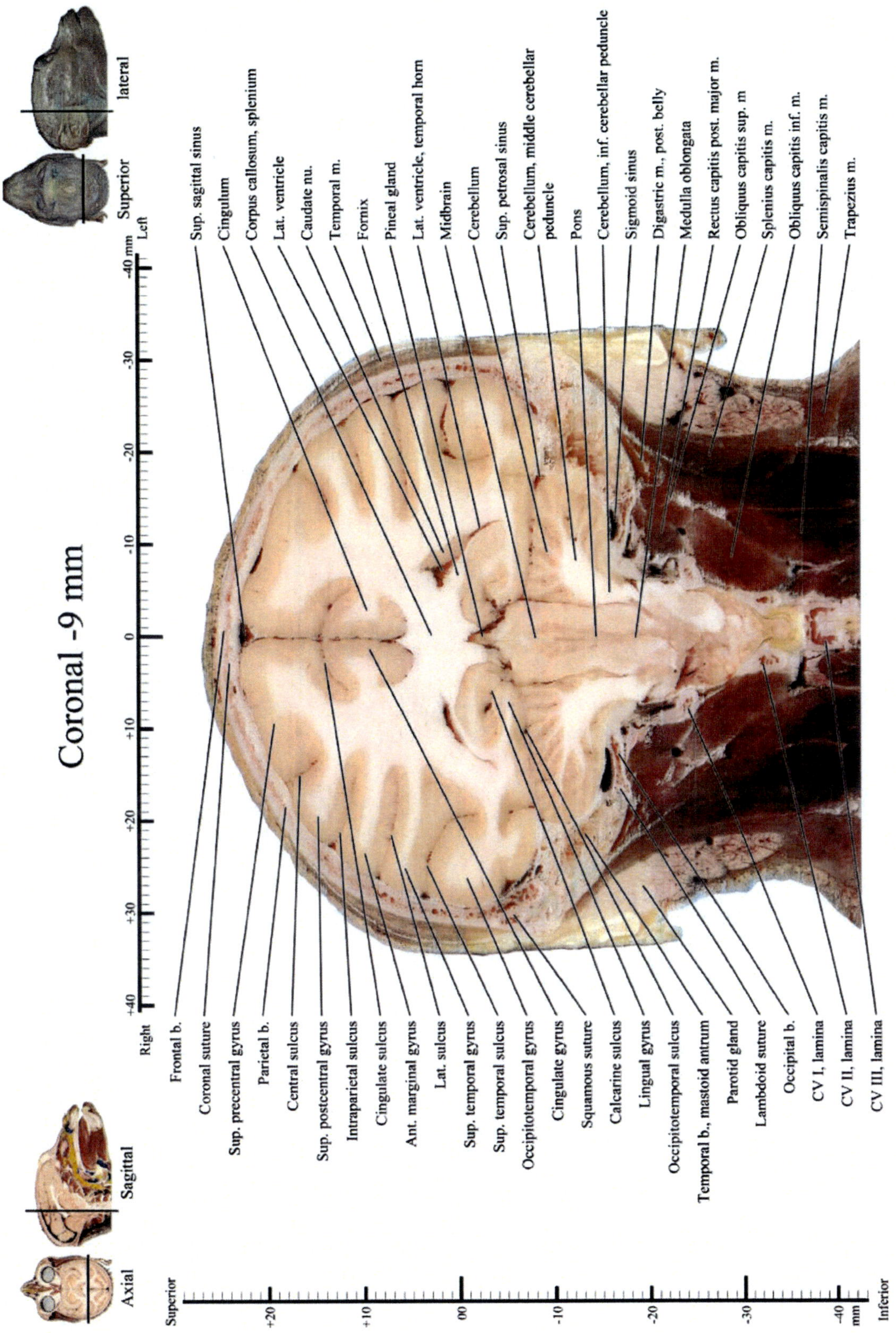
Coronal -9 mm
Superior
lateral
Axial
Sagittal
Left
Right
Sup. sagittal sinus
Cingulum
Corpus callosum, splenium
Lat. ventricle
Caudate nu.
Temporal m.
Fornix
Pineal gland
Lat. ventricle, temporal horn
Midbrain
Cerebellum
Sup. petrosal sinus
Cerebellum, middle cerebellar peduncle
Pons
Cerebellum, inf. cerebellar peduncle
Sigmoid sinus
Digastric m., post. belly
Medulla oblongata
Rectus capitis post. major m.
Obliquus capitis sup. m
Splenius capitis m.
Obliquus capitis inf. m.
Semispinalis capitis m.
Trapezius m.
Frontal b.
Coronal suture
Sup. precentral gyrus
Parietal b.
Central sulcus
Sup. postcentral gyrus
Intraparietal sulcus
Cingulate sulcus
Ant. marginal gyrus
Lat. sulcus
Sup. temporal gyrus
Sup. temporal sulcus
Occipitotemporal gyrus
Cingulate gyrus
Squamous suture
Calcarine sulcus
Lingual gyrus
Occipitotemporal sulcus
Temporal b., mastoid antrum
Parotid gland
Lambdoid suture
Occipital b.
CV I, lamina
CV II, lamina
CV III, lamina
Superior
Inferior

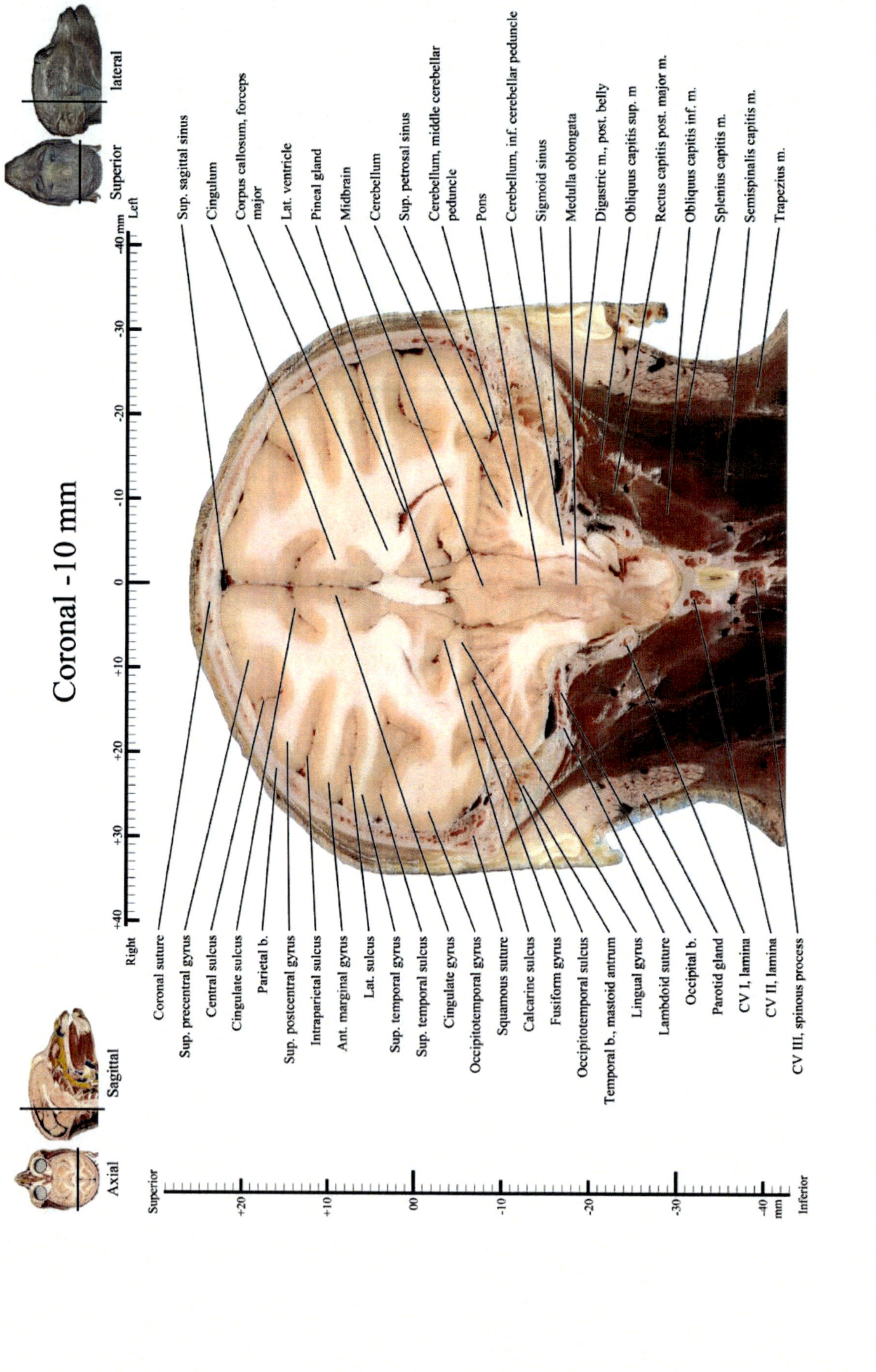
Coronal -10 mm
Axial
Sagittal
Superior
lateral
Superior
Inferior
Right
Left
mm
+40
+30
+20
+10
0
-10
-20
-30
-40
+20
+10
00
-10
-20
-30
-40
Coronal suture
Sup. precentral gyrus
Central sulcus
Cingulate sulcus
Parietal b.
Sup. postcentral gyrus
Intraparietal sulcus
Ant. marginal gyrus
Lat. sulcus
Sup. temporal gyrus
Sup. temporal sulcus
Cingulate gyrus
Occipitotemporal gyrus
Squamous suture
Calcarine sulcus
Fusiform gyrus
Occipitotemporal sulcus
Temporal b., mastoid antrum
Lingual gyrus
Lambdoid suture
Occipital b.
Parotid gland
CV I, lamina
CV II, lamina
CV III, spinous process
Sup. sagittal sinus
Cingulum
Corpus callosum, forceps major
Lat. ventricle
Pineal gland
Midbrain
Cerebellum
Sup. petrosal sinus
Cerebellum, middle cerebellar peduncle
Pons
Cerebellum, inf. cerebellar peduncle
Sigmoid sinus
Medulla oblongata
Digastric m., post. belly
Obliquus capitis sup. m
Rectus capitis post. major m.
Obliquus capitis inf. m.
Splenius capitis m.
Semispinalis capitis m.
Trapezius m.

Coronal -11 mm

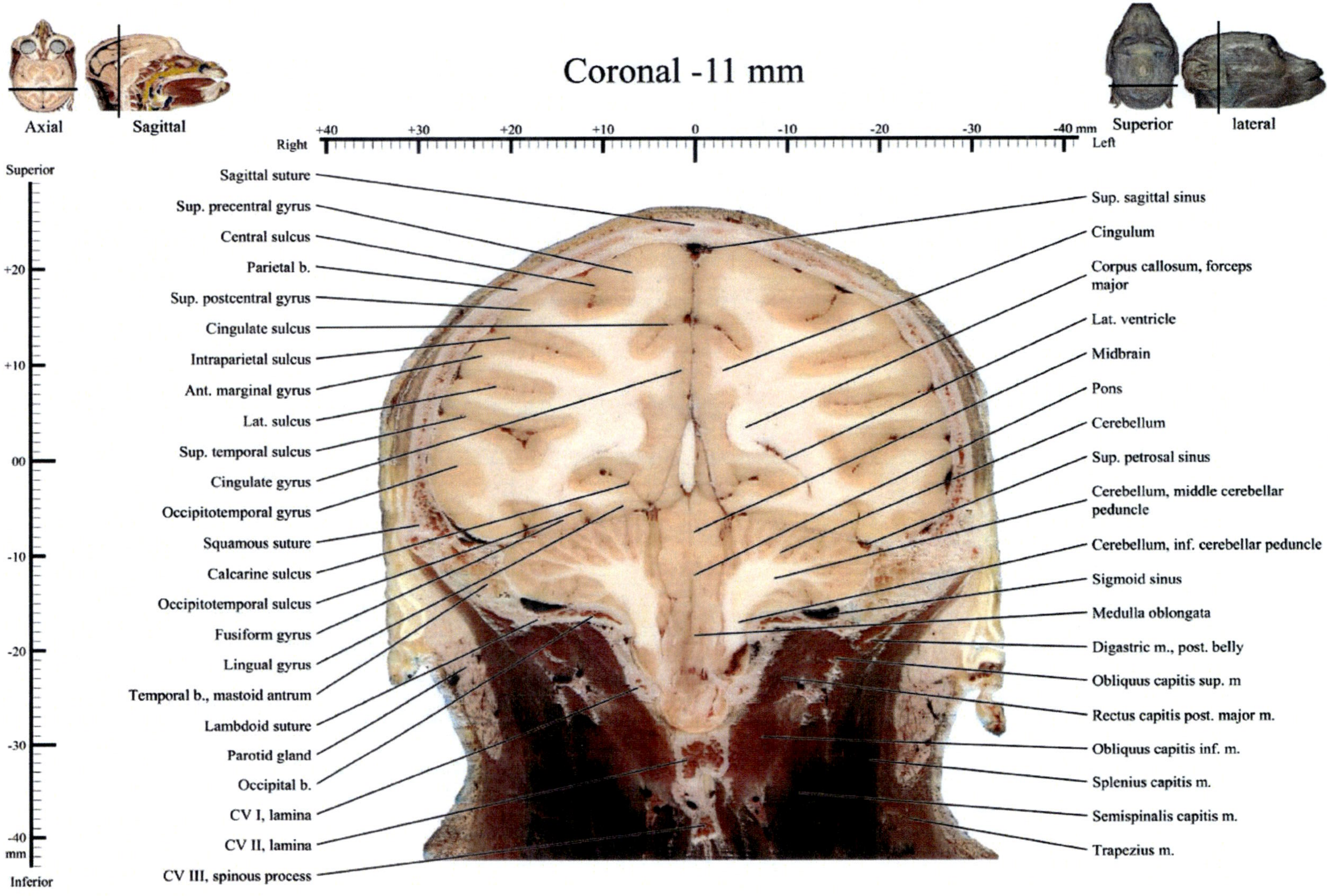

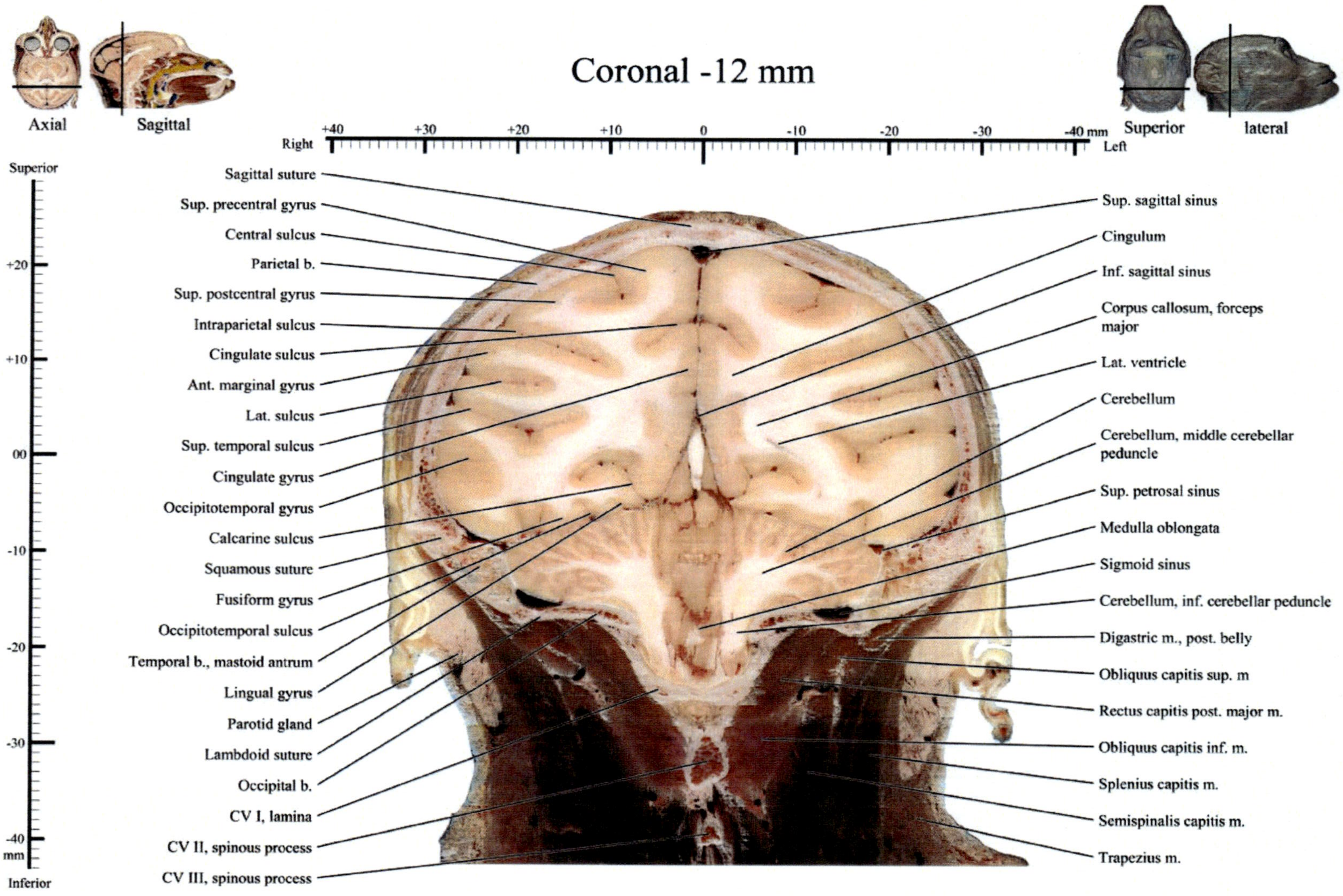
Coronal -12 mm
Axial
Sagittal
Superior
lateral
Right
Left
+40
+30
+20
+10
0
-10
-20
-30
-40 mm
Superior
+20
+10
00
-10
-20
-30
-40
mm
Inferior
Sagittal suture
Sup. precentral gyrus
Central sulcus
Parietal b.
Sup. postcentral gyrus
Intraparietal sulcus
Cingulate sulcus
Ant. marginal gyrus
Lat. sulcus
Sup. temporal sulcus
Cingulate gyrus
Occipitotemporal gyrus
Calcarine sulcus
Squamous suture
Fusiform gyrus
Occipitotemporal sulcus
Temporal b., mastoid antrum
Lingual gyrus
Parotid gland
Lambdoid suture
Occipital b.
CV I, lamina
CV II, spinous process
CV III, spinous process
Sup. sagittal sinus
Cingulum
Inf. sagittal sinus
Corpus callosum, forceps major
Lat. ventricle
Cerebellum
Cerebellum, middle cerebellar peduncle
Sup. petrosal sinus
Medulla oblongata
Sigmoid sinus
Cerebellum, inf. cerebellar peduncle
Digastric m., post. belly
Obliquus capitis sup. m
Rectus capitis post. major m.
Obliquus capitis inf. m.
Splenius capitis m.
Semispinalis capitis m.
Trapezius m.

Coronal -13 mm

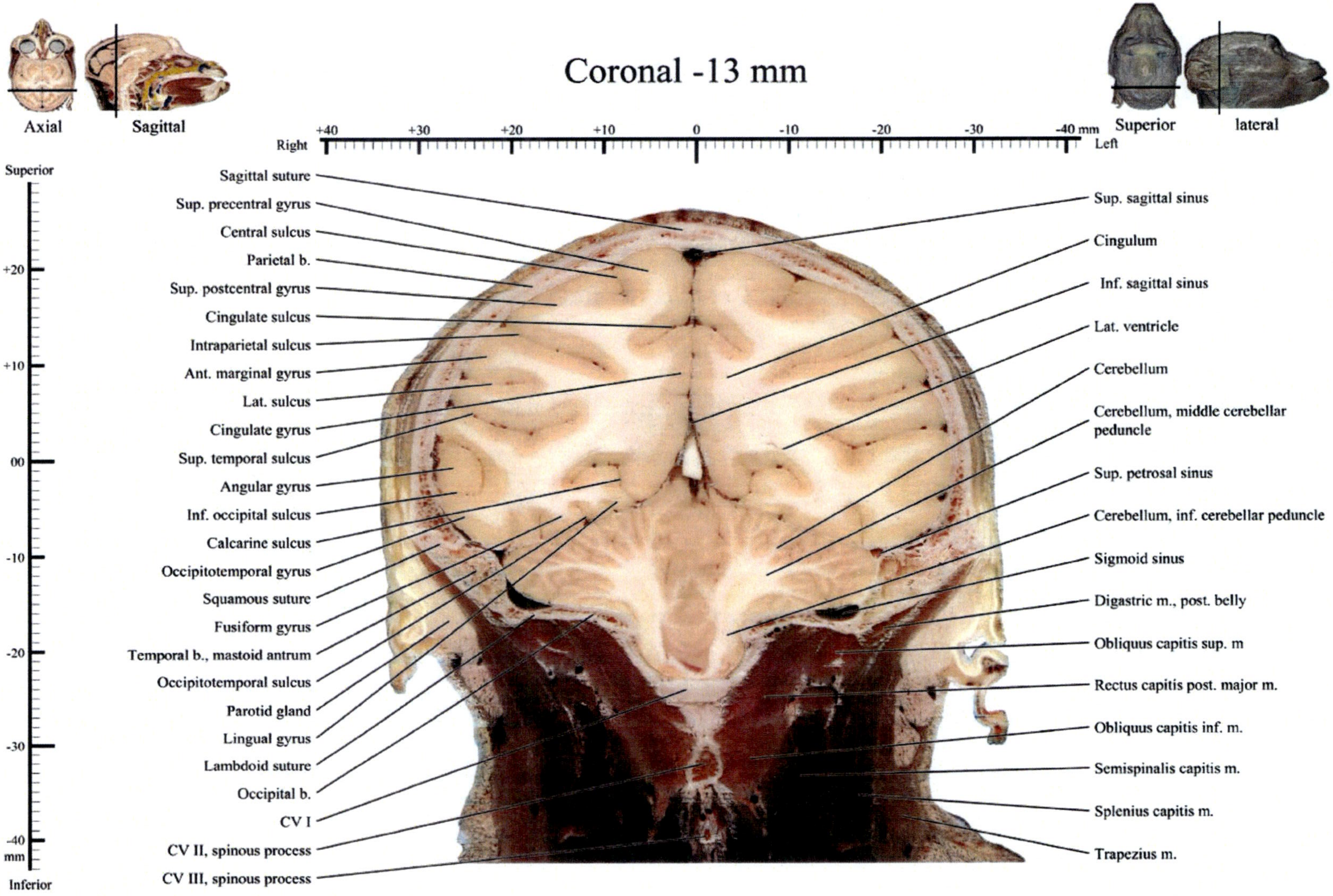

Coronal -14 mm

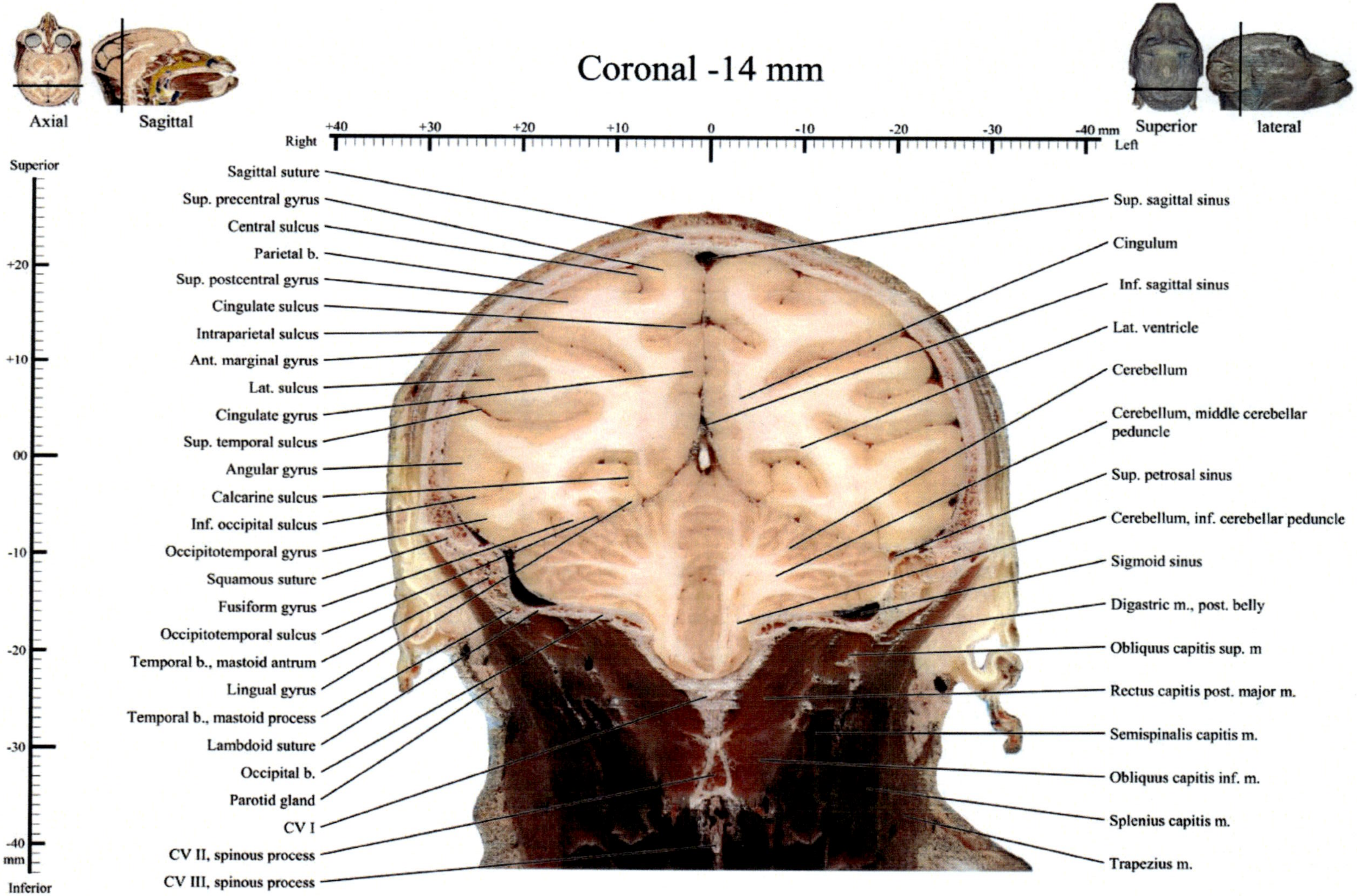

Coronal -15 mm

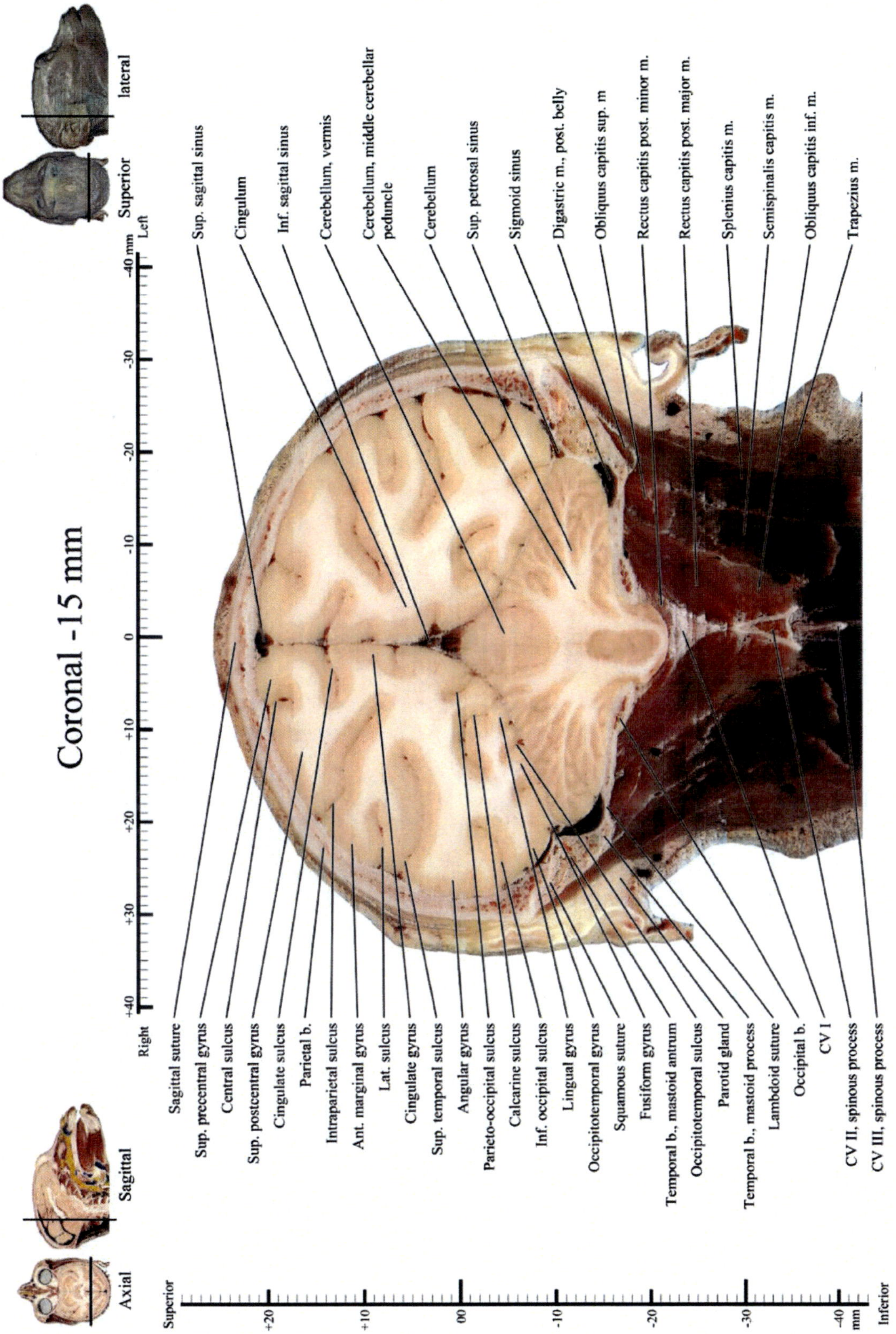

Coronal -16 mm

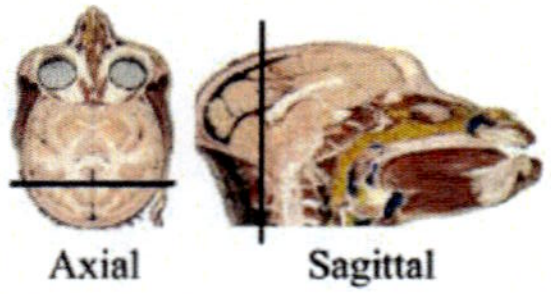

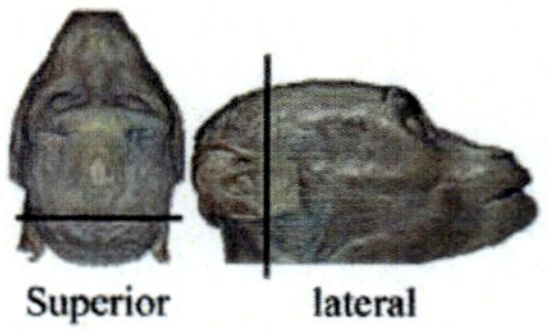

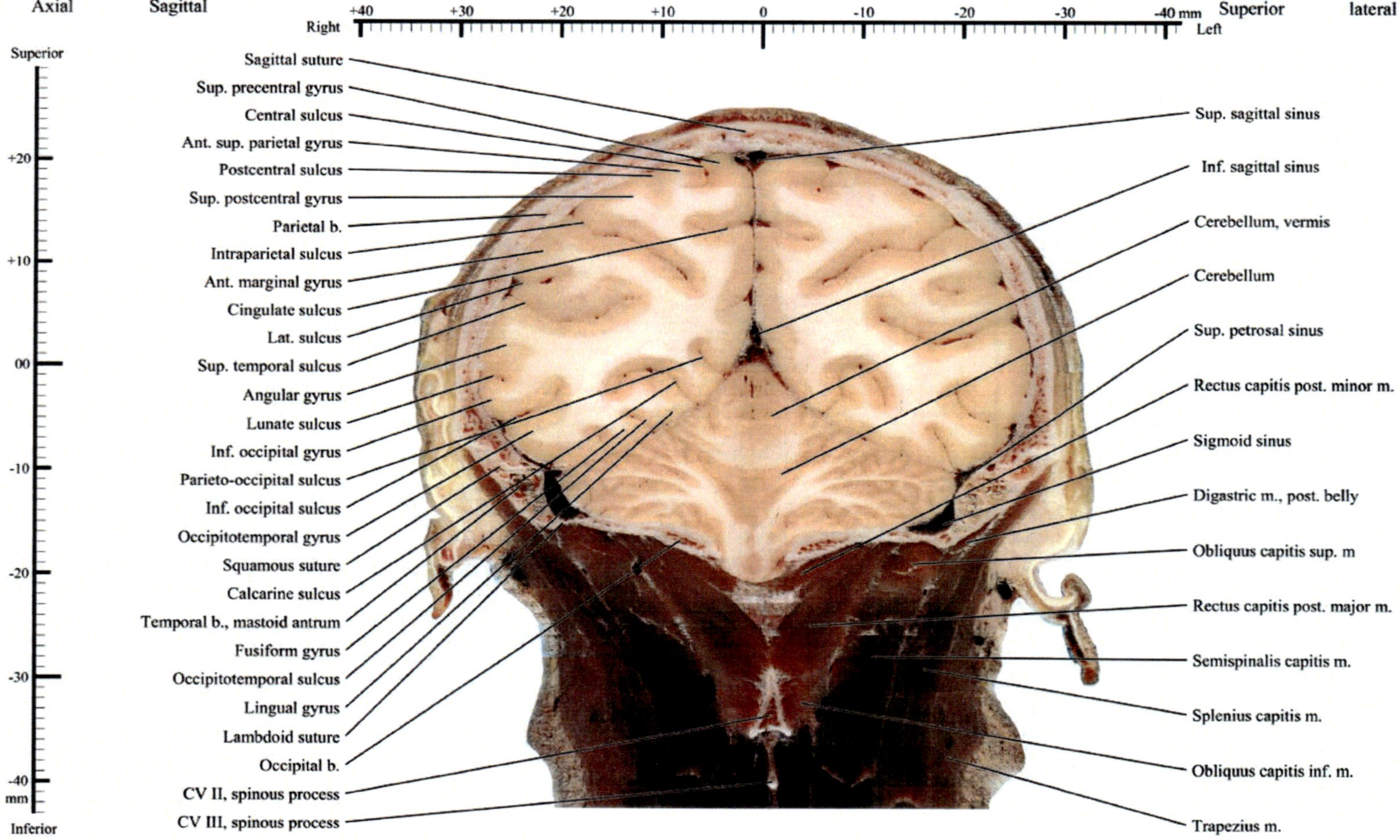

Coronal -17 mm

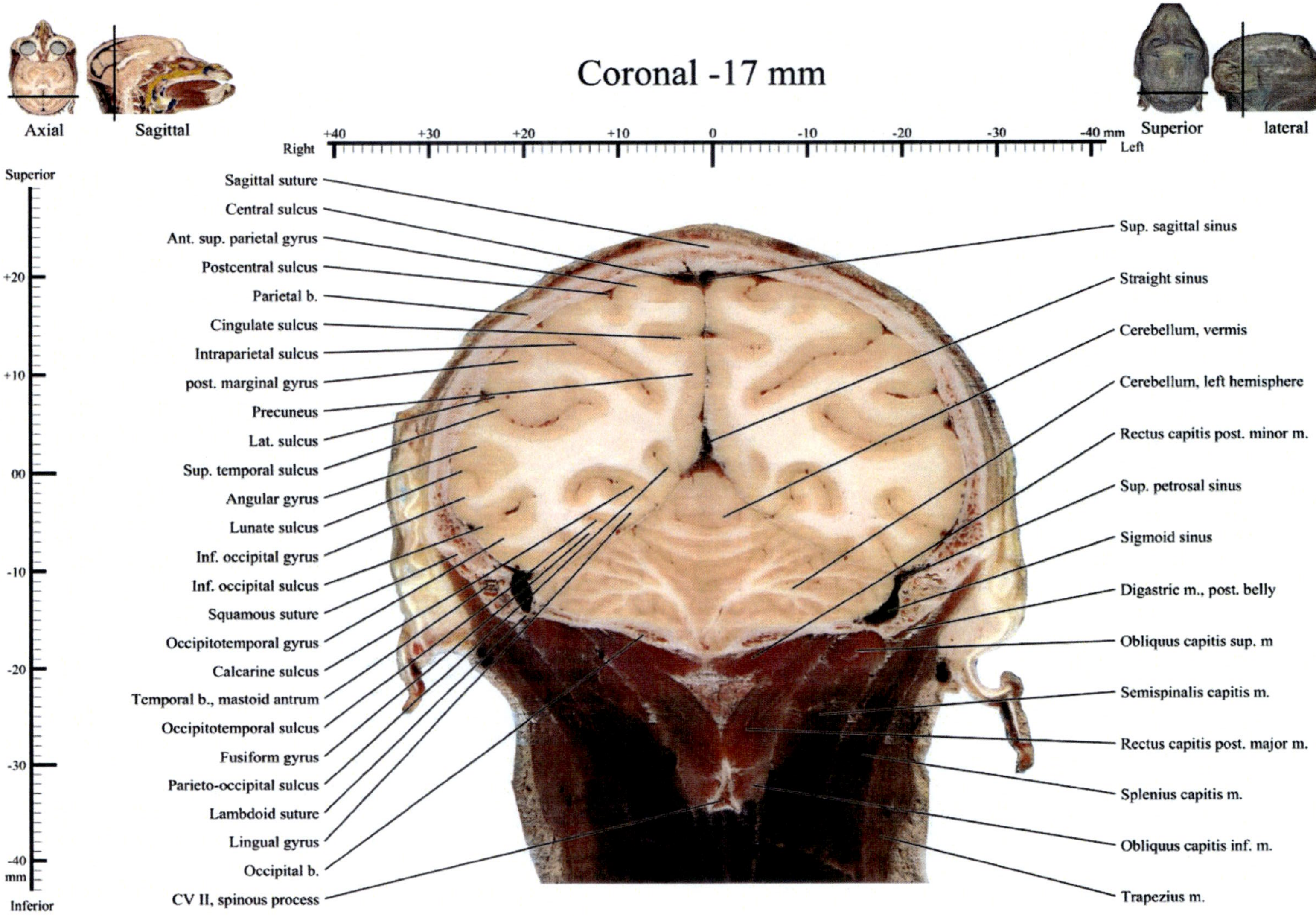

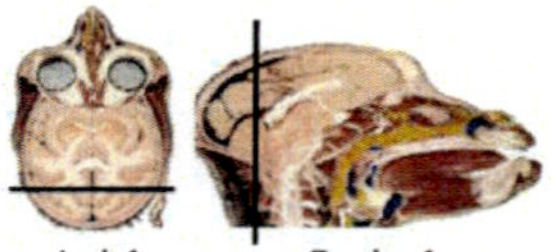

Coronal -18 mm

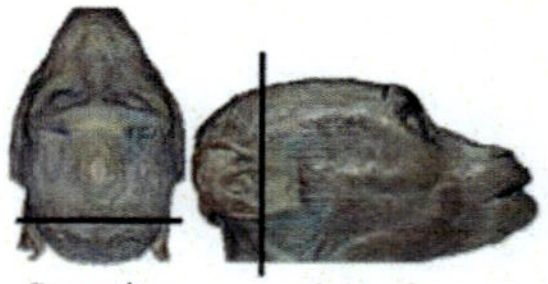

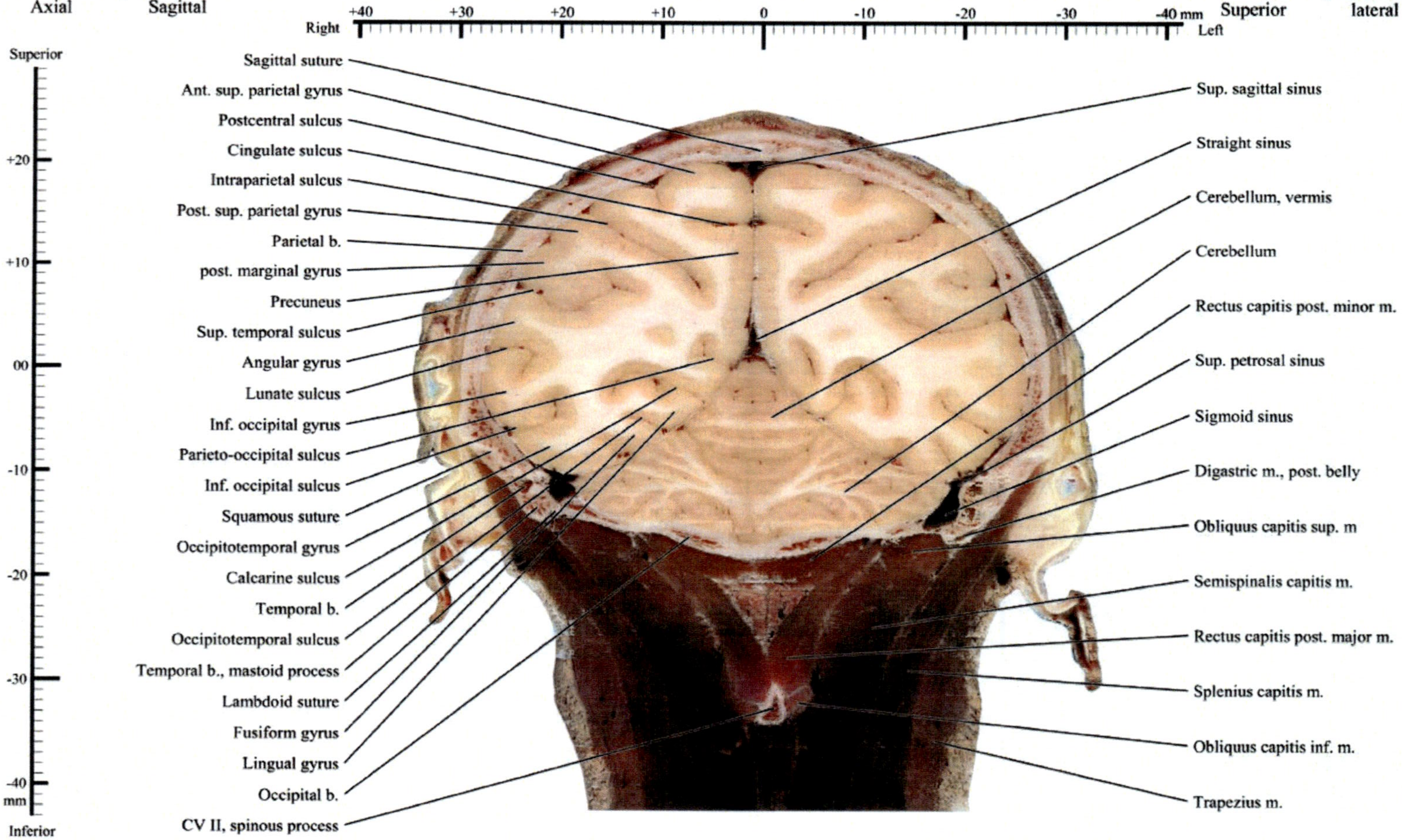

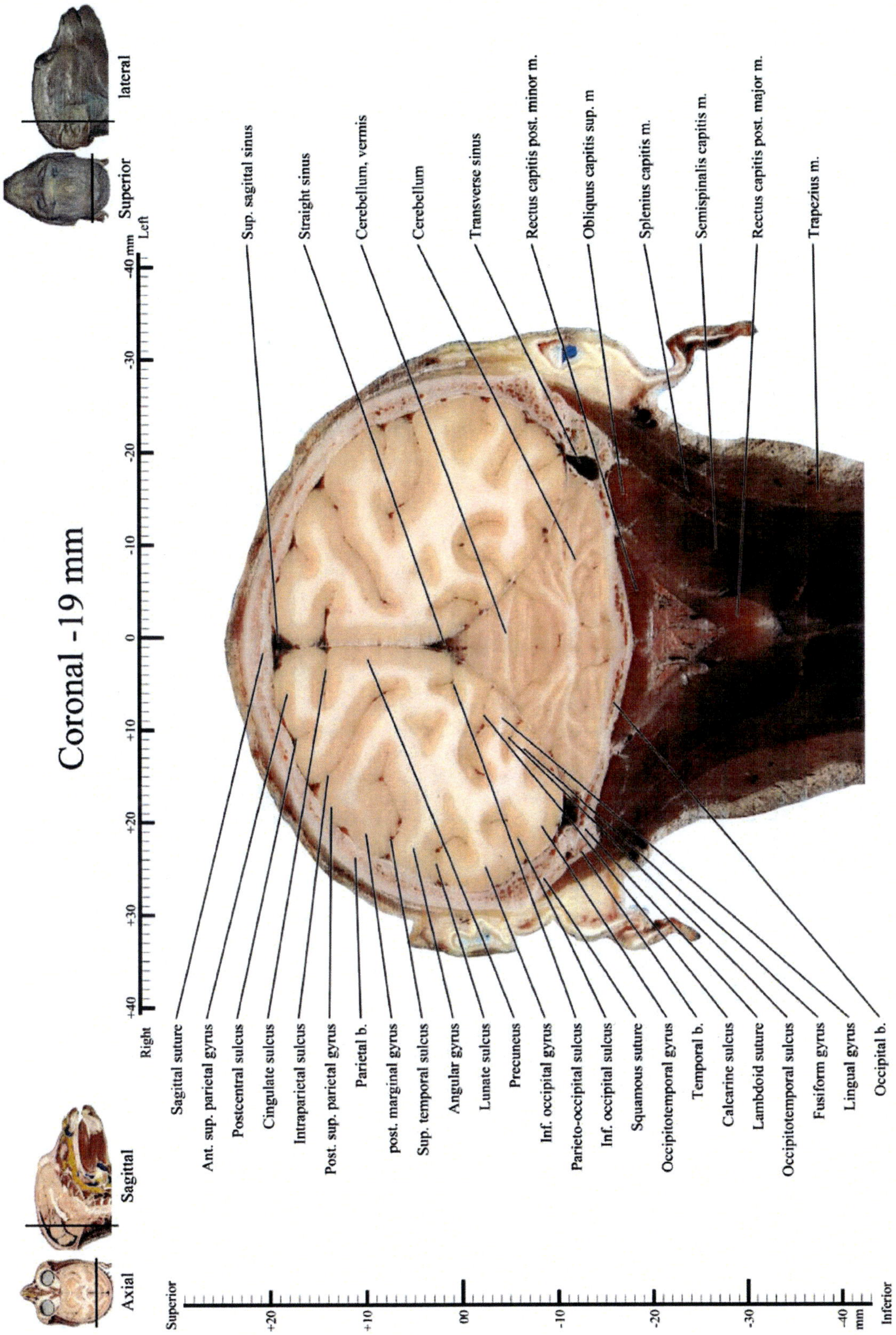
Coronal -19 mm
Superior
lateral
Sagittal
Axial
Right
Left
-40 mm
-30
-20
-10
0
+10
+20
+30
+40
Sup. sagital sinus
Straight sinus
Cerebellum, vermis
Cerebellum
Transverse sinus
Rectus capitis post. minor m.
Obliquus capitis sup. m
Splenius capitis m.
Semispinalis capitis m.
Rectus capitis post. major m.
Trapezius m.
Sagittal suture
Ant. sup. parietal gyrus
Postcentral sulcus
Cingulate sulcus
Intraparietal sulcus
Post. sup. parietal gyrus
Parietal b.
post. marginal gyrus
Sup. temporal sulcus
Angular gyrus
Lunate sulcus
Precuneus
Inf. occipital gyrus
Parieto-occipital sulcus
Inf. occipital sulcus
Squamous suture
Occipitotemporal gyrus
Temporal b.
Calcarine sulcus
Lambdoid suture
Occipitotemporal sulcus
Fusiform gyrus
Lingual gyrus
Occipital b.
Superior
+20
+10
00
-10
-20
-30
-40
mm
Inferior

Coronal -20 mm

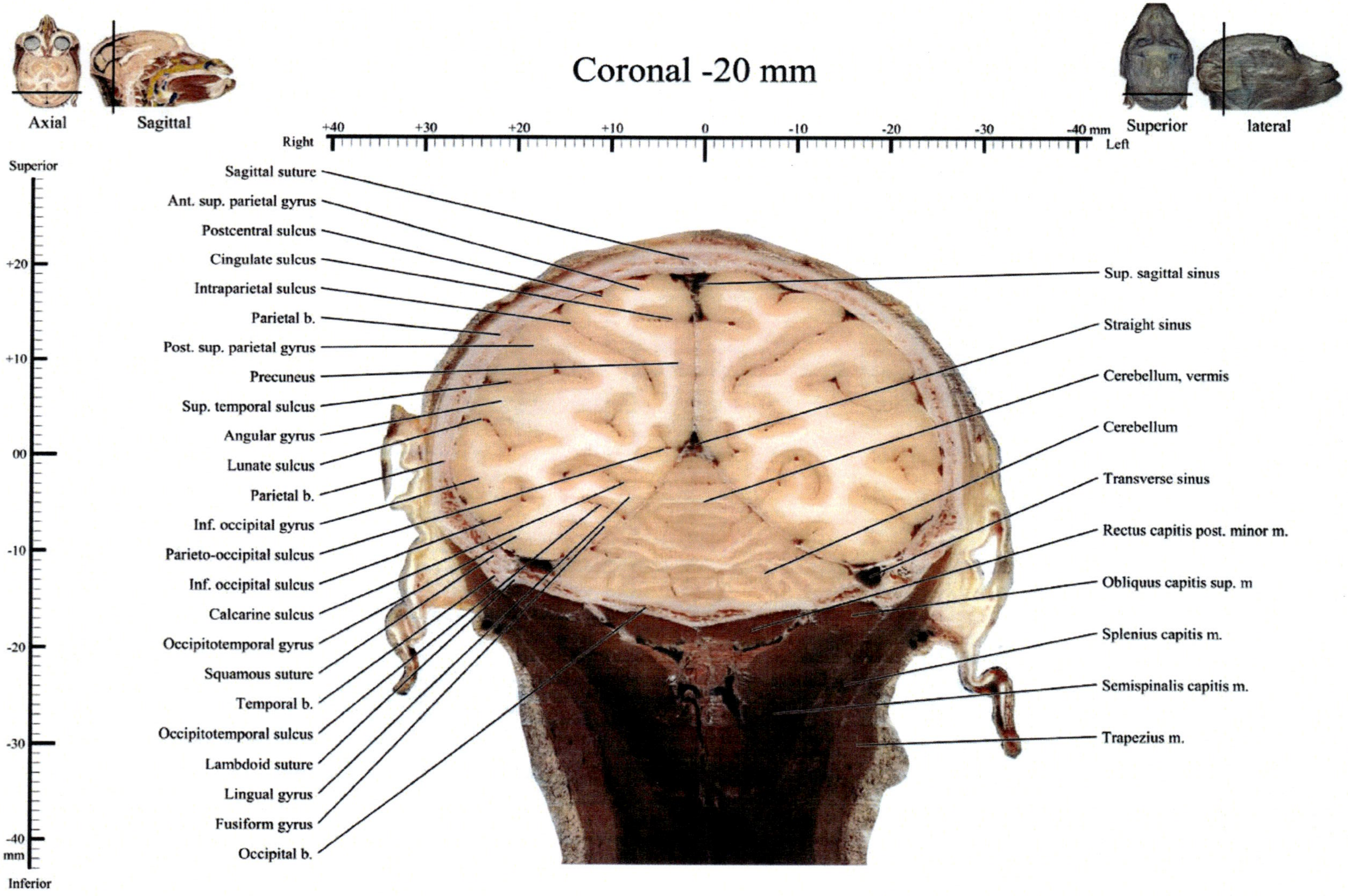

Coronal -21 mm

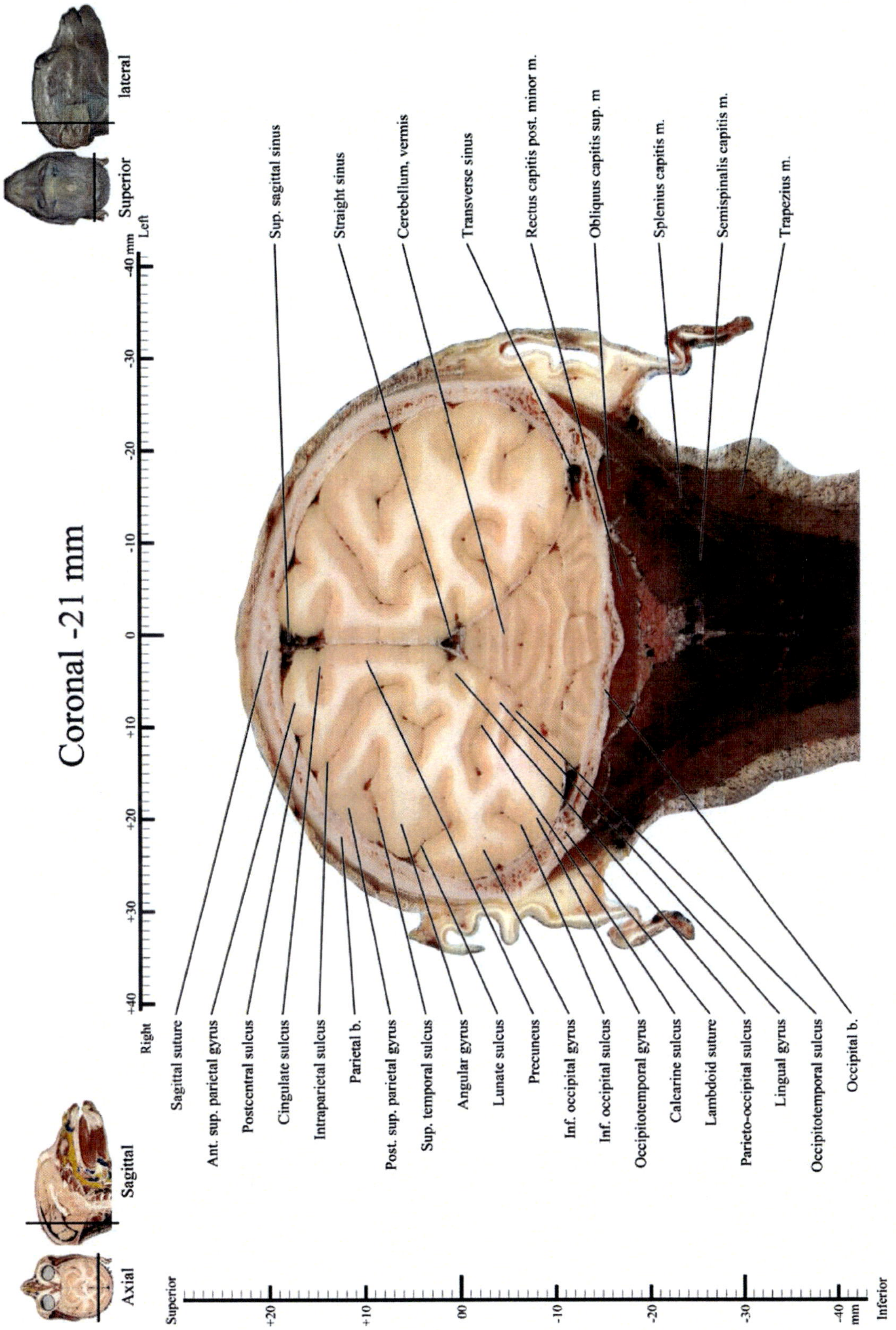

Coronal -22 mm

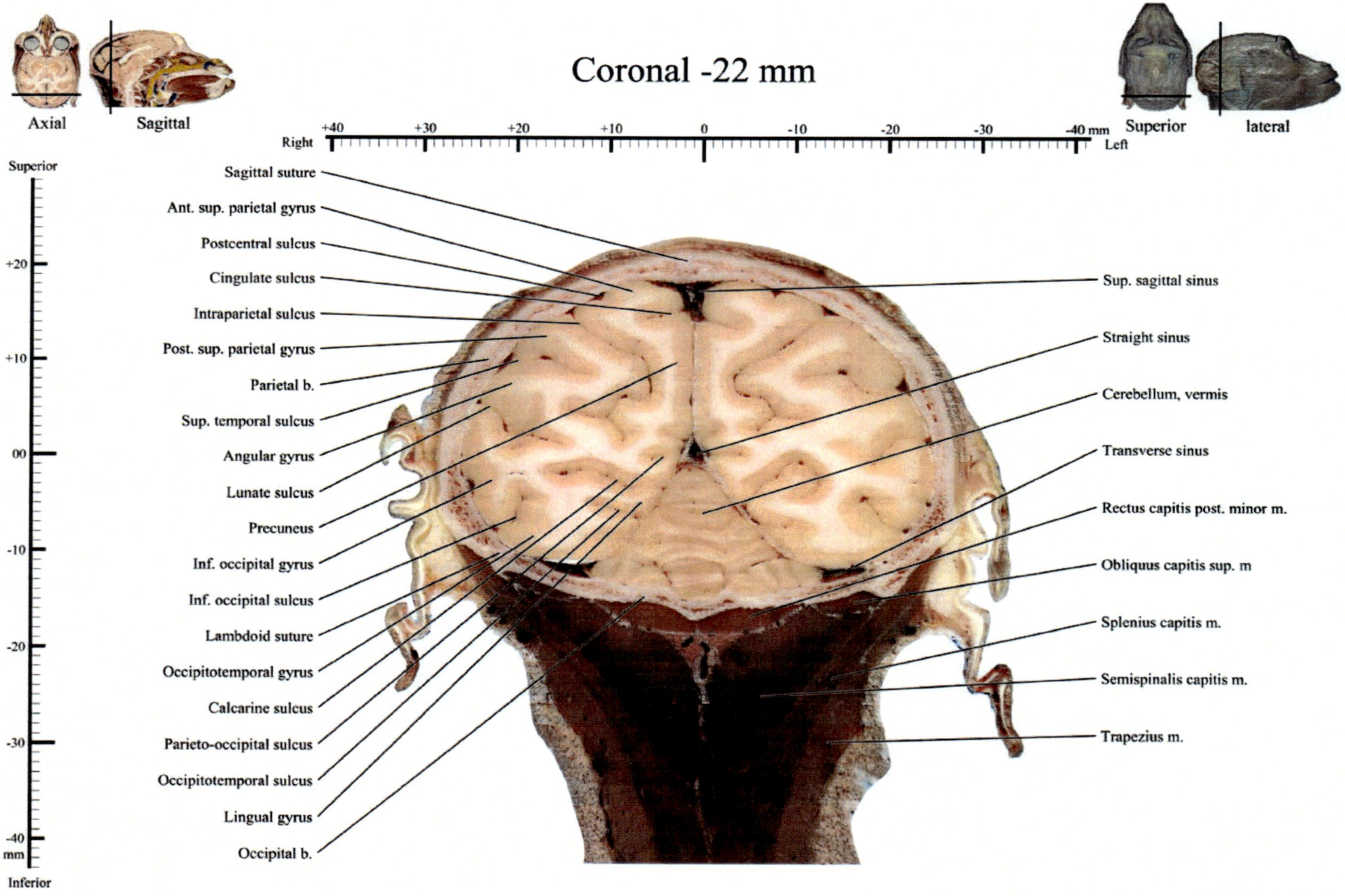

Coronal -23 mm

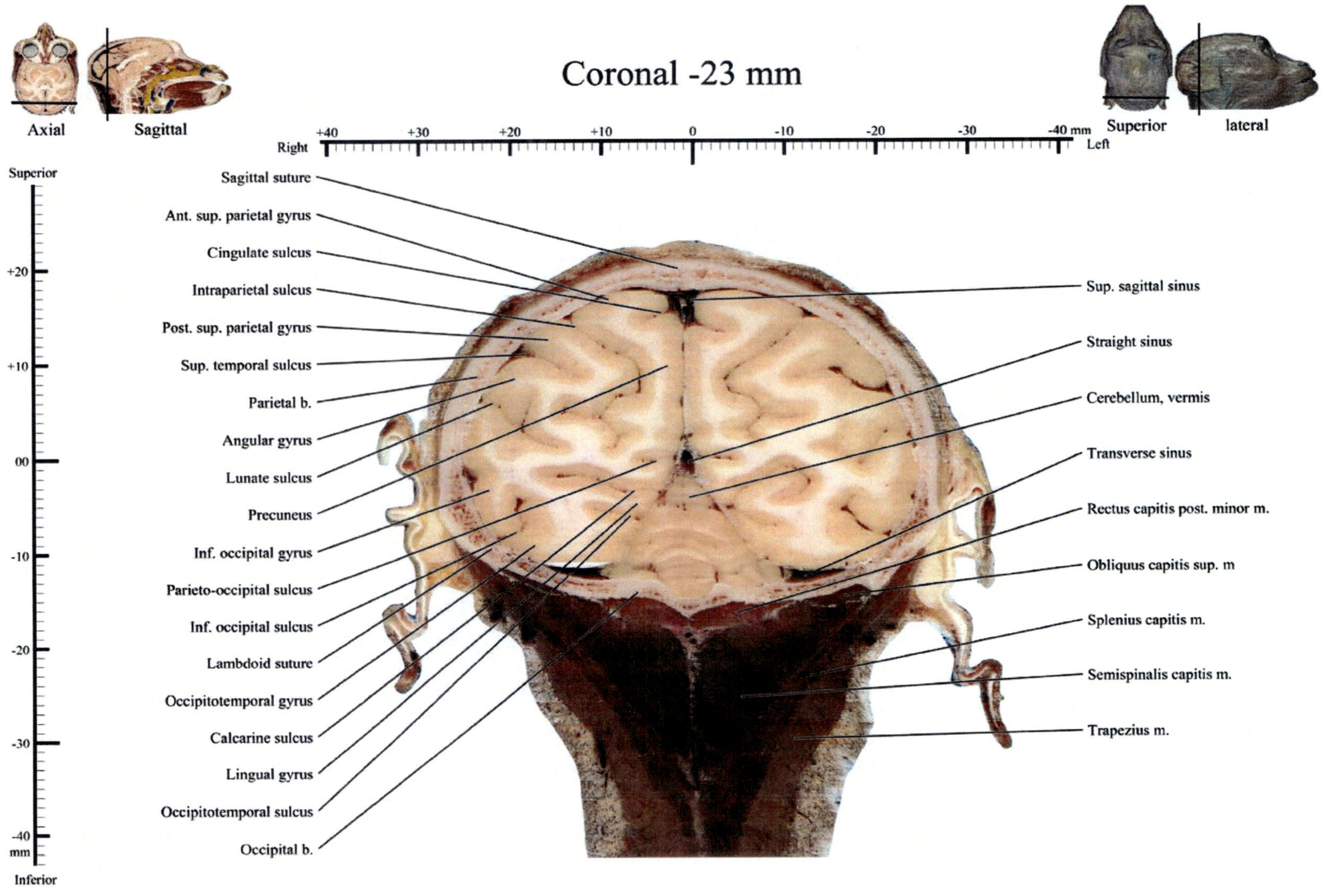

Coronal -24 mm

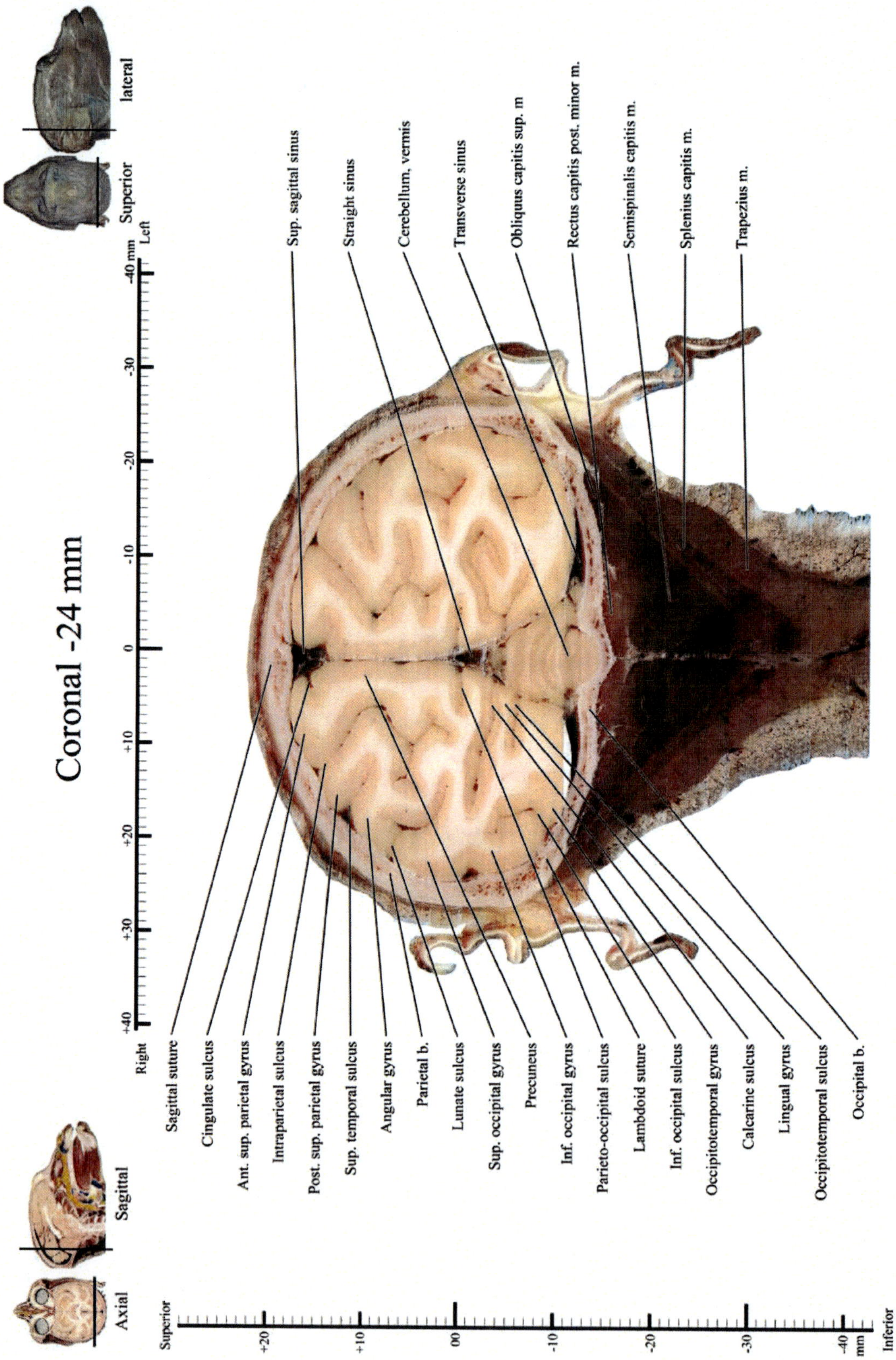

Coronal -25 mm

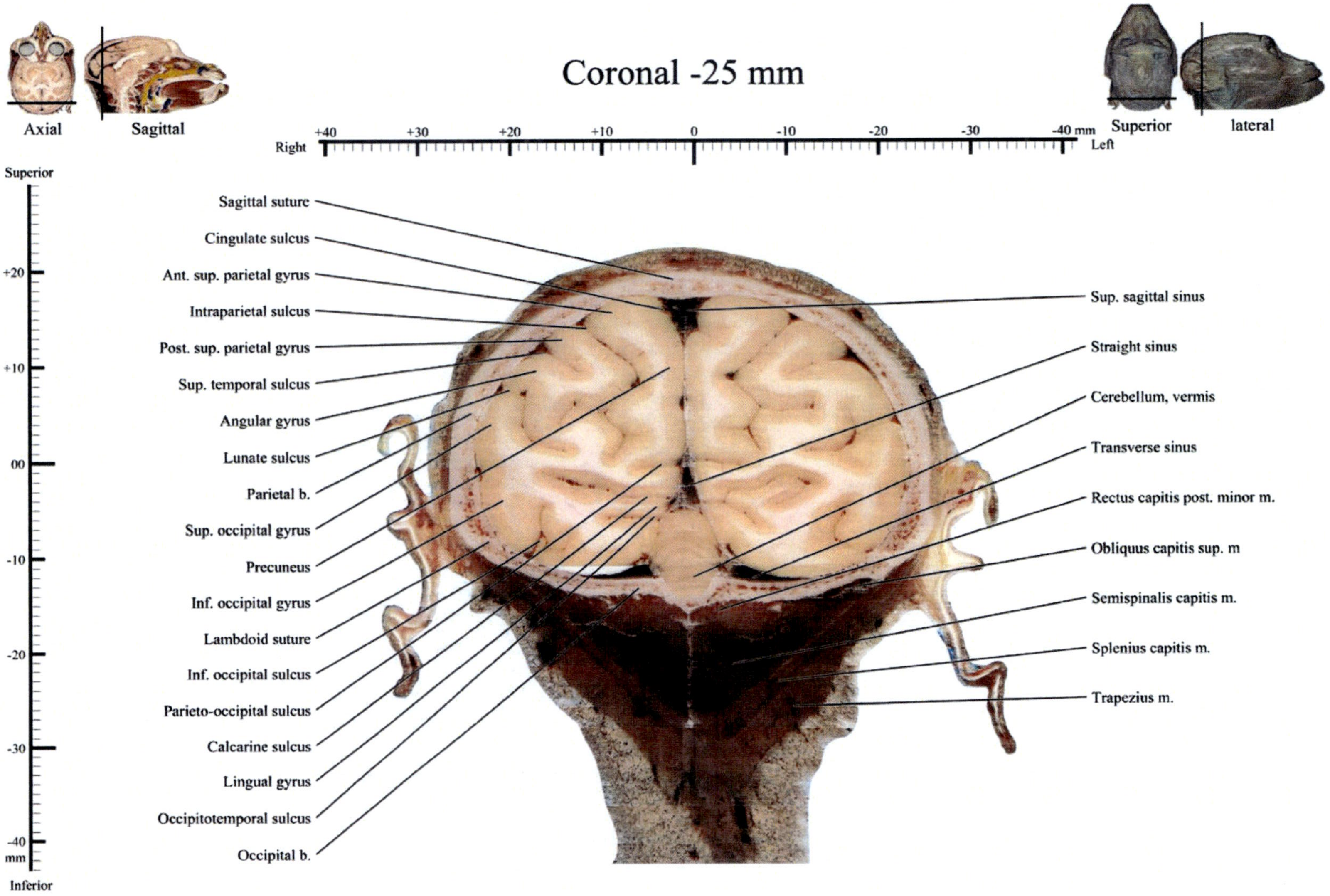

Coronal -26 mm

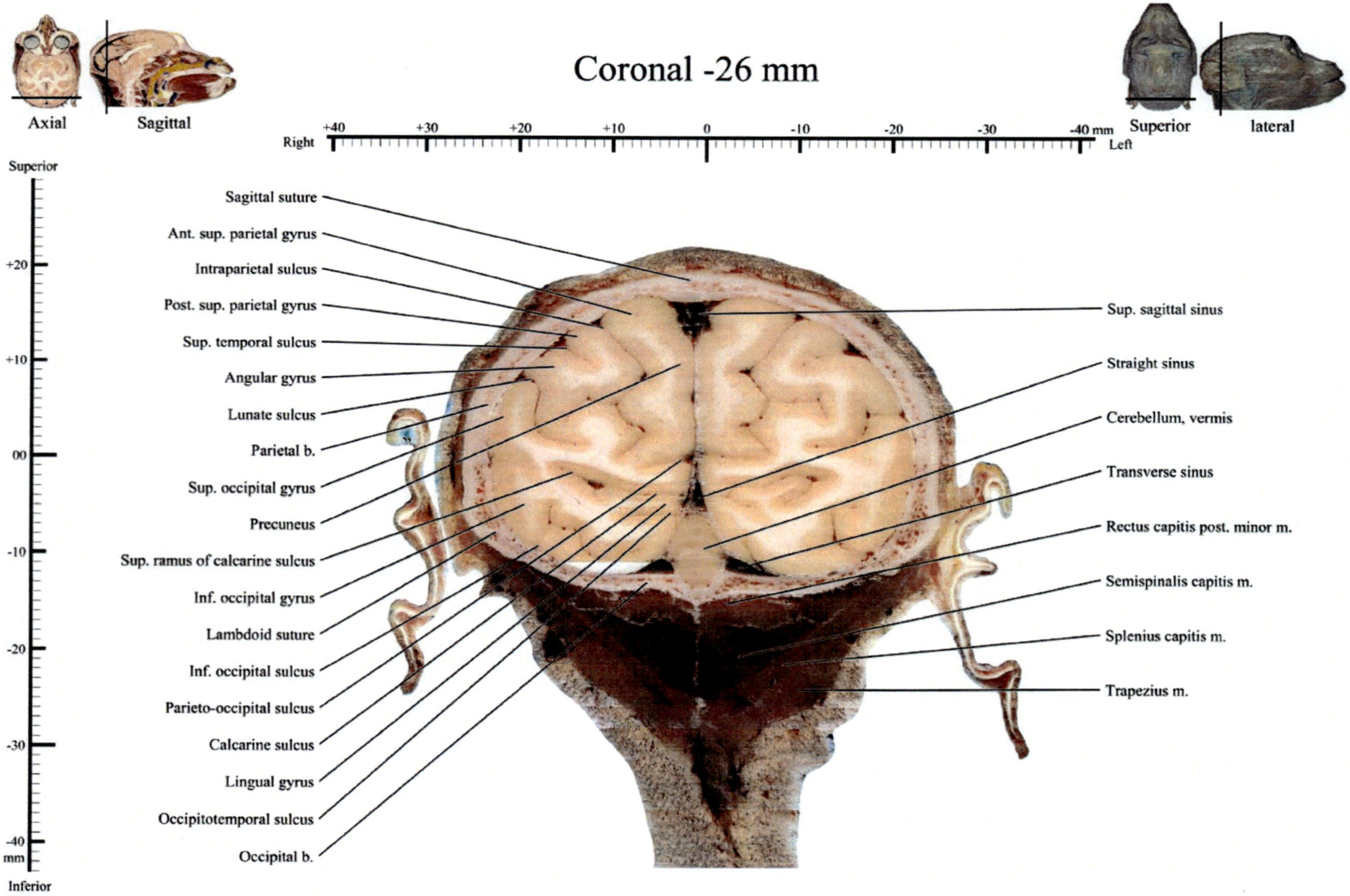

Coronal -27 mm
Axial
Sagittal
Superior
lateral
Right
Left
Superior
Inferior
mm
+40
+30
+20
+10
0
-10
-20
-30
-40 mm
Sagittal suture
Ant. sup. parietal gyrus
Intraparietal sulcus
Post. sup. parietal gyrus
Sup. temporal sulcus
Angular gyrus
Lunate sulcus
Precuneus
Parietal b.
Sup. occipital gyrus
Ectocalcarine sulcus
Sup. ramus of calcarine sulcus
Lambdoid suture
Inf. occipital gyrus
Parieto-occipital sulcus
Cuneus
Calcarine sulcus
Lingual gyrus
Occipitotemporal sulcus
Occipital b.
Sup. sagital sinus
Straight sinus
Cerebellum, vermis
Transverse sinus
Rectus capitis post. minor m.
Semispinalis capitis m.
Splenius capitis m.
Trapezius m.

Coronal -28 mm

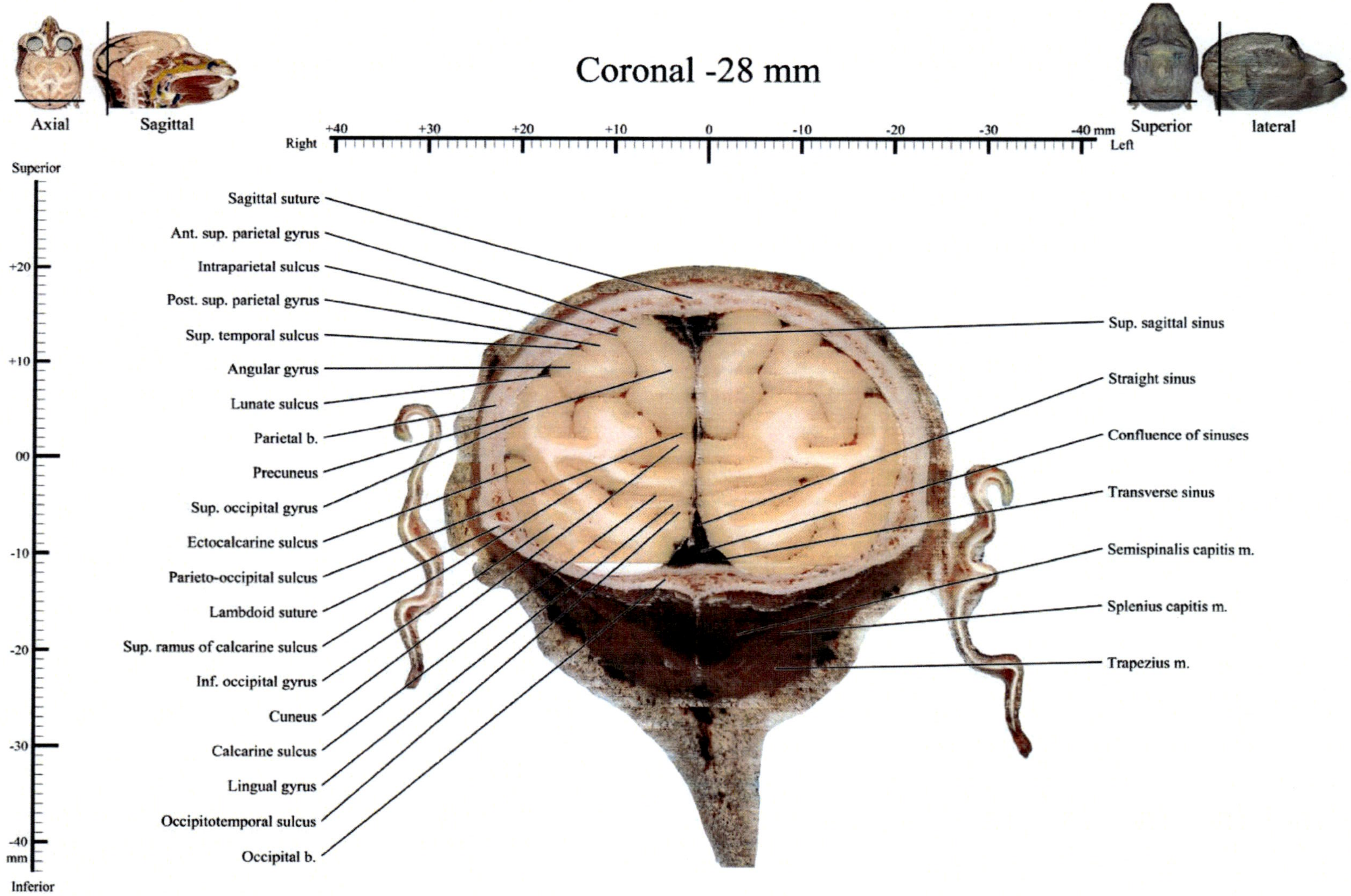

Coronal -29 mm

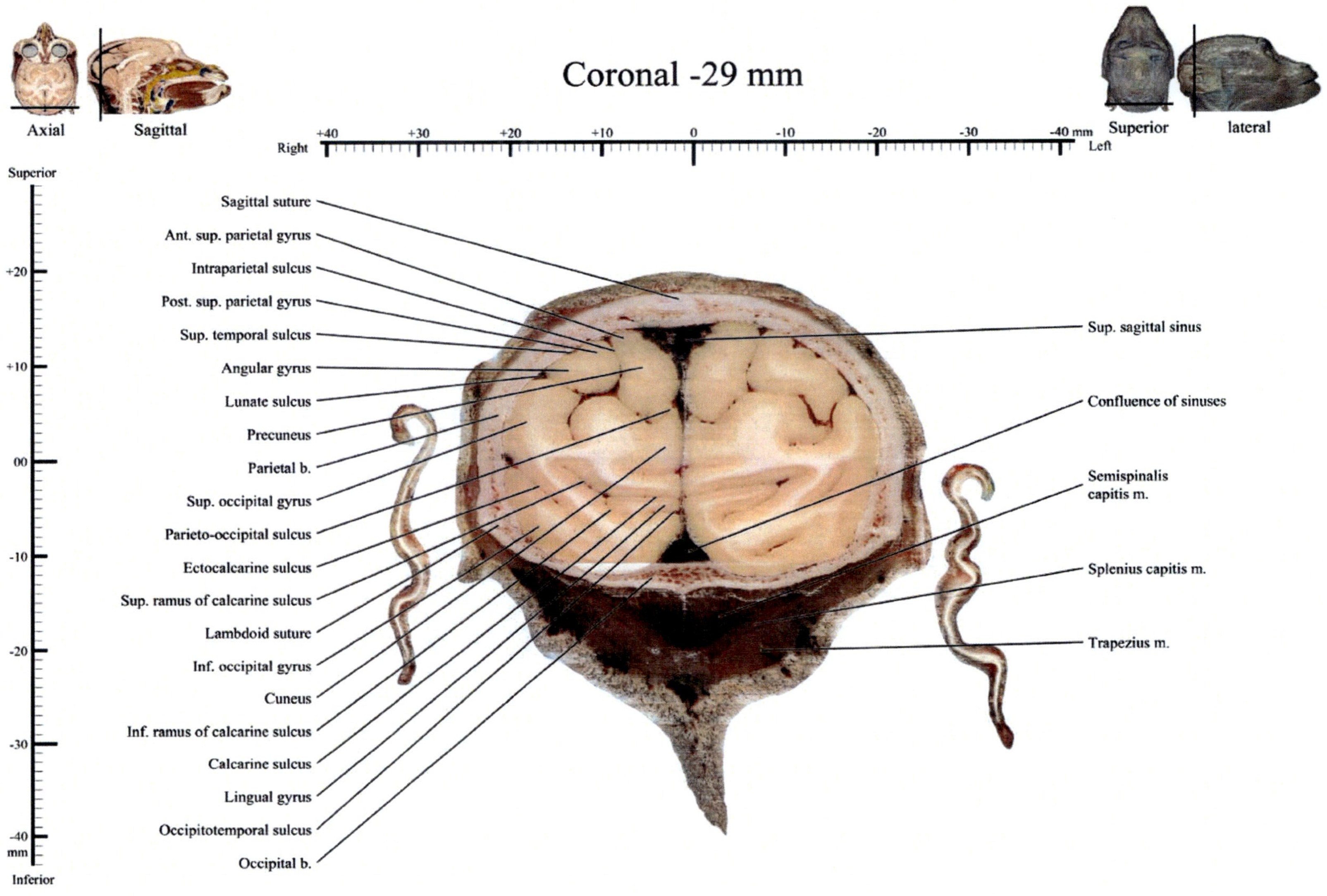

Coronal -30 mm

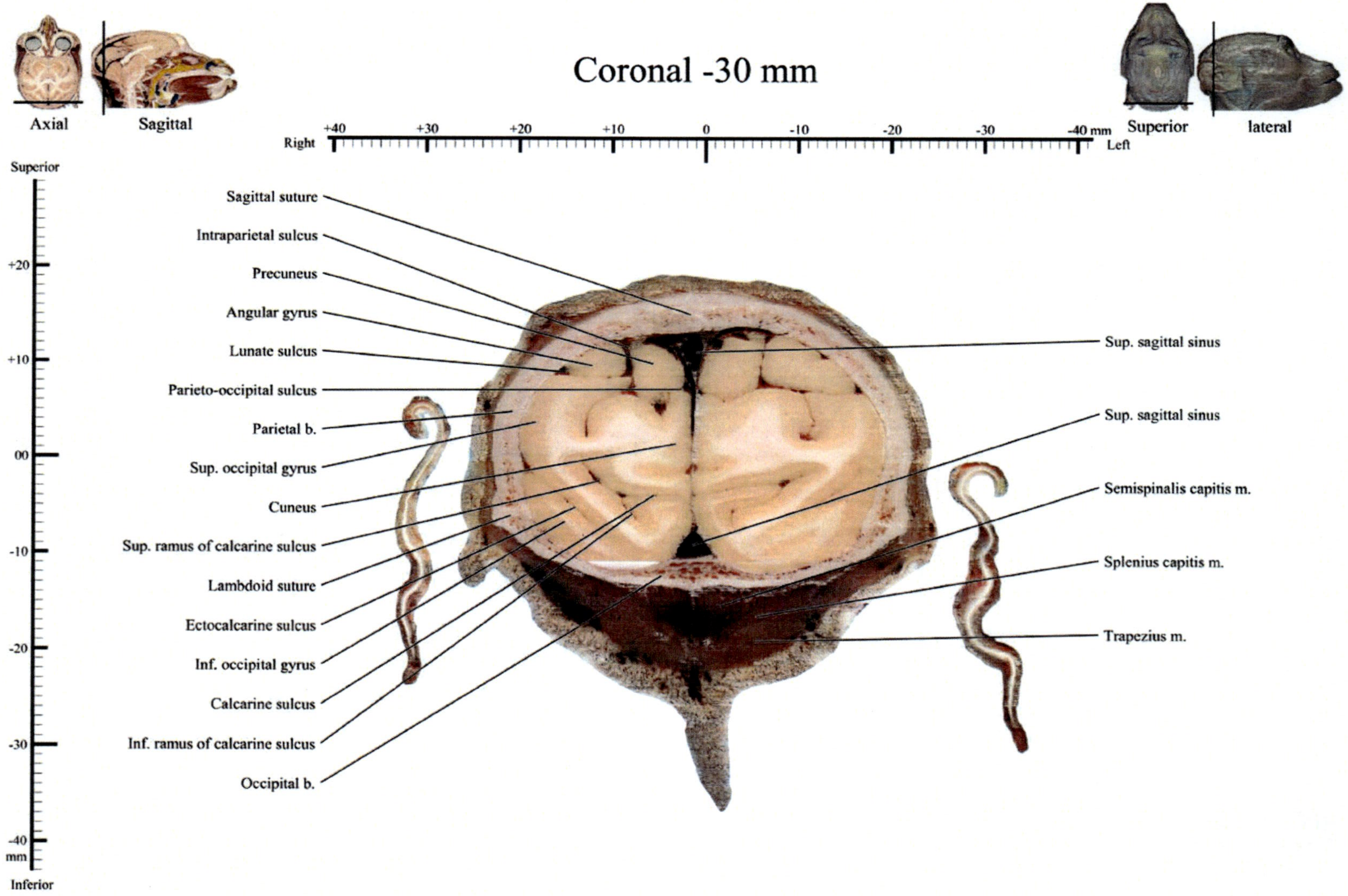

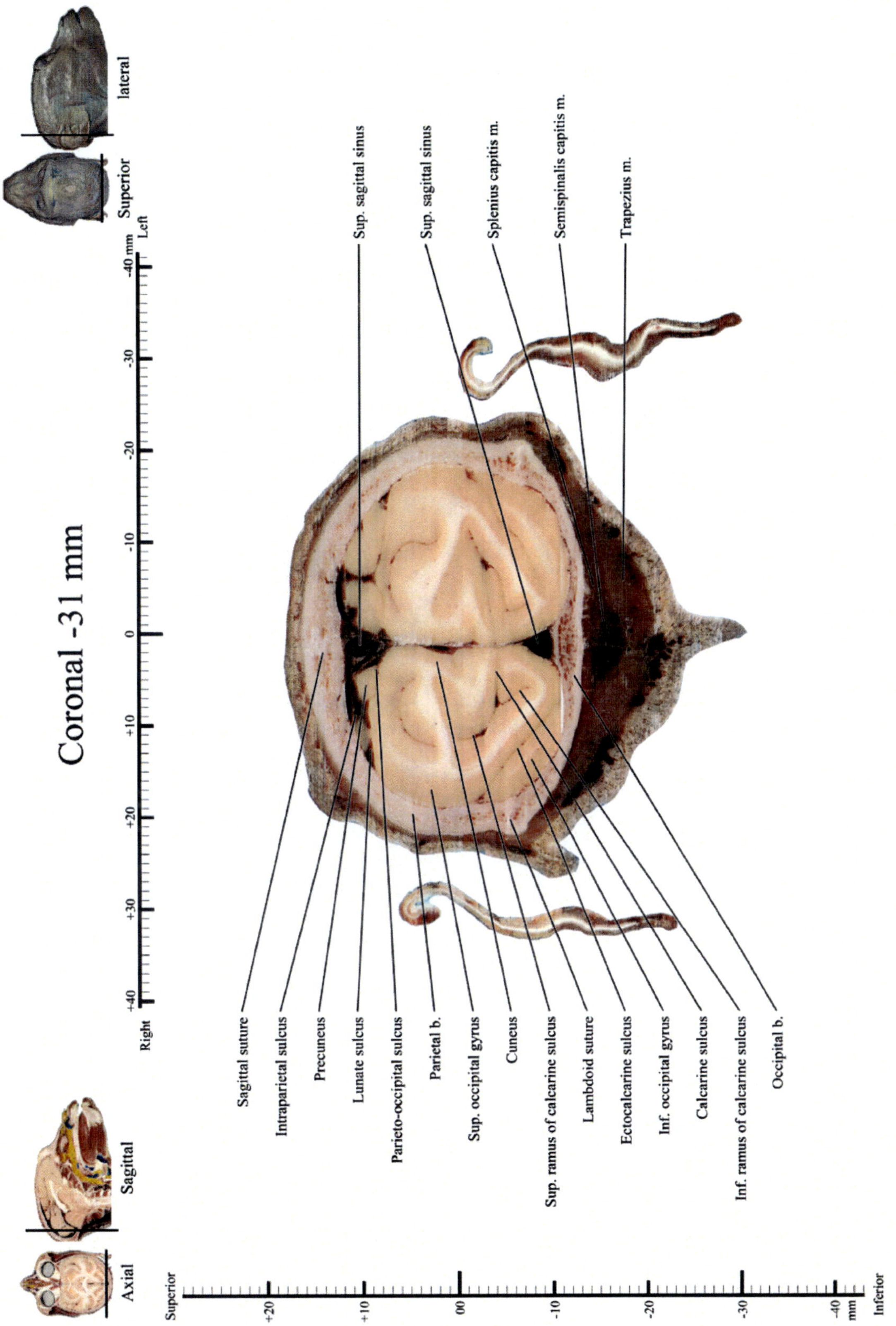
Coronal -31 mm
Superior
lateral
Axial
Sagittal
Right
Left
Superior
Inferior
mm
Sup. sagittal sinus
Sup. sagittal sinus
Splenius capitis m.
Semispinalis capitis m.
Trapezius m.
Sagittal suture
Intraparietal sulcus
Precuneus
Lunate sulcus
Parieto-occipital sulcus
Parietal b.
Sup. occipital gyrus
Cuneus
Sup. ramus of calcarine sulcus
Lambdoid suture
Ectocalcarine sulcus
Inf. occipital gyrus
Calcarine sulcus
Inf. ramus of calcarine sulcus
Occipital b.

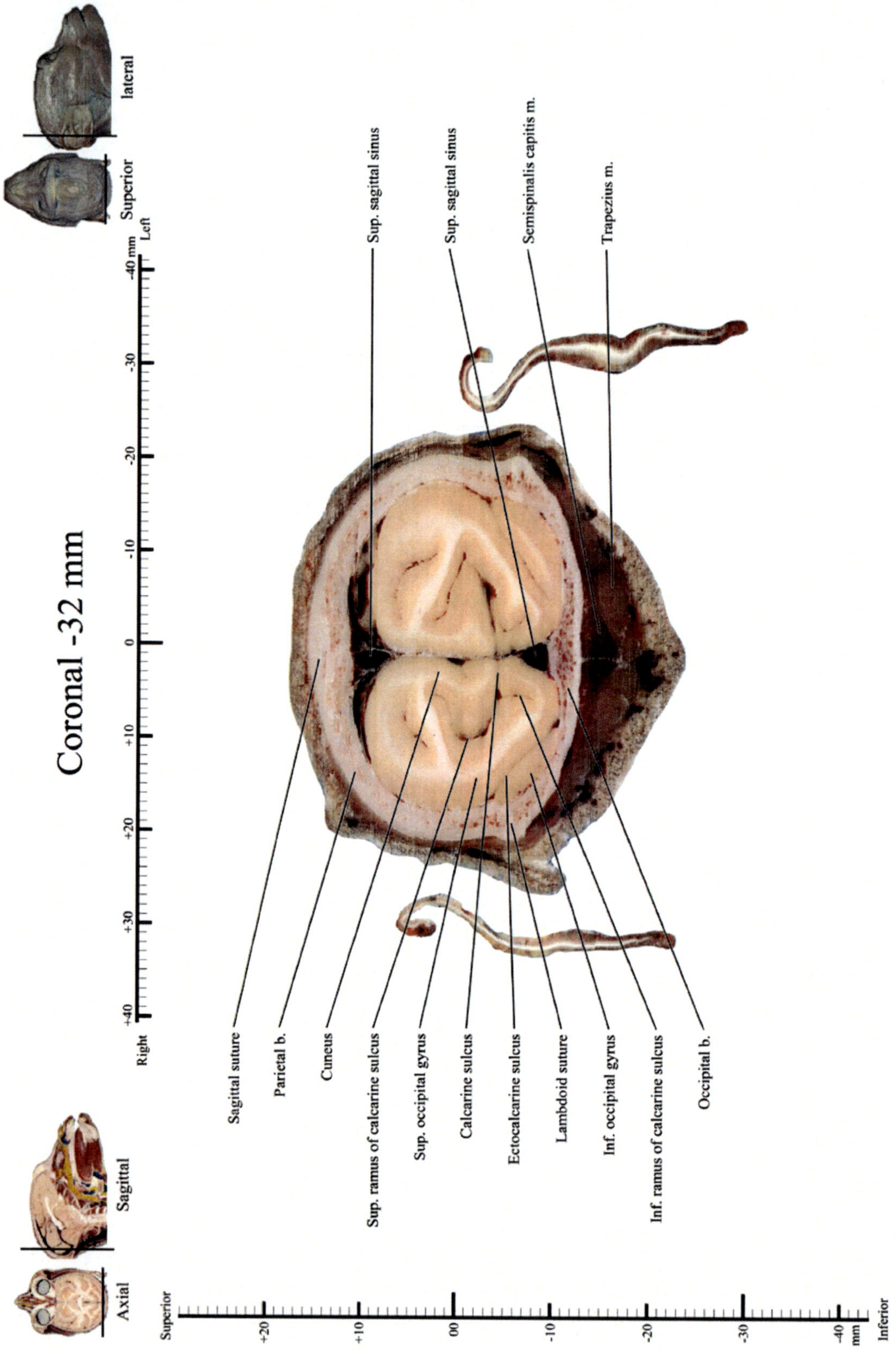
Coronal -32 mm
lateral
Superior
Sagittal
Axial
Right
Left
+40
+30
+20
+10
0
-10
-20
-30
-40 mm
Superior
Inferior
Sup. sagittal sinus
Sup. sagittal sinus
Semispinalis capitis m.
Trapezius m.
Sagittal suture
Parietal b.
Cuneus
Sup. ramus of calcarine sulcus
Sup. occipital gyrus
Calcarine sulcus
Ectocalcarine sulcus
Lambdoid suture
Inf. occipital gyrus
Inf. ramus of calcarine sulcus
Occipital b.

Coronal -33 mm

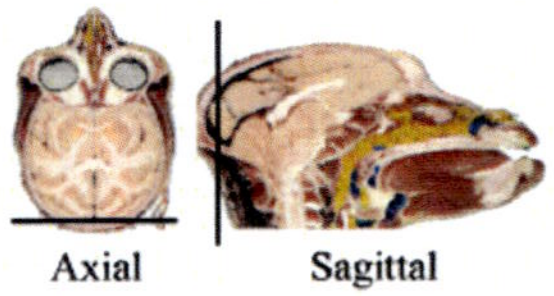

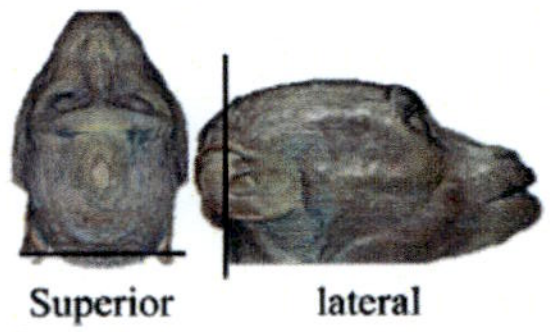

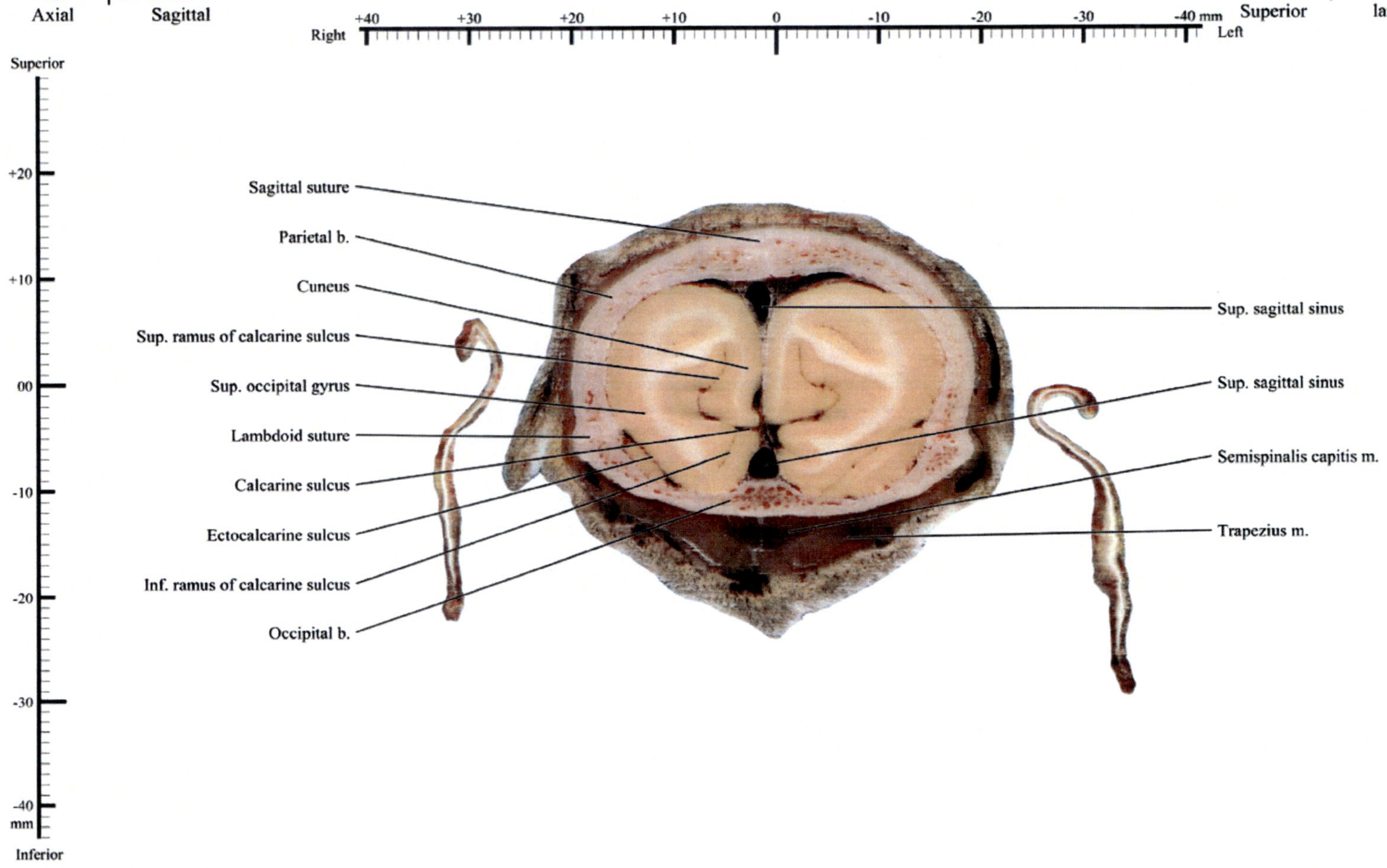

Coronal -34 mm

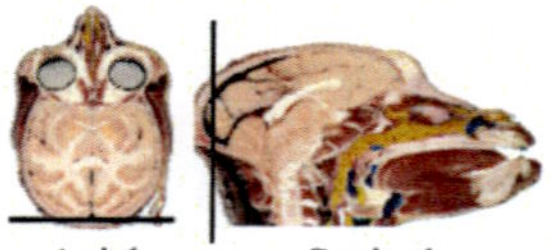

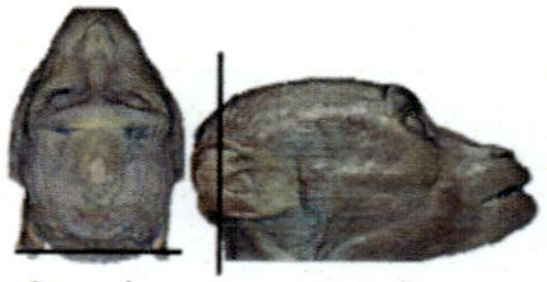

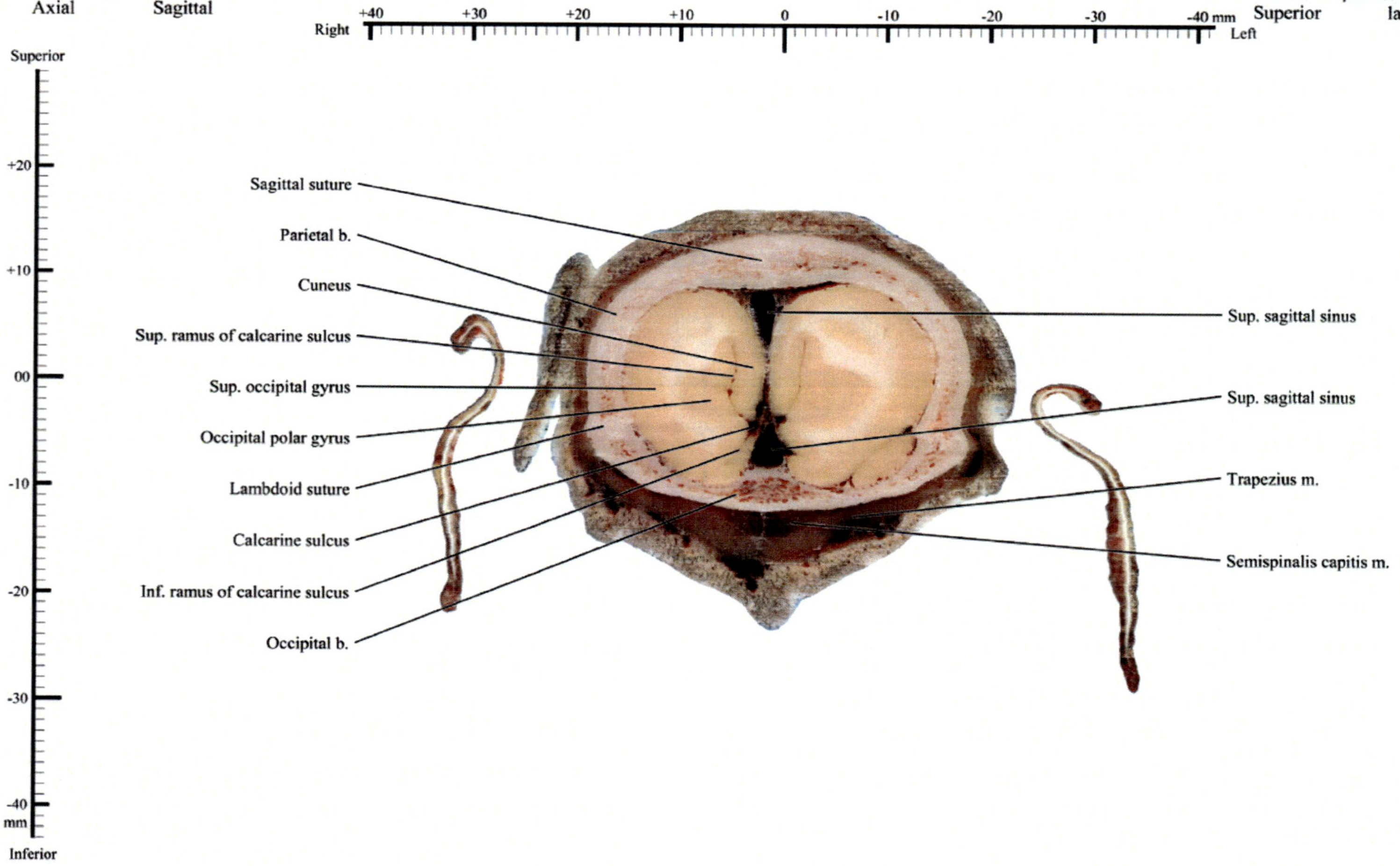

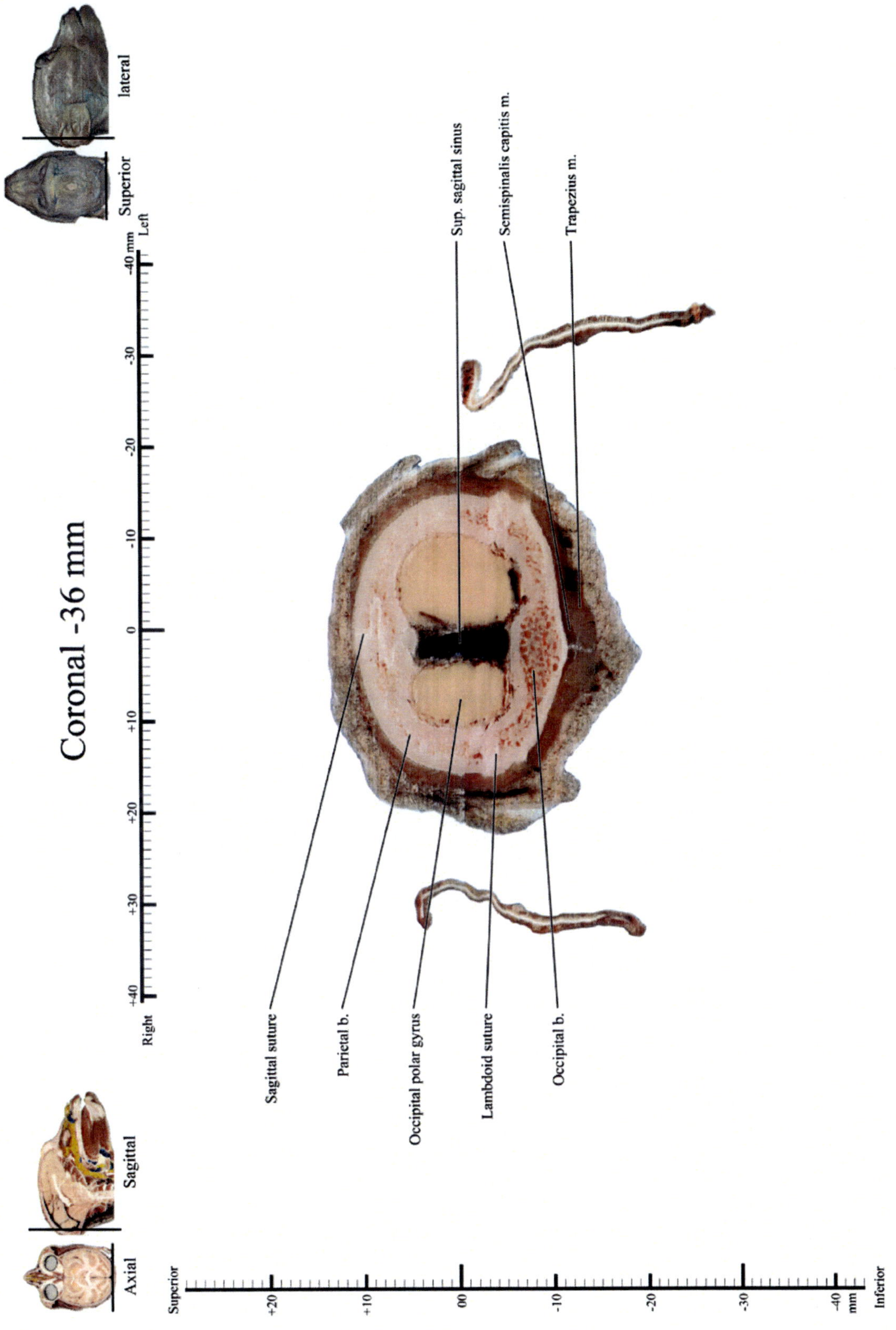
Coronal -36 mm
lateral
Superior
Sagittal
Axial
Sup. sagittal sinus
Semispinalis capitis m.
Trapezius m.
Sagittal suture
Parietal b.
Occipital polar gyrus
Lambdoid suture
Occipital b.
Right
Left
-40 mm
-30
-20
-10
0
+10
+20
+30
+40
Superior
Inferior
00
mm

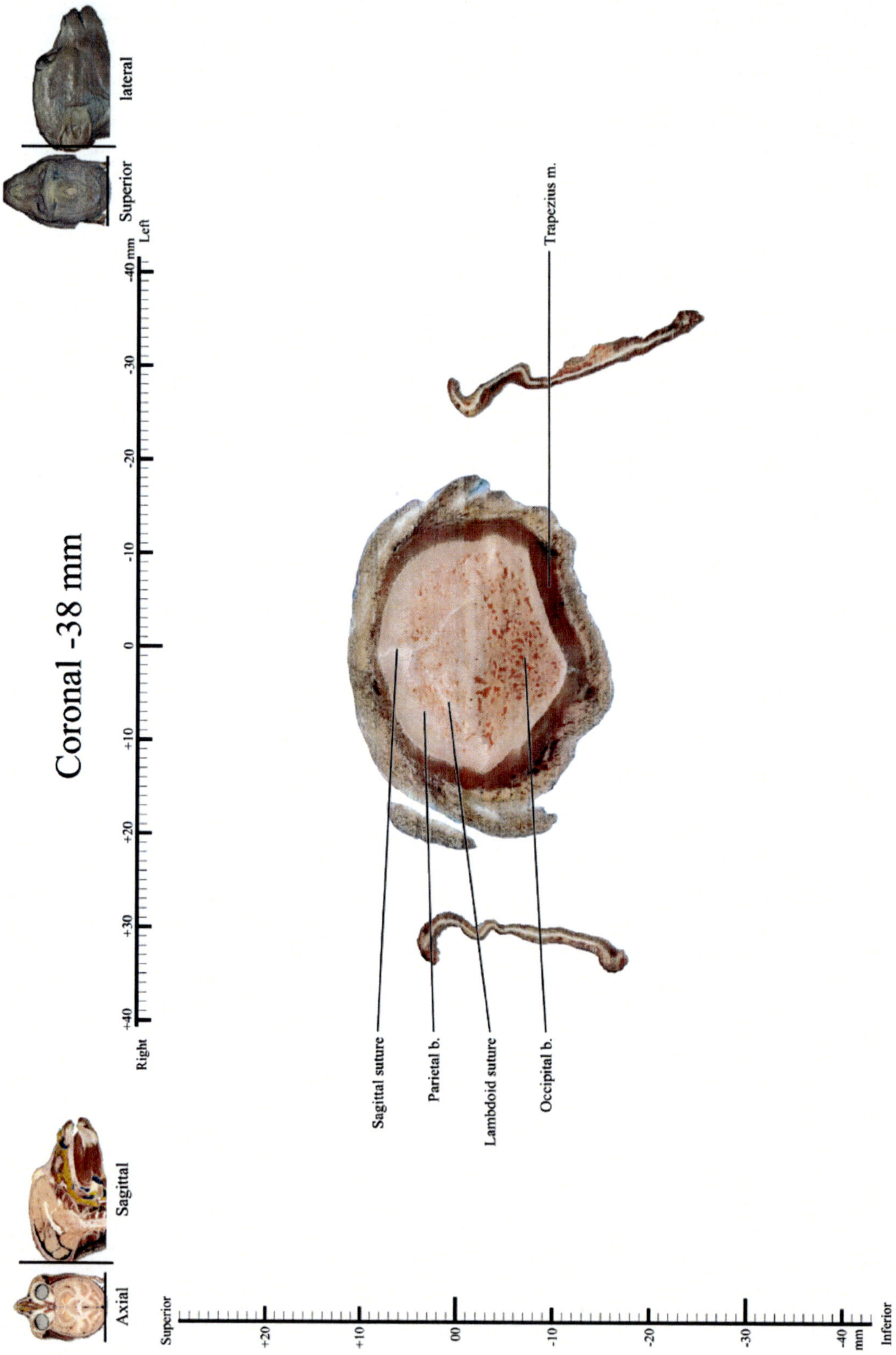

Coronal -38 mm
lateral
Superior
Sagittal
Axial
Left
Right
mm
Superior
Inferior
Trapezius m.
Sagittal suture
Parietal b.
Lambdoid suture
Occipital b.

Sagittal

4

All images, 0.024 mm × 0.024 mm sized pixel

0 mm ~ −16 mm, 0.5 mm intervals
−16 mm ~ −17 mm, 1 mm intervals
−17 mm ~ −27 mm, 2.5 mm intervals

J. S. Park, *Cross-Sectional Atlas of Rhesus Monkey Head*, https://doi.org/10.1007/978-981-19-3254-0_4

Sagittal 0 mm

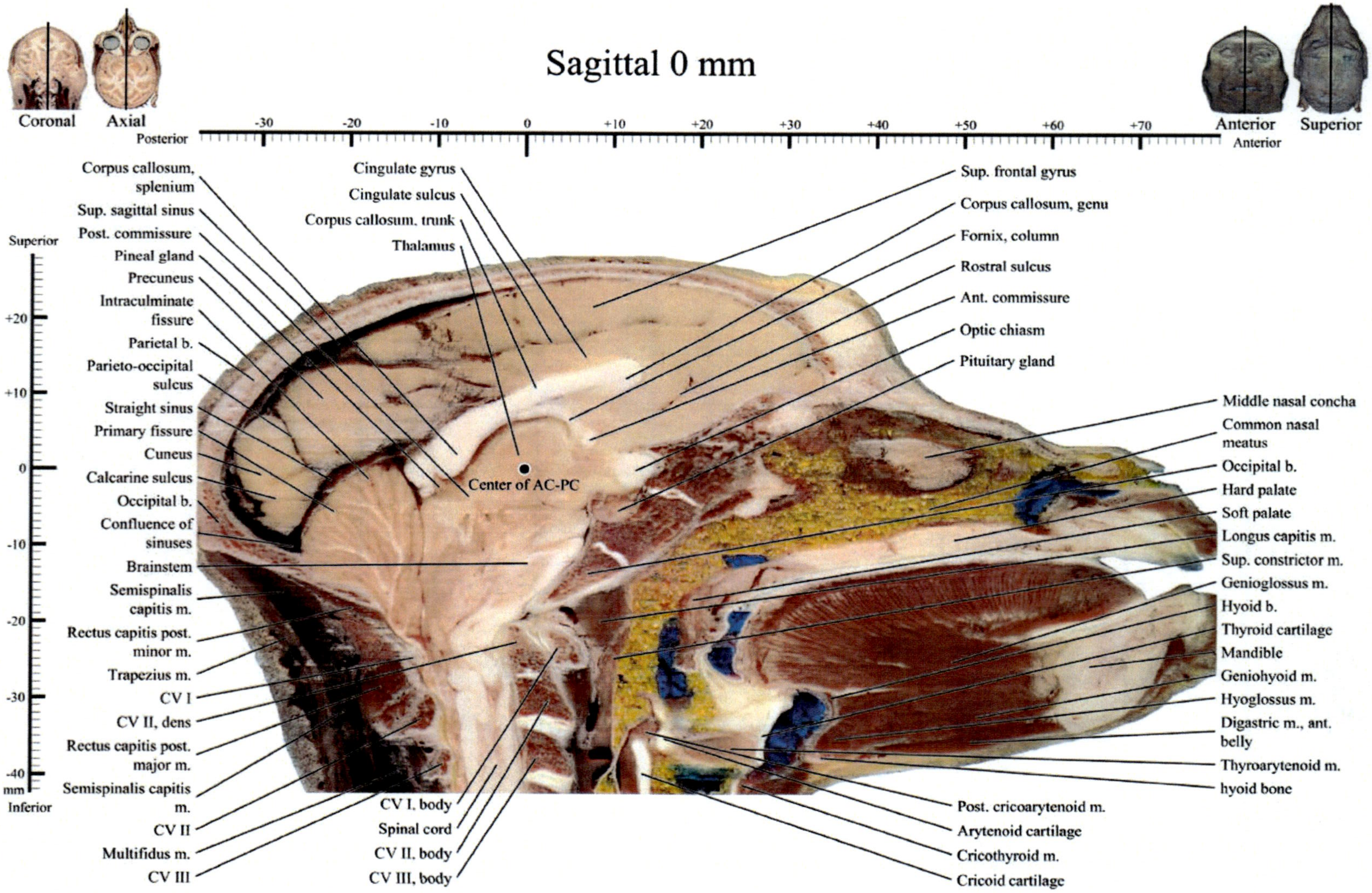

Sagittal -0.5 mm

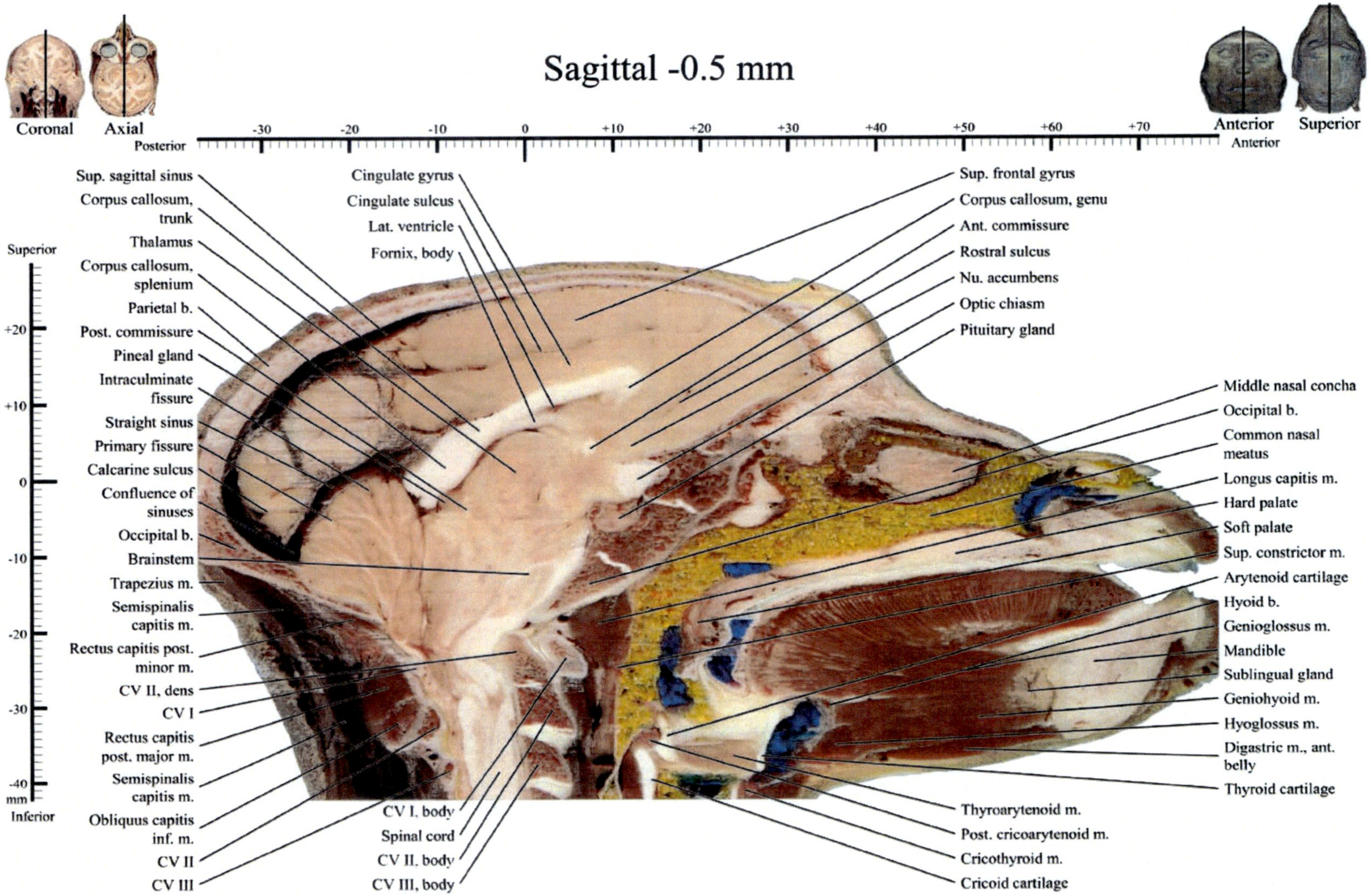

Sagittal -1 mm

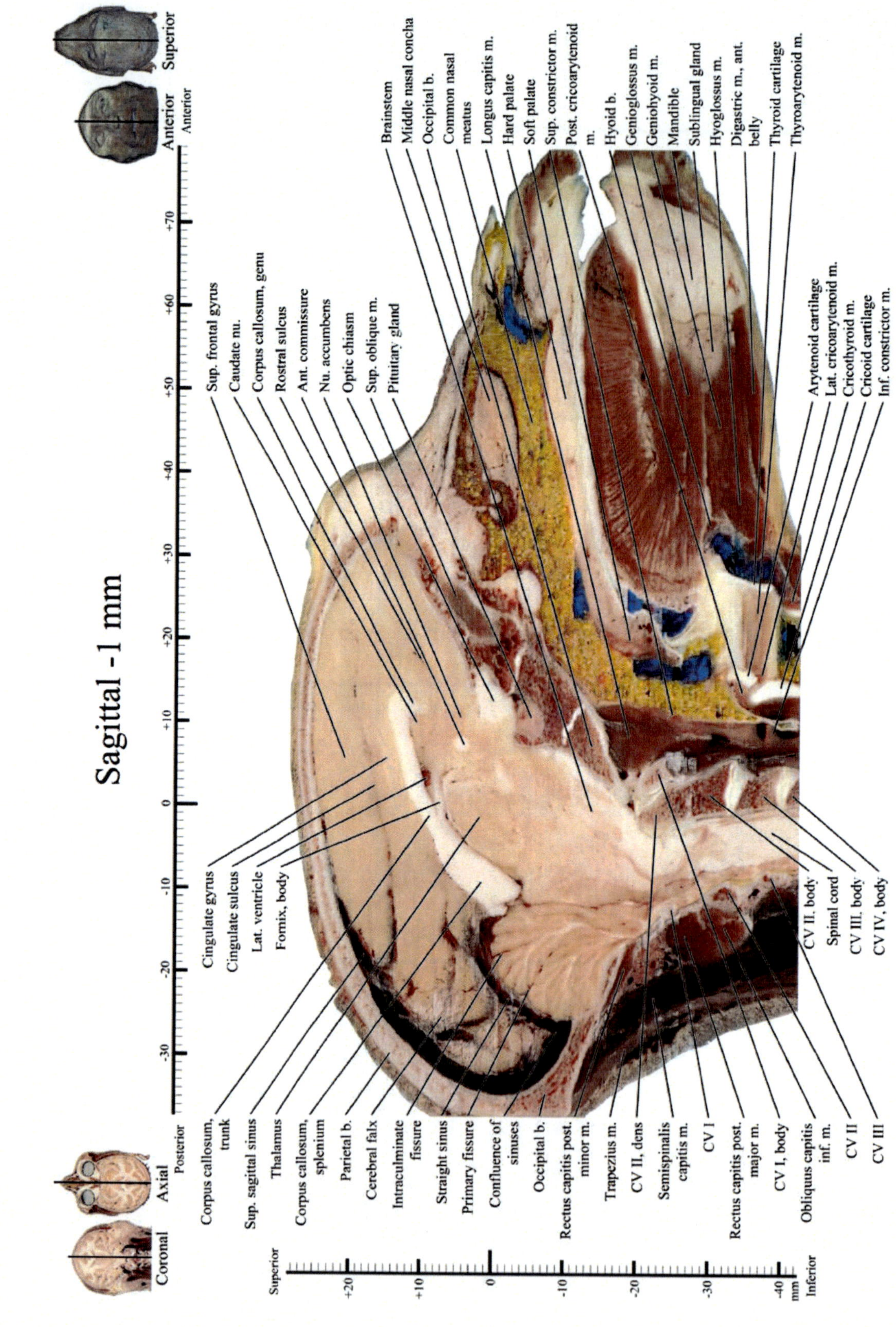

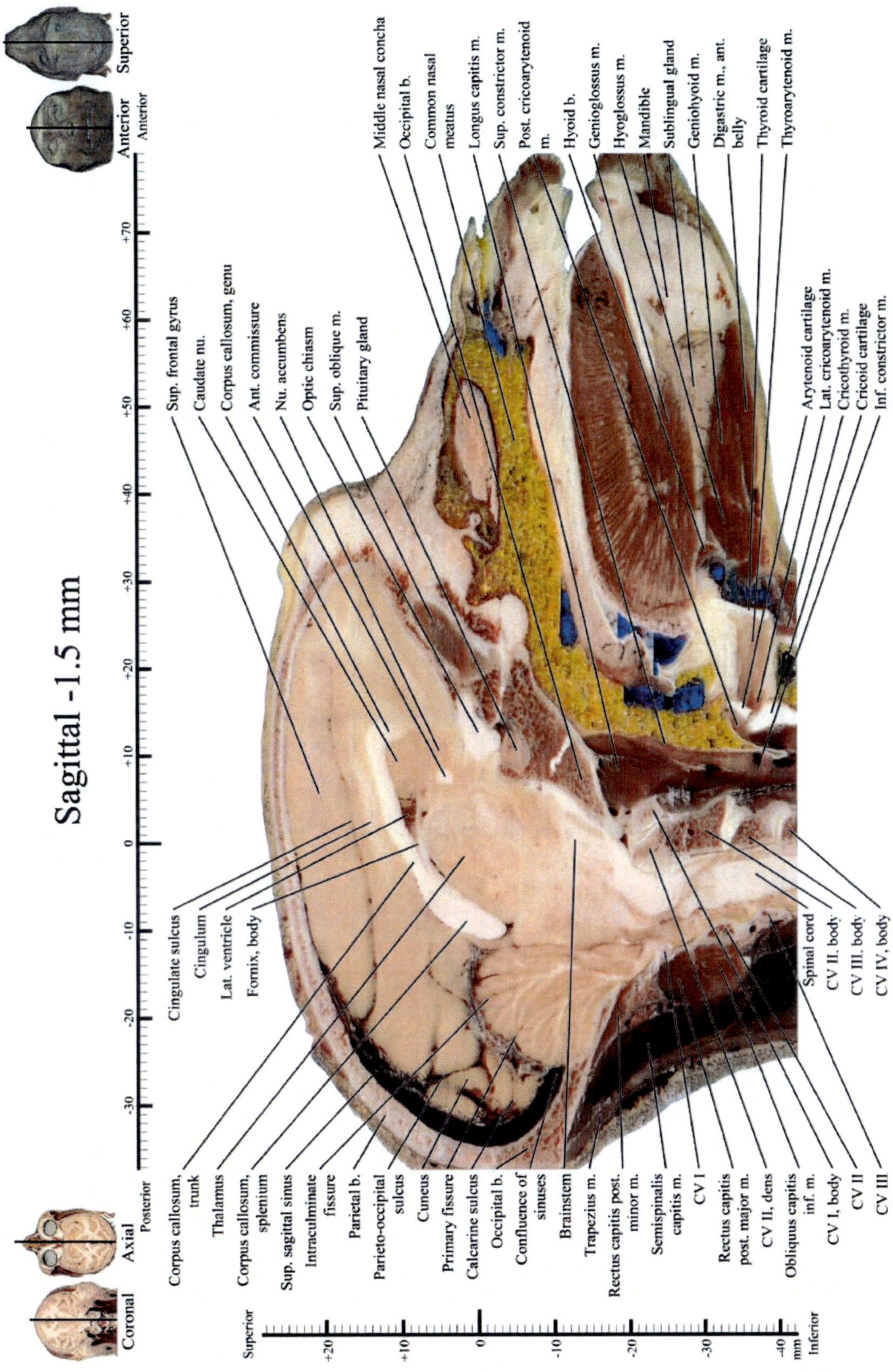

Sagittal -1.5 mm
Coronal
Axial
Anterior
Superior
Posterior
Anterior
Superior
Inferior
mm
+70
+60
+50
+40
+30
+20
+10
0
-10
-20
-30
-40
Sup. frontal gyrus
Caudate nu.
Corpus callosum, genu
Ant. commissure
Nu. accumbens
Optic chiasm
Sup. oblique m.
Pituitary gland
Middle nasal concha
Occipital b.
Common nasal meatus
Longus capitis m.
Sup. constrictor m.
Post. cricoarytenoid m.
Hyoid b.
Genioglossus m.
Hyoglossus m.
Mandible
Sublingual gland
Geniohyoid m.
Digastric m., ant. belly
Thyroid cartilage
Thyroarytenoid m.
Arytenoid cartilage
Lat. cricoarytenoid m.
Cricothyroid m.
Cricoid cartilage
Inf. constrictor m.
Cingulate sulcus
Cingulum
Lat. ventricle
Fornix, body
Spinal cord
CV II, body
CV III, body
CV IV, body
Corpus callosum, trunk
Thalamus
Corpus callosum, splenium
Sup. sagittal sinus
Intraculminate fissure
Parietal b.
Parieto-occipital sulcus
Cuneus
Primary fissure
Calcarine sulcus
Occipital b.
Confluence of sinuses
Brainstem
Trapezius m.
Rectus capitis post. minor m.
Semispinalis capitis m.
CV I
Rectus capitis post. major m.
CV II, dens
Obliquus capitis inf. m.
CV I, body
CV II
CV III

Sagittal -2 mm

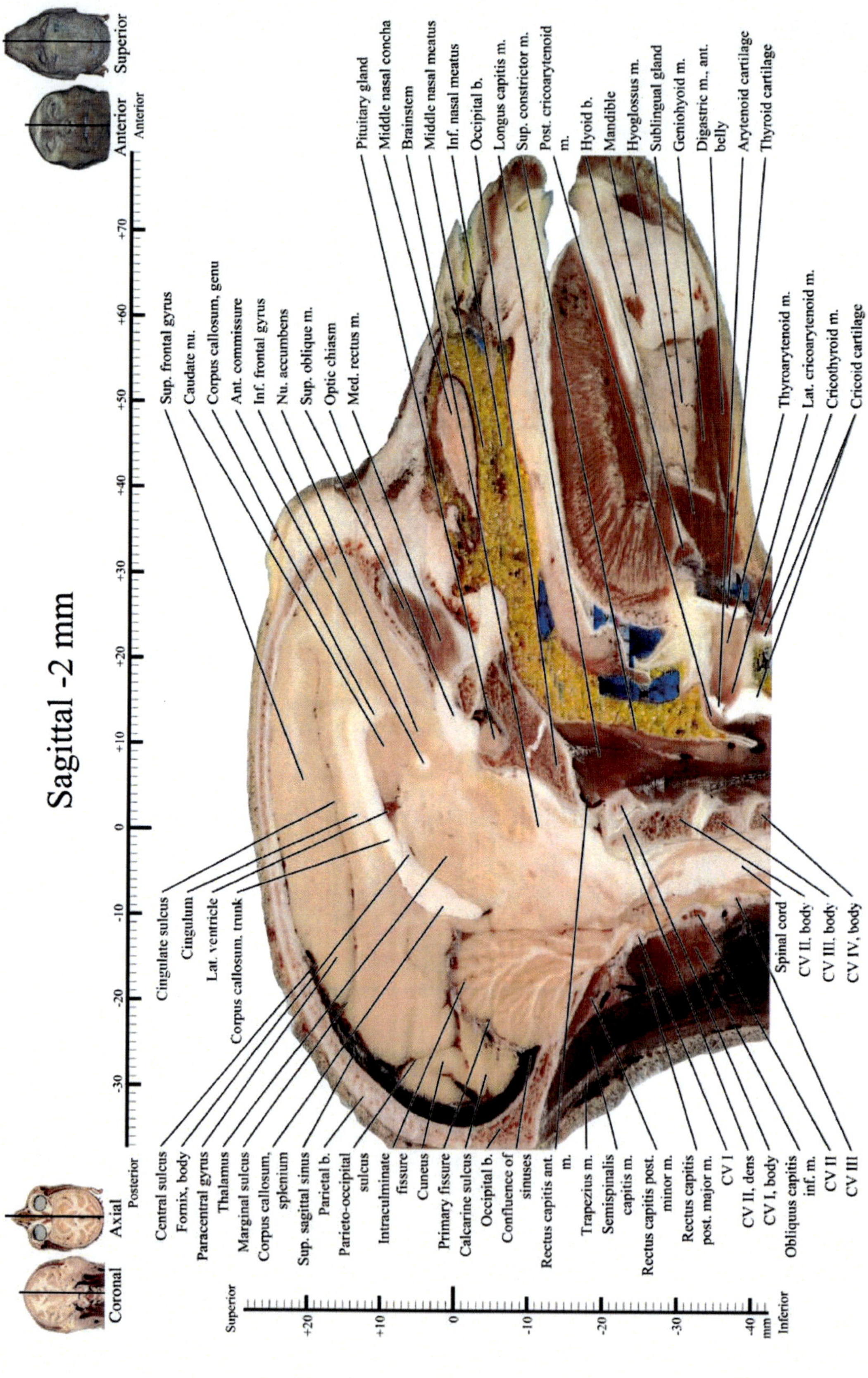

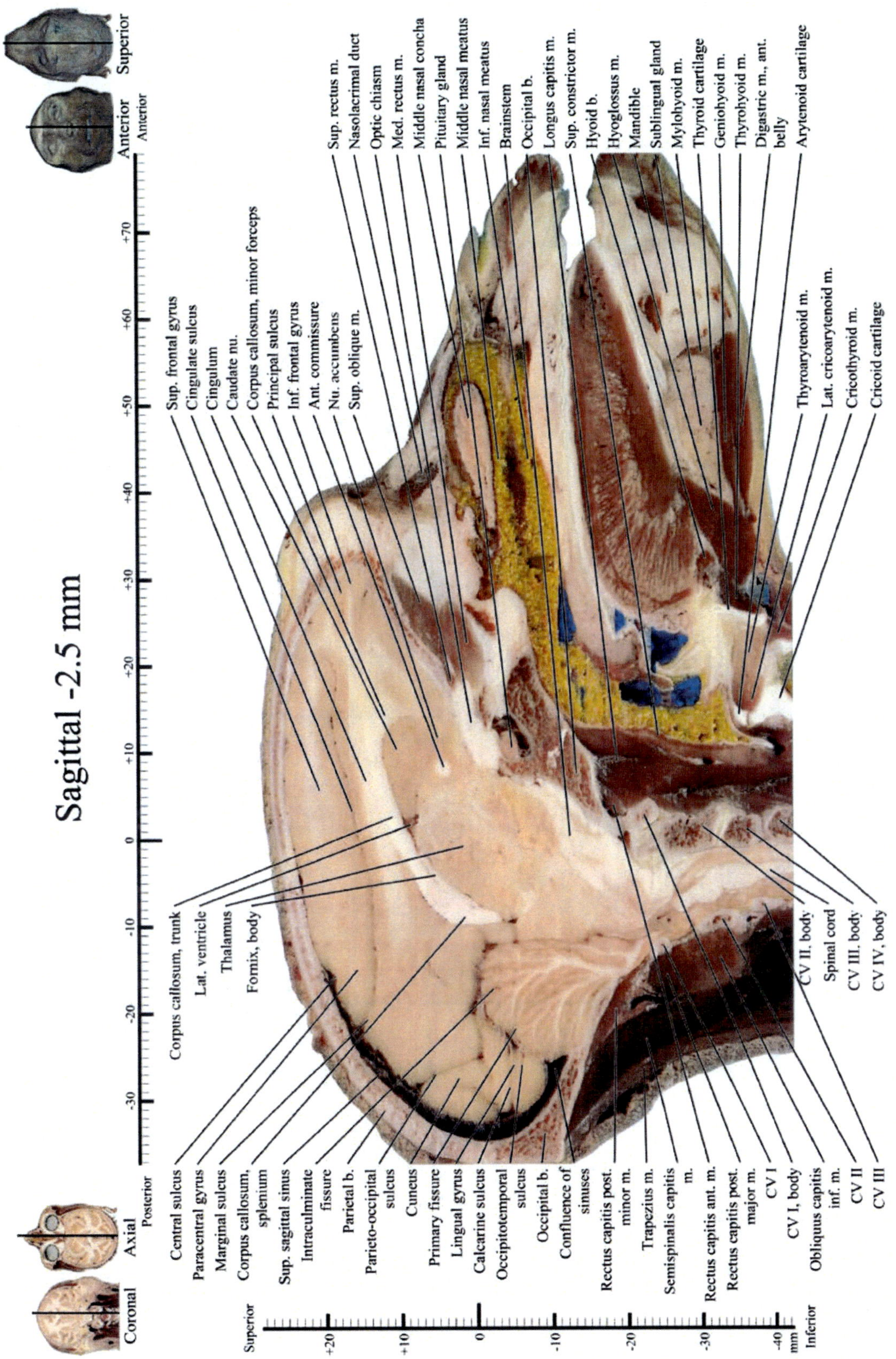
Sagittal -2.5 mm
Coronal
Axial
Posterior
Anterior
Superior
Inferior
mm
Sup. rectus m.
Nasolacrimal duct
Optic chiasm
Med. rectus m.
Middle nasal concha
Pituitary gland
Middle nasal meatus
Inf. nasal meatus
Brainstem
Occipital b.
Longus capitis m.
Sup. constrictor m.
Hyoid b.
Hyoglossus m.
Mandible
Sublingual gland
Mylohyoid m.
Thyroid cartilage
Geniohyoid m.
Thyrohyoid m.
Digastric m., ant. belly
Arytenoid cartilage
Sup. frontal gyrus
Cingulate sulcus
Cingulum
Caudate nu.
Corpus callosum, minor forceps
Principal sulcus
Inf. frontal gyrus
Ant. commissure
Nu. accumbens
Sup. oblique m.
Thyroarytenoid m.
Lat. cricoarytenoid m.
Cricothyroid m.
Cricoid cartilage
Corpus callosum, trunk
Lat. ventricle
Thalamus
Fornix, body
CV II, body
Spinal cord
CV III, body
CV IV, body
Central sulcus
Paracentral gyrus
Marginal sulcus
Corpus callosum, splenium
Sup. sagittal sinus
Intraculminate fissure
Parietal b.
Parieto-occipital sulcus
Cuneus
Primary fissure
Lingual gyrus
Calcarine sulcus
Occipitotemporal sulcus
Occipital b.
Confluence of sinuses
Rectus capitis post. minor m.
Trapezius m.
Semispinalis capitis m.
Rectus capitis ant. m.
Rectus capitis post. major m.
CV I
CV I, body
Obliquus capitis inf. m.
CV II
CV III

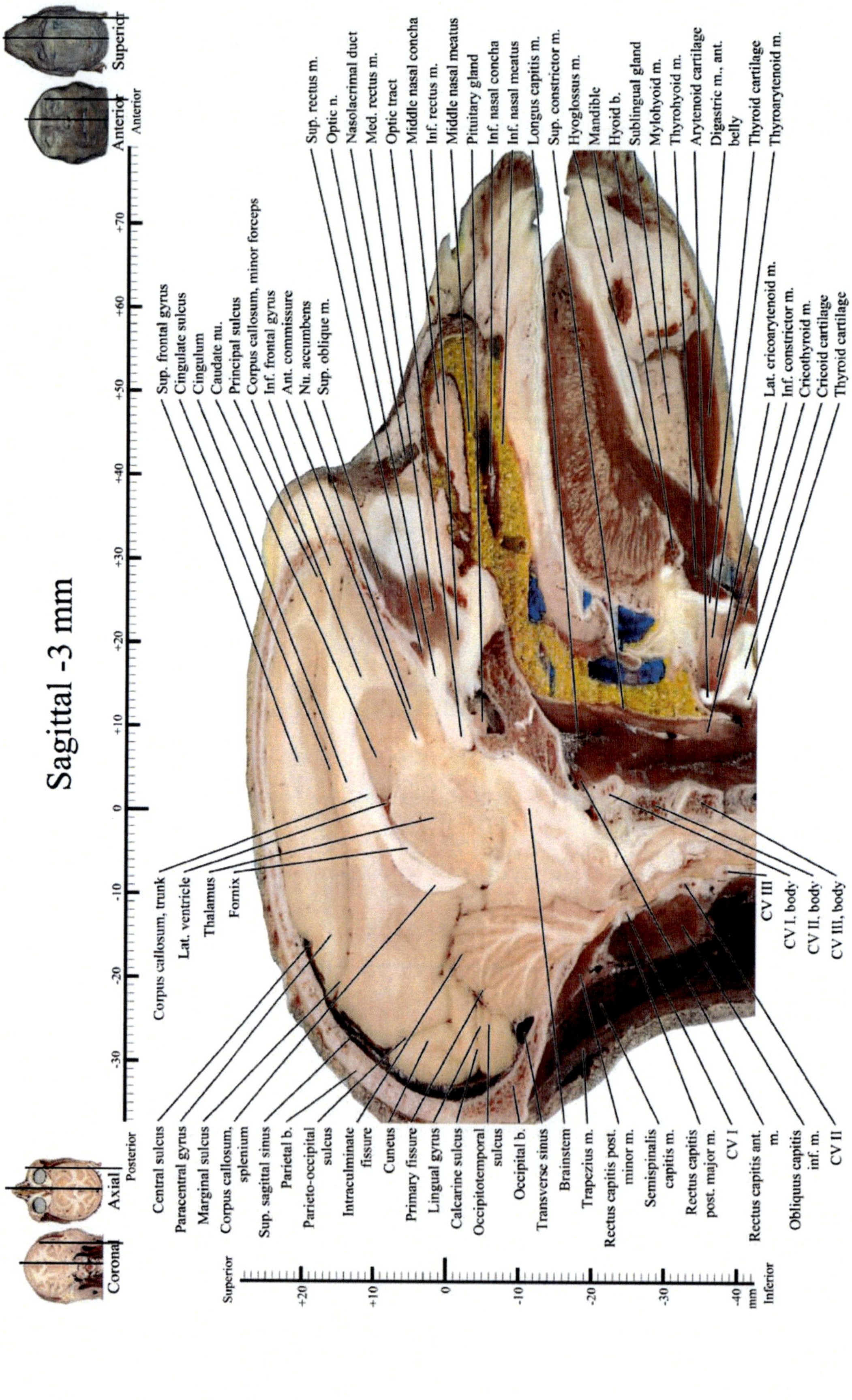

Sagittal -3 mm
Coronal
Axial
Anterior
Superior
Posterior
Inferior
mm
+20
+10
0
-10
-20
-30
-40
+70
+60
+50
+40
+30
Sup. frontal gyrus
Cingulate sulcus
Cingulum
Caudate nu.
Principal sulcus
Corpus callosum, minor forceps
Inf. frontal gyrus
Ant. commissure
Nu. accumbens
Sup. oblique m.
Sup. rectus m.
Optic n.
Nasolacrimal duct
Med. rectus m.
Optic tract
Middle nasal concha
Inf. rectus m.
Middle nasal meatus
Pituitary gland
Inf. nasal concha
Inf. nasal meatus
Longus capitis m.
Sup. constrictor m.
Hyoglossus m.
Mandible
Hyoid b.
Sublingual gland
Mylohyoid m.
Thyrohyoid m.
Arytenoid cartilage
Digastric m., ant. belly
Thyroid cartilage
Thyroarytenoid m.
Lat. cricoarytenoid m.
Inf. constrictor m.
Criothyroid m.
Cricoid cartilage
Thyroid cartilage
Corpus callosum, trunk
Lat. ventricle
Thalamus
Fornix
CV III
CV I, body
CV II, body
CV III, body
Central sulcus
Paracentral gyrus
Marginal sulcus
Corpus callosum, splenium
Sup. sagittal sinus
Parietal b.
Parieto-occipital sulcus
Intraculminate fissure
Cuneus
Primary fissure
Lingual gyrus
Calcarine sulcus
Occipitotemporal sulcus
Occipital b.
Transverse sinus
Brainstem
Trapezius m.
Rectus capitis post. minor m.
Semispinalis capitis m.
Rectus capitis post. major m.
CV I
Rectus capitis ant. m.
Obliquus capitis inf. m.
CV II

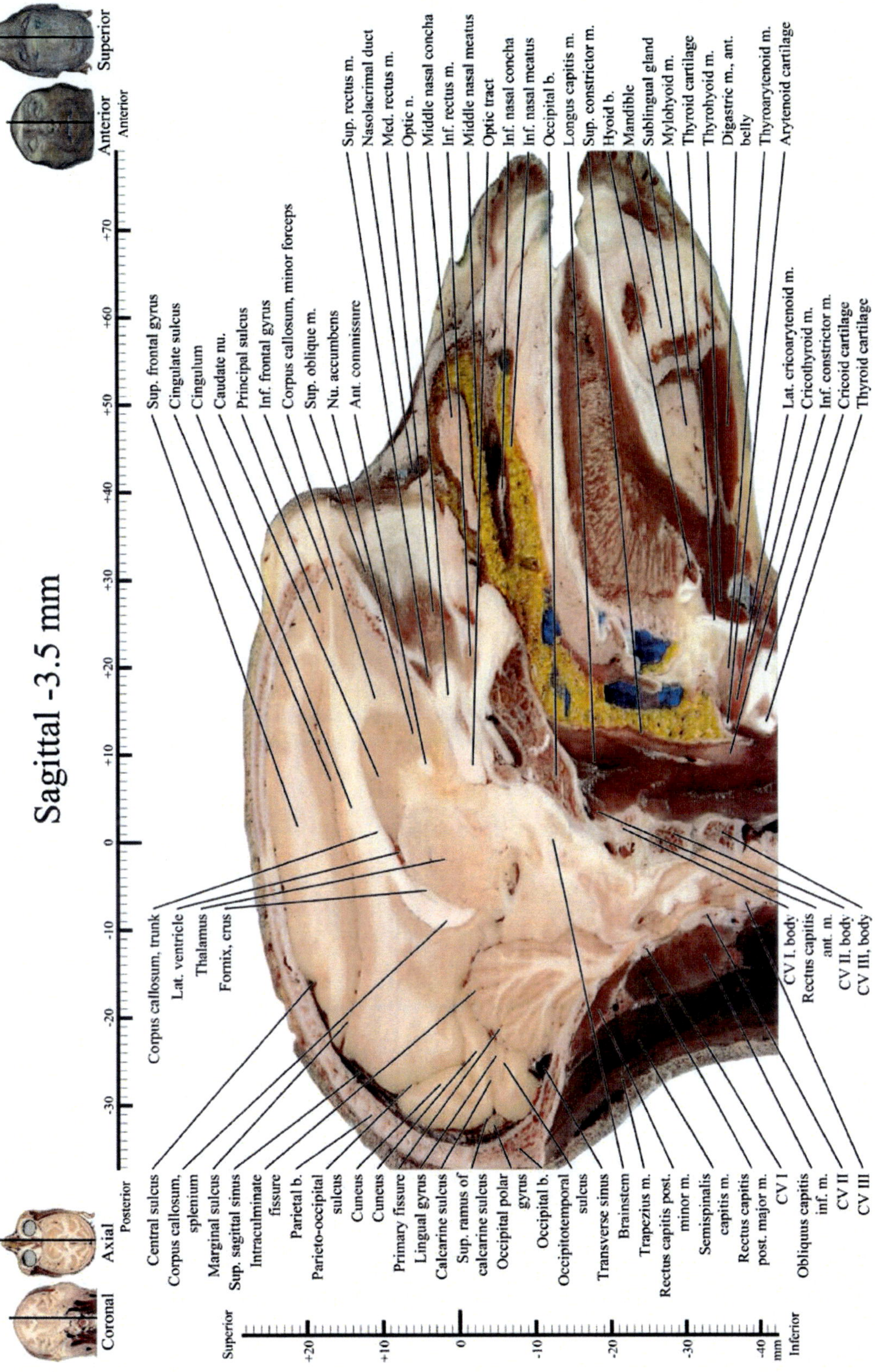
Sagittal -3.5 mm
Coronal
Axial
Anterior
Superior
Posterior
Anterior
Superior
Inferior
mm
+20
+10
0
-10
-20
-30
-40
+70
+60
+50
+40
+30
Central sulcus
Corpus callosum, splenium
Marginal sulcus
Sup. sagittal sinus
Intraculminate fissure
Parietal b.
Parieto-occipital sulcus
Cuneus
Cuneus
Primary fissure
Lingual gyrus
Calcarine sulcus
Sup. ramus of calcarine sulcus
Occipital polar gyrus
Occipital b.
Occipitotemporal sulcus
Transverse sinus
Brainstem
Trapezius m.
Rectus capitis post. minor m.
Semispinalis capitis m.
Rectus capitis post. major m.
CV I
Obliquus capitis inf. m.
CV II
CV III
Corpus callosum, trunk
Lat. ventricle
Thalamus
Fornix, crus
CV I. body
Rectus capitis ant. m.
CV II. body
CV III, body
Sup. frontal gyrus
Cingulate sulcus
Cingulum
Caudate nu.
Principal sulcus
Inf. frontal gyrus
Corpus callosum, minor forceps
Sup. oblique m.
Nu. accumbens
Ant. commissure
Lat. cricoarytenoid m.
Cricothyroid m.
Inf. constrictor m.
Cricoid cartilage
Thyroid cartilage
Sup. rectus m.
Nasolacrimal duct
Med. rectus m.
Optic n.
Middle nasal concha
Inf. rectus m.
Middle nasal meatus
Optic tract
Inf. nasal concha
Inf. nasal meatus
Occipital b.
Longus capitis m.
Sup. constrictor m.
Hyoid b.
Mandible
Sublingual gland
Mylohyoid m.
Thyroid cartilage
Thyrohyoid m.
Digastric m., ant. belly
Thyroarytenoid m.
Arytenoid cartilage

Sagittal -4 mm

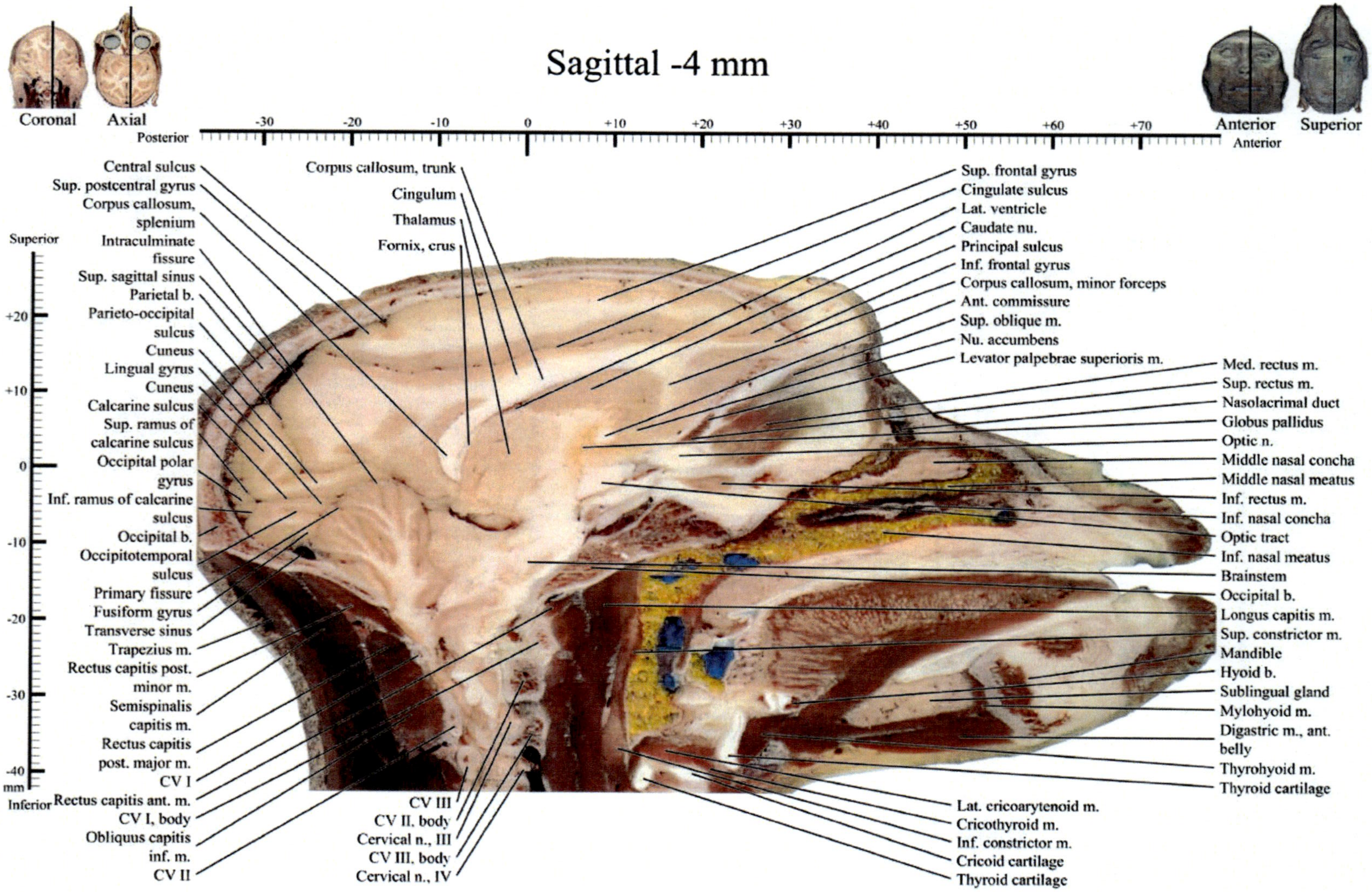

Sagittal -4.5 mm

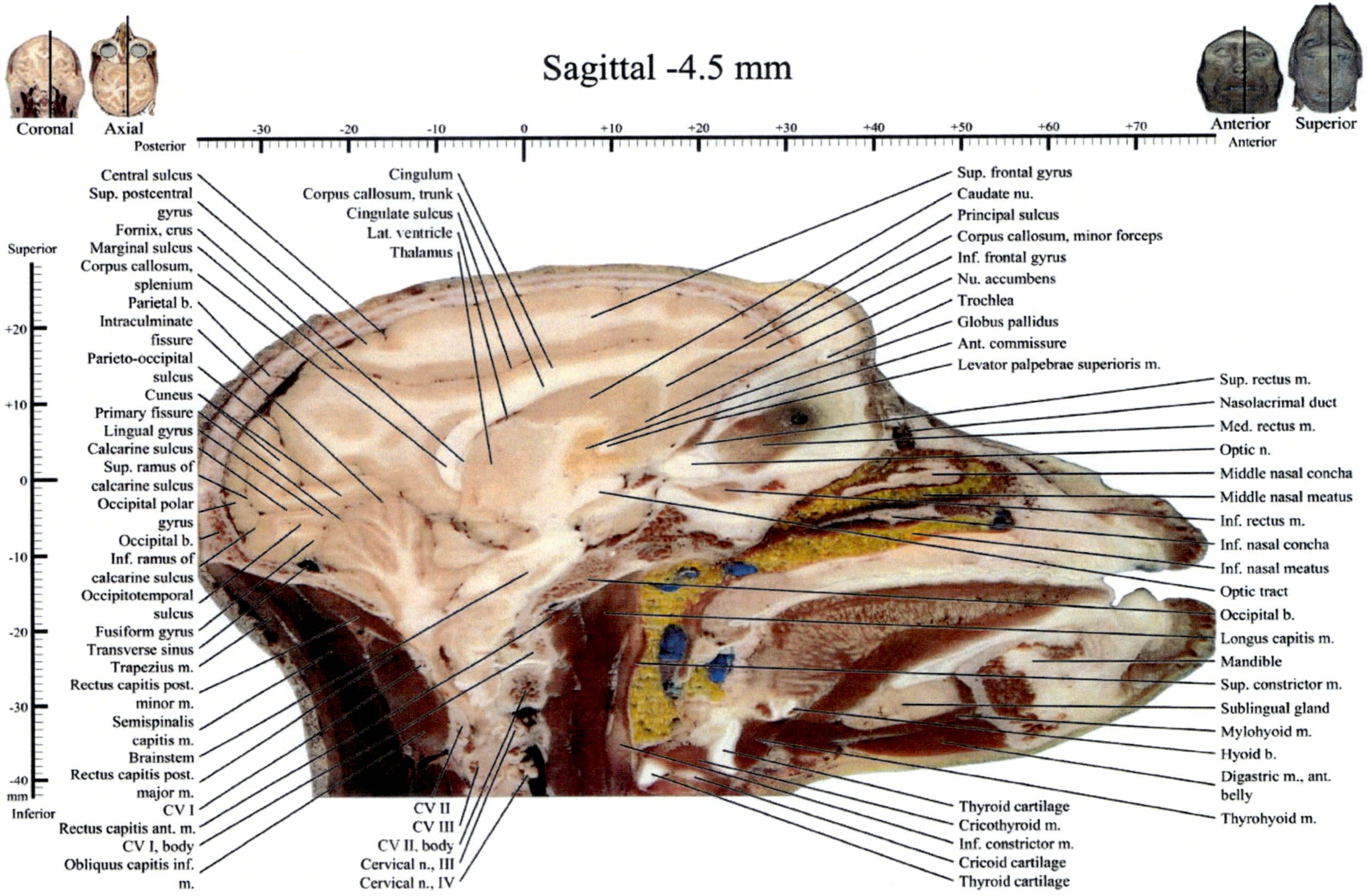

Sagittal -5 mm

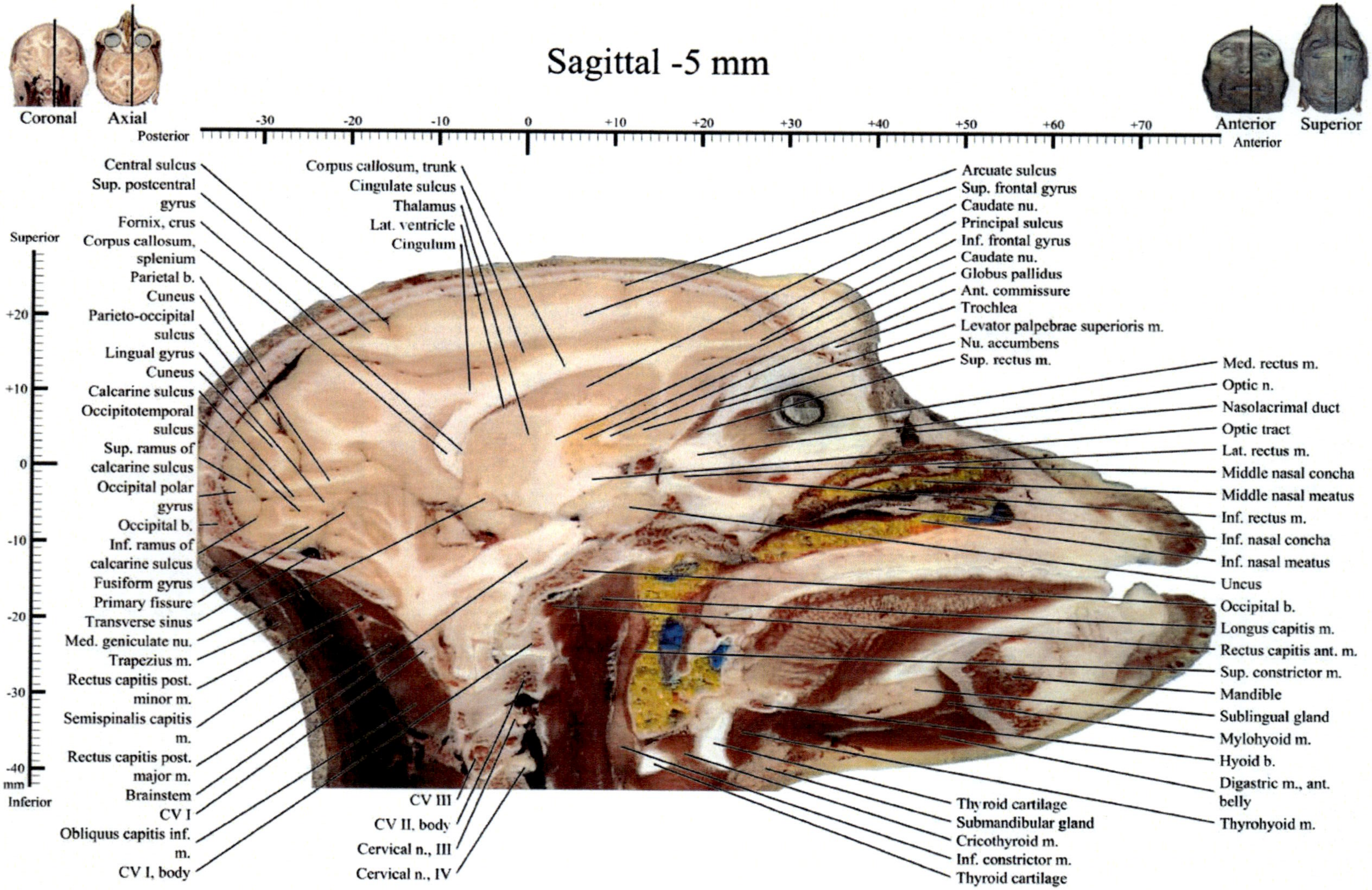

Sagittal -5.5 mm

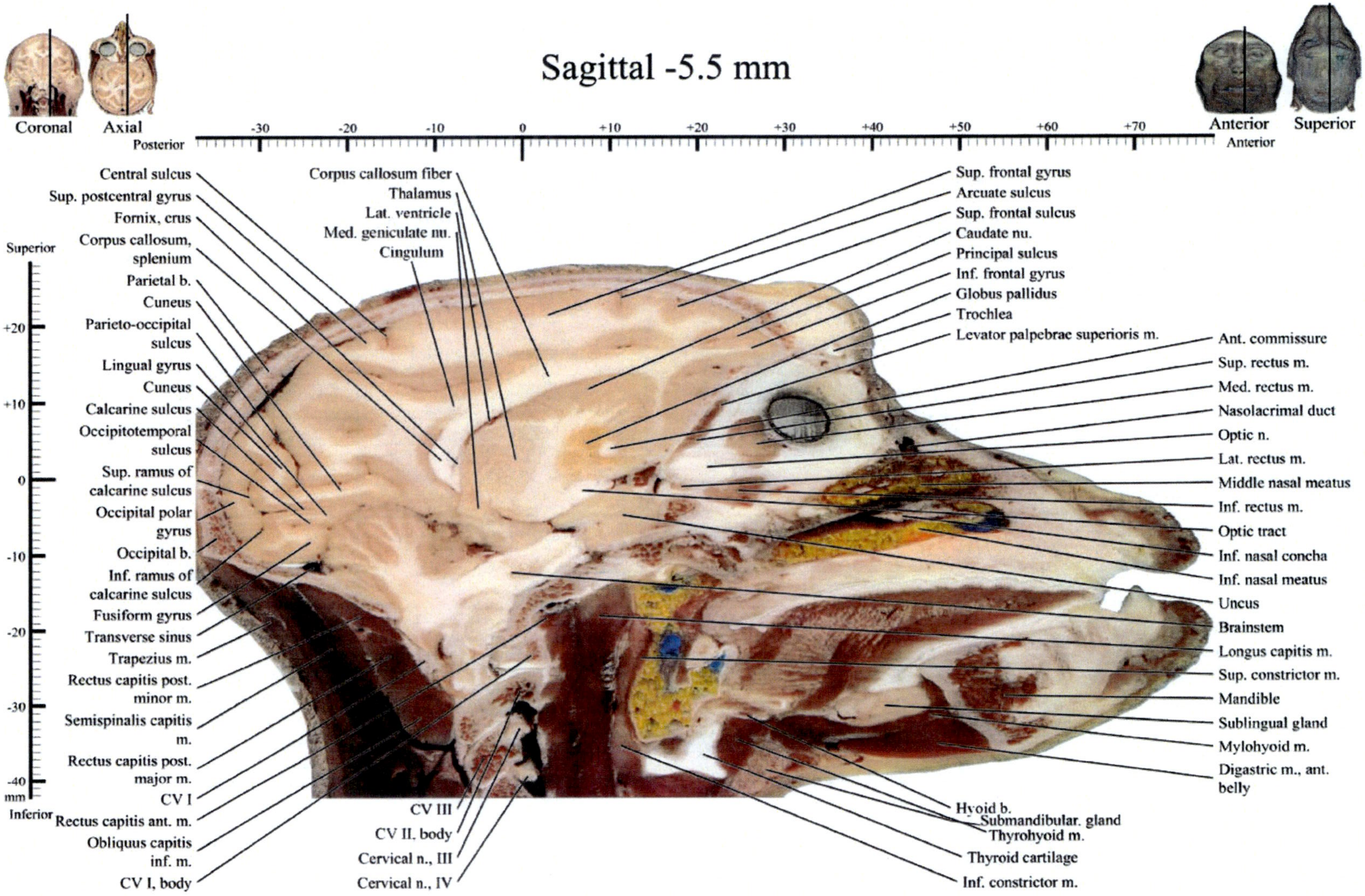

Sagittal -6 mm

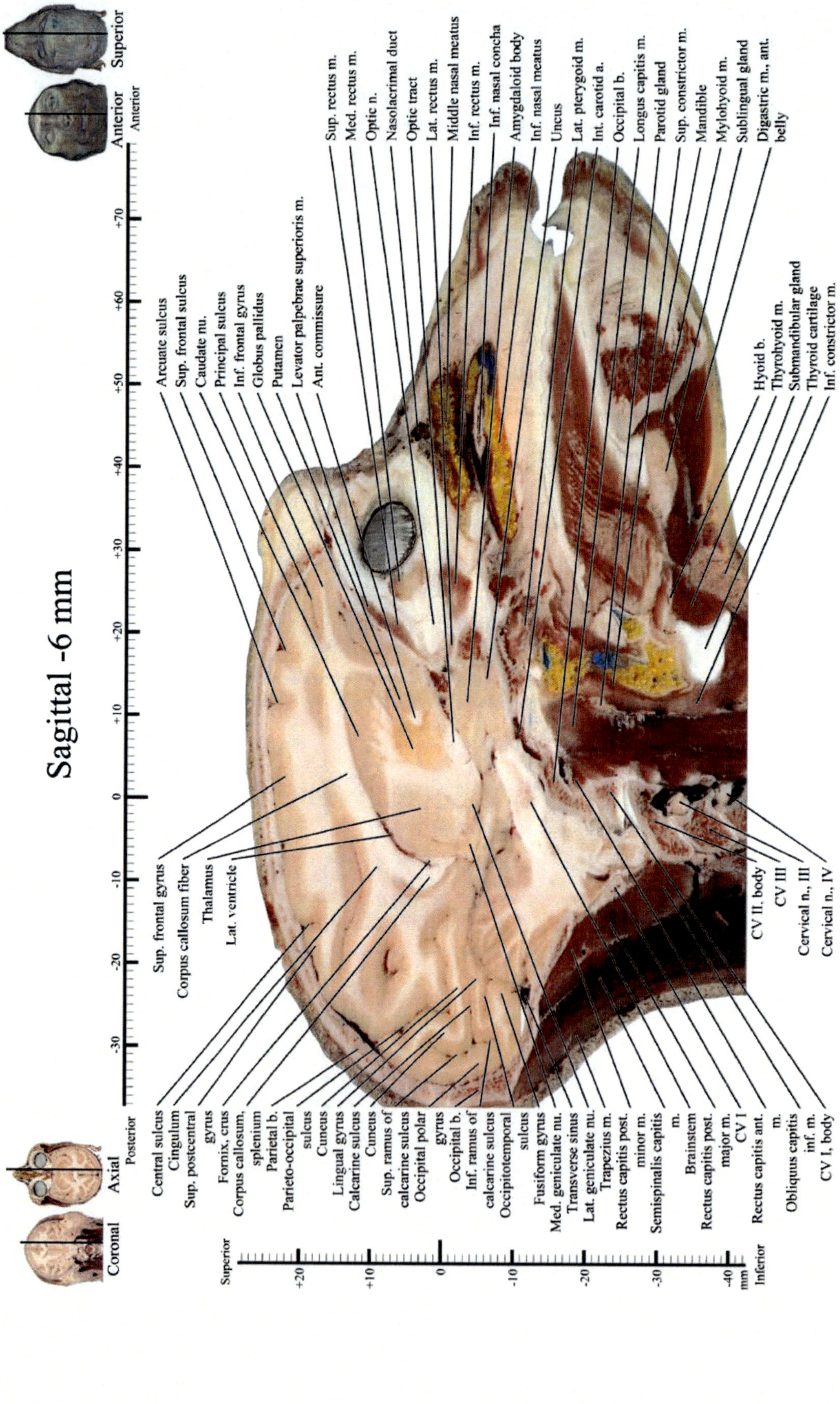

Sagittal -6.5 mm

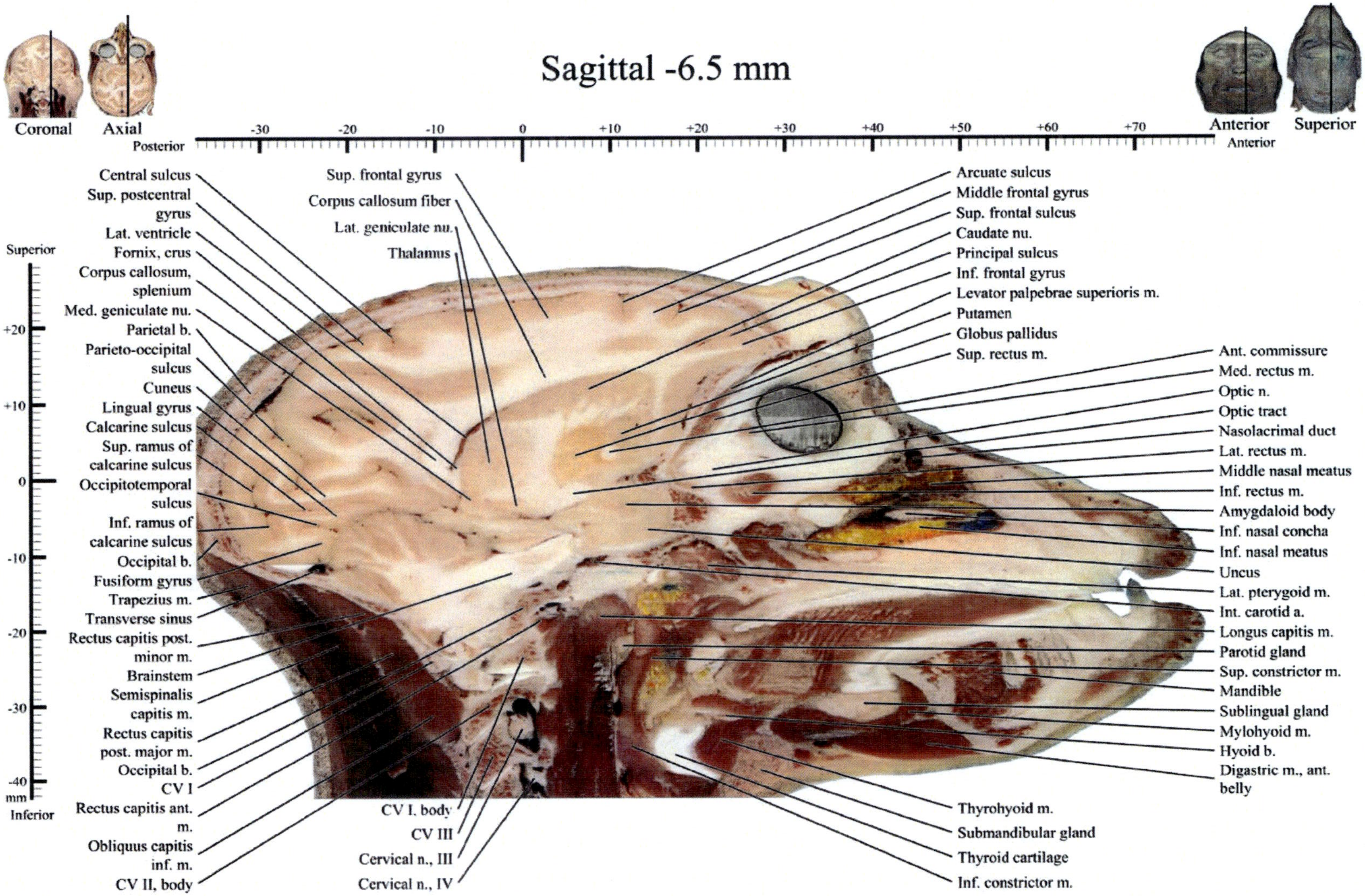

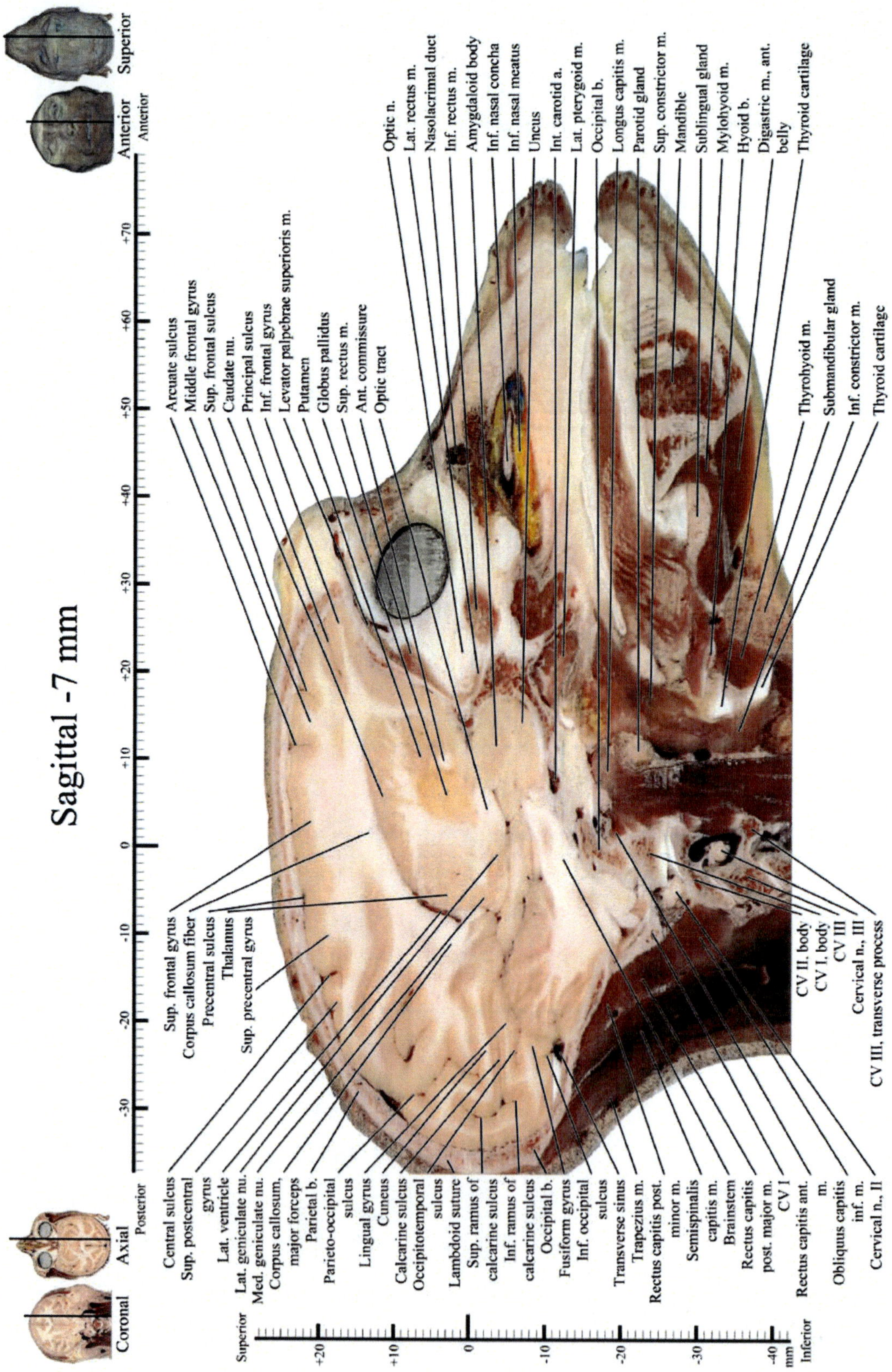
Sagittal -7 mm
Coronal
Axial
Anterior
Superior
Posterior
Anterior
Superior
Inferior
mm
Optic n.
Lat. rectus m.
Nasolacrimal duct
Inf. rectus m.
Amygdaloid body
Inf. nasal concha
Inf. nasal meatus
Uncus
Int. carotid a.
Lat. pterygoid m.
Occipital b.
Longus capitis m.
Parotid gland
Sup. constrictor m.
Mandible
Sublingual gland
Mylohyoid m.
Hyoid b.
Digastric m., ant. belly
Thyroid cartilage
Arcuate sulcus
Middle frontal gyrus
Sup. frontal sulcus
Caudate nu.
Principal sulcus
Inf. frontal gyrus
Levator palpebrae superioris m.
Putamen
Globus pallidus
Sup. rectus m.
Ant. commissure
Optic tract
Thyrohyoid m.
Submandibular gland
Inf. constrictor m.
Thyroid cartilage
Sup. frontal gyrus
Corpus callosum fiber
Precentral sulcus
Thalamus
Sup. precentral gyrus
CV II. body
CV I. body
CV III
Cervical n., III
CV III, transverse process
Central sulcus
Sup. postcentral gyrus
Lat. ventricle
Lat. geniculate nu.
Med. geniculate nu.
Corpus callosum, major forceps
Parietal b.
Parieto-occipital sulcus
Lingual gyrus
Cuneus
Calcarine sulcus
Occipitotemporal sulcus
Lambdoid suture
Sup. ramus of calcarine sulcus
Inf. ramus of calcarine sulcus
Occipital b.
Fusiform gyrus
Inf. occipital sulcus
Transverse sinus
Trapezius m.
Rectus capitis post. minor m.
Semispinalis capitis m.
Brainstem
Rectus capitis post. major m.
CV I
Rectus capitis ant. m.
Obliquus capitis inf. m.
Cervical n., II

Sagittal -7.5 mm

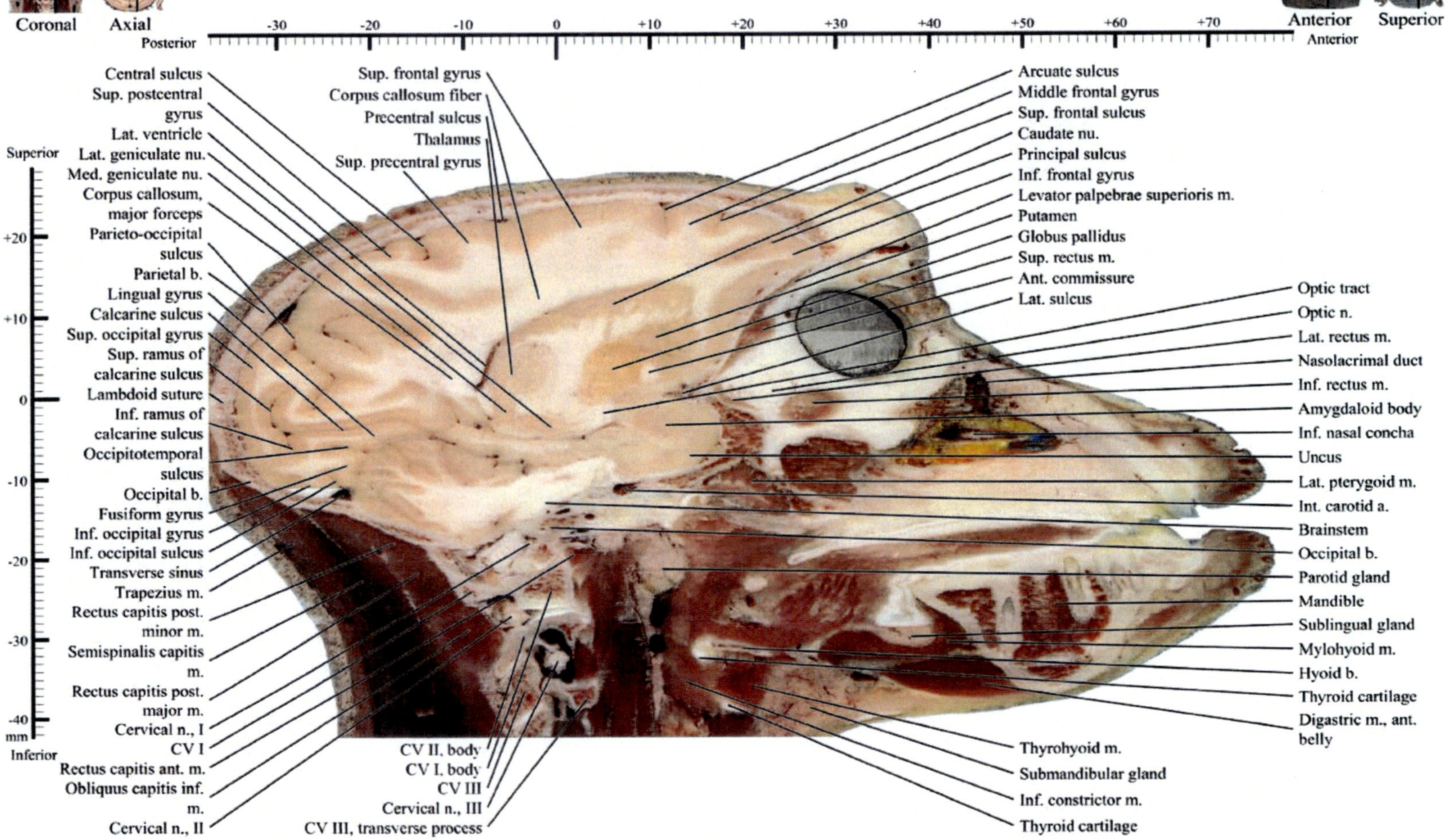

Sagittal -8 mm

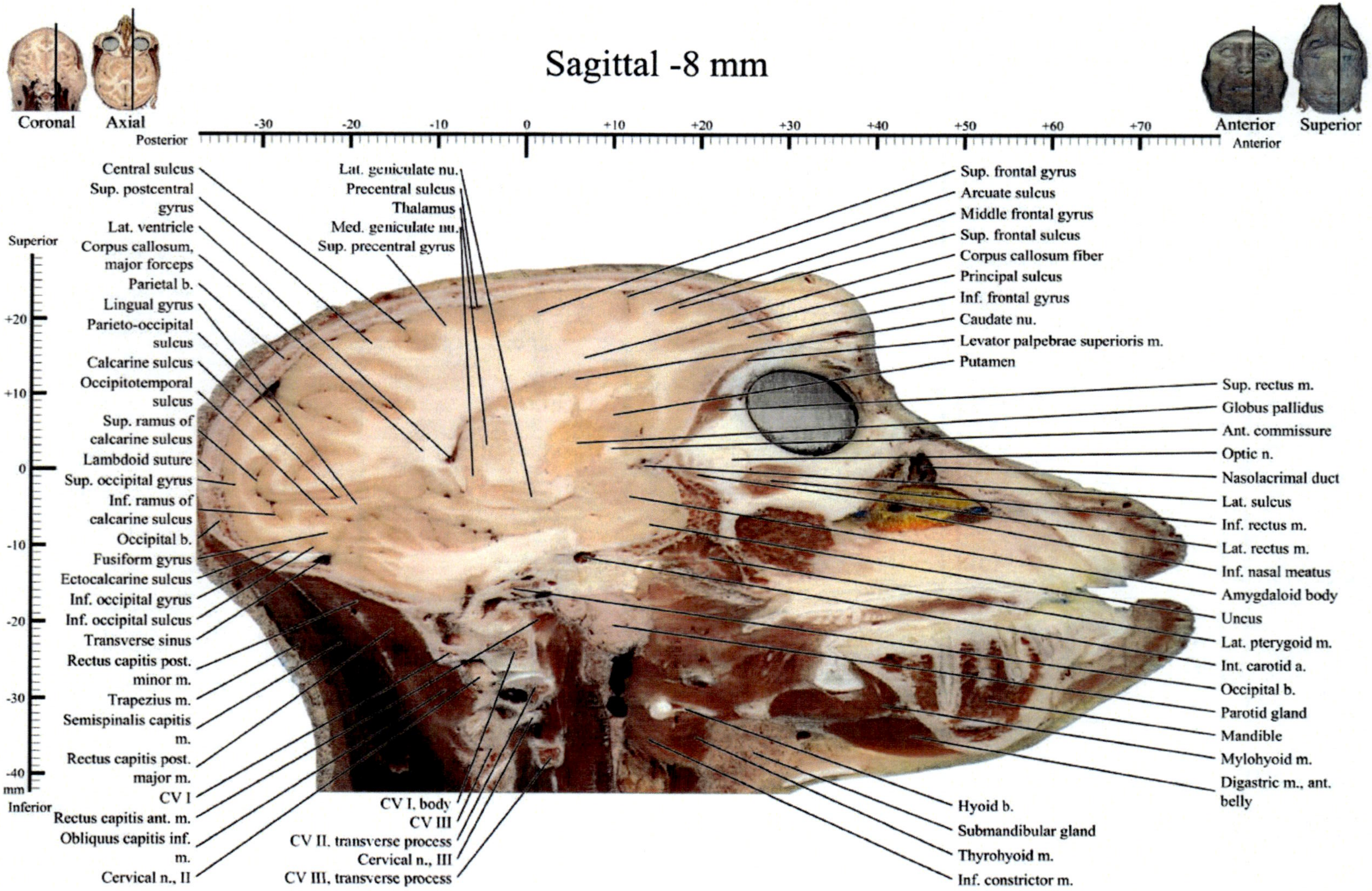

Sagittal -8.5 mm

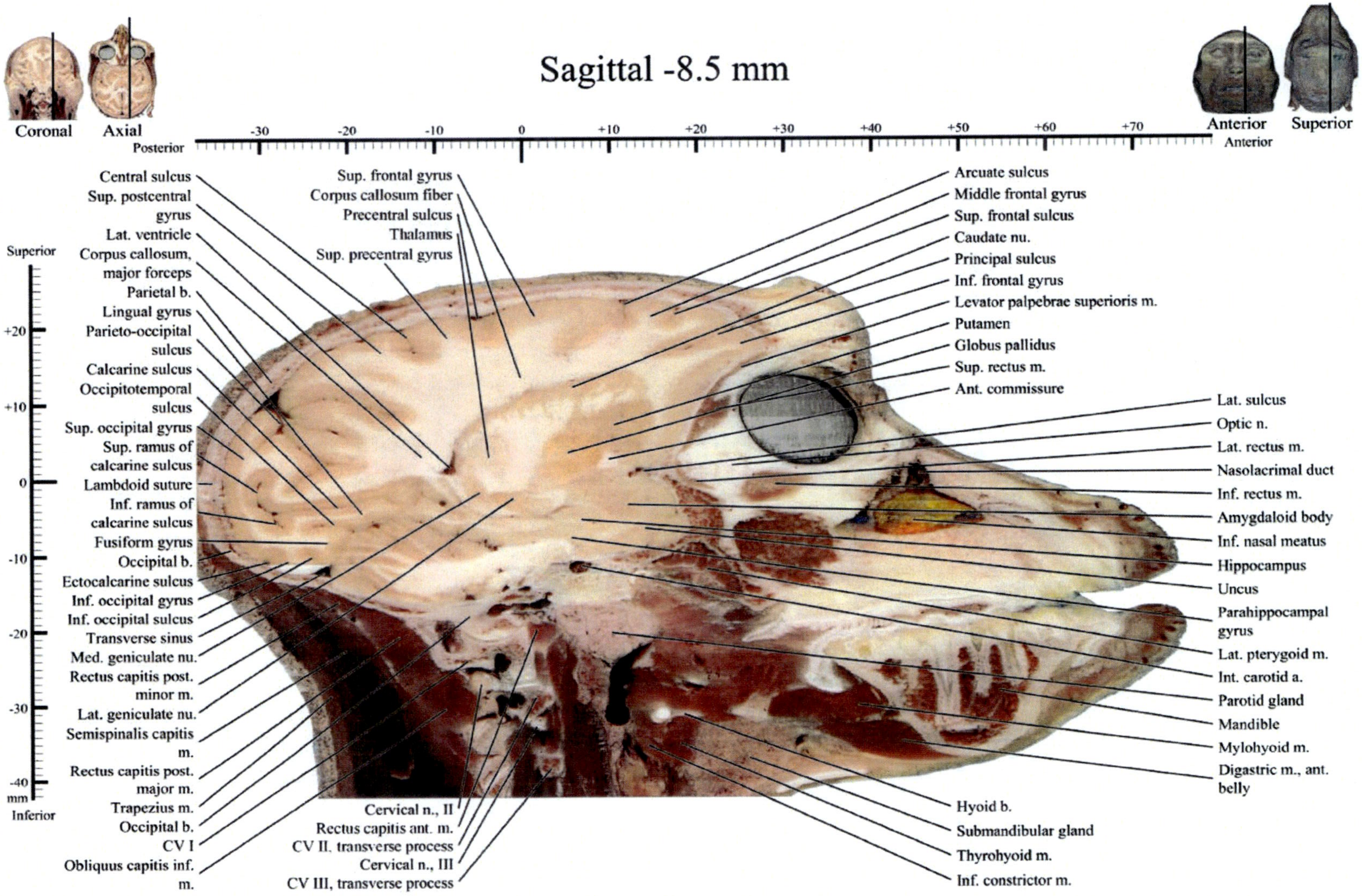

Sagittal -9 mm

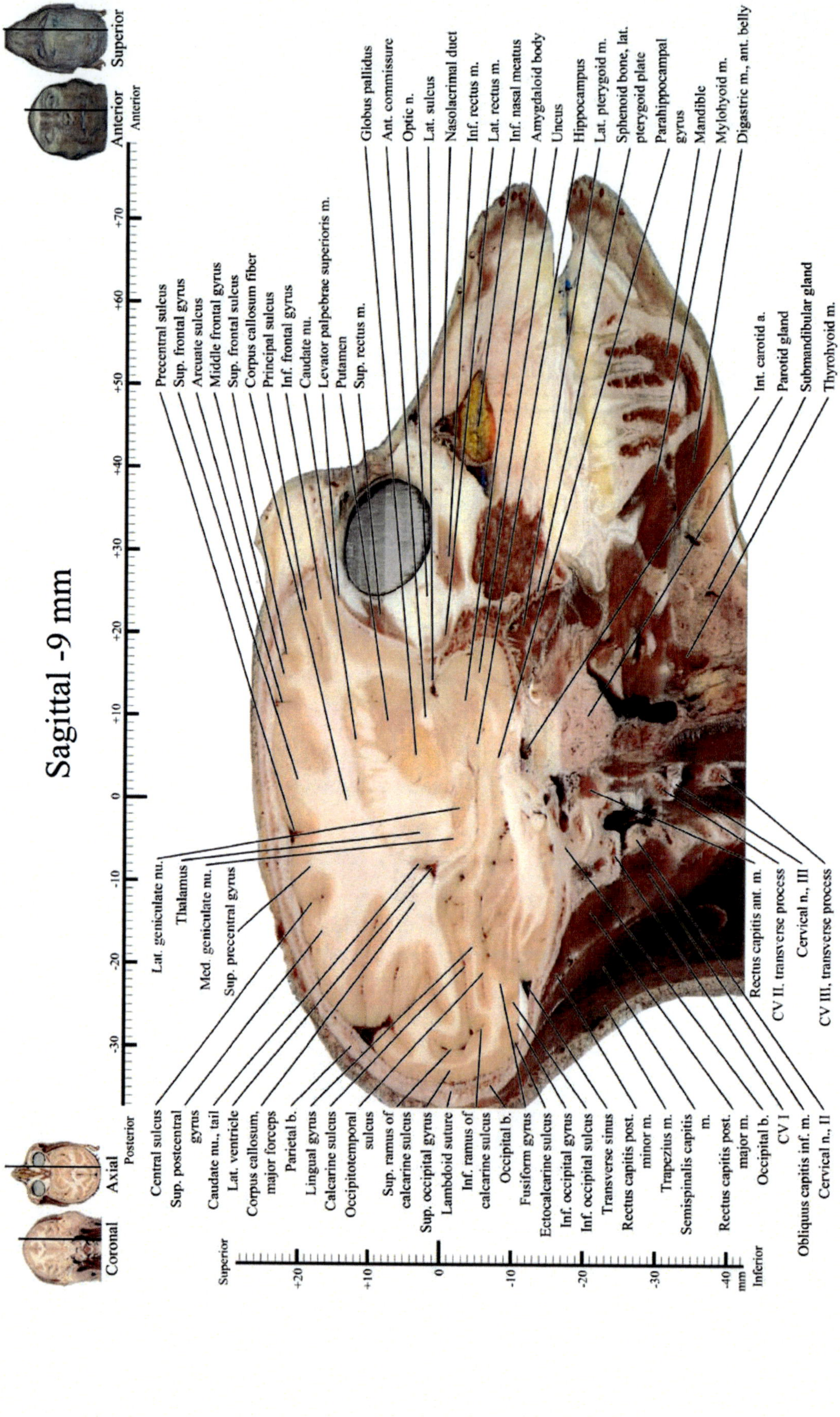

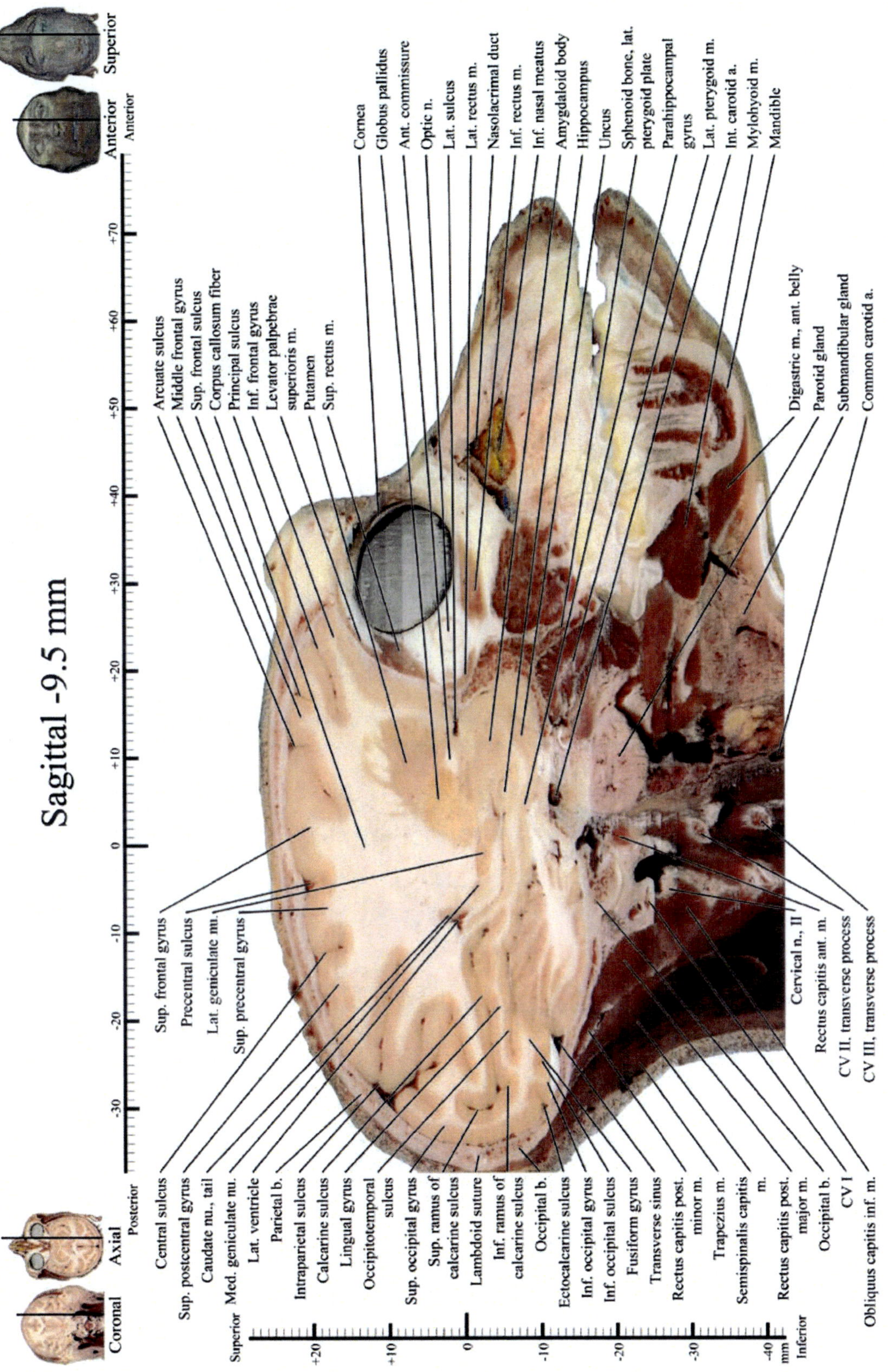

Sagittal -9.5 mm
Coronal
Axial
Anterior
Superior
Posterior
Anterior
Superior
Inferior
mm
Cornea
Globus pallidus
Ant. commissure
Optic n.
Lat. sulcus
Lat. rectus m.
Nasolacrimal duct
Inf. rectus m.
Inf. nasal meatus
Amygdaloid body
Hippocampus
Uncus
Sphenoid bone, lat. pterygoid plate
Parahippocampal gyrus
Lat. pterygoid m.
Int. carotid a.
Mylohyoid m.
Mandible
Arcuate sulcus
Middle frontal gyrus
Sup. frontal sulcus
Corpus callosum fiber
Principal sulcus
Inf. frontal gyrus
Levator palpebrae superioris m.
Putamen
Sup. rectus m.
Digastric m., ant. belly
Parotid gland
Submandibular gland
Common carotid a.
Sup. frontal gyrus
Precentral sulcus
Lat. geniculate nu.
Sup. precentral gyrus
Cervical n., II
Rectus capitis ant. m.
CV II, transverse process
CV III, transverse process
Central sulcus
Sup. postcentral gyrus
Caudate nu., tail
Med. geniculate nu.
Lat. ventricle
Parietal b.
Intraparietal sulcus
Calcarine sulcus
Lingual gyrus
Occipitotemporal sulcus
Sup. occipital gyrus
Sup. ramus of calcarine sulcus
Lambdoid suture
Inf. ramus of calcarine sulcus
Occipital b.
Ectocalcarine sulcus
Inf. occipital gyrus
Inf. occipital sulcus
Fusiform gyrus
Transverse sinus
Rectus capitis post. minor m.
Trapezius m.
Semispinalis capitis m.
Rectus capitis post. major m.
Occipital b.
CV I
Obliquus capitis inf. m.

Sagittal -10 mm

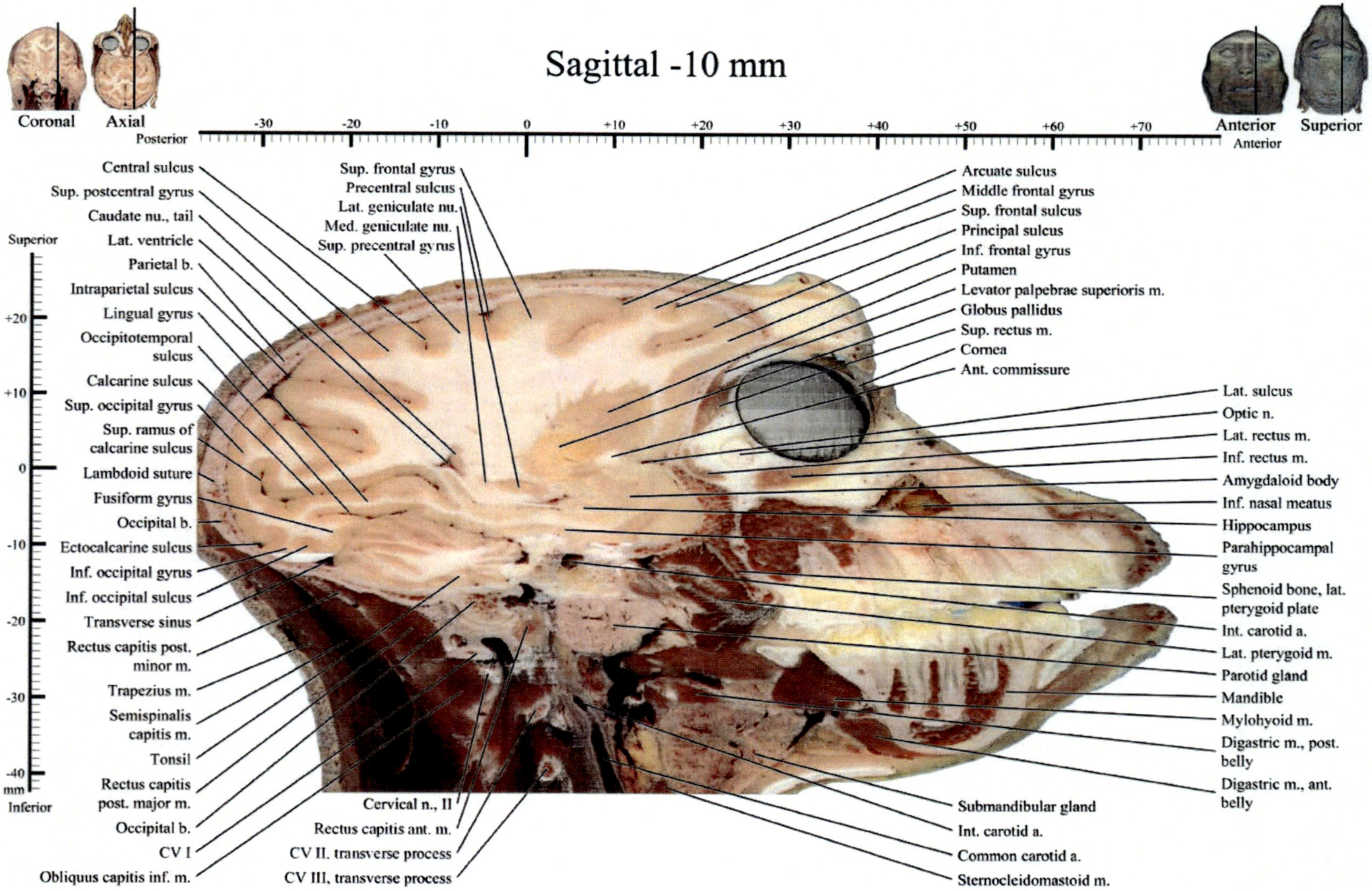

Sagittal -10.5 mm

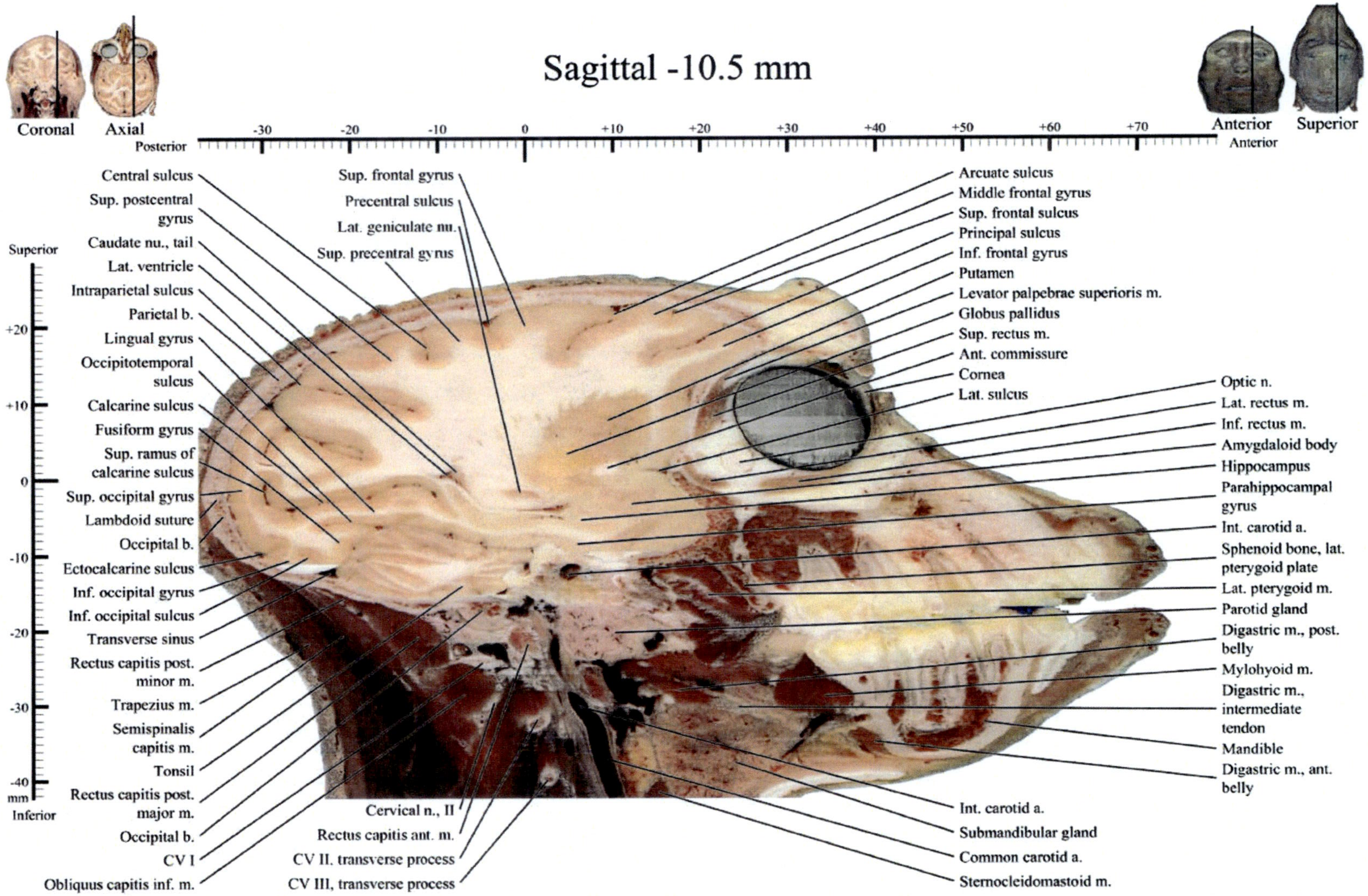

Sagittal -11 mm

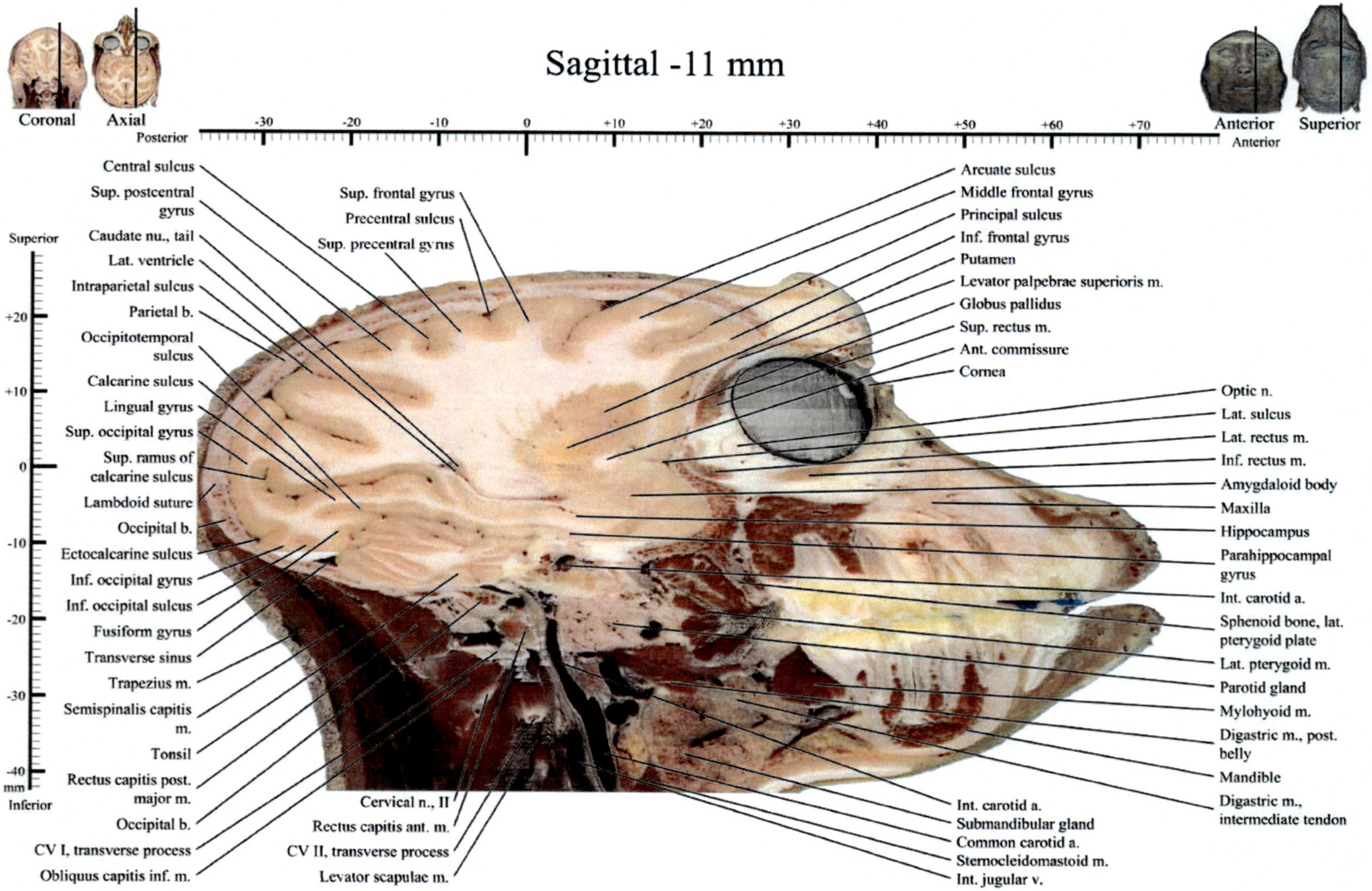

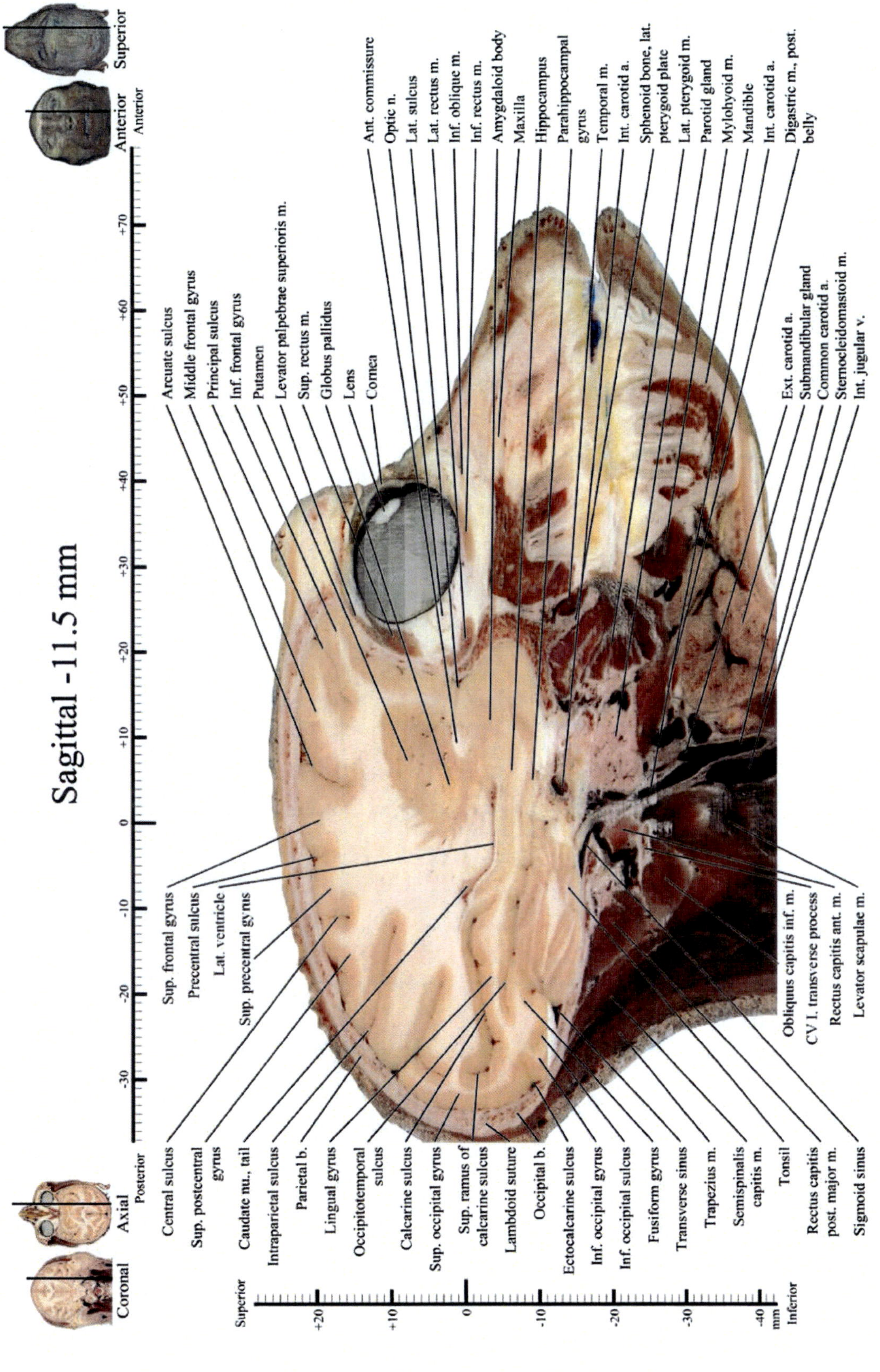
Sagittal -11.5 mm
Anterior
Superior
Anterior
Coronal
Axial
Posterior
Superior
Inferior
mm
+70
+60
+50
+40
+30
+20
+10
0
-10
-20
-30
-40
Arcuate sulcus
Middle frontal gyrus
Principal sulcus
Inf. frontal gyrus
Putamen
Levator palpebrae superioris m.
Sup. rectus m.
Globus pallidus
Lens
Cornea
Ant. commissure
Optic n.
Lat. sulcus
Lat. rectus m.
Inf. oblique m.
Inf. rectus m.
Amygdaloid body
Maxilla
Hippocampus
Parahippocampal gyrus
Temporal m.
Int. carotid a.
Sphenoid bone, lat. pterygoid plate
Lat. pterygoid m.
Parotid gland
Mylohyoid m.
Mandible
Int. carotid a.
Digastric m., post. belly
Ext. carotid a.
Submandibular gland
Common carotid a.
Sternocleidomastoid m.
Int. jugular v.
Sup. frontal gyrus
Precentral sulcus
Lat. ventricle
Sup. precentral gyrus
Obliquus capitis inf. m.
CV I, transverse process
Rectus capitis ant. m.
Levator scapulae m.
Central sulcus
Sup. postcentral gyrus
Caudate nu., tail
Intraparietal sulcus
Parietal b.
Lingual gyrus
Occipitotemporal sulcus
Calcarine sulcus
Sup. occipital gyrus
Sup. ramus of calcarine sulcus
Lambdoid suture
Occipital b.
Ectocalcarine sulcus
Inf. occipital gyrus
Inf. occipital sulcus
Fusiform gyrus
Transverse sinus
Trapezius m.
Semispinalis capitis m.
Tonsil
Rectus capitis post. major m.
Sigmoid sinus

Sagittal -12 mm

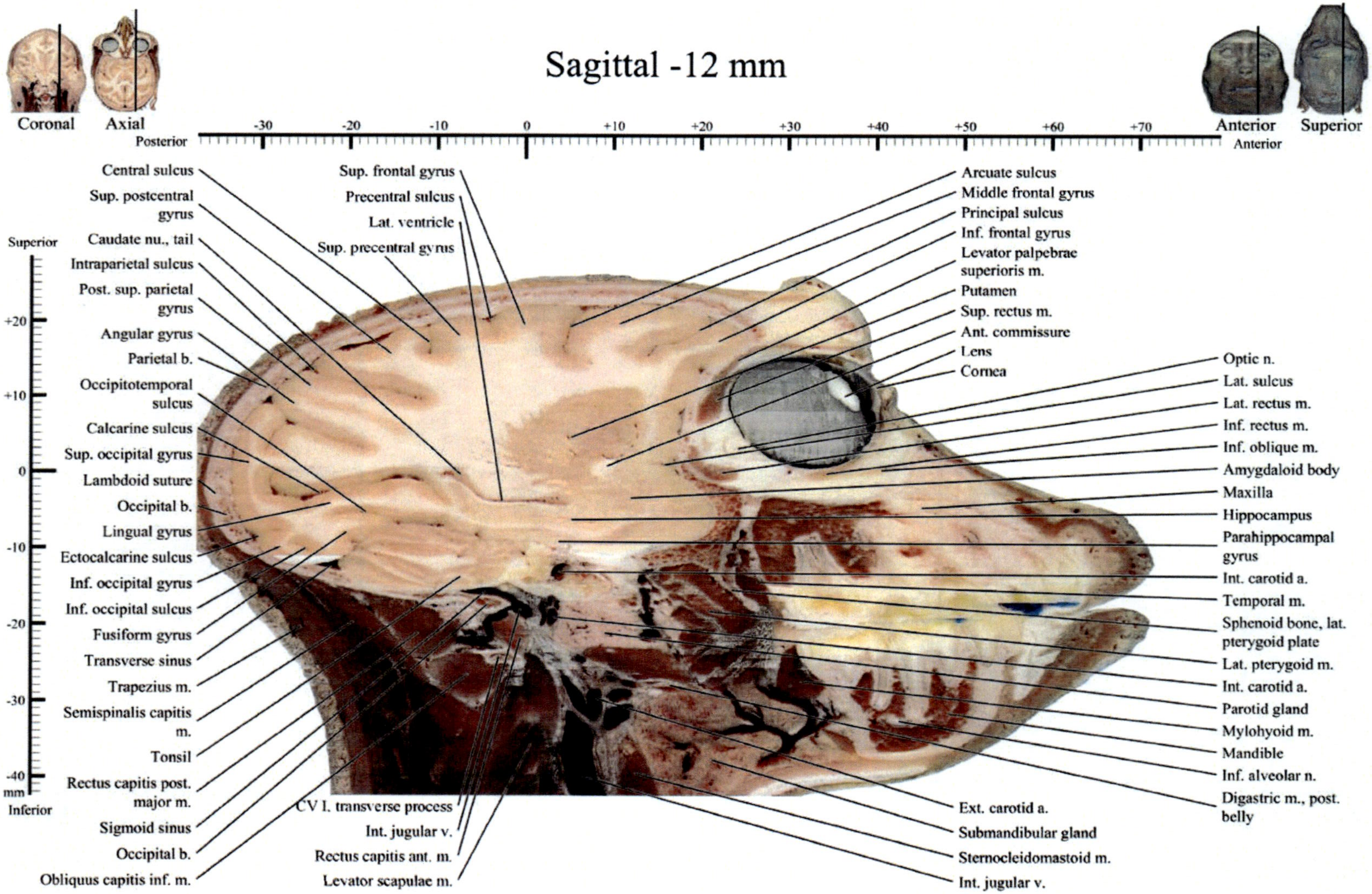

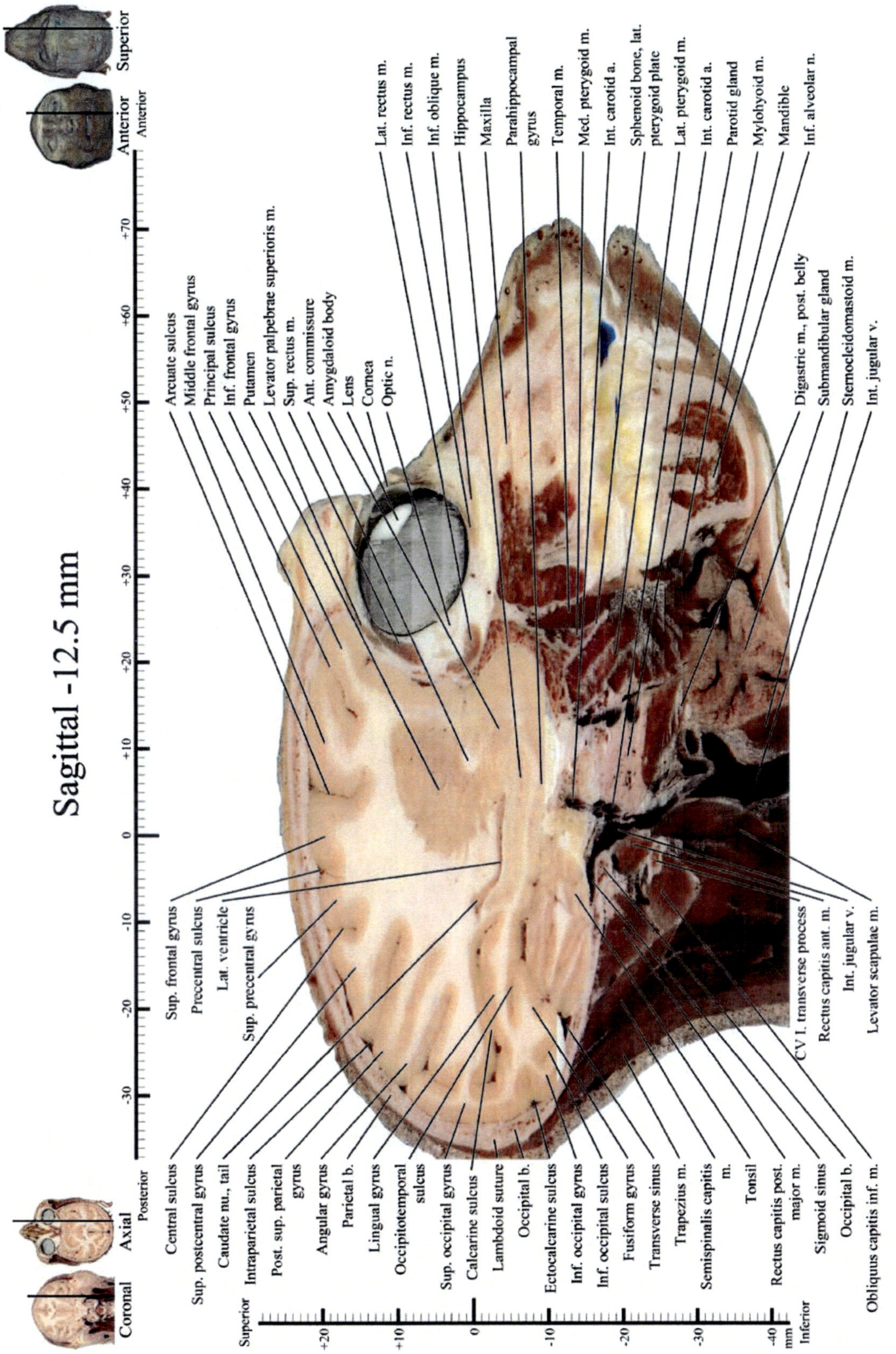
Sagittal -12.5 mm
Coronal
Axial
Anterior
Superior
Posterior
Anterior
Superior
Inferior
mm
Arcuate sulcus
Middle frontal gyrus
Principal sulcus
Inf. frontal gyrus
Putamen
Levator palpebrae superioris m.
Sup. rectus m.
Ant. commissure
Amygdaloid body
Lens
Cornea
Optic n.
Lat. rectus m.
Inf. rectus m.
Inf. oblique m.
Hippocampus
Maxilla
Parahippocampal gyrus
Temporal m.
Med. pterygoid m.
Int. carotid a.
Sphenoid bone, lat. pterygoid plate
Lat. pterygoid m.
Int. carotid a.
Parotid gland
Mylohyoid m.
Mandible
Inf. alveolar n.
Digastric m., post. belly
Submandibular gland
Sternocleidomastoid m.
Int. jugular v.
Sup. frontal gyrus
Precentral sulcus
Lat. ventricle
Sup. precentral gyrus
CV I. transverse process
Rectus capitis ant. m.
Int. jugular v.
Levator scapulae m.
Central sulcus
Sup. postcentral gyrus
Caudate nu., tail
Intraparietal sulcus
Post. sup. parietal gyrus
Angular gyrus
Parietal b.
Lingual gyrus
Occipitotemporal sulcus
Sup. occipital gyrus
Calcarine sulcus
Lambdoid suture
Occipital b.
Ectocalcarine sulcus
Inf. occipital gyrus
Inf. occipital sulcus
Fusiform gyrus
Transverse sinus
Trapezius m.
Semispinalis capitis m.
Tonsil
Rectus capitis post. major m.
Sigmoid sinus
Occipital b.
Obliquus capitis inf. m.

Sagittal -13 mm

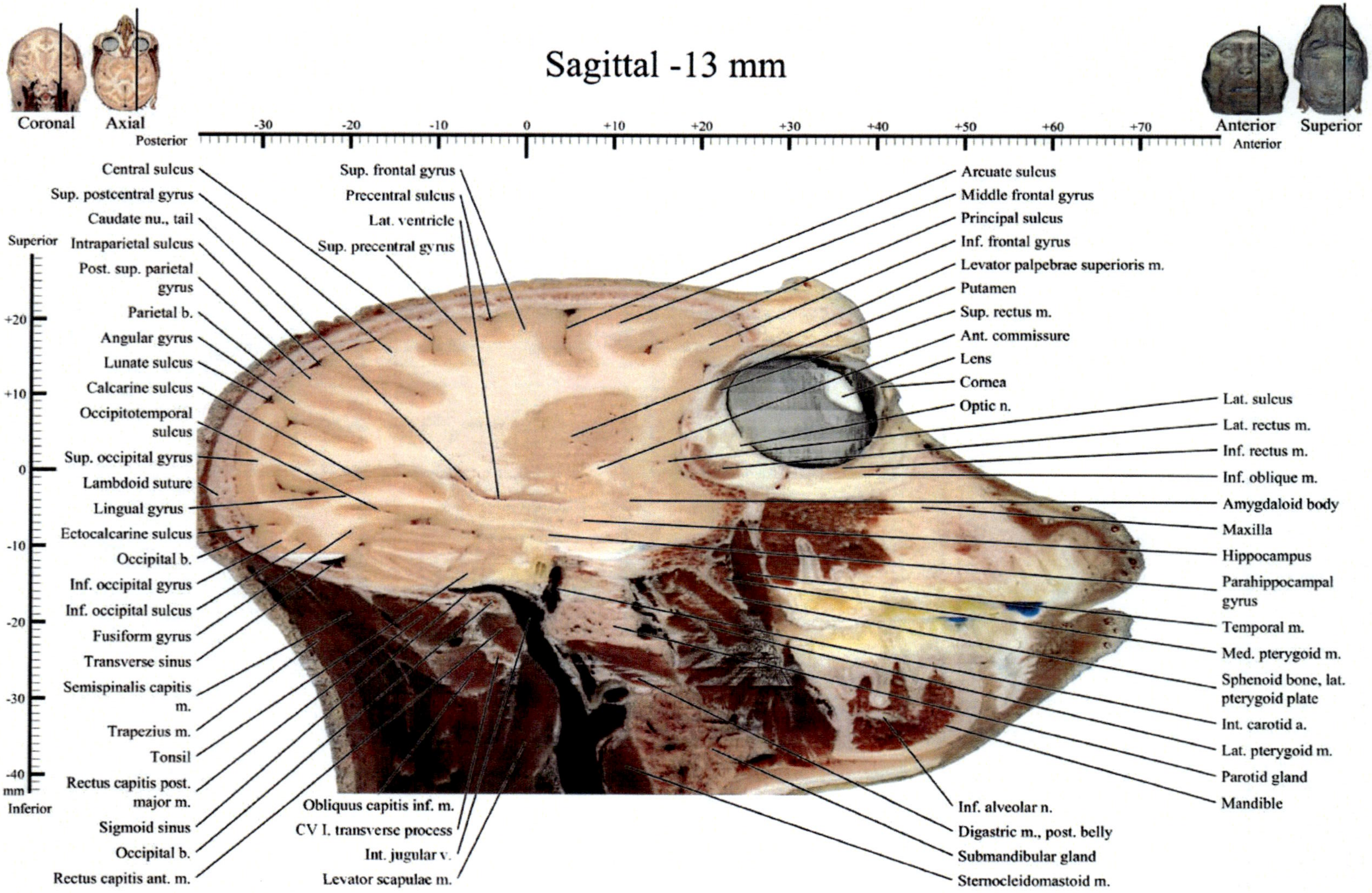

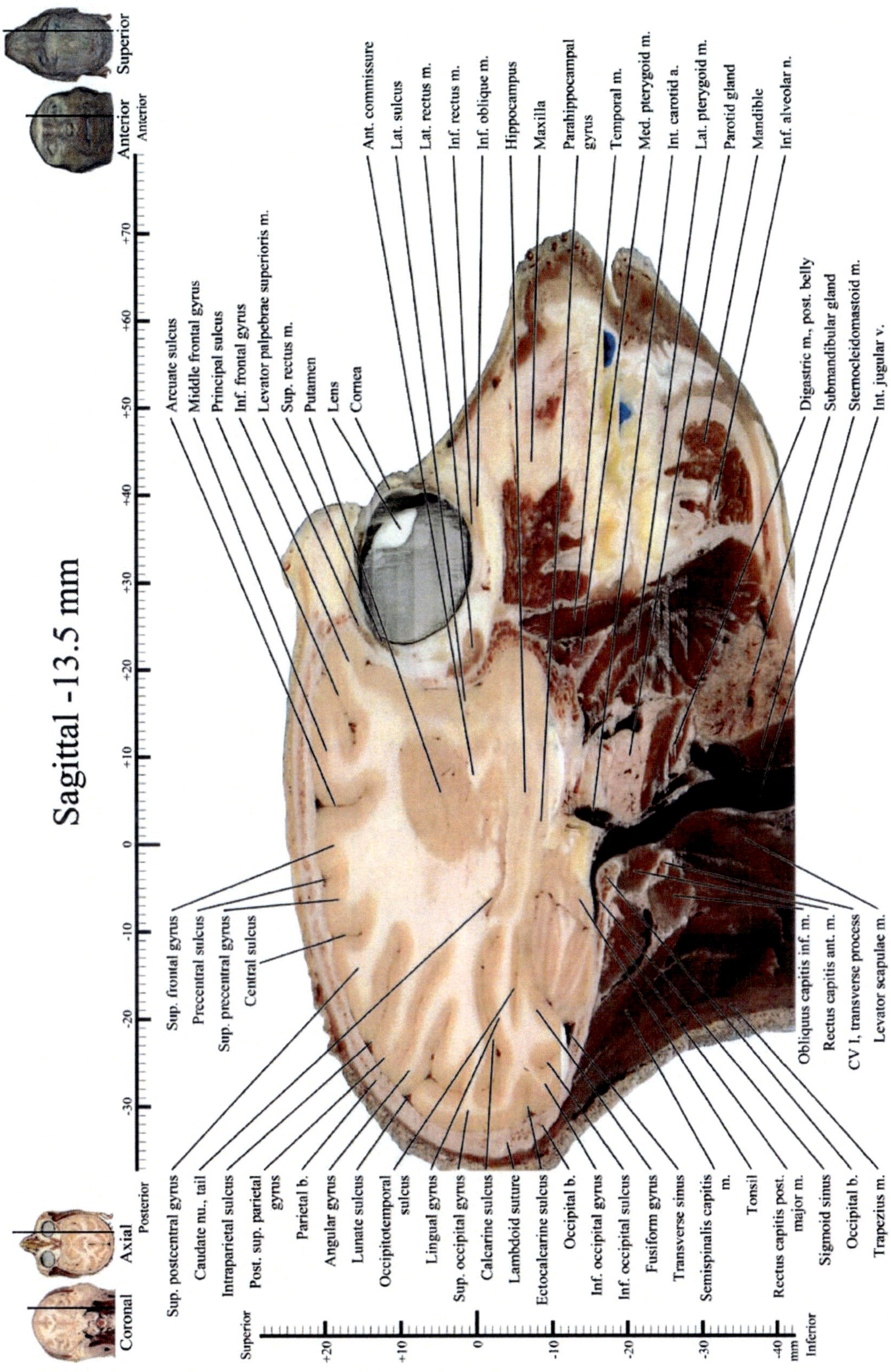
Sagittal -13.5 mm
Coronal
Axial
Anterior
Superior
Posterior
Anterior
Superior
Inferior
mm
Ant. commissure
Lat. sulcus
Lat. rectus m.
Inf. rectus m.
Inf. oblique m.
Hippocampus
Maxilla
Parahippocampal gyrus
Temporal m.
Med. pterygoid m.
Int. carotid a.
Lat. pterygoid m.
Parotid gland
Mandible
Inf. alveolar n.
Arcuate sulcus
Middle frontal gyrus
Principal sulcus
Inf. frontal gyrus
Levator palpebrae superioris m.
Sup. rectus m.
Putamen
Lens
Cornea
Digastric m., post. belly
Submandibular gland
Sternocleidomastoid m.
Int. jugular v.
Sup. frontal gyrus
Precentral sulcus
Sup. precentral gyrus
Central sulcus
Obliquus capitis inf. m.
Rectus capitis ant. m.
CV 1, transverse process
Levator scapulae m.
Sup. postcentral gyrus
Caudate nu., tail
Intraparietal sulcus
Post. sup. parietal gyrus
Parietal b.
Angular gyrus
Lunate sulcus
Occipitotemporal sulcus
Lingual gyrus
Sup. occipital gyrus
Calcarine sulcus
Lambdoid suture
Ectocalcarine sulcus
Occipital b.
Inf. occipital gyrus
Inf. occipital sulcus
Fusiform gyrus
Transverse sinus
Semispinalis capitis m.
Tonsil
Rectus capitis post. major m.
Sigmoid sinus
Occipital b.
Trapezius m.

Sagittal -14 mm

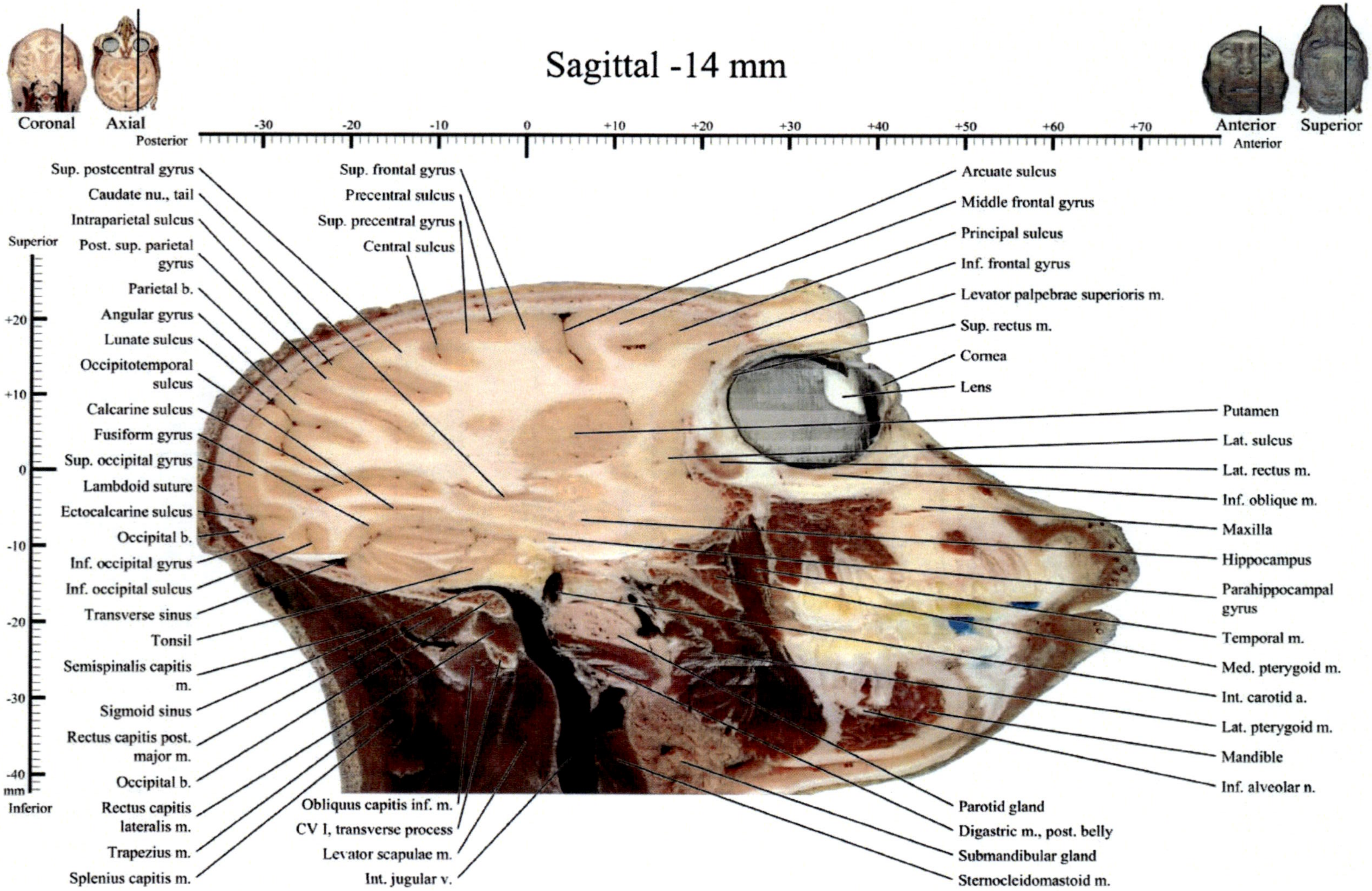

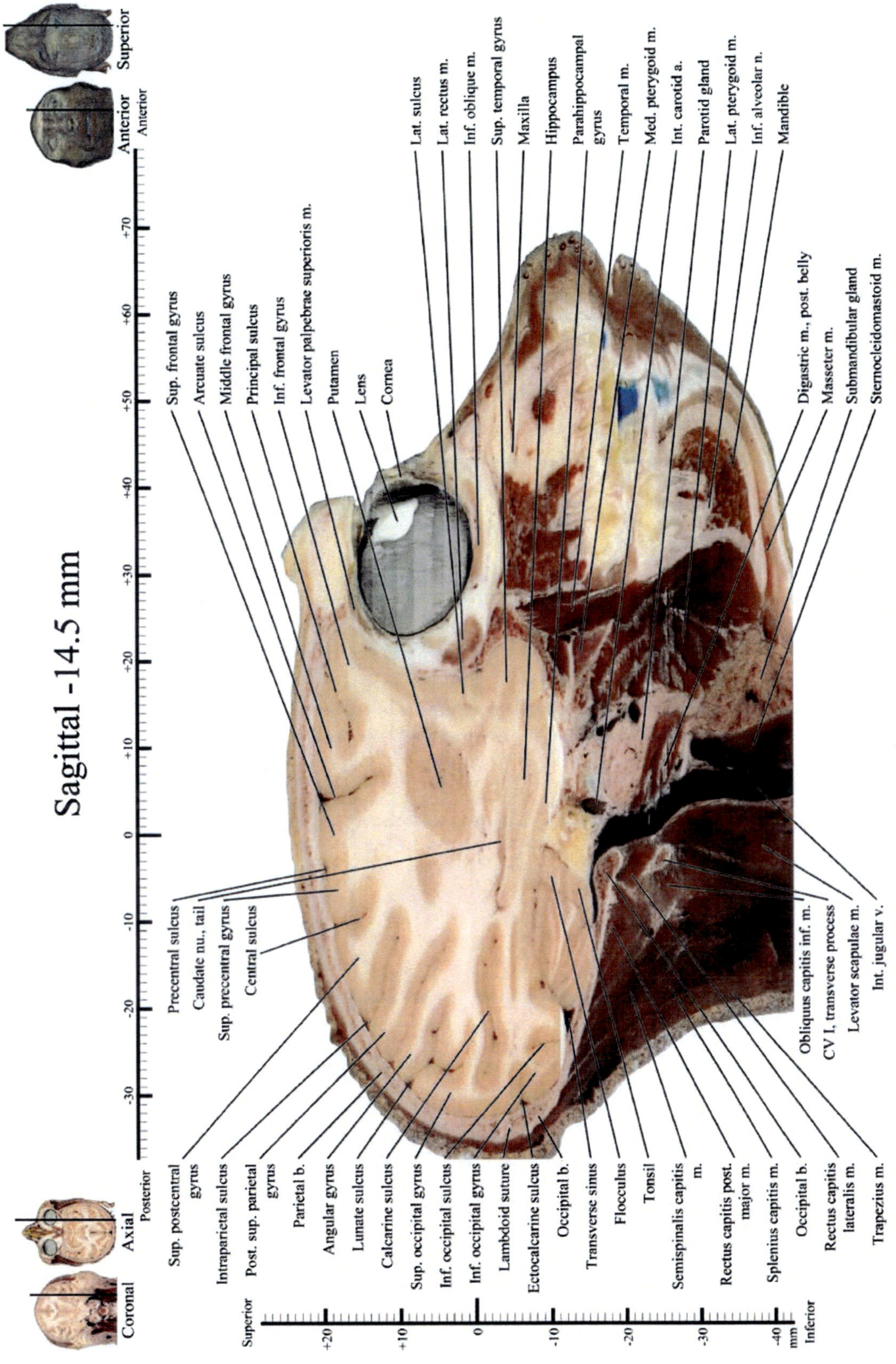

Sagittal -14.5 mm
Coronal
Axial
Anterior
Superior
Posterior
Anterior
Superior
Inferior
mm
+20
+10
0
-10
-20
-30
-40
-30
-20
-10
0
+10
+20
+30
+40
+50
+60
+70
Sup. frontal gyrus
Arcuate sulcus
Middle frontal gyrus
Principal sulcus
Inf. frontal gyrus
Levator palpebrae superioris m.
Putamen
Lens
Cornea
Lat. sulcus
Lat. rectus m.
Inf. oblique m.
Sup. temporal gyrus
Maxilla
Hippocampus
Parahippocampal gyrus
Temporal m.
Med. pterygoid m.
Int. carotid a.
Parotid gland
Lat. pterygoid m.
Inf. alveolar n.
Mandible
Digastric m., post. belly
Masseter m.
Submandibular gland
Sternocleidomastoid m.
Precentral sulcus
Caudate nu., tail
Sup. precentral gyrus
Central sulcus
Obliquus capitis inf. m.
CV I, transverse process
Levator scapulae m.
Int. jugular v.
Sup. postcentral gyrus
Intraparietal sulcus
Post. sup. parietal gyrus
Parietal b.
Angular gyrus
Lunate sulcus
Calcarine sulcus
Sup. occipital gyrus
Inf. occipital sulcus
Inf. occipital gyrus
Lambdoid suture
Ectocalcarine sulcus
Occipital b.
Transverse sinus
Flocculus
Tonsil
Semispinalis capitis m.
Rectus capitis post. major m.
Splenius capitis m.
Occipital b.
Rectus capitis lateralis m.
Trapezius m.

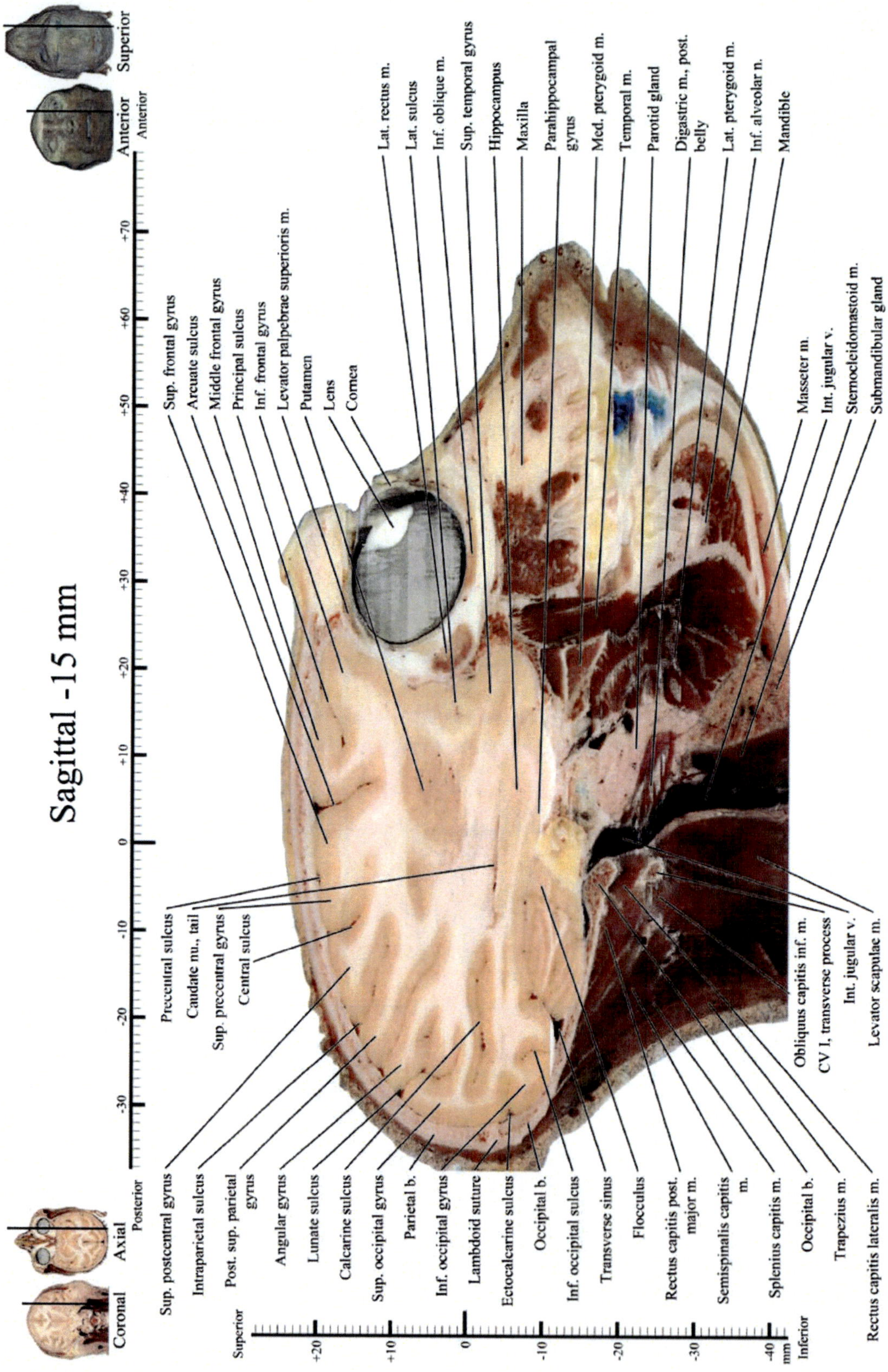
Sagittal -15 mm
Coronal
Axial
Anterior
Superior
Posterior
Anterior
Superior
Inferior
mm
-30
-20
-10
0
+10
+20
+30
+40
+50
+60
+70
Sup. postcentral gyrus
Intraparietal sulcus
Post. sup. parietal gyrus
Angular gyrus
Lunate sulcus
Calcarine sulcus
Sup. occipital gyrus
Parietal b.
Inf. occipital gyrus
Lambdoid suture
Ectocalcarine sulcus
Occipital b.
Inf. occipital sulcus
Transverse sinus
Flocculus
Rectus capitis post. major m.
Semispinalis capitis m.
Splenius capitis m.
Occipital b.
Trapezius m.
Rectus capitis lateralis m.
Precentral sulcus
Caudate nu., tail
Sup. precentral gyrus
Central sulcus
Obliquus capitis inf. m.
CV I, transverse process
Int. jugular v.
Levator scapulae m.
Sup. frontal gyrus
Arcuate sulcus
Middle frontal gyrus
Principal sulcus
Inf. frontal gyrus
Levator palpebrae superioris m.
Putamen
Lens
Cornea
Lat. rectus m.
Lat. sulcus
Inf. oblique m.
Sup. temporal gyrus
Hippocampus
Maxilla
Parahippocampal gyrus
Med. pterygoid m.
Temporal m.
Parotid gland
Digastric m., post. belly
Lat. pterygoid m.
Inf. alveolar n.
Mandible
Masseter m.
Int. jugular v.
Sternocleidomastoid m.
Submandibular gland

Sagittal -15.5 mm

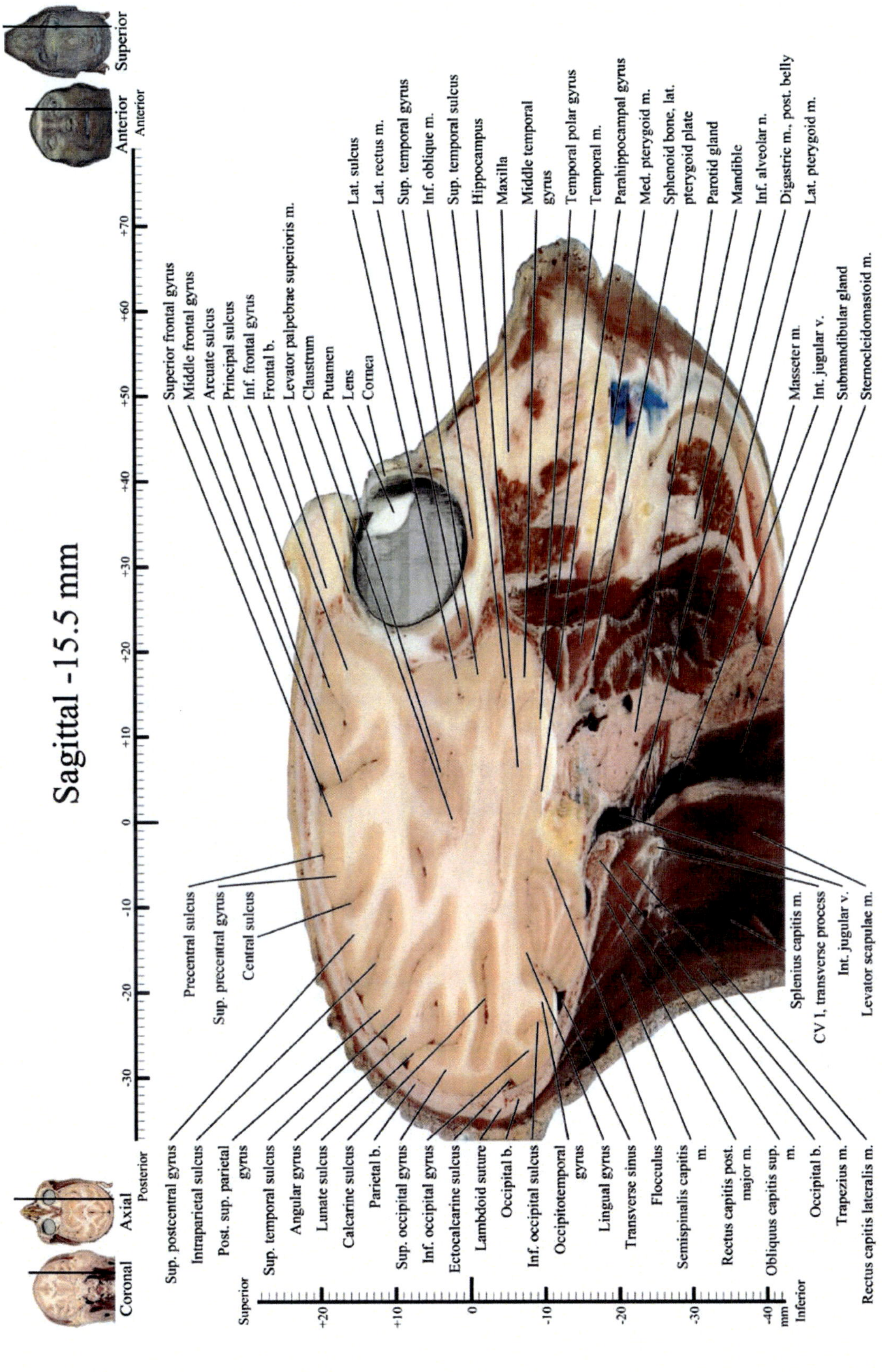

Sagittal -16 mm

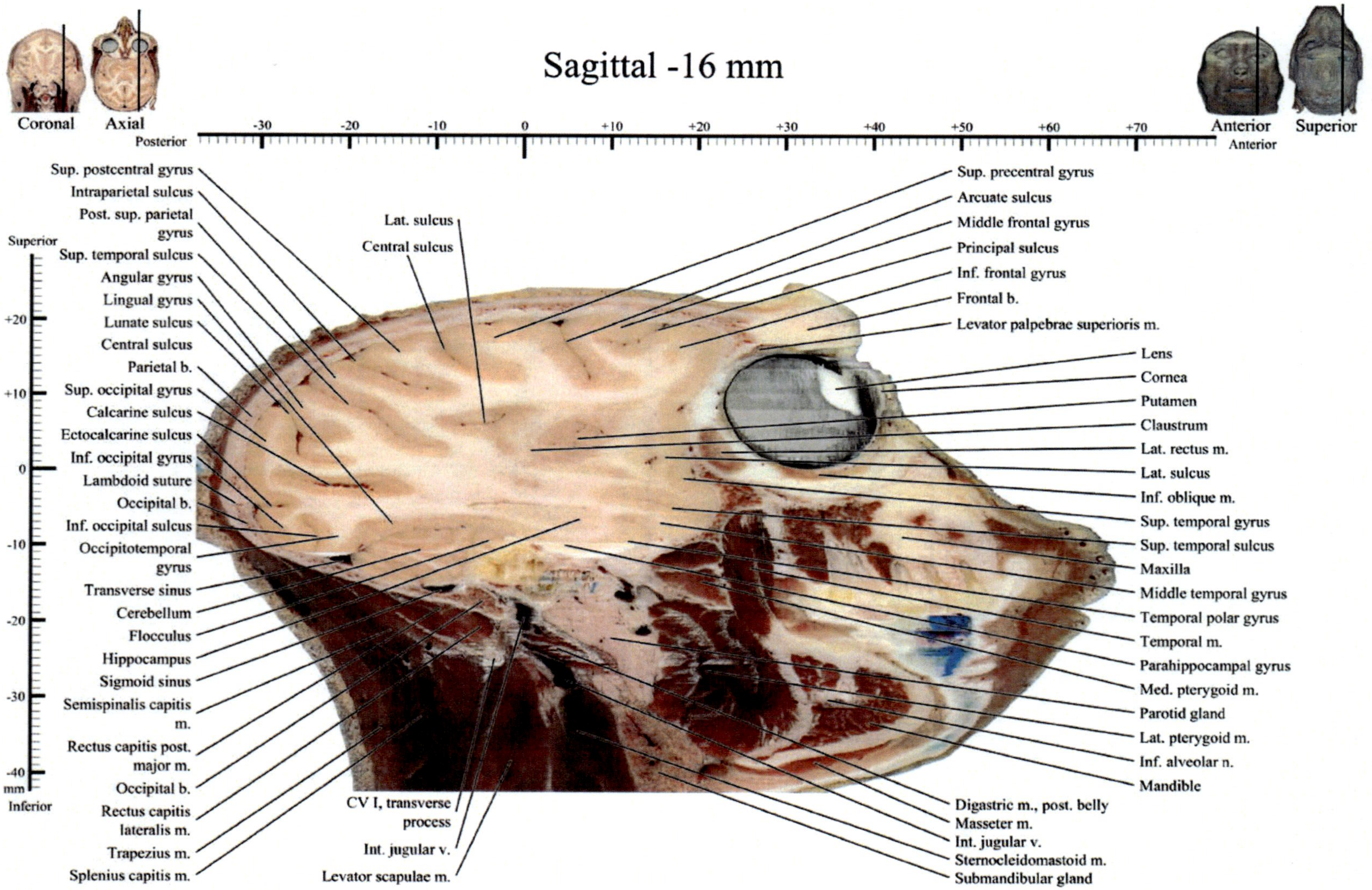

Sagittal -17 mm

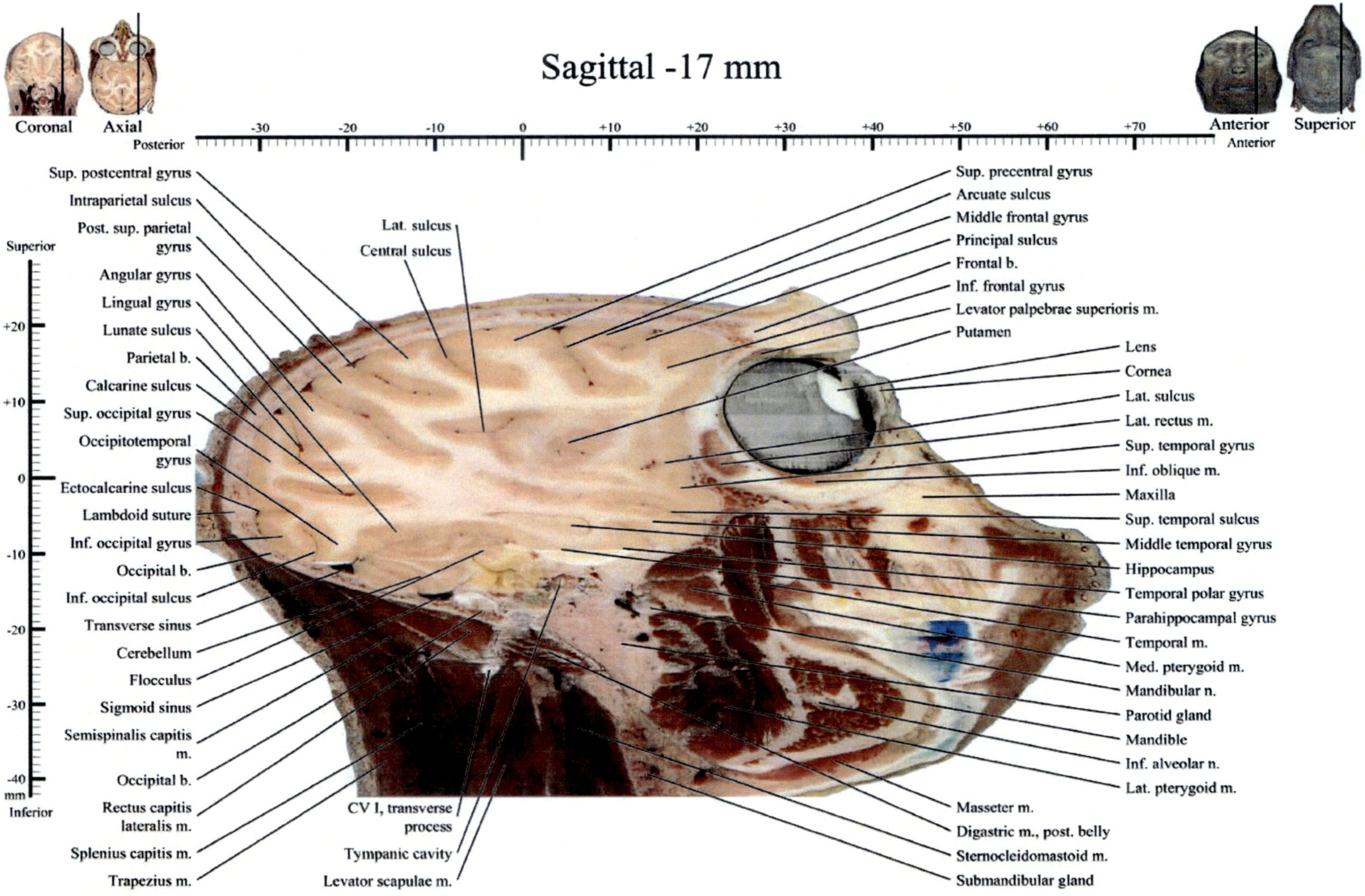

Sagittal -19.5 mm

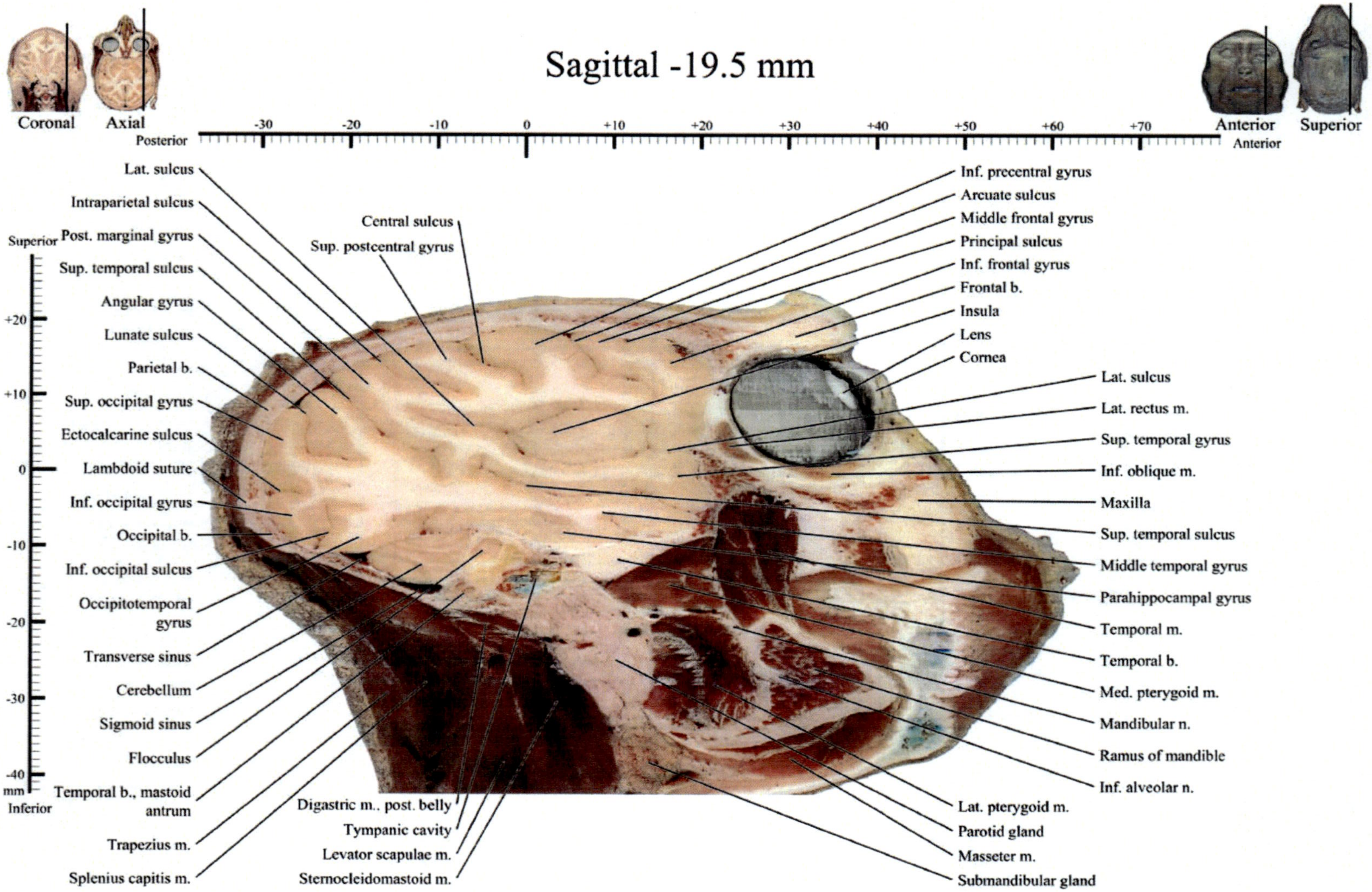

Sagittal -22 mm

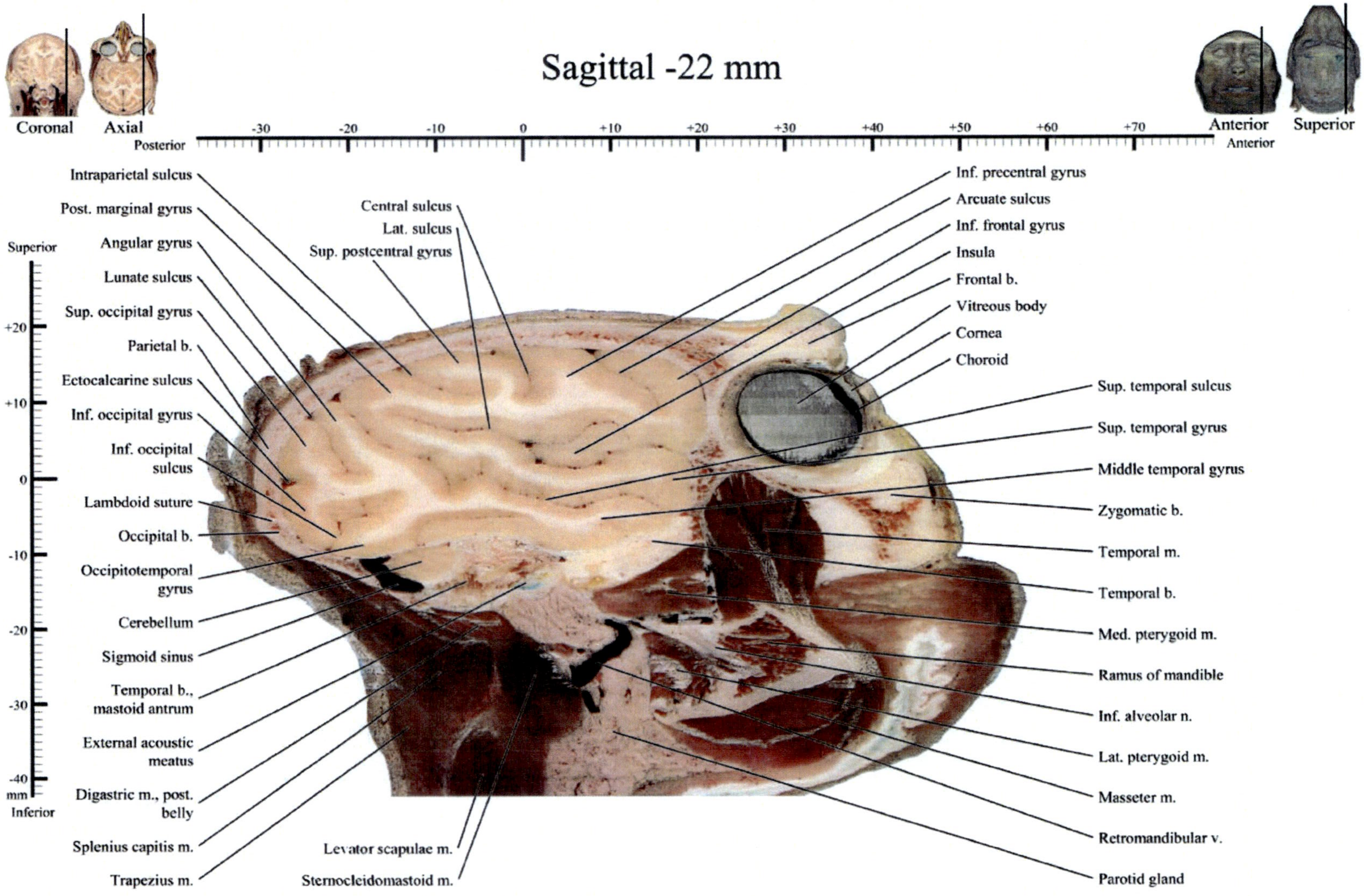

Sagittal -24.5 mm

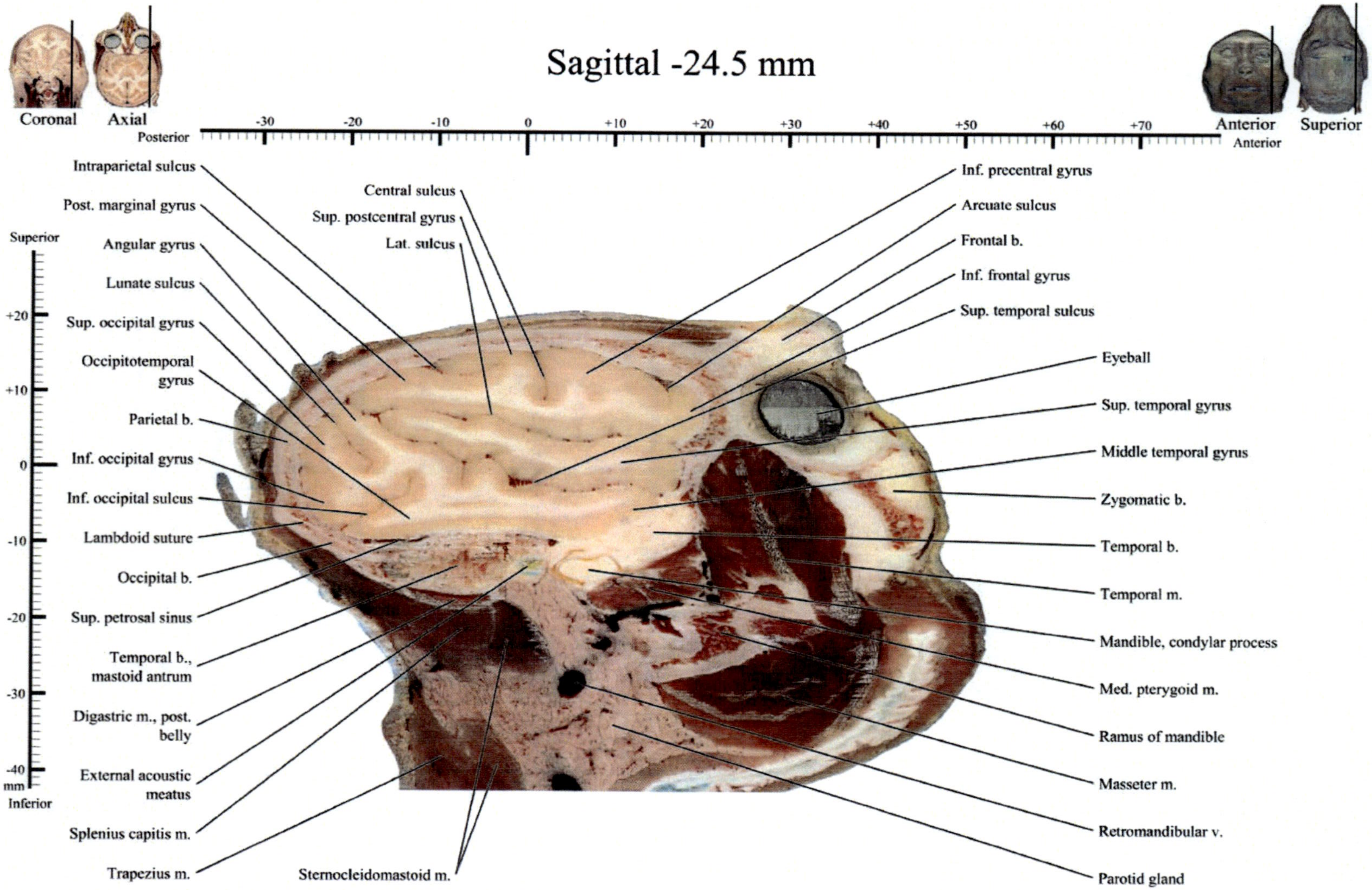

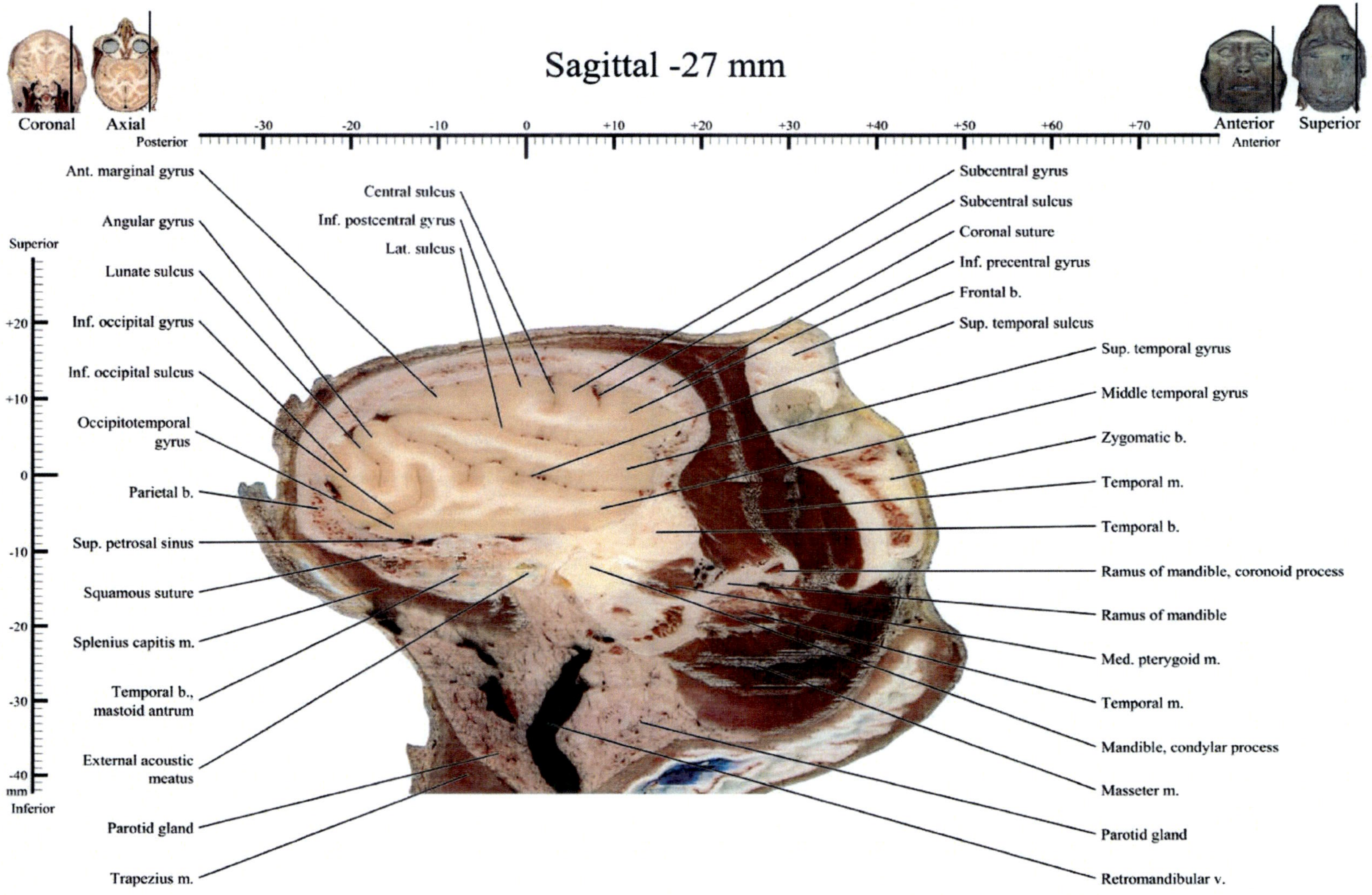
Sagittal -27 mm
Coronal
Axial
Posterior
Anterior
Superior
Anterior
Superior
Inferior
mm
-30
-20
-10
0
+10
+20
+30
+40
+50
+60
+70
+20
+10
0
-10
-20
-30
-40
Ant. marginal gyrus
Angular gyrus
Lunate sulcus
Inf. occipital gyrus
Inf. occipital sulcus
Occipitotemporal gyrus
Parietal b.
Sup. petrosal sinus
Squamous suture
Splenius capitis m.
Temporal b., mastoid antrum
External acoustic meatus
Parotid gland
Trapezius m.
Central sulcus
Inf. postcentral gyrus
Lat. sulcus
Subcentral gyrus
Subcentral sulcus
Coronal suture
Inf. precentral gyrus
Frontal b.
Sup. temporal sulcus
Sup. temporal gyrus
Middle temporal gyrus
Zygomatic b.
Temporal m.
Temporal b.
Ramus of mandible, coronoid process
Ramus of mandible
Med. pterygoid m.
Temporal m.
Mandible, condylar process
Masseter m.
Parotid gland
Retromandibular v.

Part II

Deep Head

Abbreviations

a.	Artery
ant.	Anterior
b.	Bone
CS	Cervical segment of spinal cord
CV	Cervical vertebra
ext.	External
inf.	Inferior
int.	Internal
lat.	Lateral
lig.	Ligament
m.	Muscle
med.	Medial
n.	Nerve
nu.	Nucleus
post.	Posterior
sup.	Superior
v.	Vein

5 Axial (Expanded)

All images, 0.024 mm × 0.024 mm sized pixel

+27 mm ~ +11 mm images, 4 mm intervals
+11 mm ~ 0 mm images, 1 mm intervals
0 mm ~ −5 mm images, 0.5 mm intervals
−5 mm ~ −43 mm images, 1 mm intervals

J. S. Park, *Cross-Sectional Atlas of Rhesus Monkey Head*, https://doi.org/10.1007/978-981-19-3254-0_5

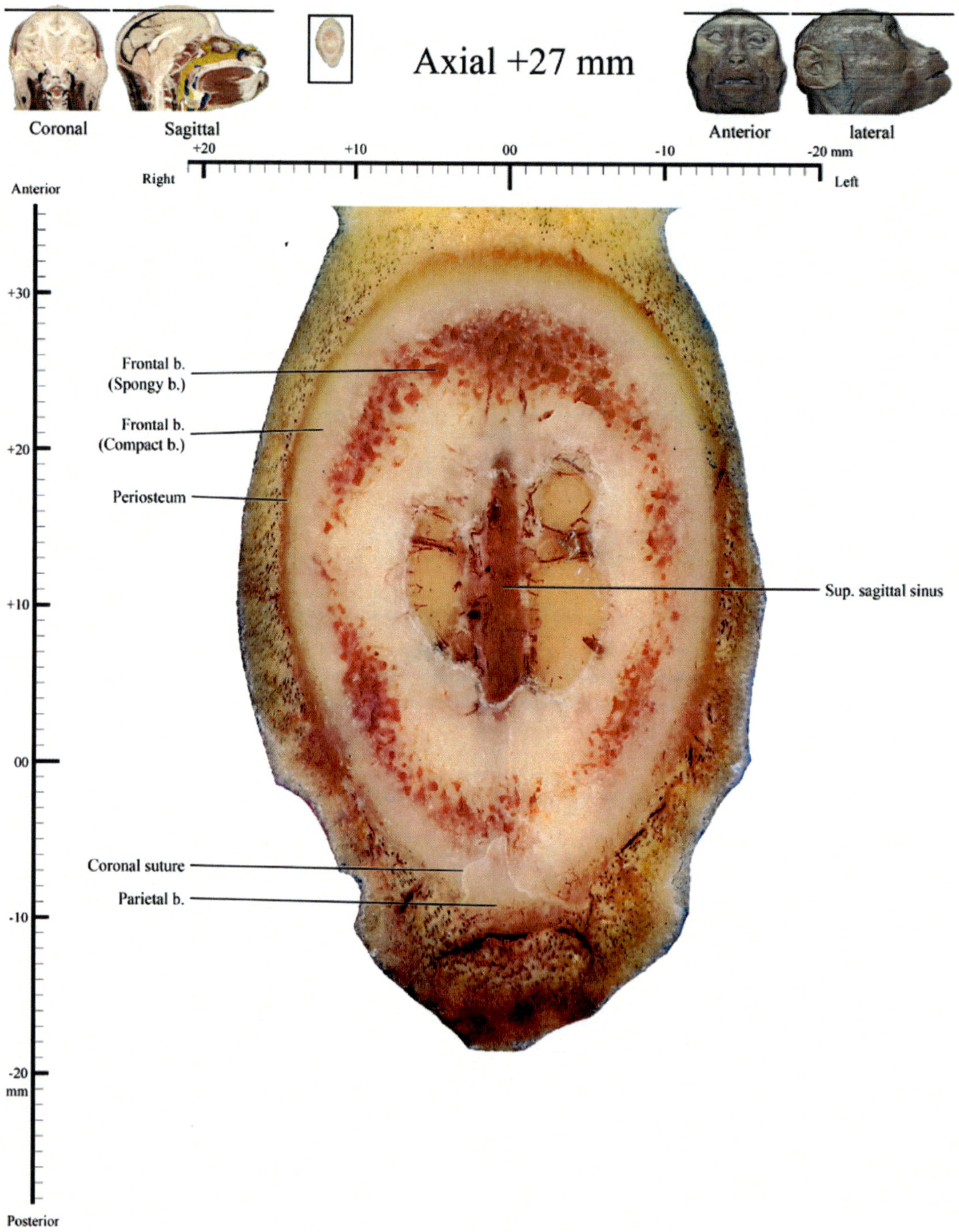

Coronal
Sagittal
Axial +27 mm
Anterior
lateral
+20
+10
00
-10
-20 mm
Right
Left
Anterior
+30
+20
+10
00
-10
-20
mm
Posterior
Frontal b. (Spongy b.)
Frontal b. (Compact b.)
Periosteum
Sup. sagittal sinus
Coronal suture
Parietal b.

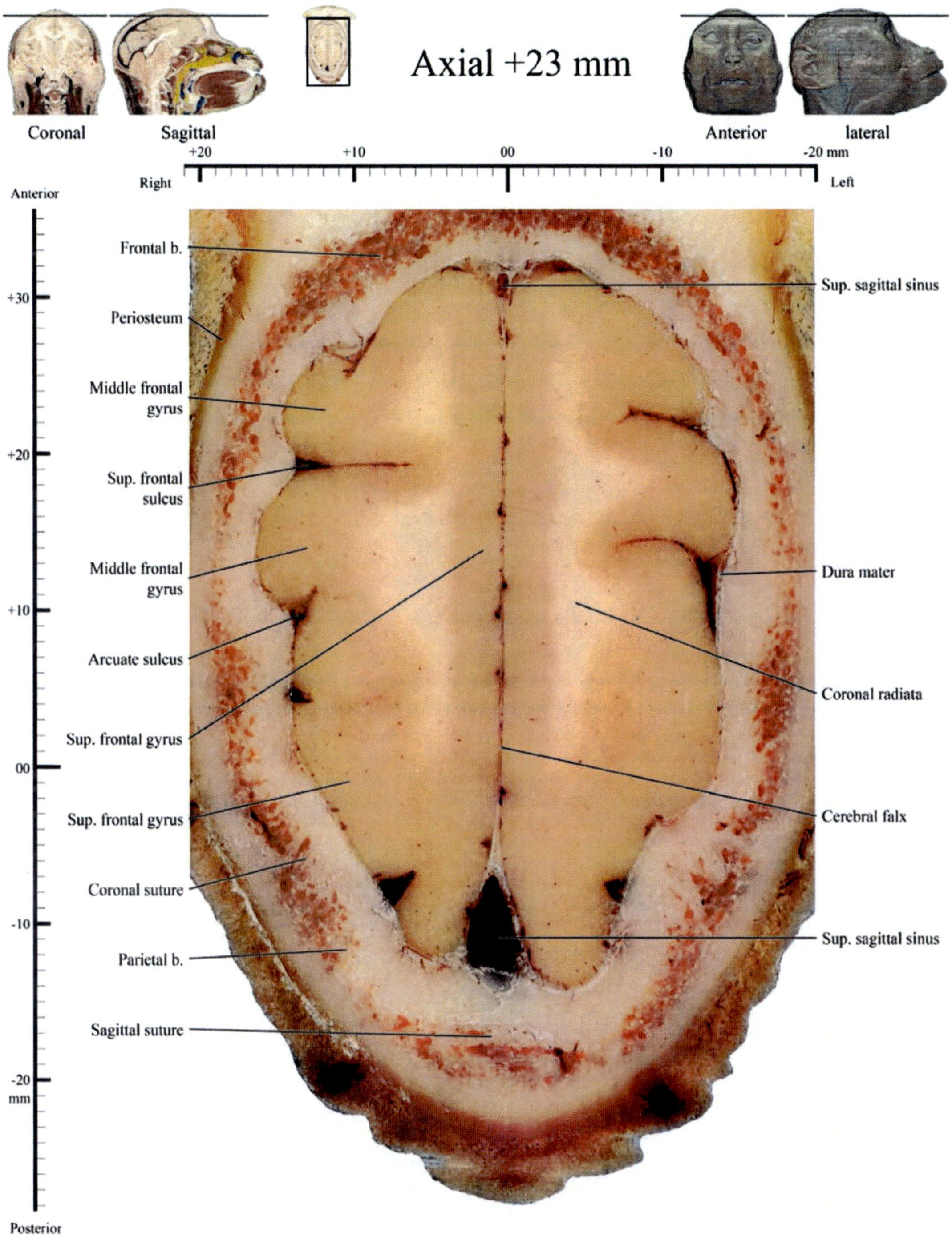
Coronal
Sagittal
Axial +23 mm
Anterior
lateral
+20
+10
00
-10
-20 mm
Right
Left
Anterior
+30
+20
+10
00
-10
-20
mm
Posterior
Frontal b.
Periosteum
Middle frontal gyrus
Sup. frontal sulcus
Middle frontal gyrus
Arcuate sulcus
Sup. frontal gyrus
Sup. frontal gyrus
Coronal suture
Parietal b.
Sagittal suture
Sup. sagittal sinus
Dura mater
Coronal radiata
Cerebral falx
Sup. sagittal sinus

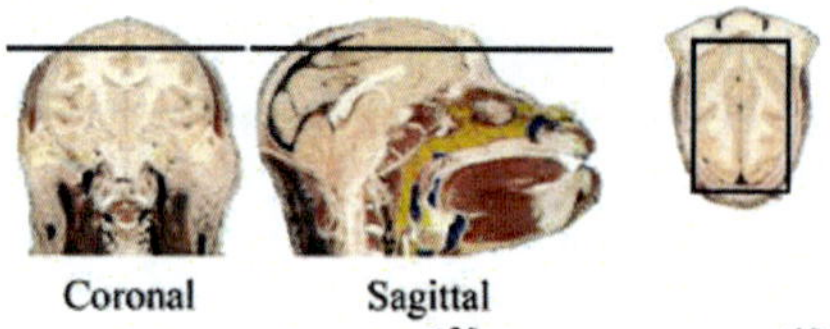
Coronal
Sagittal

Axial +19 mm

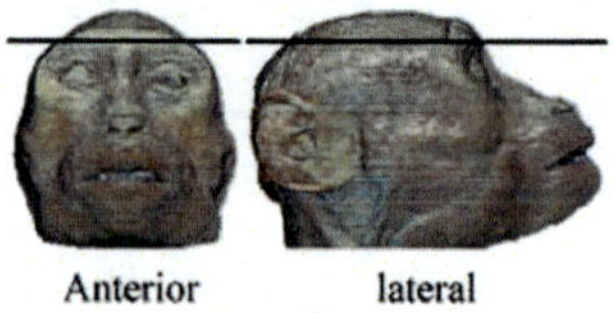
Anterior
lateral

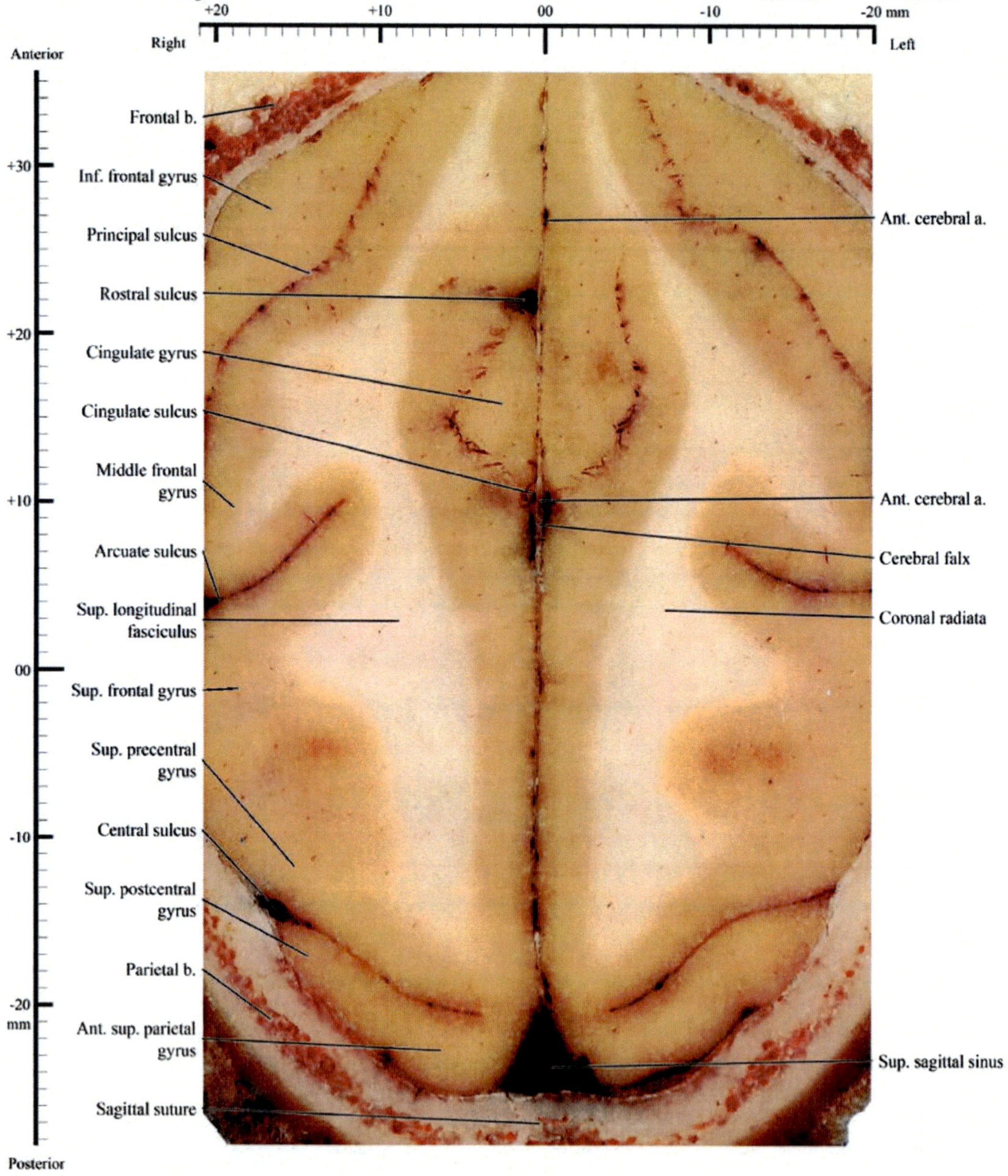
+20
+10
00
-10
-20 mm
Right
Left
Anterior
Posterior
+30
+20
+10
00
-10
-20
mm
Frontal b.
Inf. frontal gyrus
Principal sulcus
Rostral sulcus
Cingulate gyrus
Cingulate sulcus
Middle frontal gyrus
Arcuate sulcus
Sup. longitudinal fasciculus
Sup. frontal gyrus
Sup. precentral gyrus
Central sulcus
Sup. postcentral gyrus
Parietal b.
Ant. sup. parietal gyrus
Sagittal suture
Ant. cerebral a.
Ant. cerebral a.
Cerebral falx
Coronal radiata
Sup. sagittal sinus

Axial +15 mm

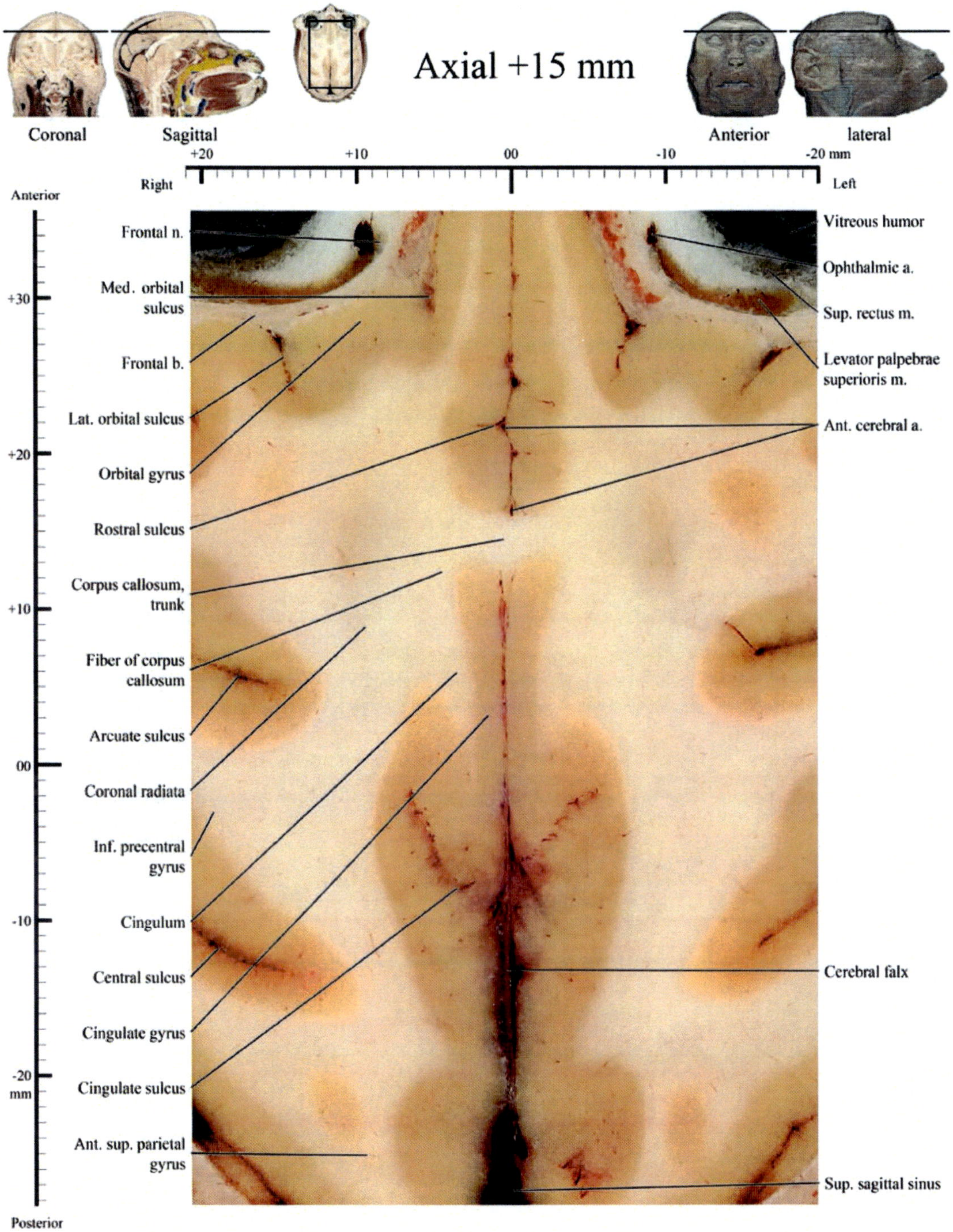
Coronal
Sagittal
Anterior
lateral
+20
+10
00
-10
-20 mm
Right
Left
Anterior
+30
+20
+10
00
-10
-20
mm
Posterior
Frontal n.
Med. orbital sulcus
Frontal b.
Lat. orbital sulcus
Orbital gyrus
Rostral sulcus
Corpus callosum, trunk
Fiber of corpus callosum
Arcuate sulcus
Coronal radiata
Inf. precentral gyrus
Cingulum
Central sulcus
Cingulate gyrus
Cingulate sulcus
Ant. sup. parietal gyrus
Vitreous humor
Ophthalmic a.
Sup. rectus m.
Levator palpebrae superioris m.
Ant. cerebral a.
Cerebral falx
Sup. sagittal sinus

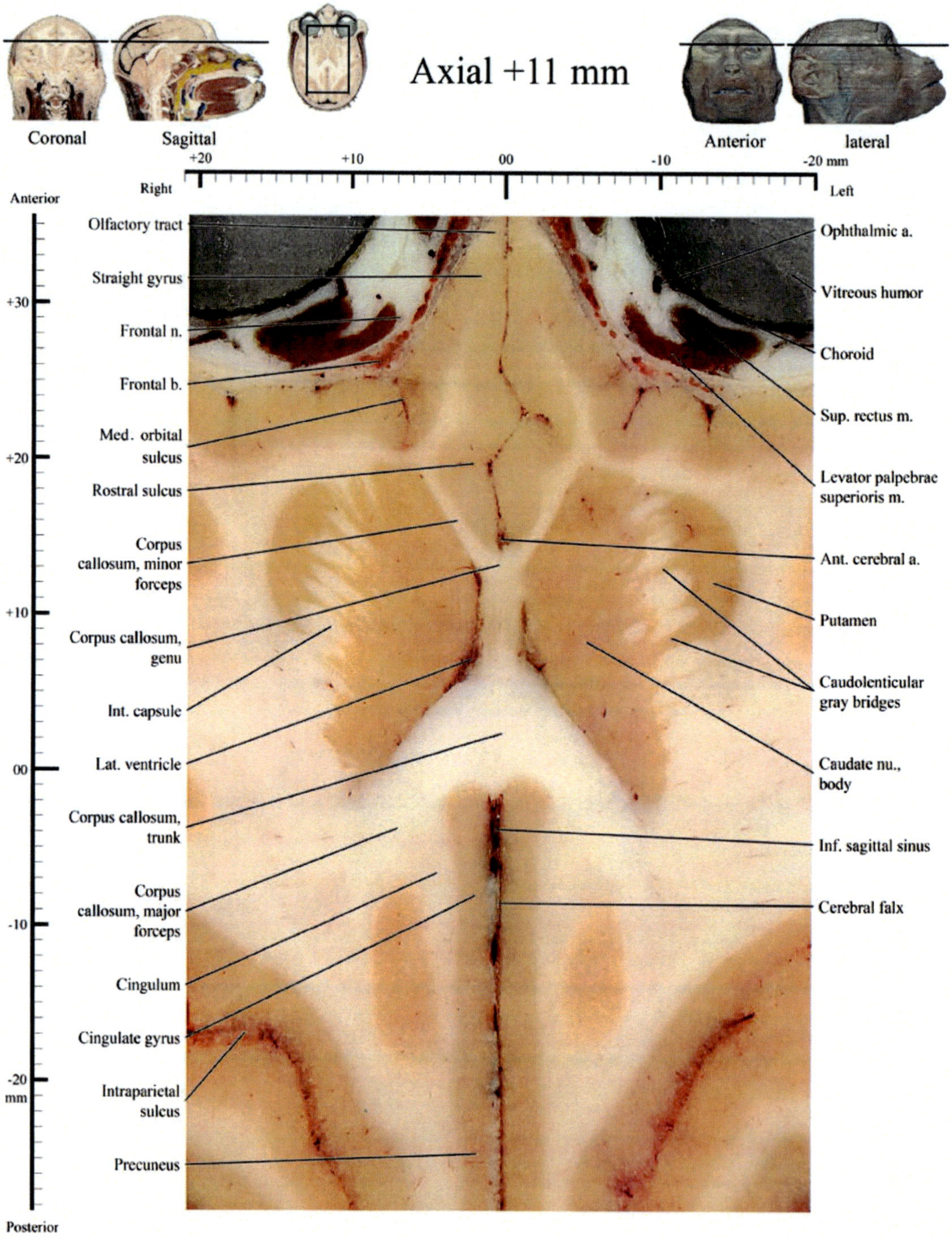
Coronal
Sagittal
Axial +11 mm
Anterior
lateral
+20
+10
00
-10
-20 mm
Right
Left
Anterior
Posterior
+30
+20
+10
00
-10
-20
mm
Olfactory tract
Straight gyrus
Frontal n.
Frontal b.
Med. orbital sulcus
Rostral sulcus
Corpus callosum, minor forceps
Corpus callosum, genu
Int. capsule
Lat. ventricle
Corpus callosum, trunk
Corpus callosum, major forceps
Cingulum
Cingulate gyrus
Intraparietal sulcus
Precuneus
Ophthalmic a.
Vitreous humor
Choroid
Sup. rectus m.
Levator palpebrae superioris m.
Ant. cerebral a.
Putamen
Caudolenticular gray bridges
Caudate nu., body
Inf. sagittal sinus
Cerebral falx

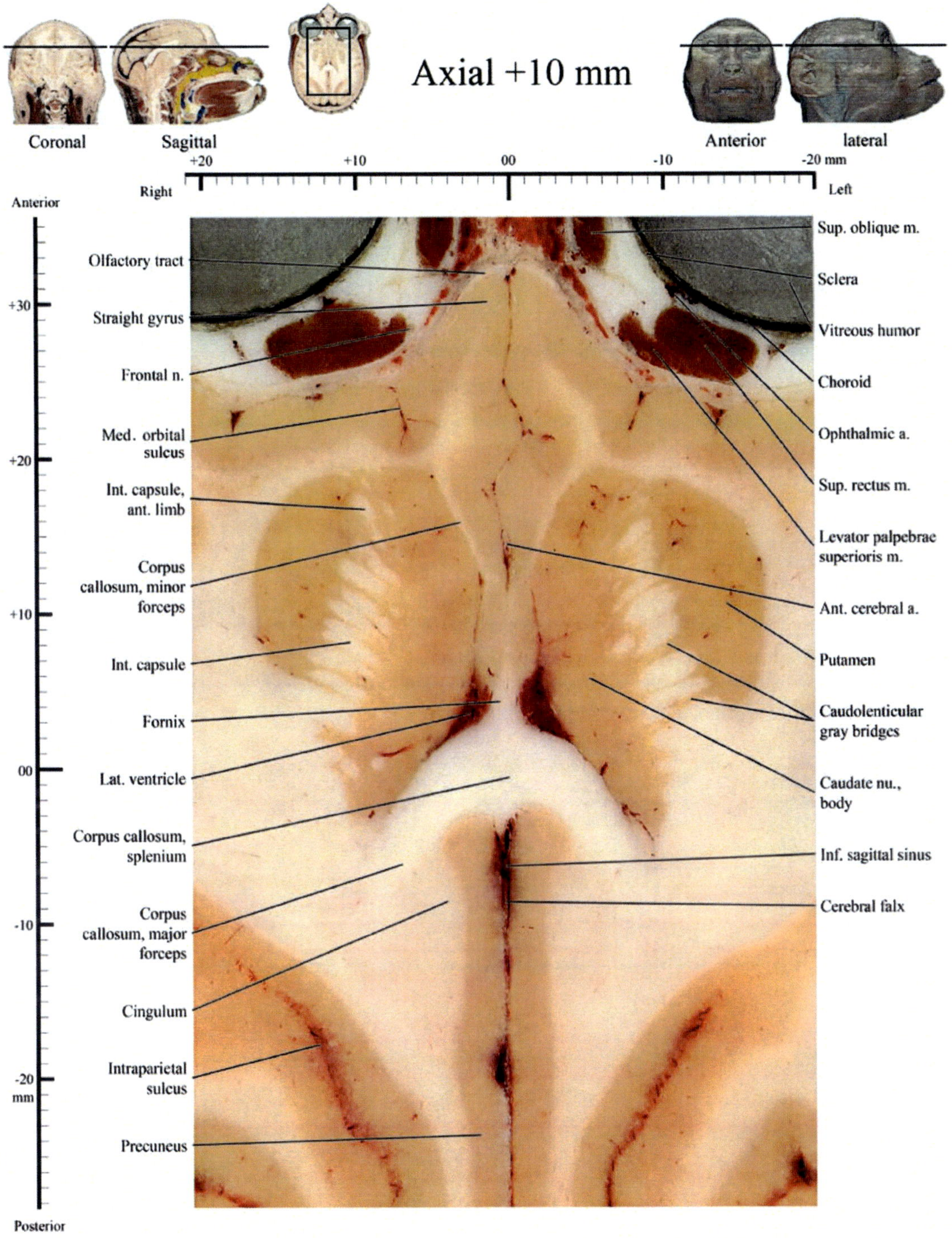
Coronal
Sagittal
Axial +10 mm
Anterior
lateral
+20
+10
00
-10
-20 mm
Right
Left
Anterior
Posterior
+30
+20
+10
00
-10
-20
mm
Olfactory tract
Straight gyrus
Frontal n.
Med. orbital sulcus
Int. capsule, ant. limb
Corpus callosum, minor forceps
Int. capsule
Fornix
Lat. ventricle
Corpus callosum, splenium
Corpus callosum, major forceps
Cingulum
Intraparietal sulcus
Precuneus
Sup. oblique m.
Sclera
Vitreous humor
Choroid
Ophthalmic a.
Sup. rectus m.
Levator palpebrae superioris m.
Ant. cerebral a.
Putamen
Caudolenticular gray bridges
Caudate nu., body
Inf. sagittal sinus
Cerebral falx

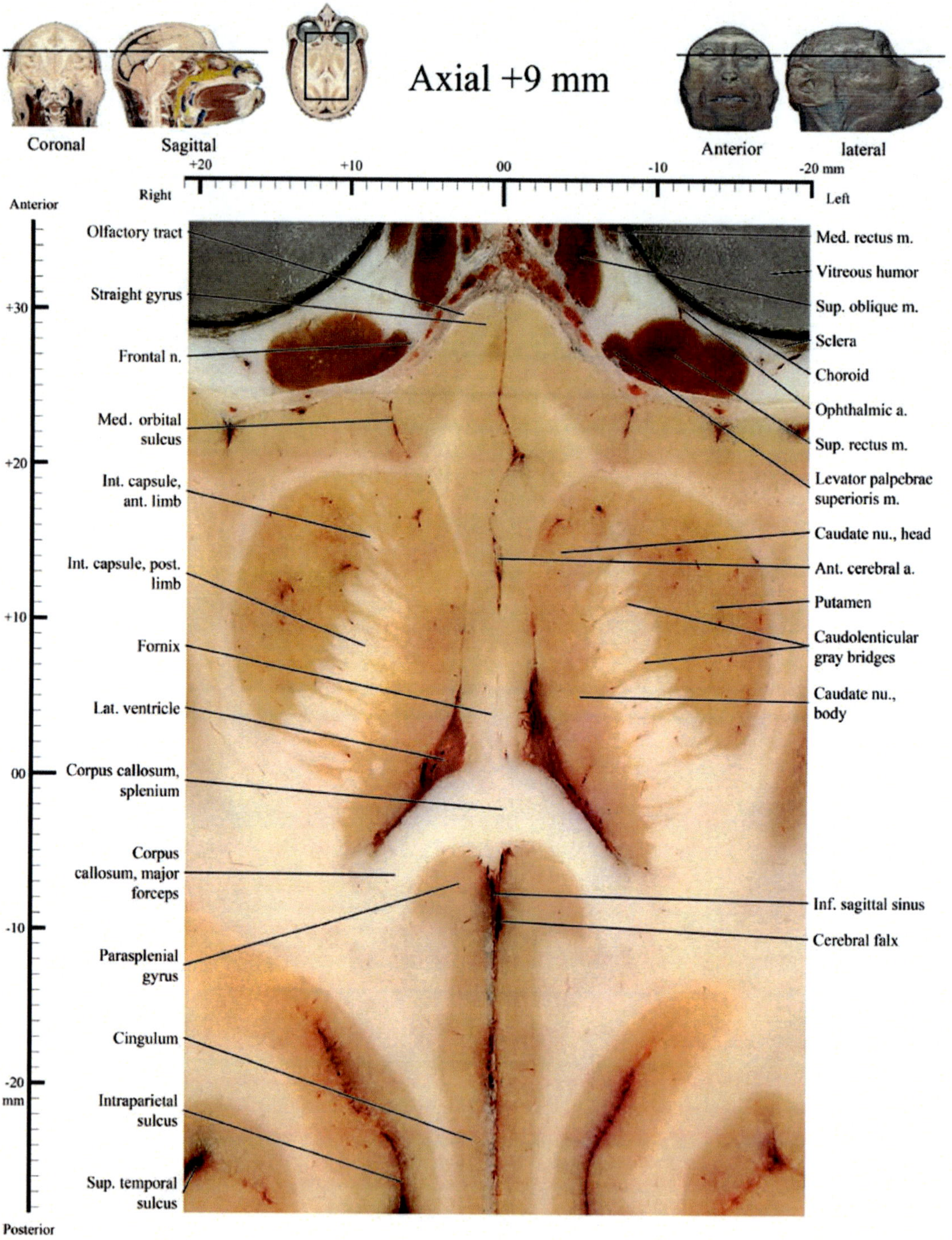
Axial +9 mm
Coronal
Sagittal
Anterior
lateral
+20
+10
00
-10
-20 mm
Right
Left
Anterior
Posterior
+30
+20
+10
00
-10
-20
mm
Olfactory tract
Straight gyrus
Frontal n.
Med. orbital sulcus
Int. capsule, ant. limb
Int. capsule, post. limb
Fornix
Lat. ventricle
Corpus callosum, splenium
Corpus callosum, major forceps
Parasplenial gyrus
Cingulum
Intraparietal sulcus
Sup. temporal sulcus
Med. rectus m.
Vitreous humor
Sup. oblique m.
Sclera
Choroid
Ophthalmic a.
Sup. rectus m.
Levator palpebrae superioris m.
Caudate nu., head
Ant. cerebral a.
Putamen
Caudolenticular gray bridges
Caudate nu., body
Inf. sagittal sinus
Cerebral falx

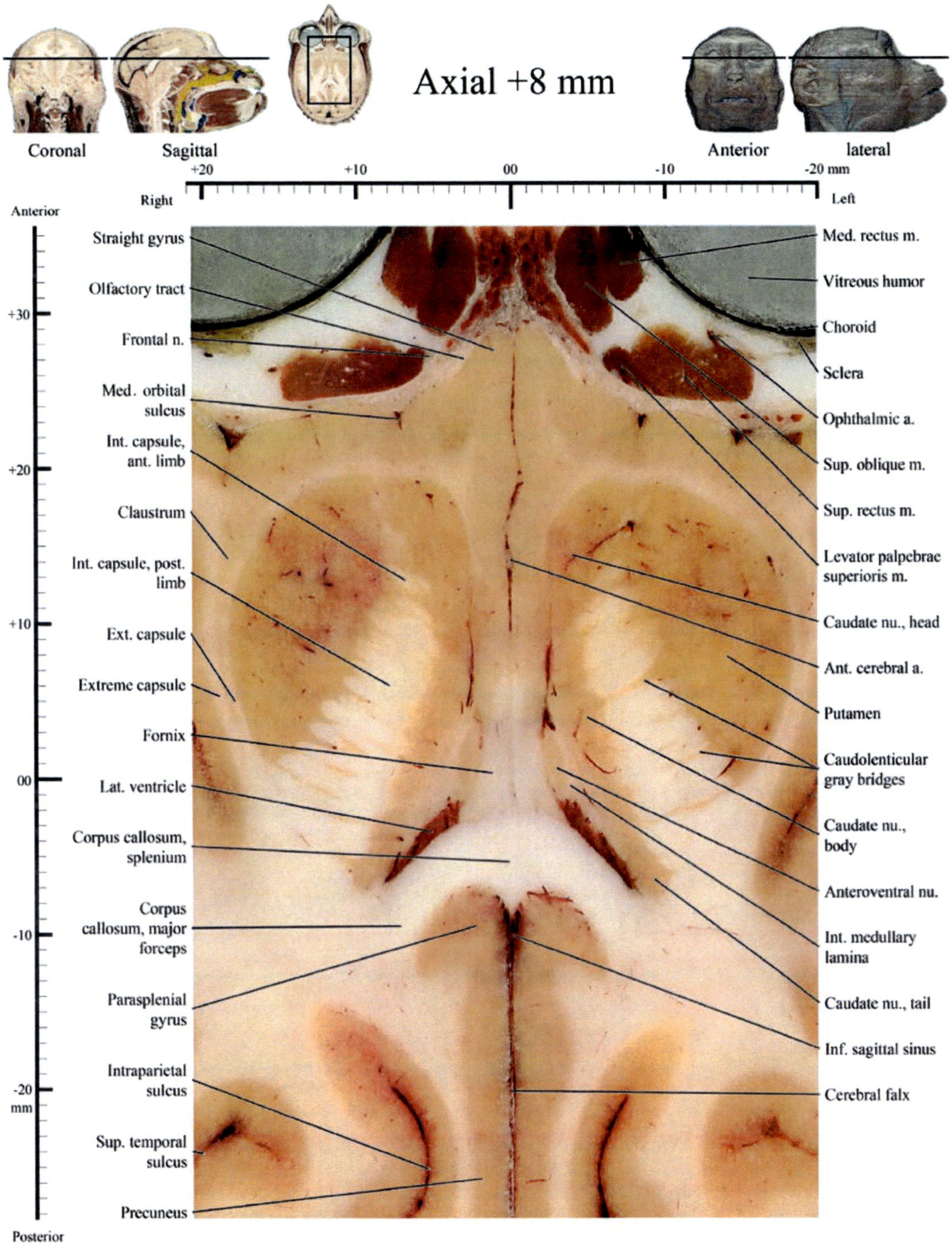
Axial +8 mm
Coronal
Sagittal
Anterior
lateral
+20
+10
00
-10
-20 mm
Right
Left
Anterior
Posterior
+30
+20
+10
00
-10
-20
mm
Straight gyrus
Olfactory tract
Frontal n.
Med. orbital sulcus
Int. capsule, ant. limb
Claustrum
Int. capsule, post. limb
Ext. capsule
Extreme capsule
Fornix
Lat. ventricle
Corpus callosum, splenium
Corpus callosum, major forceps
Parasplenial gyrus
Intraparietal sulcus
Sup. temporal sulcus
Precuneus
Med. rectus m.
Vitreous humor
Choroid
Sclera
Ophthalmic a.
Sup. oblique m.
Sup. rectus m.
Levator palpebrae superioris m.
Caudate nu., head
Ant. cerebral a.
Putamen
Caudolenticular gray bridges
Caudate nu., body
Anteroventral nu.
Int. medullary lamina
Caudate nu., tail
Inf. sagittal sinus
Cerebral falx

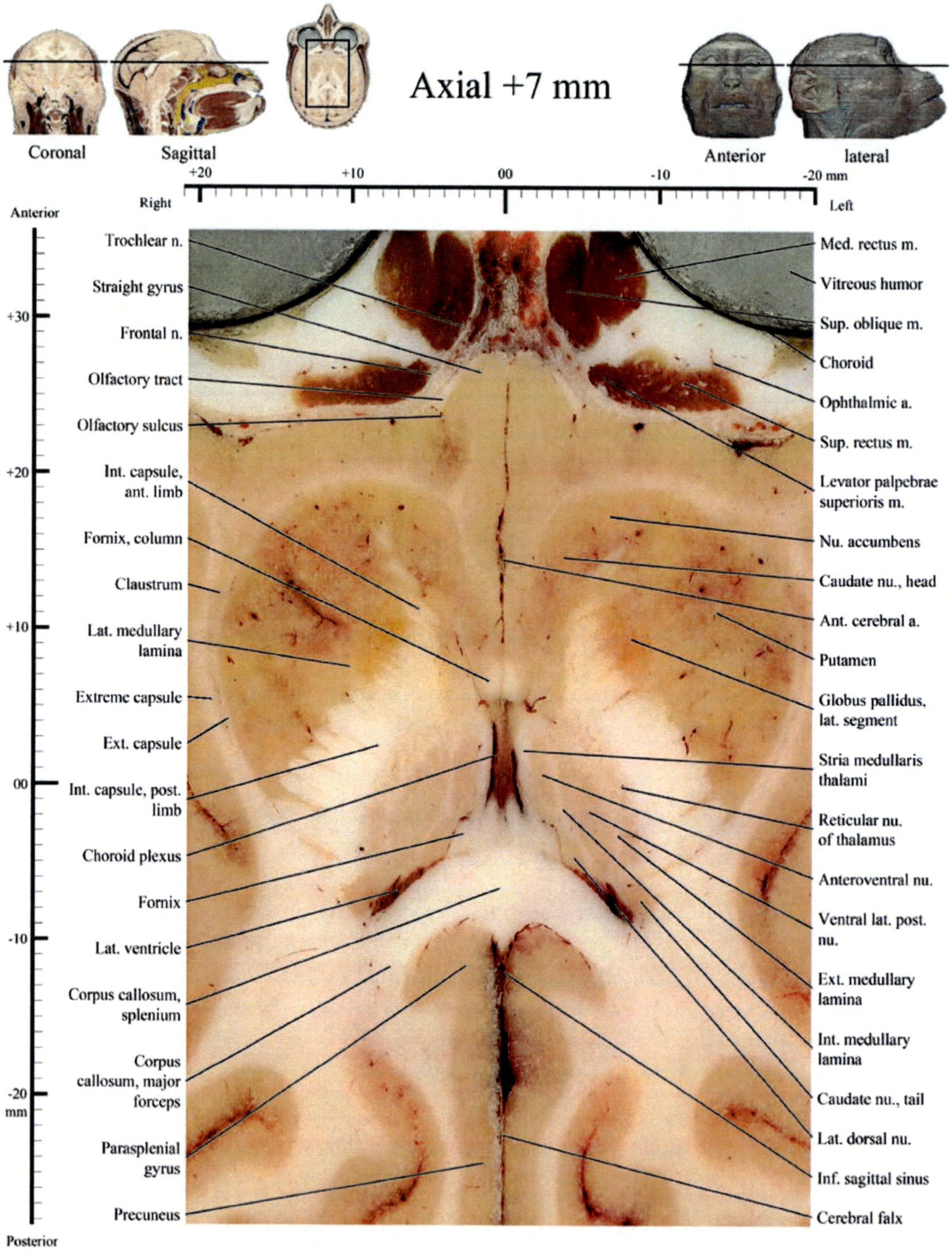
Coronal
Sagittal
Axial +7 mm
Anterior
lateral
+20
+10
00
-10
-20 mm
Right
Left
Anterior
Posterior
+30
+20
+10
00
-10
-20 mm
Trochlear n.
Straight gyrus
Frontal n.
Olfactory tract
Olfactory sulcus
Int. capsule, ant. limb
Fornix, column
Claustrum
Lat. medullary lamina
Extreme capsule
Ext. capsule
Int. capsule, post. limb
Choroid plexus
Fornix
Lat. ventricle
Corpus callosum, splenium
Corpus callosum, major forceps
Parasplenial gyrus
Precuneus
Med. rectus m.
Vitreous humor
Sup. oblique m.
Choroid
Ophthalmic a.
Sup. rectus m.
Levator palpebrae superioris m.
Nu. accumbens
Caudate nu., head
Ant. cerebral a.
Putamen
Globus pallidus, lat. segment
Stria medullaris thalami
Reticular nu. of thalamus
Anteroventral nu.
Ventral lat. post. nu.
Ext. medullary lamina
Int. medullary lamina
Caudate nu., tail
Lat. dorsal nu.
Inf. sagittal sinus
Cerebral falx

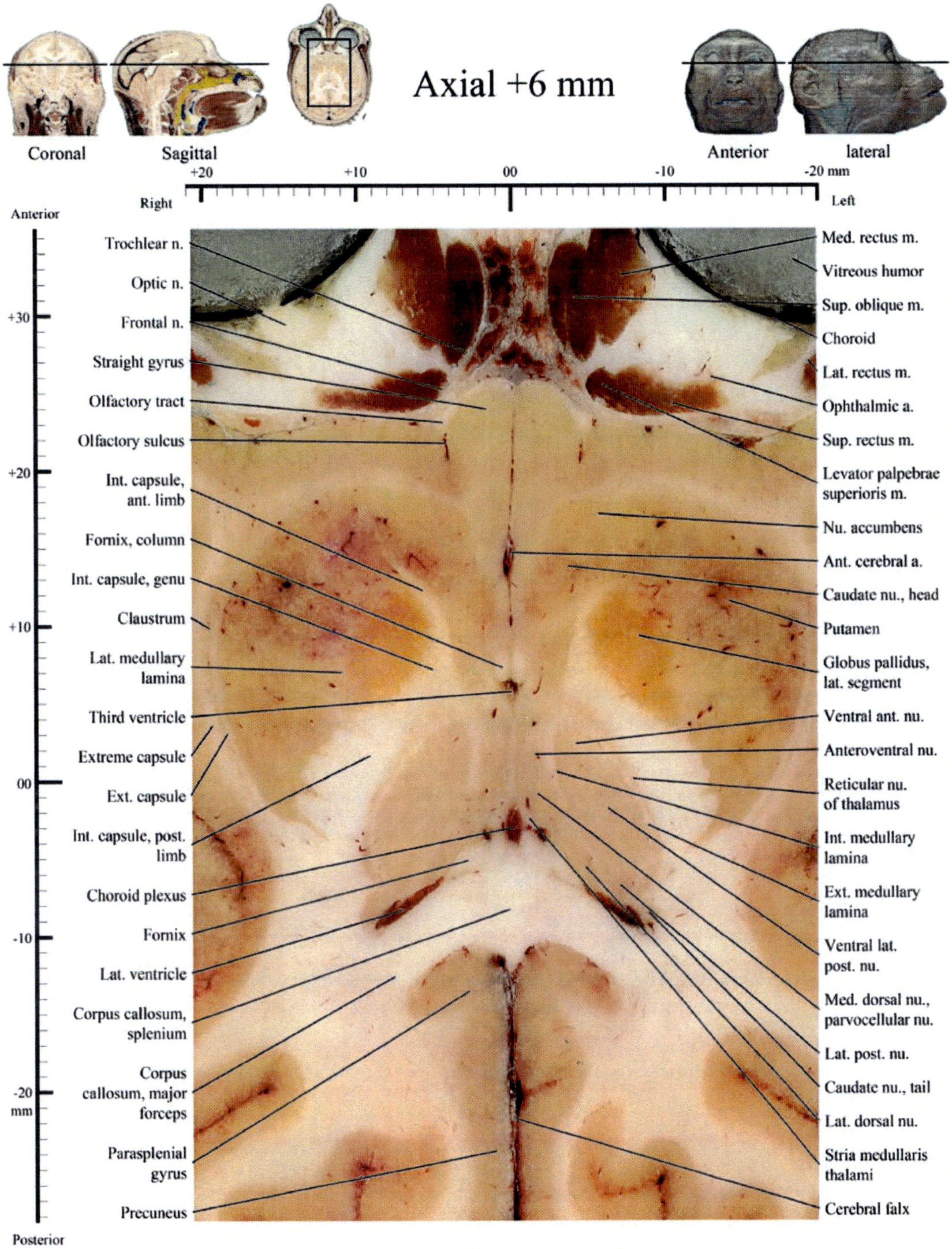
Coronal
Sagittal
Axial +6 mm
Anterior
lateral
+20
+10
00
-10
-20 mm
Right
Left
Anterior
Posterior
+30
+20
+10
00
-10
-20 mm
Trochlear n.
Optic n.
Frontal n.
Straight gyrus
Olfactory tract
Olfactory sulcus
Int. capsule, ant. limb
Fornix, column
Int. capsule, genu
Claustrum
Lat. medullary lamina
Third ventricle
Extreme capsule
Ext. capsule
Int. capsule, post. limb
Choroid plexus
Fornix
Lat. ventricle
Corpus callosum, splenium
Corpus callosum, major forceps
Parasplenial gyrus
Precuneus
Med. rectus m.
Vitreous humor
Sup. oblique m.
Choroid
Lat. rectus m.
Ophthalmic a.
Sup. rectus m.
Levator palpebrae superioris m.
Nu. accumbens
Ant. cerebral a.
Caudate nu., head
Putamen
Globus pallidus, lat. segment
Ventral ant. nu.
Anteroventral nu.
Reticular nu. of thalamus
Int. medullary lamina
Ext. medullary lamina
Ventral lat. post. nu.
Med. dorsal nu., parvocellular nu.
Lat. post. nu.
Caudate nu., tail
Lat. dorsal nu.
Stria medullaris thalami
Cerebral falx

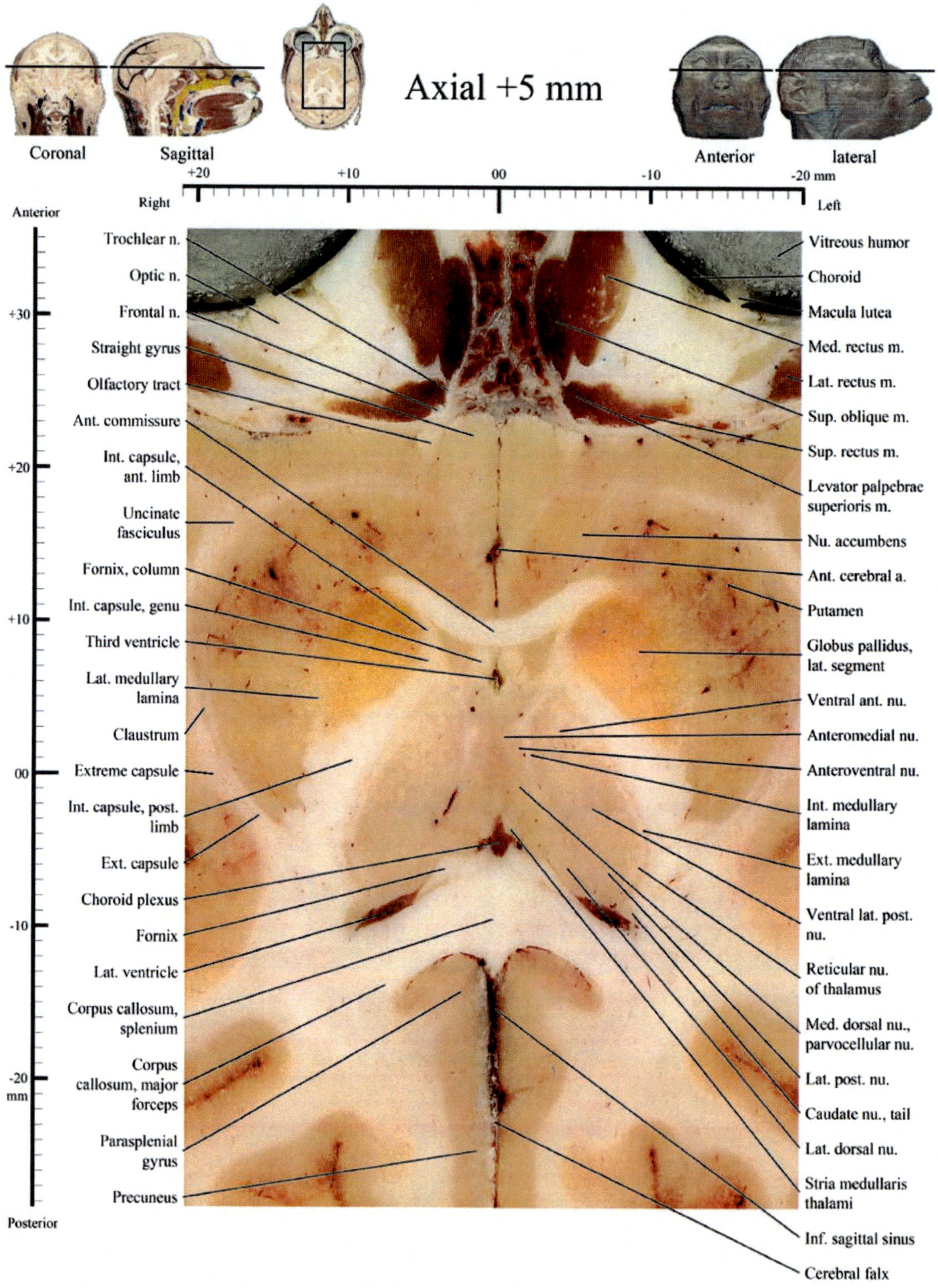
Axial +5 mm
Coronal
Sagittal
Anterior
lateral
+20
+10
00
-10
-20 mm
Right
Left
Anterior
Posterior
+30
+20
+10
00
-10
-20
mm
Trochlear n.
Optic n.
Frontal n.
Straight gyrus
Olfactory tract
Ant. commissure
Int. capsule, ant. limb
Uncinate fasciculus
Fornix, column
Int. capsule, genu
Third ventricle
Lat. medullary lamina
Claustrum
Extreme capsule
Int. capsule, post. limb
Ext. capsule
Choroid plexus
Fornix
Lat. ventricle
Corpus callosum, splenium
Corpus callosum, major forceps
Parasplenial gyrus
Precuneus
Vitreous humor
Choroid
Macula lutea
Med. rectus m.
Lat. rectus m.
Sup. oblique m.
Sup. rectus m.
Levator palpebrae superioris m.
Nu. accumbens
Ant. cerebral a.
Putamen
Globus pallidus, lat. segment
Ventral ant. nu.
Anteromedial nu.
Anteroventral nu.
Int. medullary lamina
Ext. medullary lamina
Ventral lat. post. nu.
Reticular nu. of thalamus
Med. dorsal nu., parvocellular nu.
Lat. post. nu.
Caudate nu., tail
Lat. dorsal nu.
Stria medullaris thalami
Inf. sagittal sinus
Cerebral falx

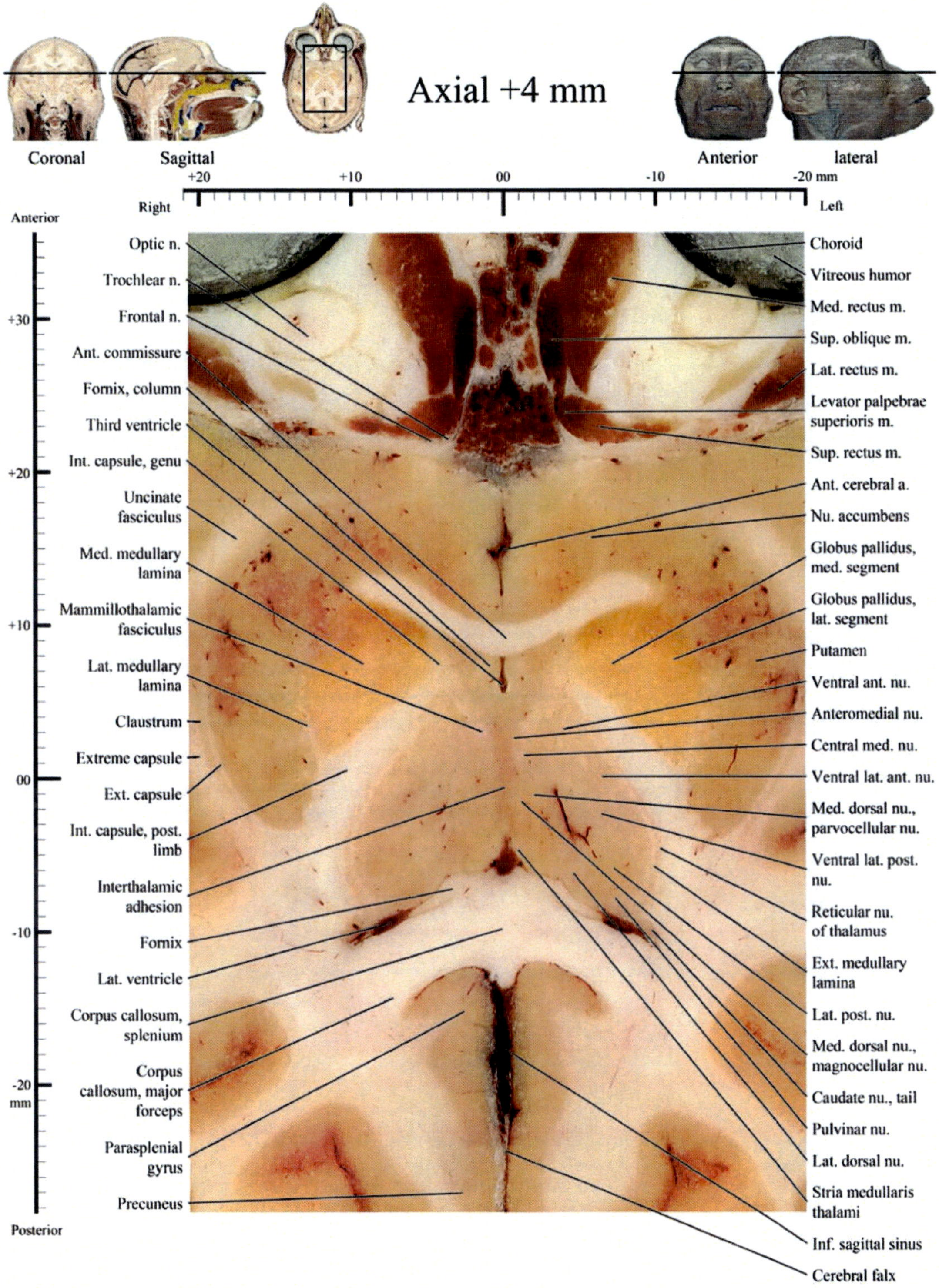
Coronal
Sagittal
Axial +4 mm
Anterior
lateral
+20
+10
00
-10
-20 mm
Right
Left
Anterior
Posterior
+30
+20
+10
00
-10
-20
mm
Optic n.
Trochlear n.
Frontal n.
Ant. commissure
Fornix, column
Third ventricle
Int. capsule, genu
Uncinate fasciculus
Med. medullary lamina
Mammillothalamic fasciculus
Lat. medullary lamina
Claustrum
Extreme capsule
Ext. capsule
Int. capsule, post. limb
Interthalamic adhesion
Fornix
Lat. ventricle
Corpus callosum, splenium
Corpus callosum, major forceps
Parasplenial gyrus
Precuneus
Choroid
Vitreous humor
Med. rectus m.
Sup. oblique m.
Lat. rectus m.
Levator palpebrae superioris m.
Sup. rectus m.
Ant. cerebral a.
Nu. accumbens
Globus pallidus, med. segment
Globus pallidus, lat. segment
Putamen
Ventral ant. nu.
Anteromedial nu.
Central med. nu.
Ventral lat. ant. nu.
Med. dorsal nu., parvocellular nu.
Ventral lat. post. nu.
Reticular nu. of thalamus
Ext. medullary lamina
Lat. post. nu.
Med. dorsal nu., magnocellular nu.
Caudate nu., tail
Pulvinar nu.
Lat. dorsal nu.
Stria medullaris thalami
Inf. sagittal sinus
Cerebral falx

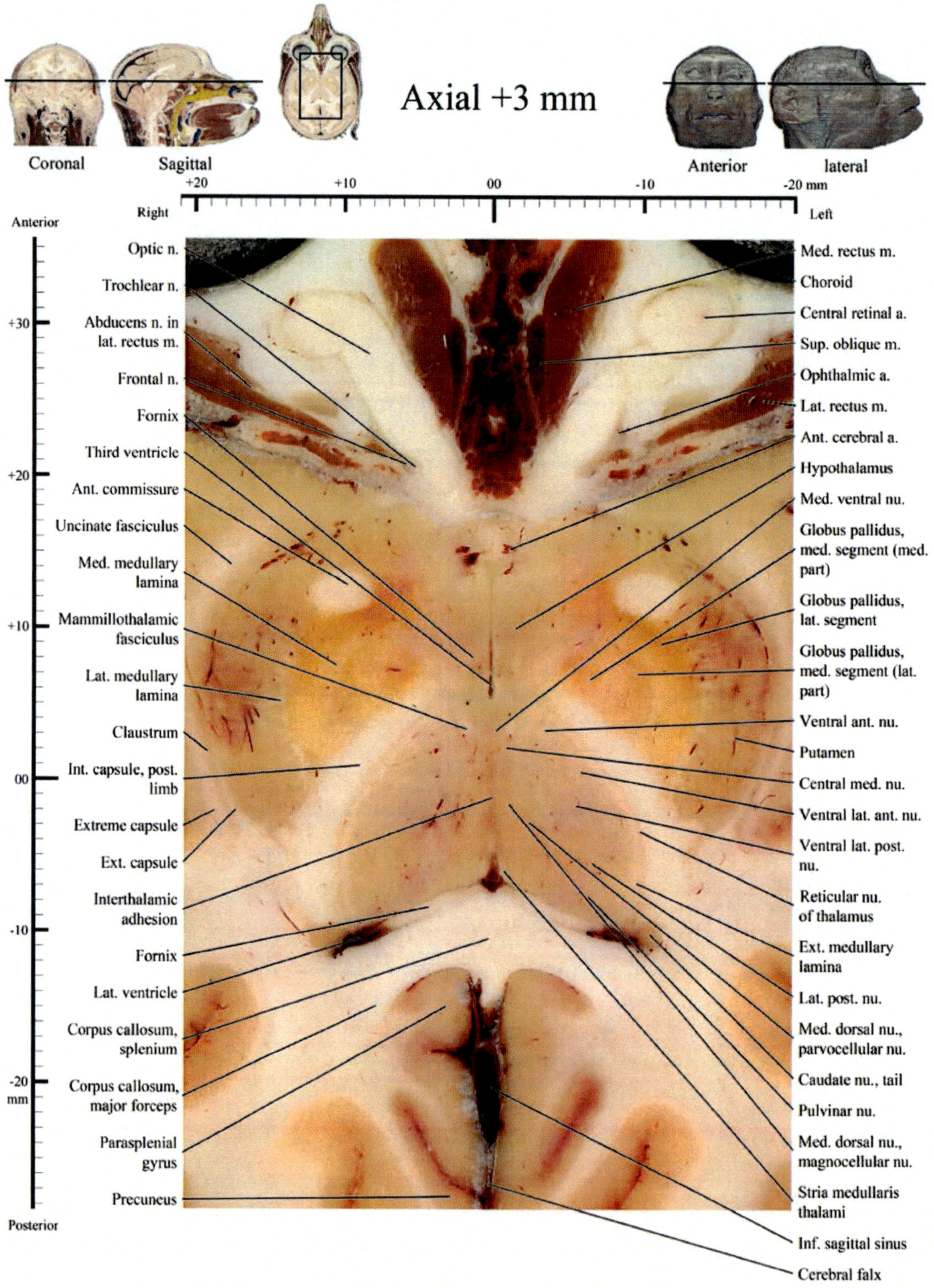
Axial +3 mm
Coronal
Sagittal
Anterior
lateral
+20
+10
00
-10
-20 mm
Right
Left
Anterior
Posterior
+30
+20
+10
00
-10
-20
mm
Optic n.
Trochlear n.
Abducens n. in lat. rectus m.
Frontal n.
Fornix
Third ventricle
Ant. commissure
Uncinate fasciculus
Med. medullary lamina
Mammillothalamic fasciculus
Lat. medullary lamina
Claustrum
Int. capsule, post. limb
Extreme capsule
Ext. capsule
Interthalamic adhesion
Fornix
Lat. ventricle
Corpus callosum, splenium
Corpus callosum, major forceps
Parasplenial gyrus
Precuneus
Med. rectus m.
Choroid
Central retinal a.
Sup. oblique m.
Ophthalmic a.
Lat. rectus m.
Ant. cerebral a.
Hypothalamus
Med. ventral nu.
Globus pallidus, med. segment (med. part)
Globus pallidus, lat. segment
Globus pallidus, med. segment (lat. part)
Ventral ant. nu.
Putamen
Central med. nu.
Ventral lat. ant. nu.
Ventral lat. post. nu.
Reticular nu. of thalamus
Ext. medullary lamina
Lat. post. nu.
Med. dorsal nu., parvocellular nu.
Caudate nu., tail
Pulvinar nu.
Med. dorsal nu., magnocellular nu.
Stria medullaris thalami
Inf. sagittal sinus
Cerebral falx

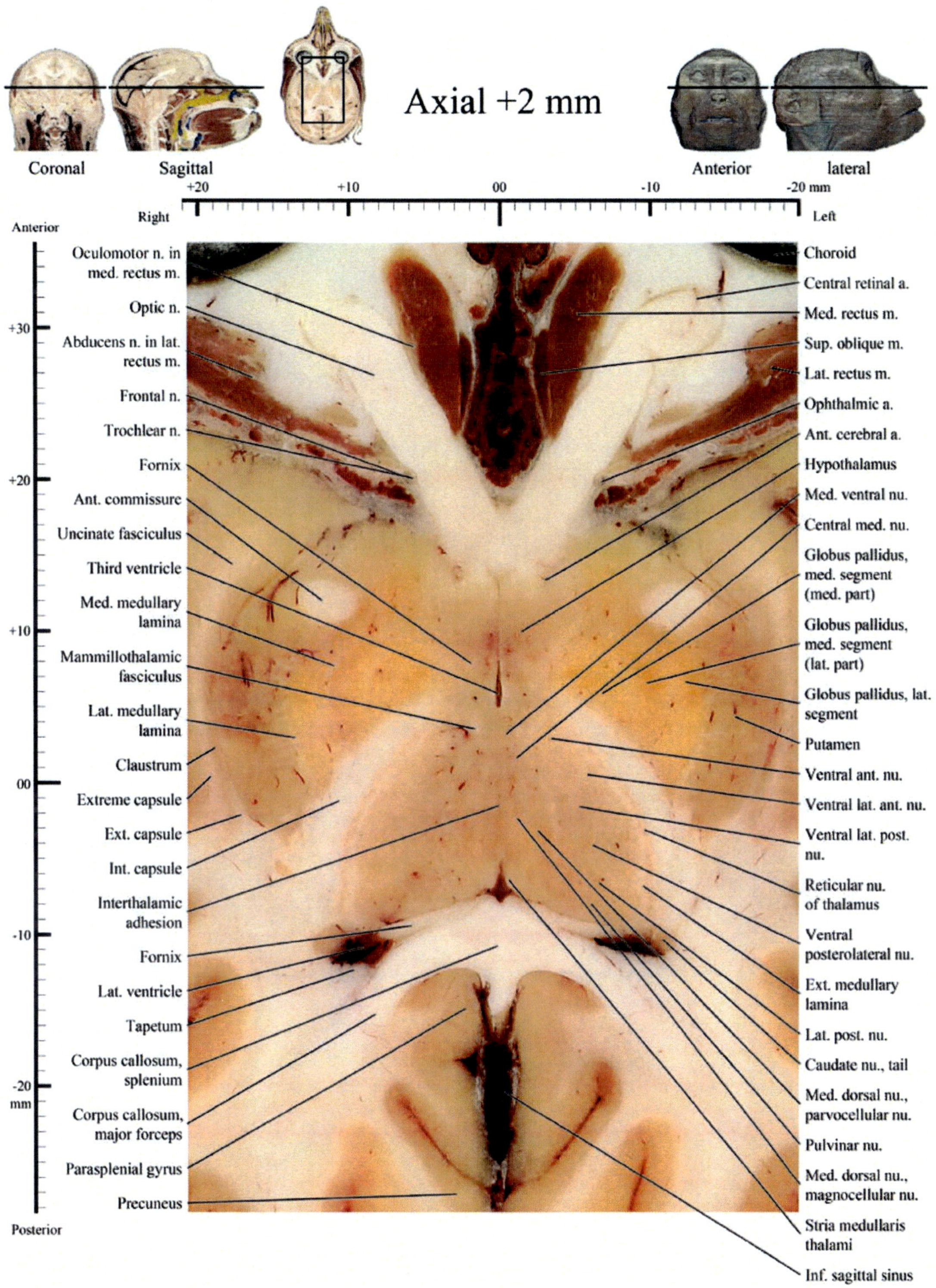

Axial +2 mm
Coronal
Sagittal
Anterior
lateral
+20
+10
00
-10
-20 mm
Right
Left
Anterior
Posterior
+30
+20
+10
00
-10
-20
mm
Oculomotor n. in med. rectus m.
Optic n.
Abducens n. in lat. rectus m.
Frontal n.
Trochlear n.
Fornix
Ant. commissure
Uncinate fasciculus
Third ventricle
Med. medullary lamina
Mammillothalamic fasciculus
Lat. medullary lamina
Claustrum
Extreme capsule
Ext. capsule
Int. capsule
Interthalamic adhesion
Fornix
Lat. ventricle
Tapetum
Corpus callosum, splenium
Corpus callosum, major forceps
Parasplenial gyrus
Precuneus
Choroid
Central retinal a.
Med. rectus m.
Sup. oblique m.
Lat. rectus m.
Ophthalmic a.
Ant. cerebral a.
Hypothalamus
Med. ventral nu.
Central med. nu.
Globus pallidus, med. segment (med. part)
Globus pallidus, med. segment (lat. part)
Globus pallidus, lat. segment
Putamen
Ventral ant. nu.
Ventral lat. ant. nu.
Ventral lat. post. nu.
Reticular nu. of thalamus
Ventral posterolateral nu.
Ext. medullary lamina
Lat. post. nu.
Caudate nu., tail
Med. dorsal nu., parvocellular nu.
Pulvinar nu.
Med. dorsal nu., magnocellular nu.
Stria medullaris thalami
Inf. sagittal sinus

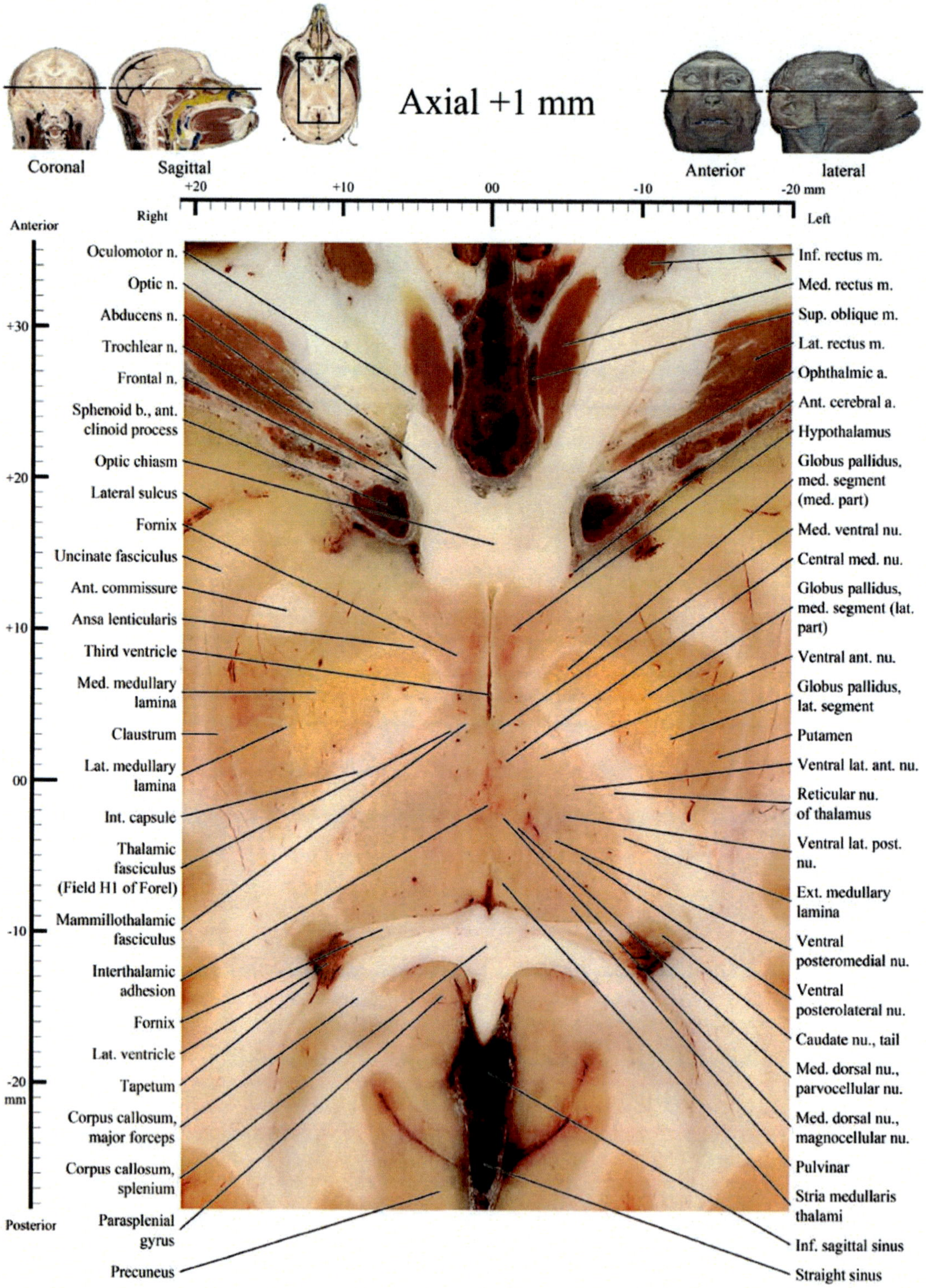
Coronal
Sagittal
Axial +1 mm
Anterior
lateral
+20
+10
00
-10
-20 mm
Right
Left
Anterior
Posterior
+30
+20
+10
00
-10
-20
mm
Oculomotor n.
Optic n.
Abducens n.
Trochlear n.
Frontal n.
Sphenoid b., ant. clinoid process
Optic chiasm
Lateral sulcus
Fornix
Uncinate fasciculus
Ant. commissure
Ansa lenticularis
Third ventricle
Med. medullary lamina
Claustrum
Lat. medullary lamina
Int. capsule
Thalamic fasciculus (Field H1 of Forel)
Mammillothalamic fasciculus
Interthalamic adhesion
Fornix
Lat. ventricle
Tapetum
Corpus callosum, major forceps
Corpus callosum, splenium
Parasplenial gyrus
Precuneus
Inf. rectus m.
Med. rectus m.
Sup. oblique m.
Lat. rectus m.
Ophthalmic a.
Ant. cerebral a.
Hypothalamus
Globus pallidus, med. segment (med. part)
Med. ventral nu.
Central med. nu.
Globus pallidus, med. segment (lat. part)
Ventral ant. nu.
Globus pallidus, lat. segment
Putamen
Ventral lat. ant. nu.
Reticular nu. of thalamus
Ventral lat. post. nu.
Ext. medullary lamina
Ventral posteromedial nu.
Ventral posterolateral nu.
Caudate nu., tail
Med. dorsal nu., parvocellular nu.
Med. dorsal nu., magnocellular nu.
Pulvinar
Stria medullaris thalami
Inf. sagittal sinus
Straight sinus

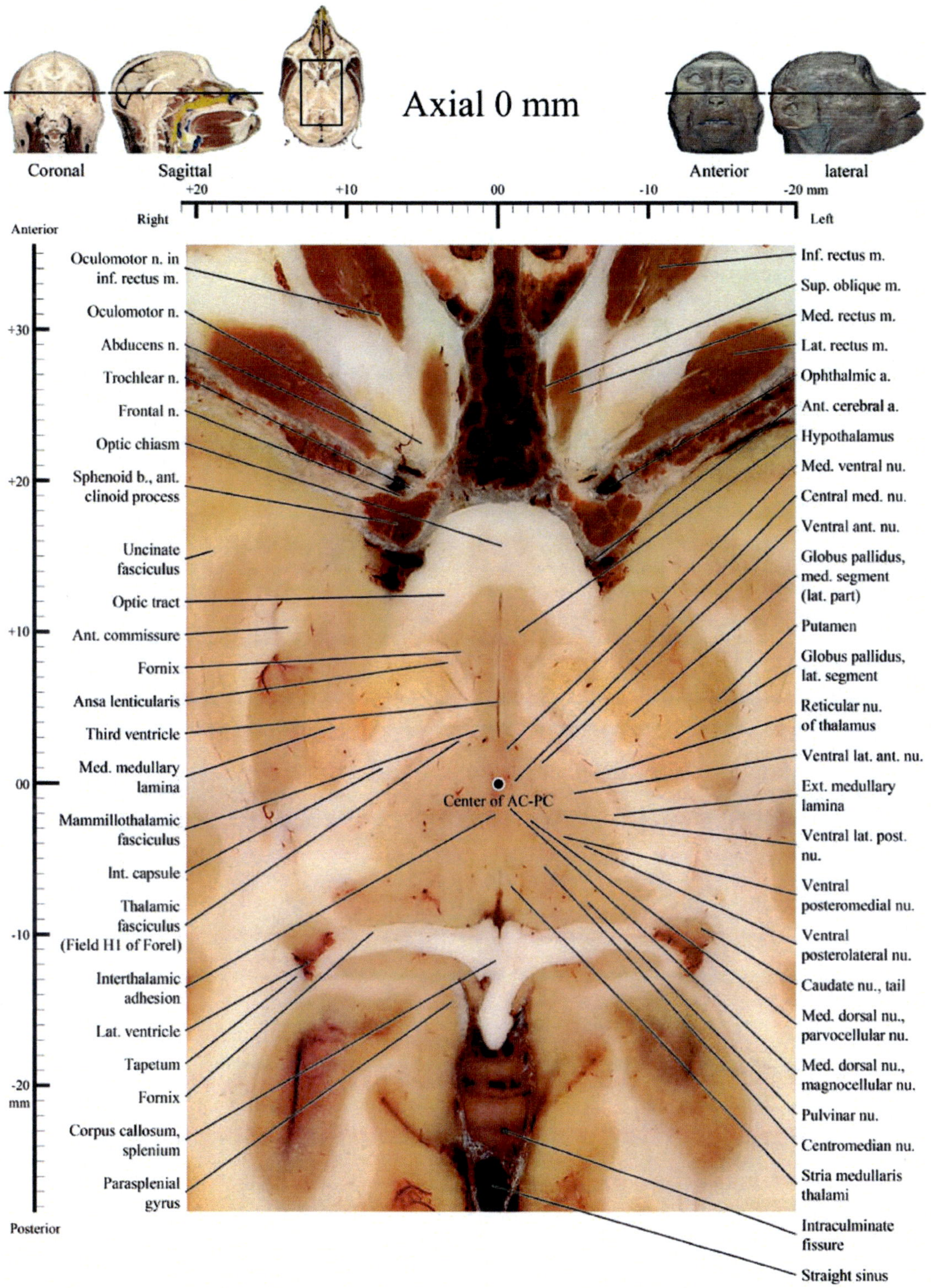
Axial 0 mm
Coronal
Sagittal
Anterior
lateral
+20
+10
00
-10
-20 mm
Right
Left
Anterior
Posterior
+30
+20
+10
00
-10
-20
mm
Oculomotor n. in inf. rectus m.
Oculomotor n.
Abducens n.
Trochlear n.
Frontal n.
Optic chiasm
Sphenoid b., ant. clinoid process
Uncinate fasciculus
Optic tract
Ant. commissure
Fornix
Ansa lenticularis
Third ventricle
Med. medullary lamina
Mammillothalamic fasciculus
Int. capsule
Thalamic fasciculus (Field H1 of Forel)
Interthalamic adhesion
Lat. ventricle
Tapetum
Fornix
Corpus callosum, splenium
Parasplenial gyrus
Inf. rectus m.
Sup. oblique m.
Med. rectus m.
Lat. rectus m.
Ophthalmic a.
Ant. cerebral a.
Hypothalamus
Med. ventral nu.
Central med. nu.
Ventral ant. nu.
Globus pallidus, med. segment (lat. part)
Putamen
Globus pallidus, lat. segment
Reticular nu. of thalamus
Ventral lat. ant. nu.
Ext. medullary lamina
Ventral lat. post. nu.
Ventral posteromedial nu.
Ventral posterolateral nu.
Caudate nu., tail
Med. dorsal nu., parvocellular nu.
Med. dorsal nu., magnocellular nu.
Pulvinar nu.
Centromedian nu.
Stria medullaris thalami
Intraculminate fissure
Straight sinus
Center of AC-PC

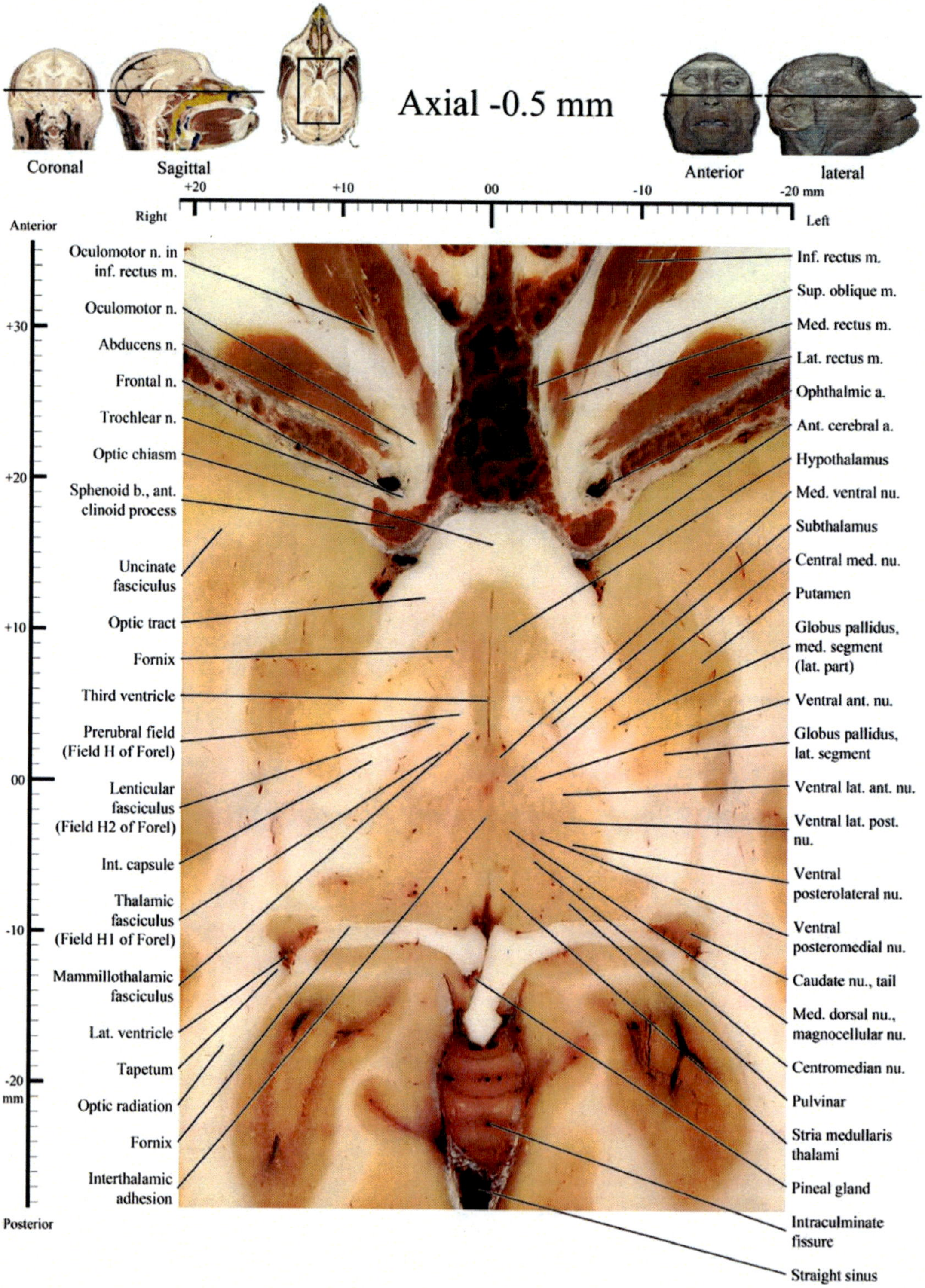
Coronal
Sagittal
Axial -0.5 mm
Anterior
lateral
+20
+10
00
-10
-20 mm
Right
Left
Anterior
Posterior
+30
+20
+10
00
-10
-20
mm
Oculomotor n. in inf. rectus m.
Oculomotor n.
Abducens n.
Frontal n.
Trochlear n.
Optic chiasm
Sphenoid b., ant. clinoid process
Uncinate fasciculus
Optic tract
Fornix
Third ventricle
Prerubral field (Field H of Forel)
Lenticular fasciculus (Field H2 of Forel)
Int. capsule
Thalamic fasciculus (Field H1 of Forel)
Mammillothalamic fasciculus
Lat. ventricle
Tapetum
Optic radiation
Fornix
Interthalamic adhesion
Inf. rectus m.
Sup. oblique m.
Med. rectus m.
Lat. rectus m.
Ophthalmic a.
Ant. cerebral a.
Hypothalamus
Med. ventral nu.
Subthalamus
Central med. nu.
Putamen
Globus pallidus, med. segment (lat. part)
Ventral ant. nu.
Globus pallidus, lat. segment
Ventral lat. ant. nu.
Ventral lat. post. nu.
Ventral posterolateral nu.
Ventral posteromedial nu.
Caudate nu., tail
Med. dorsal nu., magnocellular nu.
Centromedian nu.
Pulvinar
Stria medullaris thalami
Pineal gland
Intraculminate fissure
Straight sinus

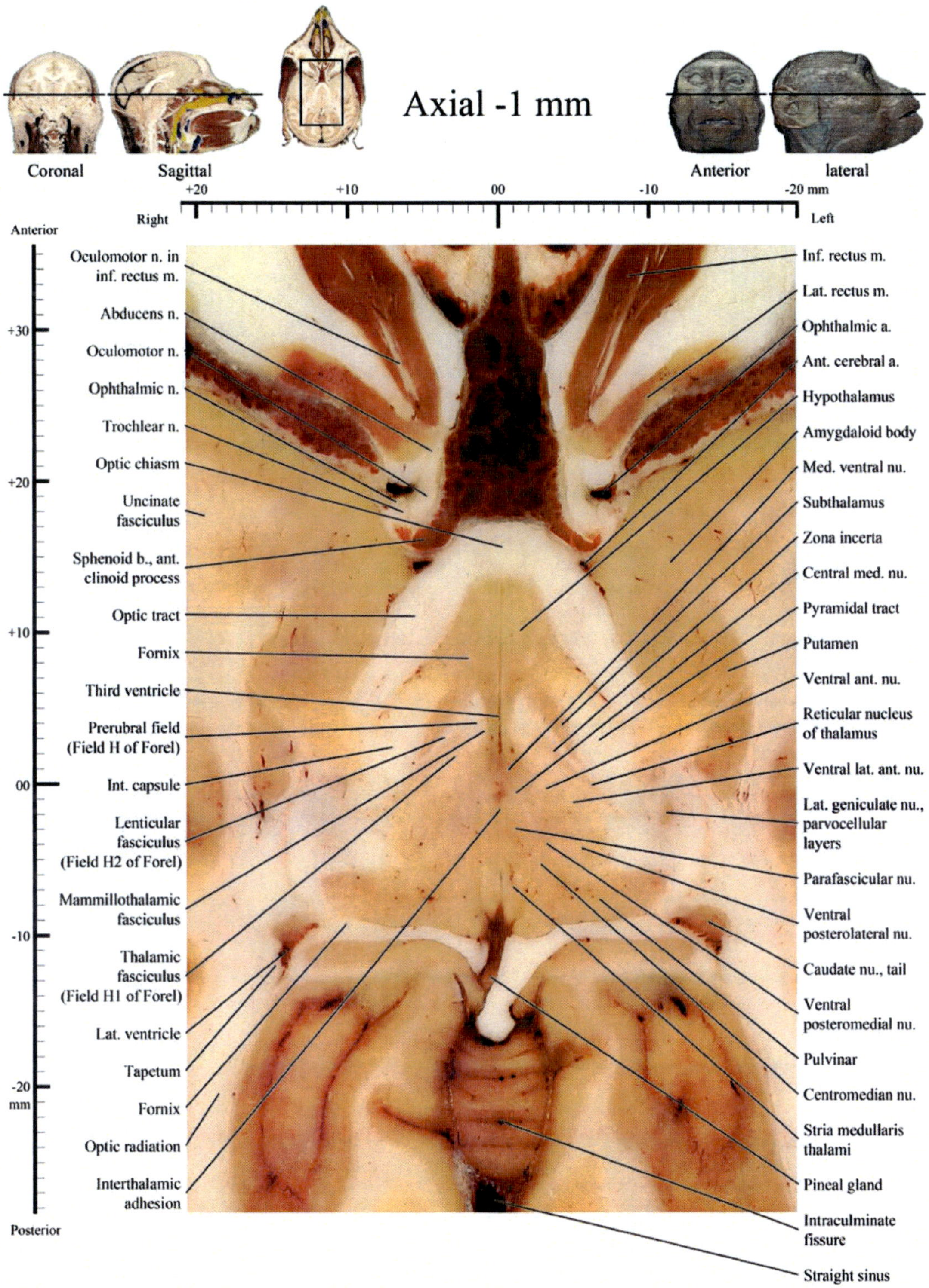
Axial -1 mm
Coronal
Sagittal
Anterior
lateral
+20
+10
00
-10
-20 mm
Right
Left
Anterior
Posterior
+30
+20
+10
00
-10
-20
mm
Oculomotor n. in inf. rectus m.
Abducens n.
Oculomotor n.
Ophthalmic n.
Trochlear n.
Optic chiasm
Uncinate fasciculus
Sphenoid b., ant. clinoid process
Optic tract
Fornix
Third ventricle
Prerubral field (Field H of Forel)
Int. capsule
Lenticular fasciculus (Field H2 of Forel)
Mammillothalamic fasciculus
Thalamic fasciculus (Field H1 of Forel)
Lat. ventricle
Tapetum
Fornix
Optic radiation
Interthalamic adhesion
Inf. rectus m.
Lat. rectus m.
Ophthalmic a.
Ant. cerebral a.
Hypothalamus
Amygdaloid body
Med. ventral nu.
Subthalamus
Zona incerta
Central med. nu.
Pyramidal tract
Putamen
Ventral ant. nu.
Reticular nucleus of thalamus
Ventral lat. ant. nu.
Lat. geniculate nu., parvocellular layers
Parafascicular nu.
Ventral posterolateral nu.
Caudate nu., tail
Ventral posteromedial nu.
Pulvinar
Centromedian nu.
Stria medullaris thalami
Pineal gland
Intraculminate fissure
Straight sinus

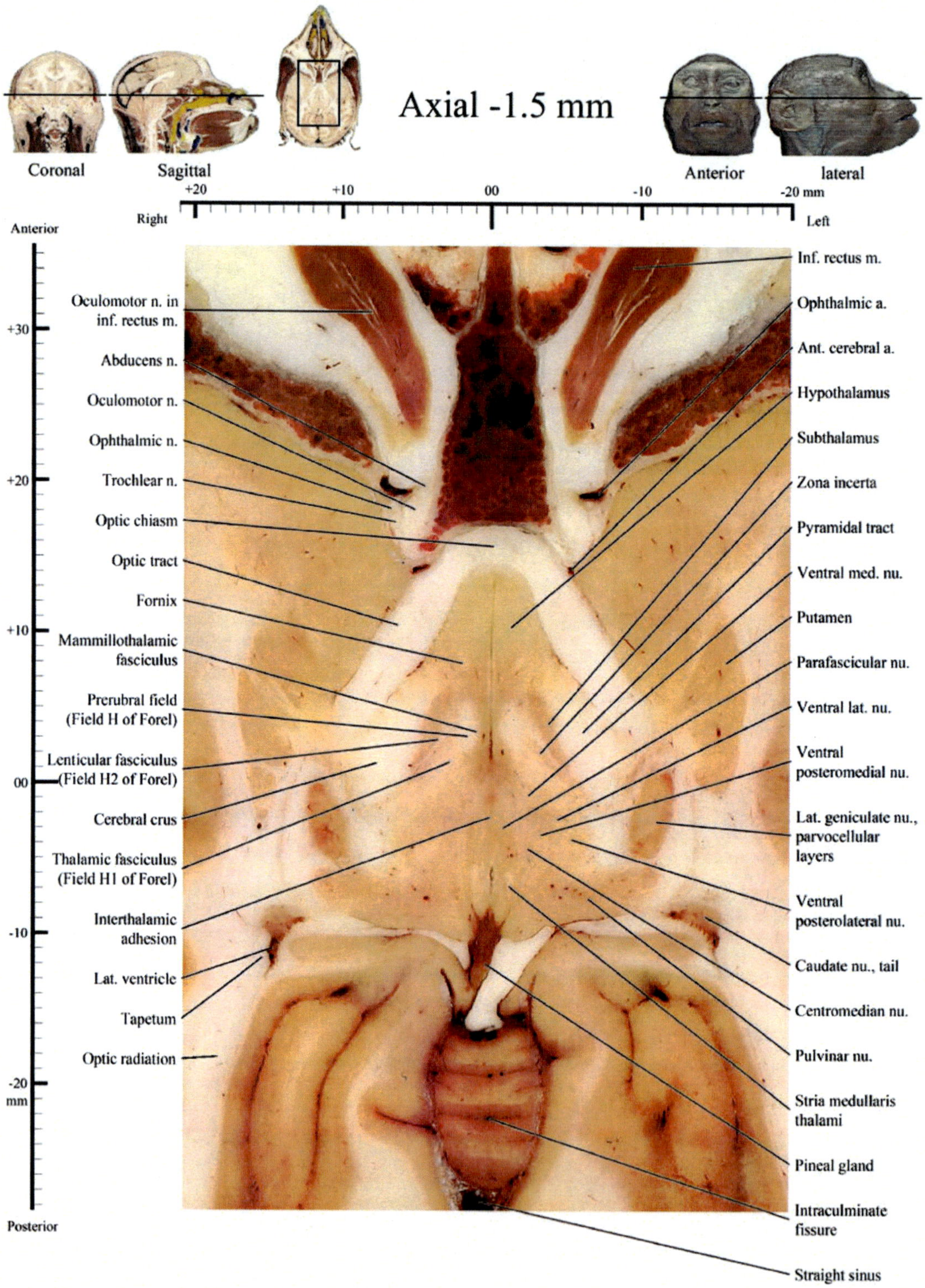

Coronal
Sagittal
Axial -1.5 mm
Anterior
lateral
+20
+10
00
-10
-20 mm
Right
Left
Anterior
Posterior
+30
+20
+10
00
-10
-20
mm
Oculomotor n. in inf. rectus m.
Abducens n.
Oculomotor n.
Ophthalmic n.
Trochlear n.
Optic chiasm
Optic tract
Fornix
Mammillothalamic fasciculus
Prerubral field (Field H of Forel)
Lenticular fasciculus (Field H2 of Forel)
Cerebral crus
Thalamic fasciculus (Field H1 of Forel)
Interthalamic adhesion
Lat. ventricle
Tapetum
Optic radiation
Inf. rectus m.
Ophthalmic a.
Ant. cerebral a.
Hypothalamus
Subthalamus
Zona incerta
Pyramidal tract
Ventral med. nu.
Putamen
Parafascicular nu.
Ventral lat. nu.
Ventral posteromedial nu.
Lat. geniculate nu., parvocellular layers
Ventral posterolateral nu.
Caudate nu., tail
Centromedian nu.
Pulvinar nu.
Stria medullaris thalami
Pineal gland
Intraculminate fissure
Straight sinus

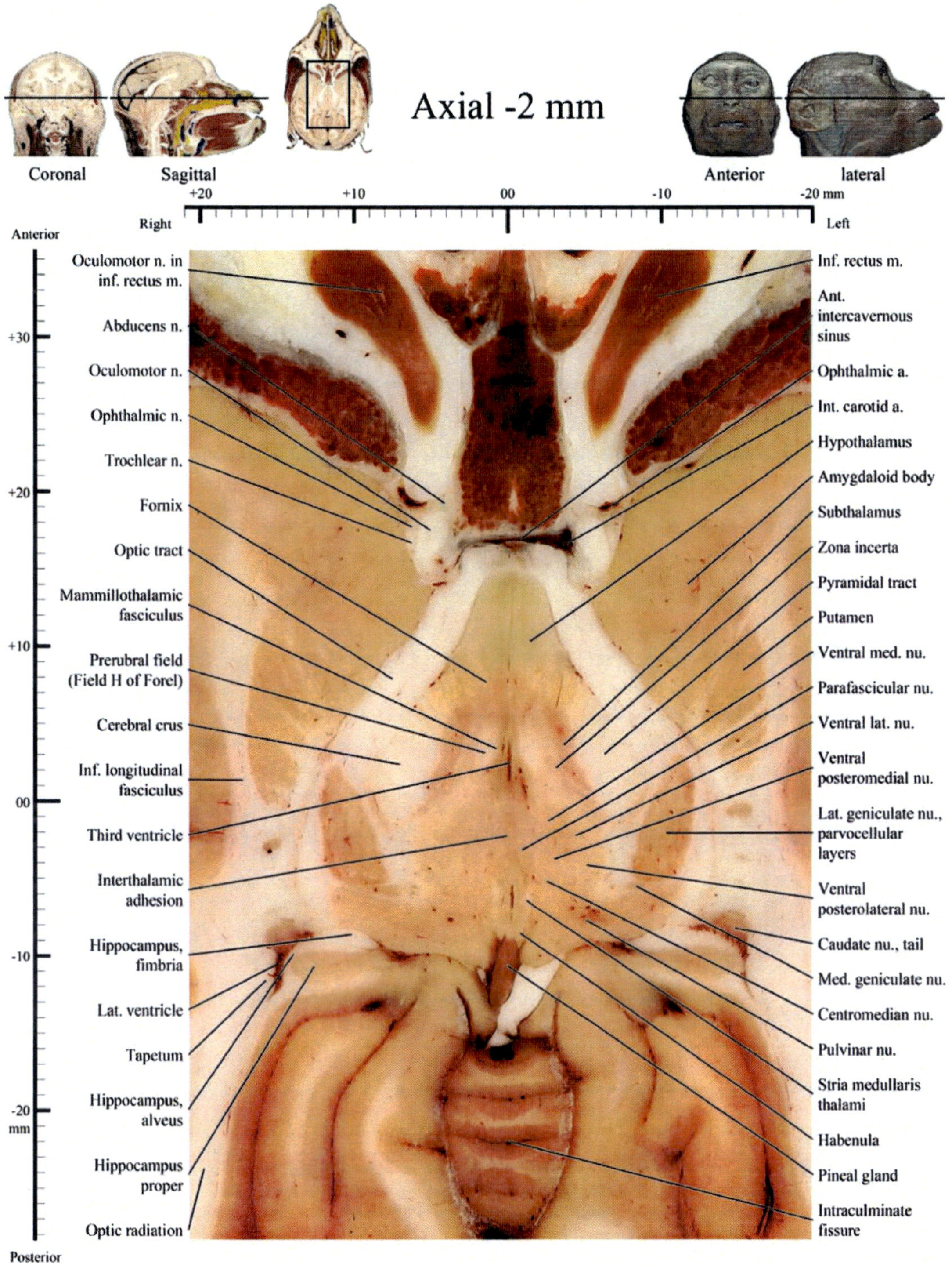
Coronal
Sagittal
Axial -2 mm
Anterior
lateral
+20
+10
00
-10
-20 mm
Right
Left
Anterior
+30
+20
+10
00
-10
-20
mm
Posterior
Oculomotor n. in inf. rectus m.
Abducens n.
Oculomotor n.
Ophthalmic n.
Trochlear n.
Fornix
Optic tract
Mammillothalamic fasciculus
Prerubral field (Field H of Forel)
Cerebral crus
Inf. longitudinal fasciculus
Third ventricle
Interthalamic adhesion
Hippocampus, fimbria
Lat. ventricle
Tapetum
Hippocampus, alveus
Hippocampus proper
Optic radiation
Inf. rectus m.
Ant. intercavernous sinus
Ophthalmic a.
Int. carotid a.
Hypothalamus
Amygdaloid body
Subthalamus
Zona incerta
Pyramidal tract
Putamen
Ventral med. nu.
Parafascicular nu.
Ventral lat. nu.
Ventral posteromedial nu.
Lat. geniculate nu., parvocellular layers
Ventral posterolateral nu.
Caudate nu., tail
Med. geniculate nu.
Centromedian nu.
Pulvinar nu.
Stria medullaris thalami
Habenula
Pineal gland
Intraculminate fissure

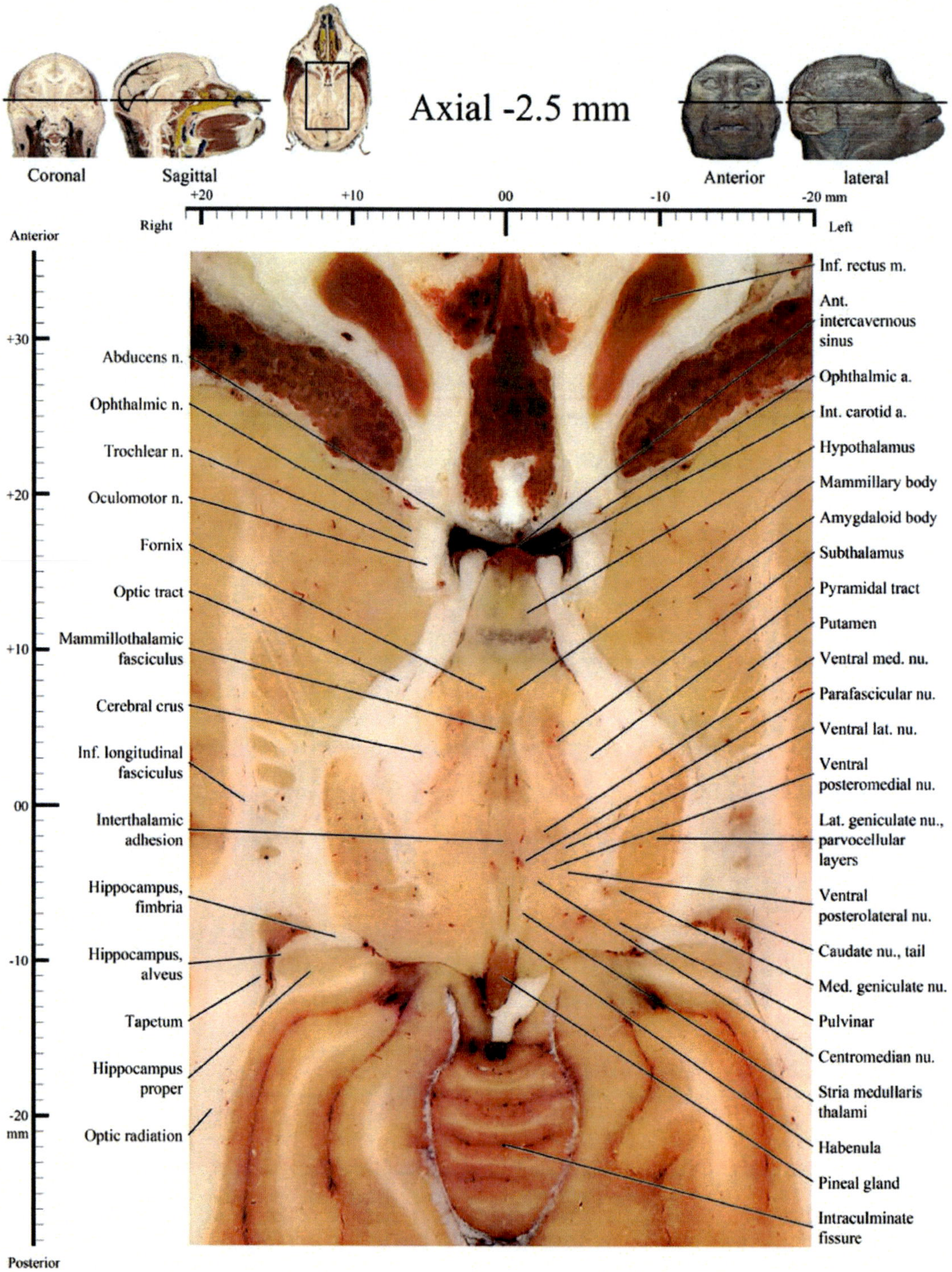
Axial -2.5 mm
Coronal
Sagittal
Anterior
lateral
+20
+10
00
-10
-20 mm
Right
Left
Anterior
Posterior
+30
+20
+10
00
-10
-20
mm
Abducens n.
Ophthalmic n.
Trochlear n.
Oculomotor n.
Fornix
Optic tract
Mammillothalamic fasciculus
Cerebral crus
Inf. longitudinal fasciculus
Interthalamic adhesion
Hippocampus, fimbria
Hippocampus, alveus
Tapetum
Hippocampus proper
Optic radiation
Inf. rectus m.
Ant. intercavernous sinus
Ophthalmic a.
Int. carotid a.
Hypothalamus
Mammillary body
Amygdaloid body
Subthalamus
Pyramidal tract
Putamen
Ventral med. nu.
Parafascicular nu.
Ventral lat. nu.
Ventral posteromedial nu.
Lat. geniculate nu., parvocellular layers
Ventral posterolateral nu.
Caudate nu., tail
Med. geniculate nu.
Pulvinar
Centromedian nu.
Stria medullaris thalami
Habenula
Pineal gland
Intraculminate fissure

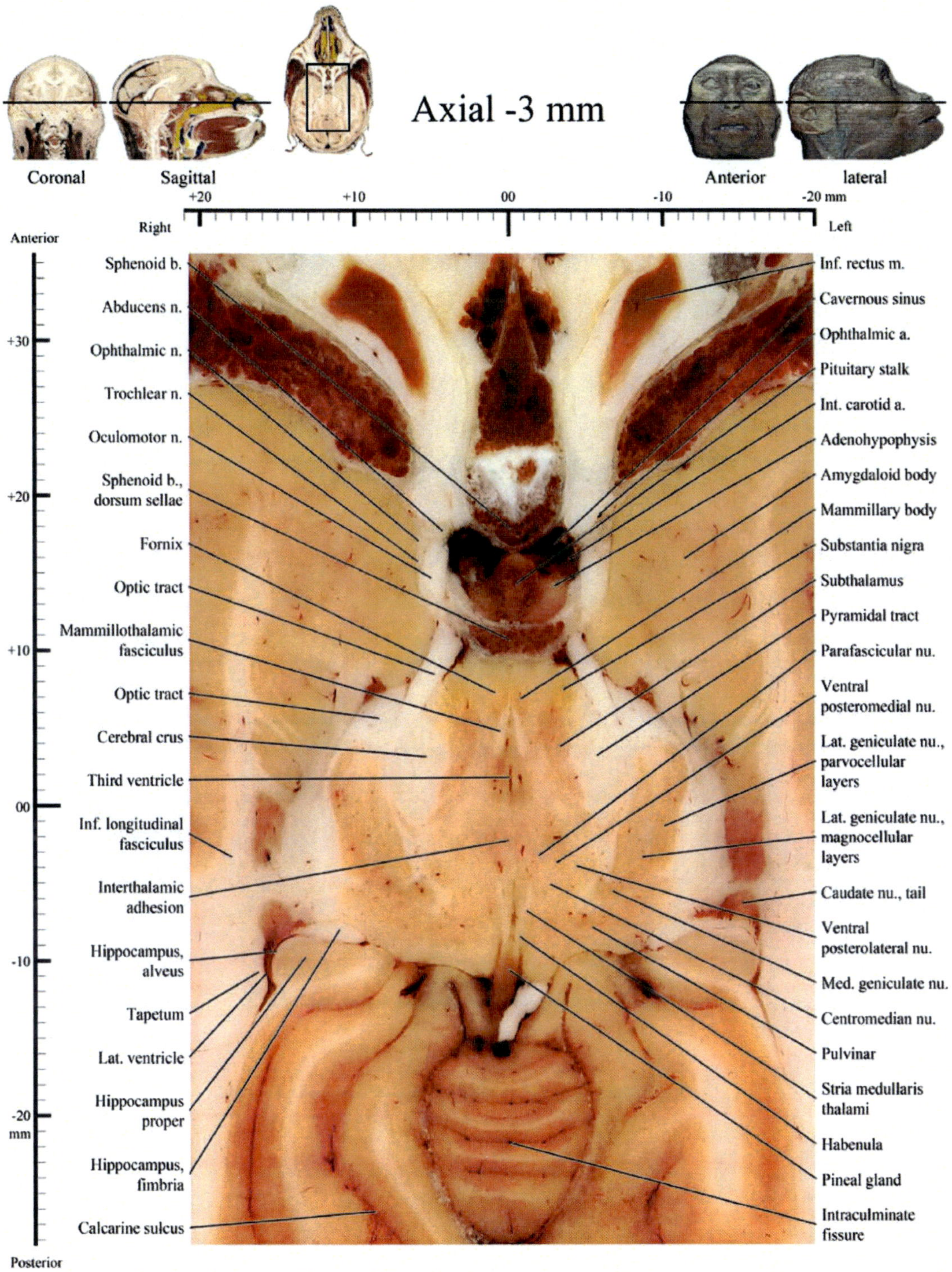
Axial -3 mm
Coronal
Sagittal
Anterior
lateral
+20
+10
00
-10
-20 mm
Right
Left
Anterior
Posterior
+30
+20
+10
00
-10
-20
mm
Sphenoid b.
Abducens n.
Ophthalmic n.
Trochlear n.
Oculomotor n.
Sphenoid b., dorsum sellae
Fornix
Optic tract
Mammillothalamic fasciculus
Optic tract
Cerebral crus
Third ventricle
Inf. longitudinal fasciculus
Interthalamic adhesion
Hippocampus, alveus
Tapetum
Lat. ventricle
Hippocampus proper
Hippocampus, fimbria
Calcarine sulcus
Inf. rectus m.
Cavernous sinus
Ophthalmic a.
Pituitary stalk
Int. carotid a.
Adenohypophysis
Amygdaloid body
Mammillary body
Substantia nigra
Subthalamus
Pyramidal tract
Parafascicular nu.
Ventral posteromedial nu.
Lat. geniculate nu., parvocellular layers
Lat. geniculate nu., magnocellular layers
Caudate nu., tail
Ventral posterolateral nu.
Med. geniculate nu.
Centromedian nu.
Pulvinar
Stria medullaris thalami
Habenula
Pineal gland
Intraculminate fissure

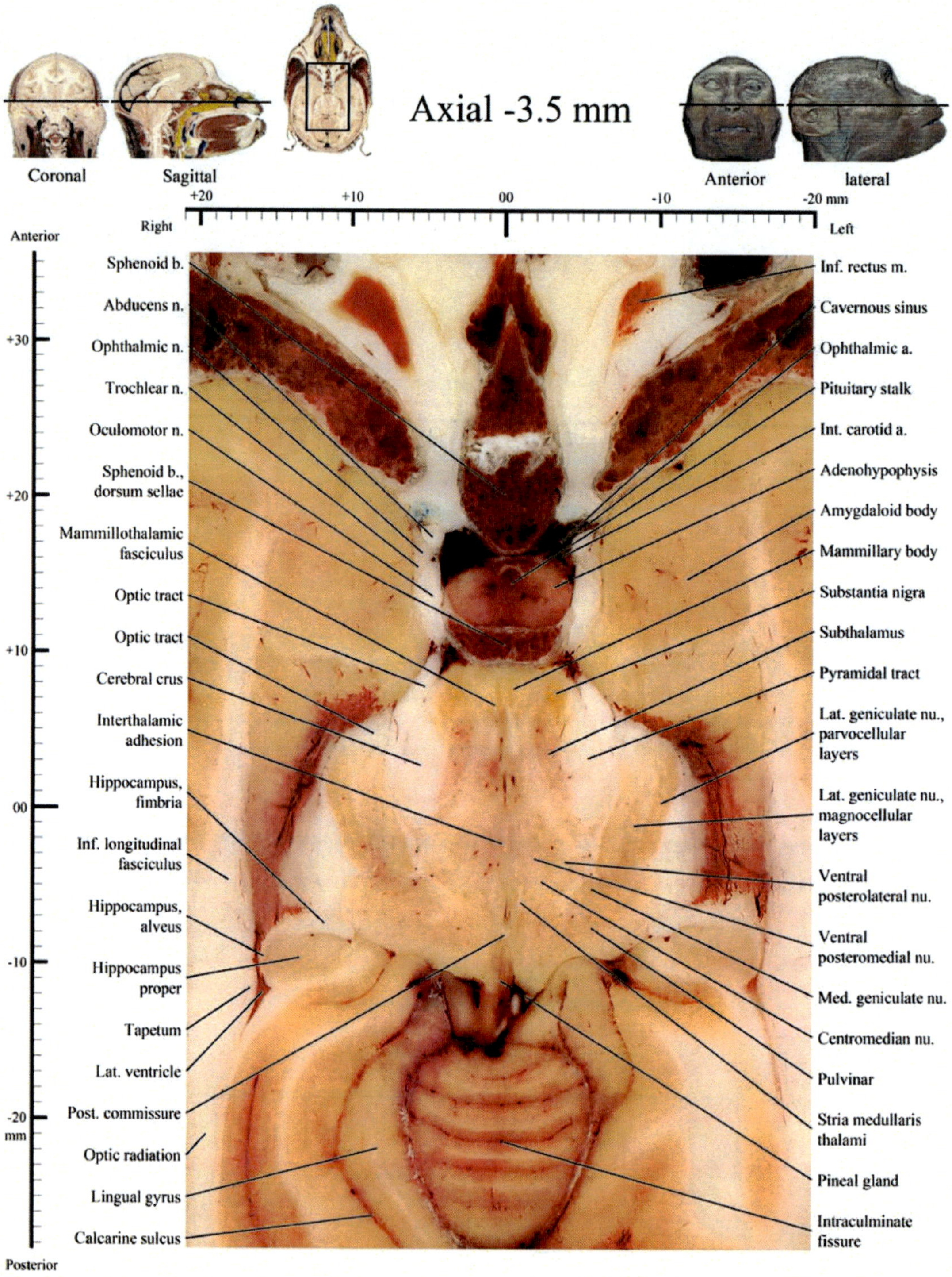

Axial -3.5 mm
Coronal
Sagittal
Anterior
lateral
+20
+10
00
-10
-20 mm
Right
Left
Anterior
Posterior
+30
+20
+10
00
-10
-20
mm
Sphenoid b.
Abducens n.
Ophthalmic n.
Trochlear n.
Oculomotor n.
Sphenoid b., dorsum sellae
Mammillothalamic fasciculus
Optic tract
Optic tract
Cerebral crus
Interthalamic adhesion
Hippocampus, fimbria
Inf. longitudinal fasciculus
Hippocampus, alveus
Hippocampus proper
Tapetum
Lat. ventricle
Post. commissure
Optic radiation
Lingual gyrus
Calcarine sulcus
Inf. rectus m.
Cavernous sinus
Ophthalmic a.
Pituitary stalk
Int. carotid a.
Adenohypophysis
Amygdaloid body
Mammillary body
Substantia nigra
Subthalamus
Pyramidal tract
Lat. geniculate nu., parvocellular layers
Lat. geniculate nu., magnocellular layers
Ventral posterolateral nu.
Ventral posteromedial nu.
Med. geniculate nu.
Centromedian nu.
Pulvinar
Stria medullaris thalami
Pineal gland
Intraculminate fissure

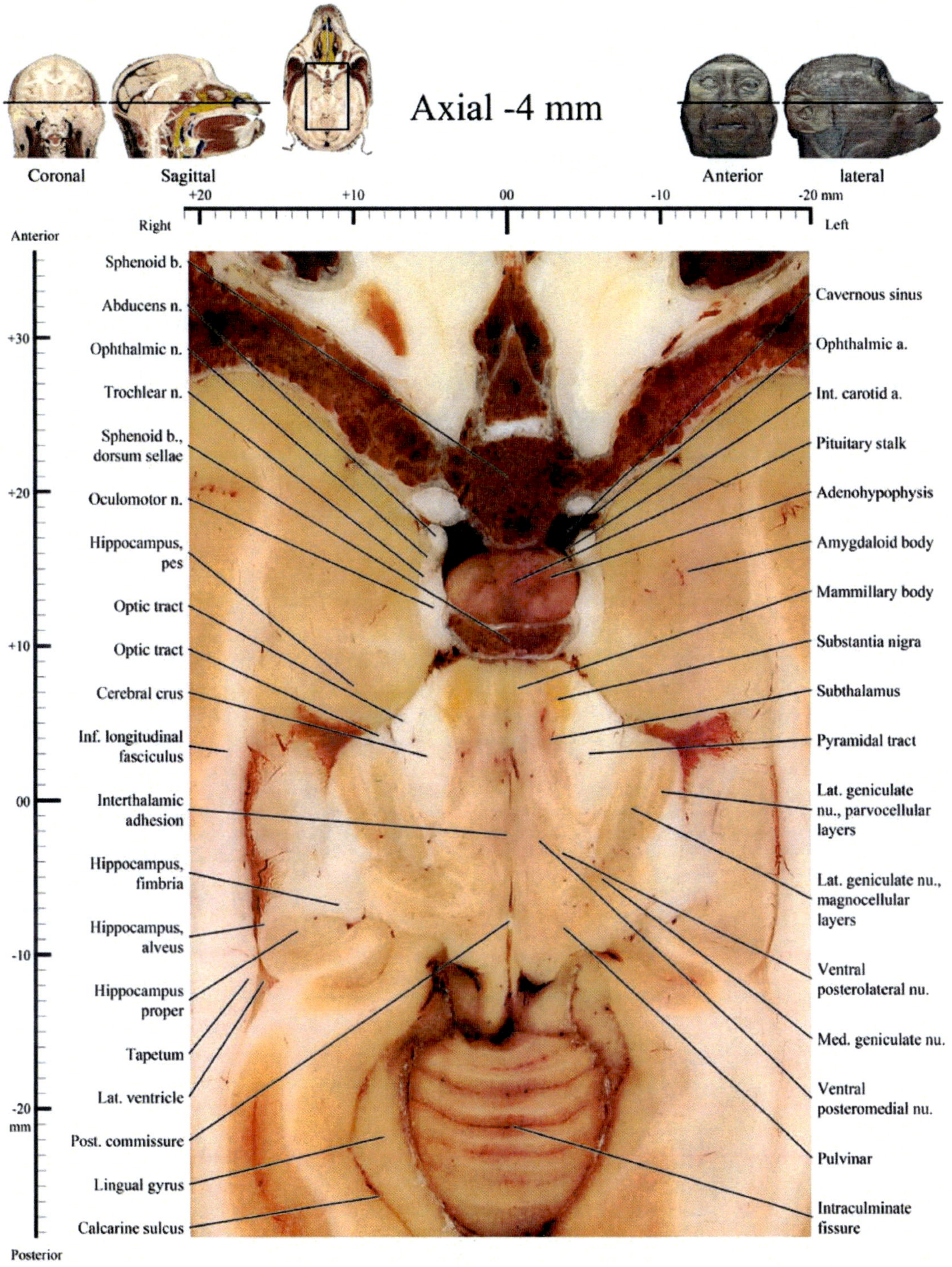
Axial -4 mm
Coronal
Sagittal
Anterior
lateral
+20
+10
00
-10
-20 mm
Right
Left
Anterior
Posterior
+30
+20
+10
00
-10
-20
mm
Sphenoid b.
Abducens n.
Ophthalmic n.
Trochlear n.
Sphenoid b., dorsum sellae
Oculomotor n.
Hippocampus, pes
Optic tract
Optic tract
Cerebral crus
Inf. longitudinal fasciculus
Interthalamic adhesion
Hippocampus, fimbria
Hippocampus, alveus
Hippocampus proper
Tapetum
Lat. ventricle
Post. commissure
Lingual gyrus
Calcarine sulcus
Cavernous sinus
Ophthalmic a.
Int. carotid a.
Pituitary stalk
Adenohypophysis
Amygdaloid body
Mammillary body
Substantia nigra
Subthalamus
Pyramidal tract
Lat. geniculate nu., parvocellular layers
Lat. geniculate nu., magnocellular layers
Ventral posterolateral nu.
Med. geniculate nu.
Ventral posteromedial nu.
Pulvinar
Intraculminate fissure

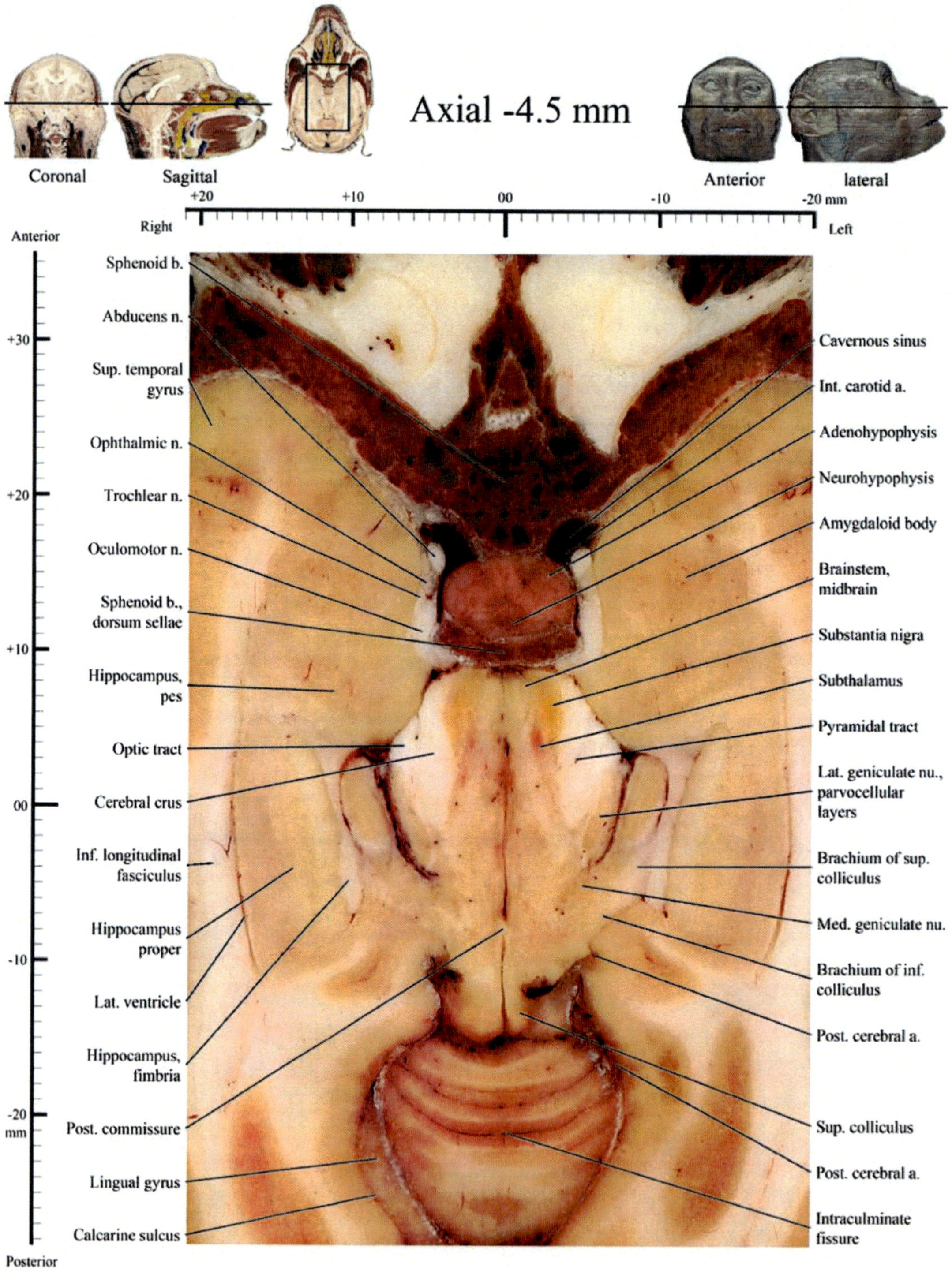
Coronal
Sagittal
Axial -4.5 mm
Anterior
lateral
+20
+10
00
-10
-20 mm
Right
Left
Anterior
+30
+20
+10
00
-10
-20 mm
Posterior
Sphenoid b.
Abducens n.
Sup. temporal gyrus
Ophthalmic n.
Trochlear n.
Oculomotor n.
Sphenoid b., dorsum sellae
Hippocampus, pes
Optic tract
Cerebral crus
Inf. longitudinal fasciculus
Hippocampus proper
Lat. ventricle
Hippocampus, fimbria
Post. commissure
Lingual gyrus
Calcarine sulcus
Cavernous sinus
Int. carotid a.
Adenohypophysis
Neurohypophysis
Amygdaloid body
Brainstem, midbrain
Substantia nigra
Subthalamus
Pyramidal tract
Lat. geniculate nu., parvocellular layers
Brachium of sup. colliculus
Med. geniculate nu.
Brachium of inf. colliculus
Post. cerebral a.
Sup. colliculus
Post. cerebral a.
Intraculminate fissure

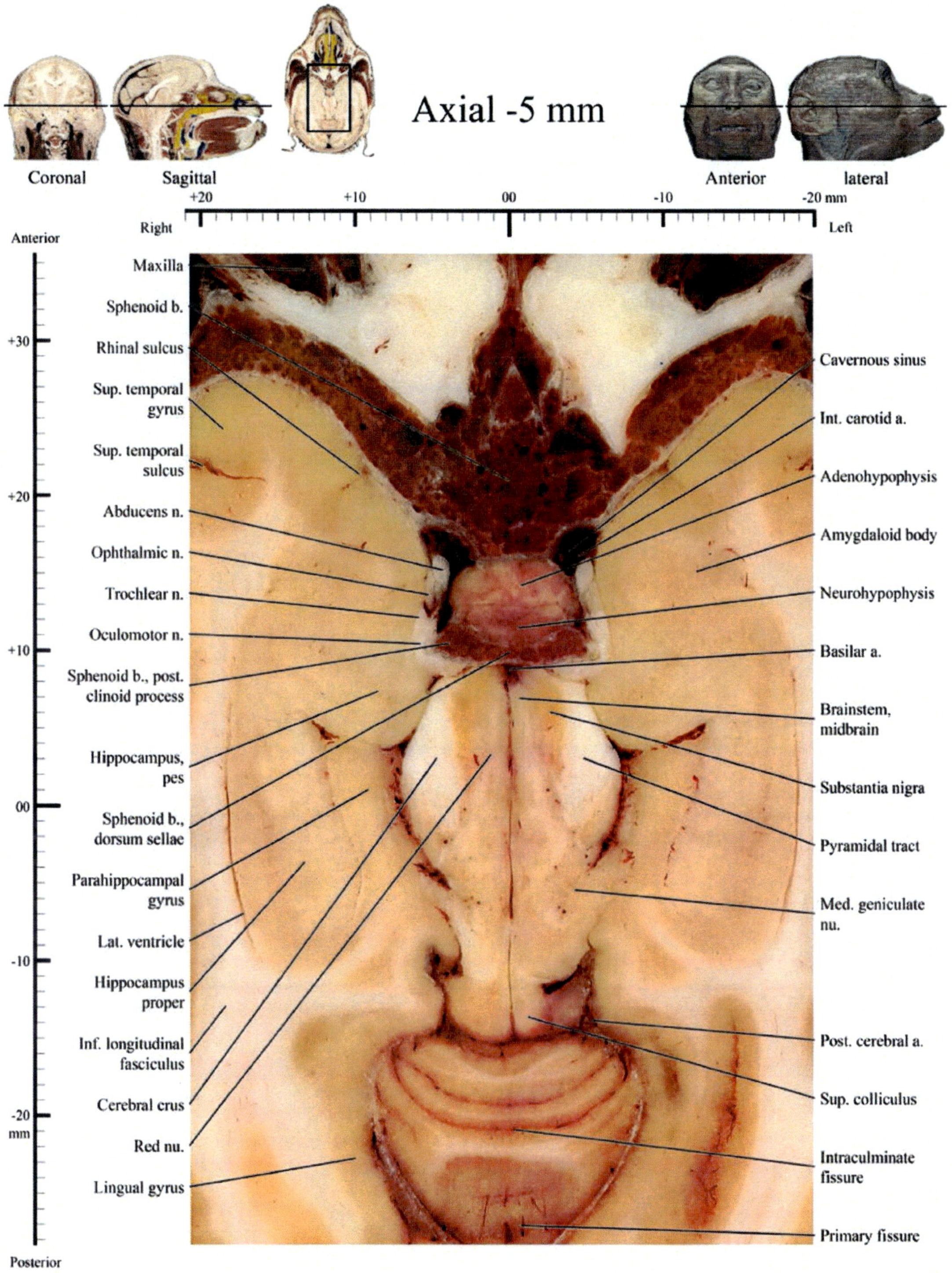

Axial -5 mm
Coronal
Sagittal
Anterior
lateral
+20
+10
00
-10
-20 mm
Right
Left
Anterior
Posterior
+30
+20
+10
00
-10
-20
mm
Maxilla
Sphenoid b.
Rhinal sulcus
Sup. temporal gyrus
Sup. temporal sulcus
Abducens n.
Ophthalmic n.
Trochlear n.
Oculomotor n.
Sphenoid b., post. clinoid process
Hippocampus, pes
Sphenoid b., dorsum sellae
Parahippocampal gyrus
Lat. ventricle
Hippocampus proper
Inf. longitudinal fasciculus
Cerebral crus
Red nu.
Lingual gyrus
Cavernous sinus
Int. carotid a.
Adenohypophysis
Amygdaloid body
Neurohypophysis
Basilar a.
Brainstem, midbrain
Substantia nigra
Pyramidal tract
Med. geniculate nu.
Post. cerebral a.
Sup. colliculus
Intraculminate fissure
Primary fissure

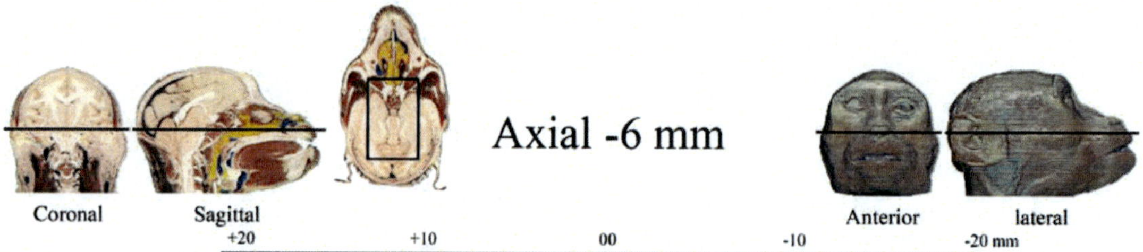

Coronal
Sagittal
Axial -6 mm
Anterior
lateral

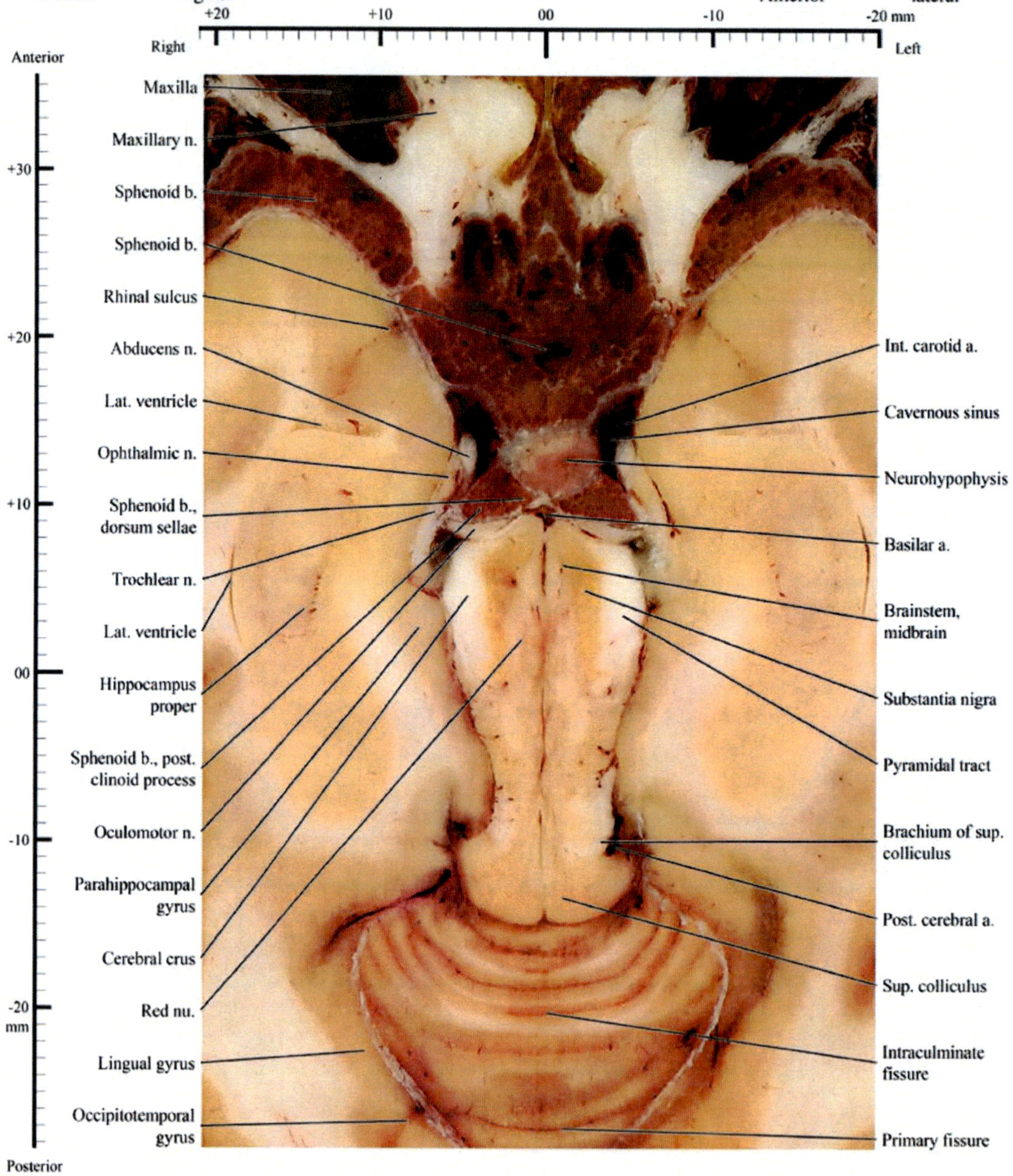

+20
+10
00
-10
-20 mm
Right
Left
Anterior
+30
+20
+10
00
-10
-20
mm
Posterior
Maxilla
Maxillary n.
Sphenoid b.
Sphenoid b.
Rhinal sulcus
Abducens n.
Lat. ventricle
Ophthalmic n.
Sphenoid b., dorsum sellae
Trochlear n.
Lat. ventricle
Hippocampus proper
Sphenoid b., post. clinoid process
Oculomotor n.
Parahippocampal gyrus
Cerebral crus
Red nu.
Lingual gyrus
Occipitotemporal gyrus
Int. carotid a.
Cavernous sinus
Neurohypophysis
Basilar a.
Brainstem, midbrain
Substantia nigra
Pyramidal tract
Brachium of sup. colliculus
Post. cerebral a.
Sup. colliculus
Intraculminate fissure
Primary fissure

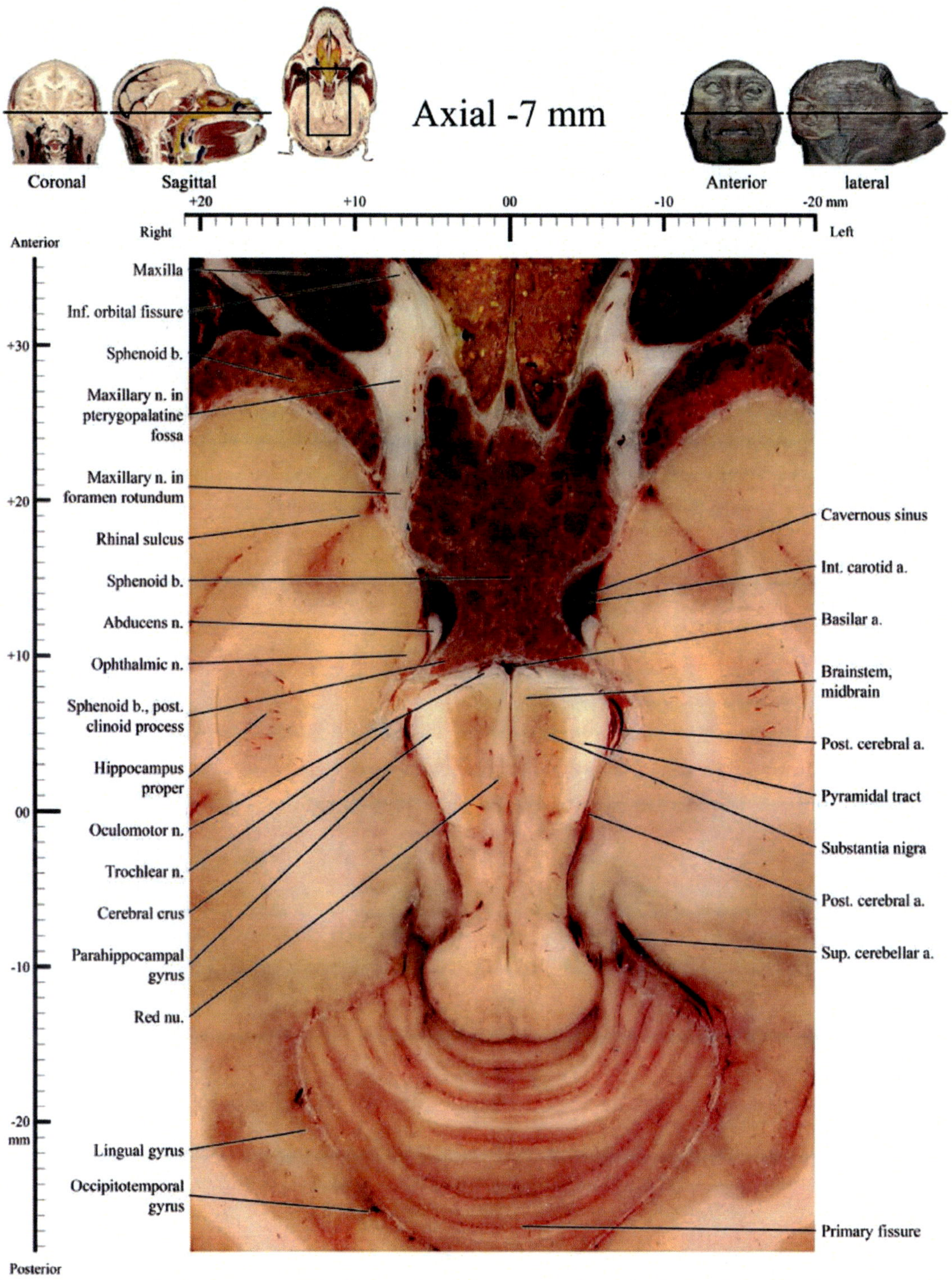
Axial -7 mm
Coronal
Sagittal
Anterior
lateral
+20
+10
00
-10
-20 mm
Right
Left
Anterior
Posterior
+30
+20
+10
00
-10
-20
mm
Maxilla
Inf. orbital fissure
Sphenoid b.
Maxillary n. in pterygopalatine fossa
Maxillary n. in foramen rotundum
Rhinal sulcus
Sphenoid b.
Abducens n.
Ophthalmic n.
Sphenoid b., post. clinoid process
Hippocampus proper
Oculomotor n.
Trochlear n.
Cerebral crus
Parahippocampal gyrus
Red nu.
Lingual gyrus
Occipitotemporal gyrus
Cavernous sinus
Int. carotid a.
Basilar a.
Brainstem, midbrain
Post. cerebral a.
Pyramidal tract
Substantia nigra
Post. cerebral a.
Sup. cerebellar a.
Primary fissure

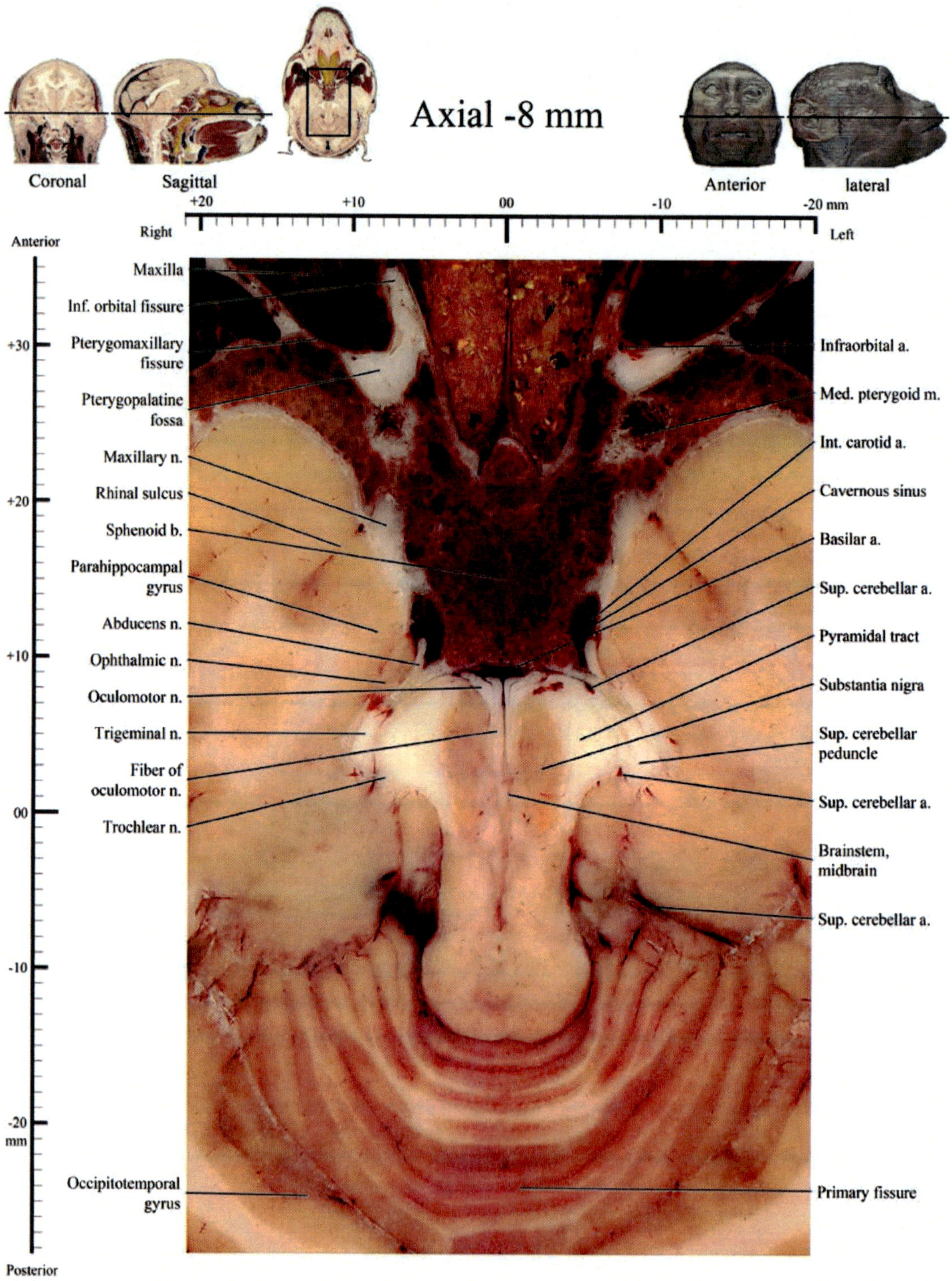
Coronal
Sagittal
Axial -8 mm
Anterior
lateral
+20
+10
00
-10
-20 mm
Right
Left
Anterior
Posterior
+30
+20
+10
00
-10
-20
mm
Maxilla
Inf. orbital fissure
Pterygomaxillary fissure
Pterygopalatine fossa
Maxillary n.
Rhinal sulcus
Sphenoid b.
Parahippocampal gyrus
Abducens n.
Ophthalmic n.
Oculomotor n.
Trigeminal n.
Fiber of oculomotor n.
Trochlear n.
Occipitotemporal gyrus
Infraorbital a.
Med. pterygoid m.
Int. carotid a.
Cavernous sinus
Basilar a.
Sup. cerebellar a.
Pyramidal tract
Substantia nigra
Sup. cerebellar peduncle
Sup. cerebellar a.
Brainstem, midbrain
Sup. cerebellar a.
Primary fissure

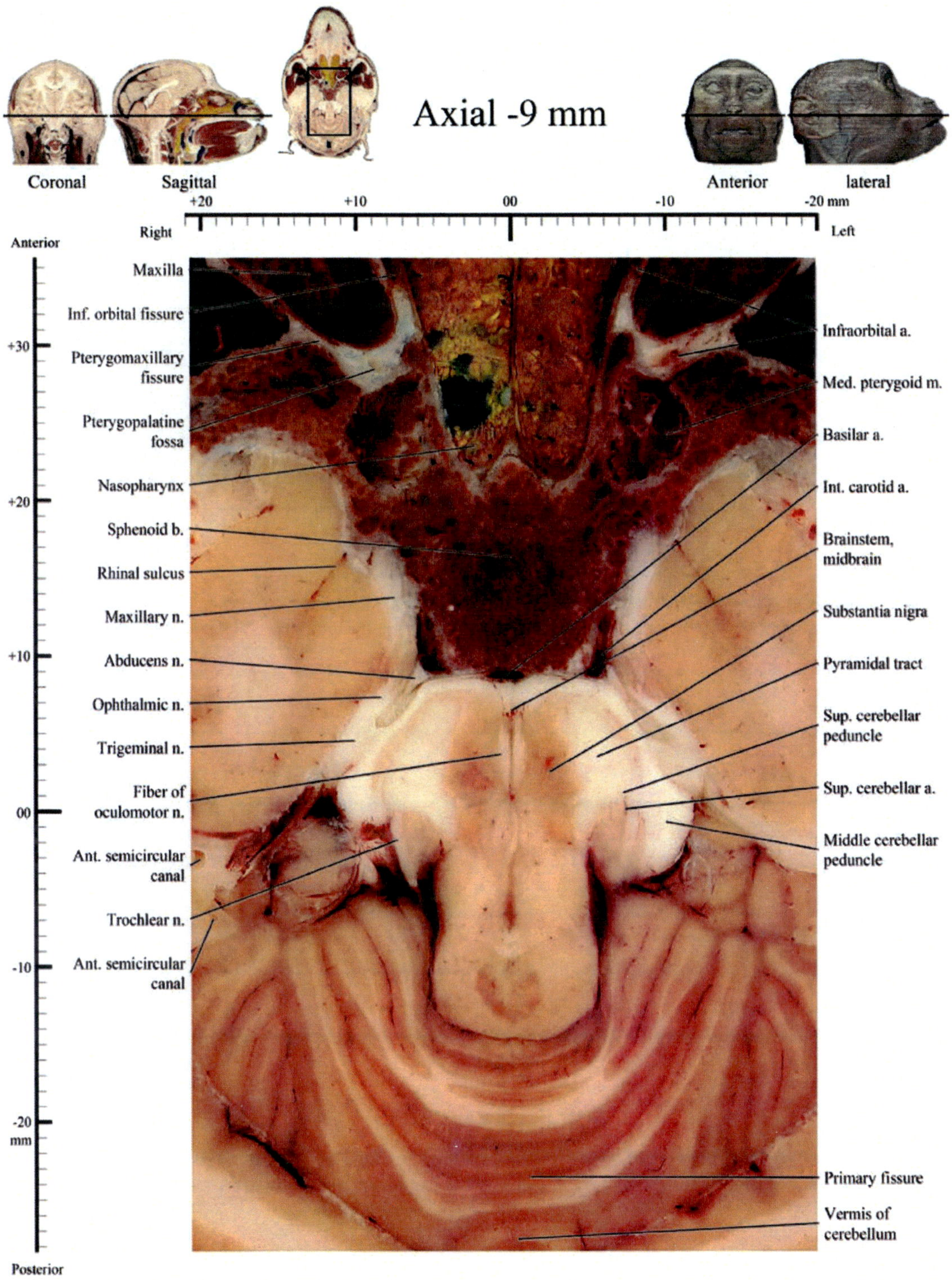
Axial -9 mm
Coronal
Sagittal
Anterior
lateral
+20
+10
00
-10
-20 mm
Right
Left
Anterior
Posterior
+30
+20
+10
00
-10
-20
mm
Maxilla
Inf. orbital fissure
Pterygomaxillary fissure
Pterygopalatine fossa
Nasopharynx
Sphenoid b.
Rhinal sulcus
Maxillary n.
Abducens n.
Ophthalmic n.
Trigeminal n.
Fiber of oculomotor n.
Ant. semicircular canal
Trochlear n.
Ant. semicircular canal
Infraorbital a.
Med. pterygoid m.
Basilar a.
Int. carotid a.
Brainstem, midbrain
Substantia nigra
Pyramidal tract
Sup. cerebellar peduncle
Sup. cerebellar a.
Middle cerebellar peduncle
Primary fissure
Vermis of cerebellum

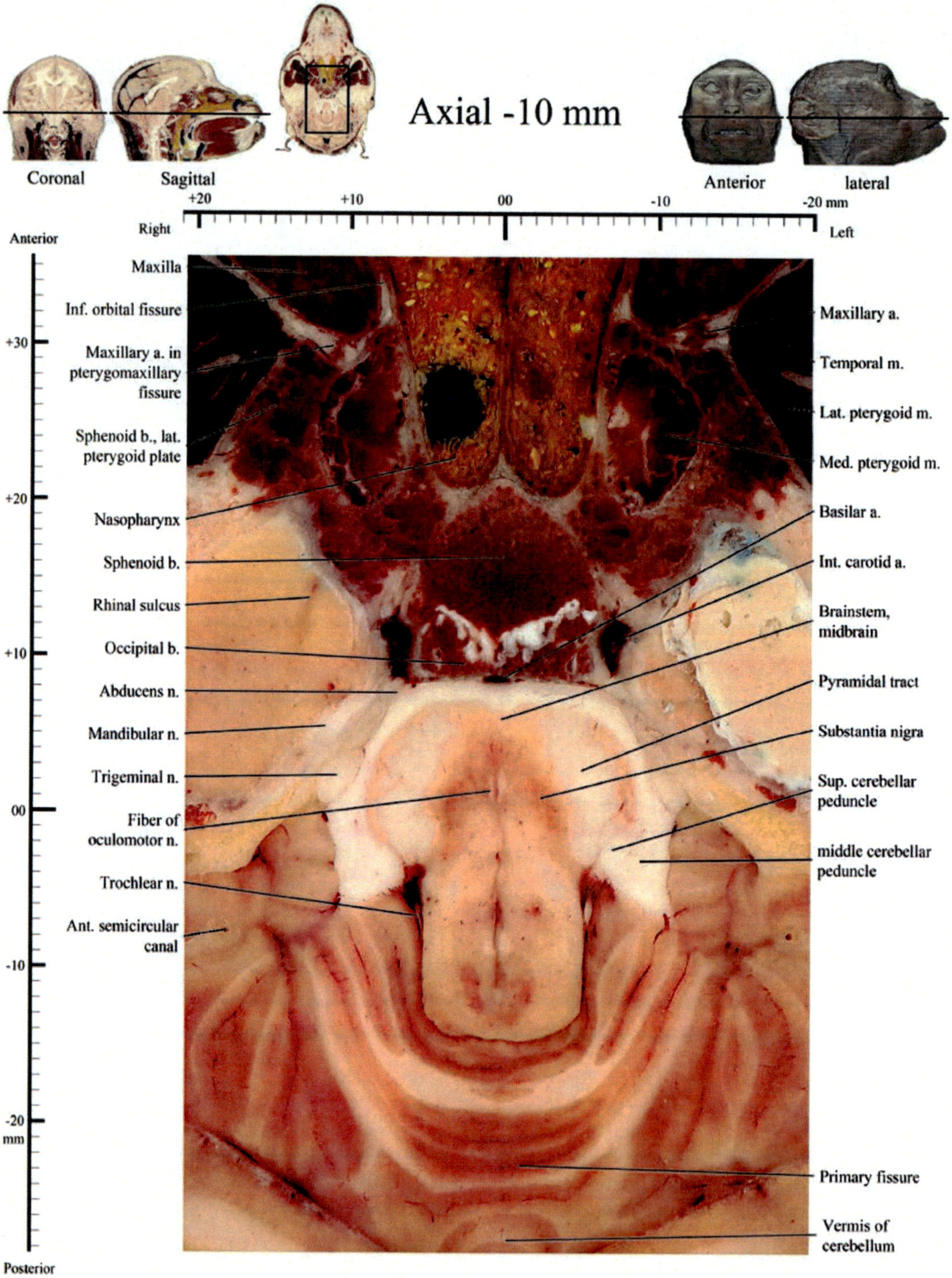

Axial -10 mm
Coronal
Sagittal
Anterior
lateral
+20
+10
00
-10
-20 mm
Right
Left
Anterior
Posterior
+30
+20
+10
00
-10
-20
mm
Maxilla
Inf. orbital fissure
Maxillary a. in pterygomaxillary fissure
Sphenoid b., lat. pterygoid plate
Nasopharynx
Sphenoid b.
Rhinal sulcus
Occipital b.
Abducens n.
Mandibular n.
Trigeminal n.
Fiber of oculomotor n.
Trochlear n.
Ant. semicircular canal
Maxillary a.
Temporal m.
Lat. pterygoid m.
Med. pterygoid m.
Basilar a.
Int. carotid a.
Brainstem, midbrain
Pyramidal tract
Substantia nigra
Sup. cerebellar peduncle
middle cerebellar peduncle
Primary fissure
Vermis of cerebellum

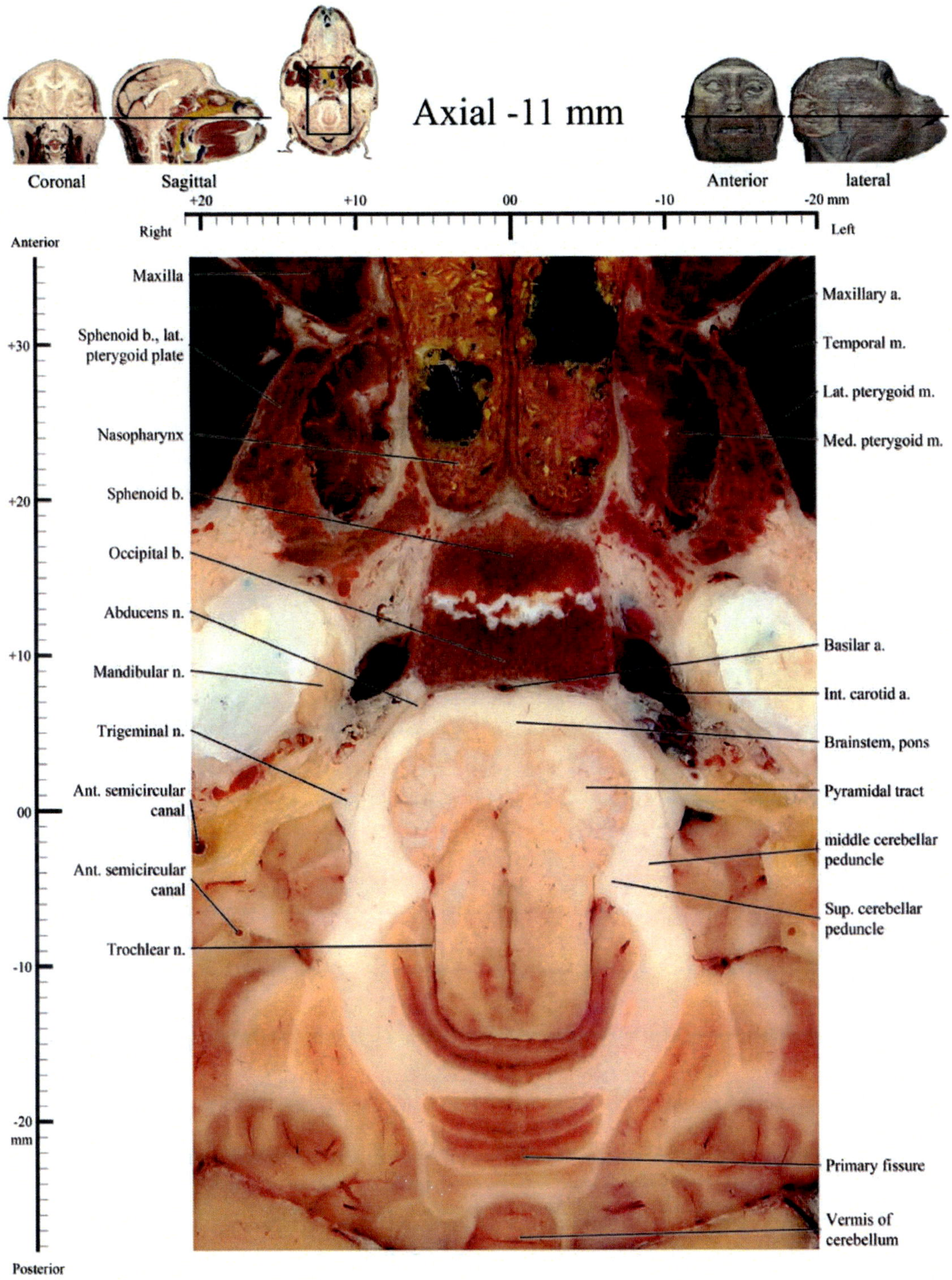
Coronal
Sagittal
Axial -11 mm
Anterior
lateral
+20
+10
00
-10
-20 mm
Right
Left
Anterior
Posterior
+30
+20
+10
00
-10
-20
mm
Maxilla
Sphenoid b., lat. pterygoid plate
Nasopharynx
Sphenoid b.
Occipital b.
Abducens n.
Mandibular n.
Trigeminal n.
Ant. semicircular canal
Ant. semicircular canal
Trochlear n.
Maxillary a.
Temporal m.
Lat. pterygoid m.
Med. pterygoid m.
Basilar a.
Int. carotid a.
Brainstem, pons
Pyramidal tract
middle cerebellar peduncle
Sup. cerebellar peduncle
Primary fissure
Vermis of cerebellum

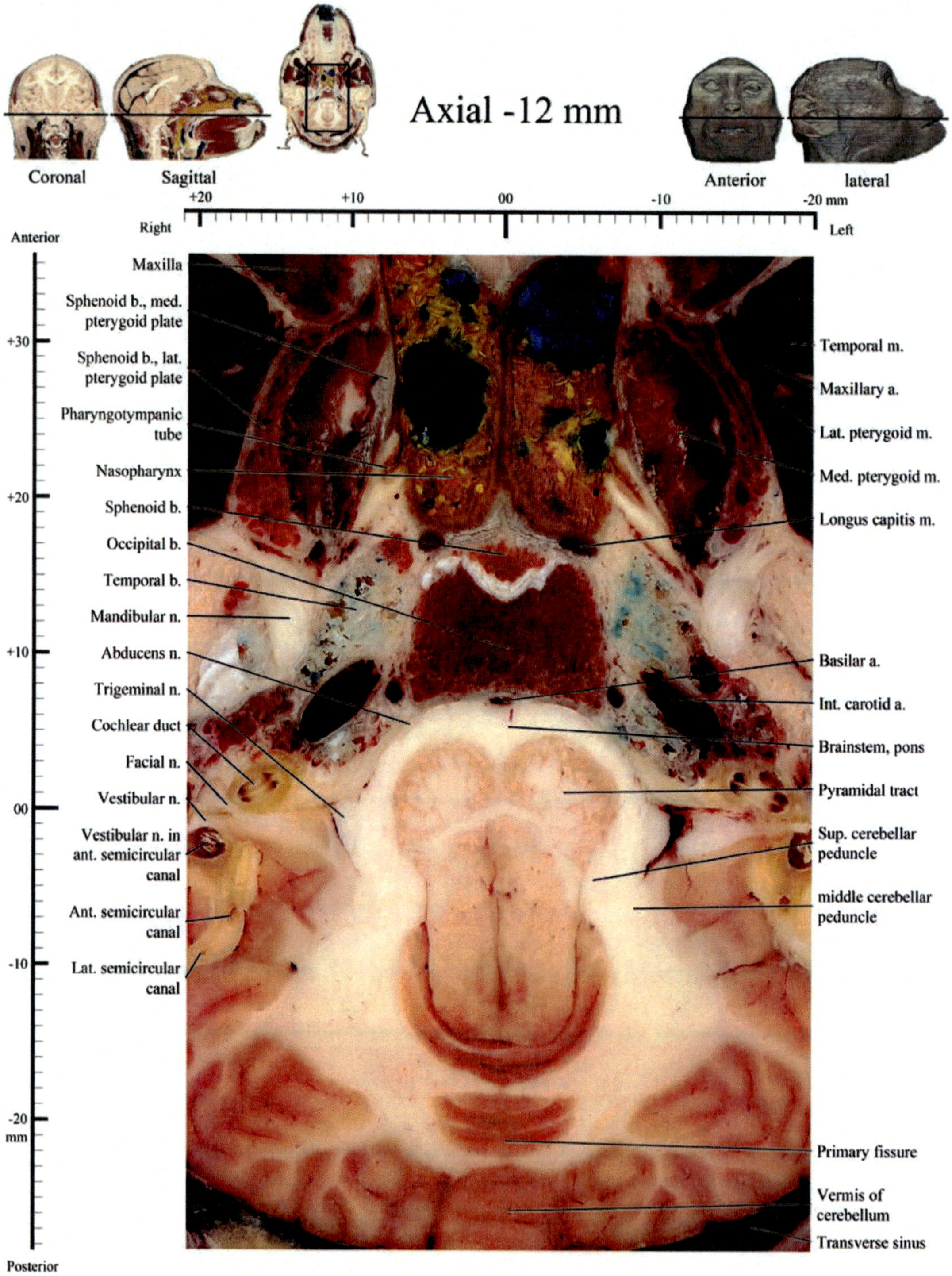
Axial -12 mm
Coronal
Sagittal
Anterior
lateral
+20
+10
00
-10
-20 mm
Right
Left
Anterior
Posterior
+30
+20
+10
00
-10
-20
mm
Maxilla
Sphenoid b., med. pterygoid plate
Sphenoid b., lat. pterygoid plate
Pharyngotympanic tube
Nasopharynx
Sphenoid b.
Occipital b.
Temporal b.
Mandibular n.
Abducens n.
Trigeminal n.
Cochlear duct
Facial n.
Vestibular n.
Vestibular n. in ant. semicircular canal
Ant. semicircular canal
Lat. semicircular canal
Temporal m.
Maxillary a.
Lat. pterygoid m.
Med. pterygoid m.
Longus capitis m.
Basilar a.
Int. carotid a.
Brainstem, pons
Pyramidal tract
Sup. cerebellar peduncle
middle cerebellar peduncle
Primary fissure
Vermis of cerebellum
Transverse sinus

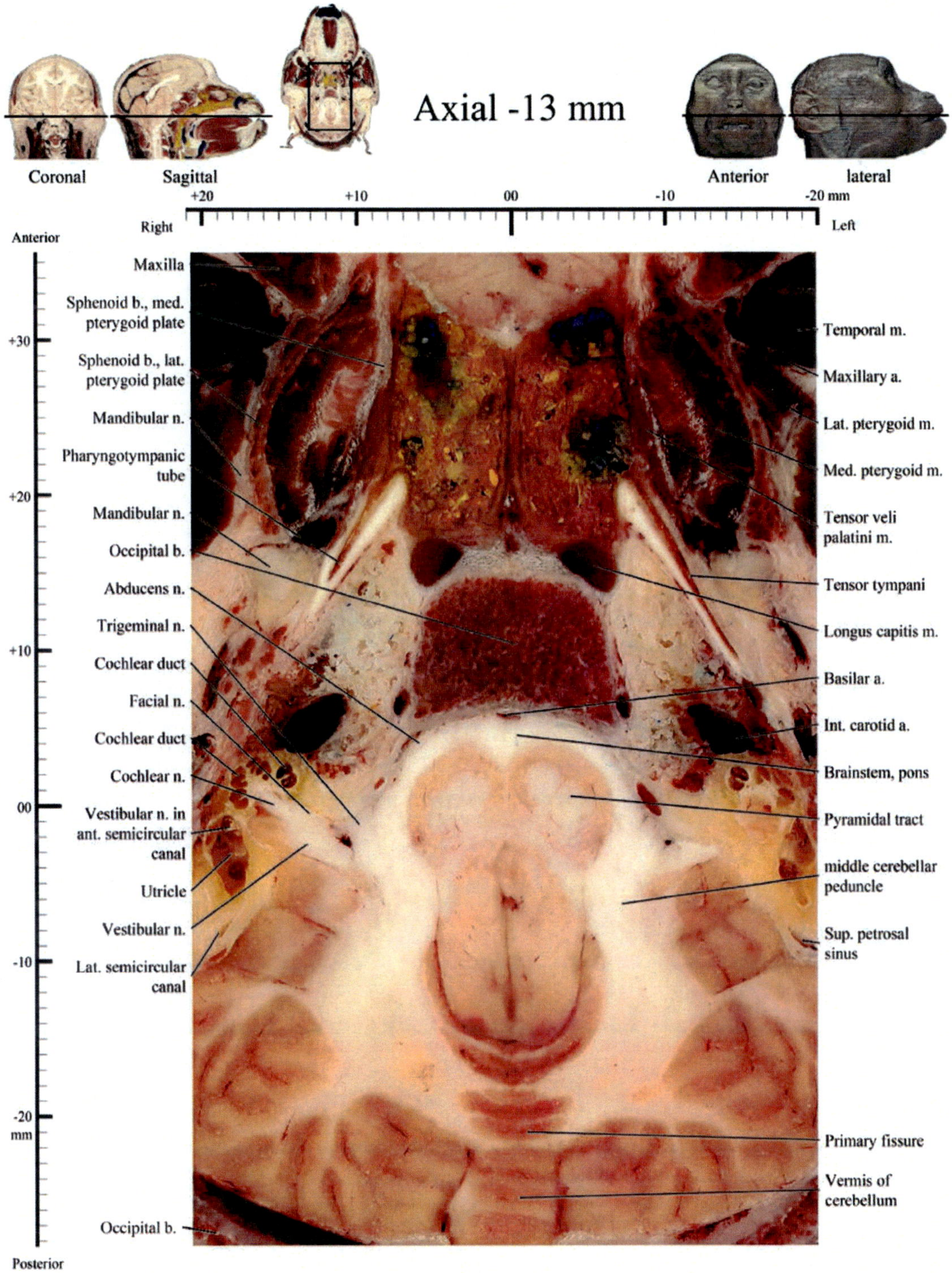

Axial -13 mm
Coronal
Sagittal
Anterior
lateral
+20
+10
00
-10
-20 mm
Right
Left
Anterior
Posterior
+30
+20
+10
00
-10
-20
mm
Maxilla
Sphenoid b., med. pterygoid plate
Sphenoid b., lat. pterygoid plate
Mandibular n.
Pharyngotympanic tube
Mandibular n.
Occipital b.
Abducens n.
Trigeminal n.
Cochlear duct
Facial n.
Cochlear duct
Cochlear n.
Vestibular n. in ant. semicircular canal
Utricle
Vestibular n.
Lat. semicircular canal
Occipital b.
Temporal m.
Maxillary a.
Lat. pterygoid m.
Med. pterygoid m.
Tensor veli palatini m.
Tensor tympani
Longus capitis m.
Basilar a.
Int. carotid a.
Brainstem, pons
Pyramidal tract
middle cerebellar peduncle
Sup. petrosal sinus
Primary fissure
Vermis of cerebellum

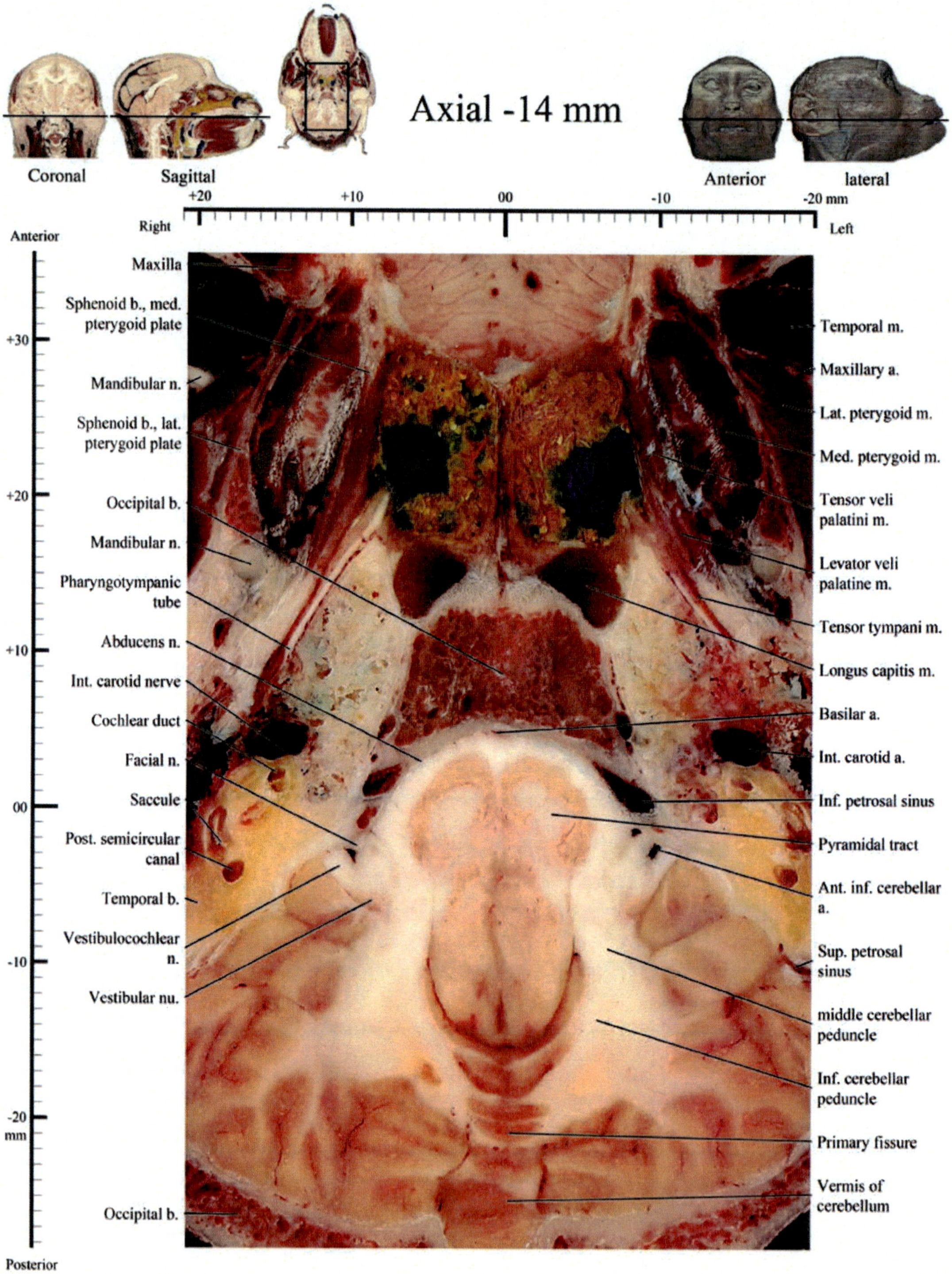
Axial -14 mm
Coronal
Sagittal
Anterior
lateral
+20
+10
00
-10
-20 mm
Right
Left
Anterior
Posterior
+30
+20
+10
00
-10
-20
mm
Maxilla
Sphenoid b., med. pterygoid plate
Mandibular n.
Sphenoid b., lat. pterygoid plate
Occipital b.
Mandibular n.
Pharyngotympanic tube
Abducens n.
Int. carotid nerve
Cochlear duct
Facial n.
Saccule
Post. semicircular canal
Temporal b.
Vestibulocochlear n.
Vestibular nu.
Occipital b.
Temporal m.
Maxillary a.
Lat. pterygoid m.
Med. pterygoid m.
Tensor veli palatini m.
Levator veli palatine m.
Tensor tympani m.
Longus capitis m.
Basilar a.
Int. carotid a.
Inf. petrosal sinus
Pyramidal tract
Ant. inf. cerebellar a.
Sup. petrosal sinus
middle cerebellar peduncle
Inf. cerebellar peduncle
Primary fissure
Vermis of cerebellum

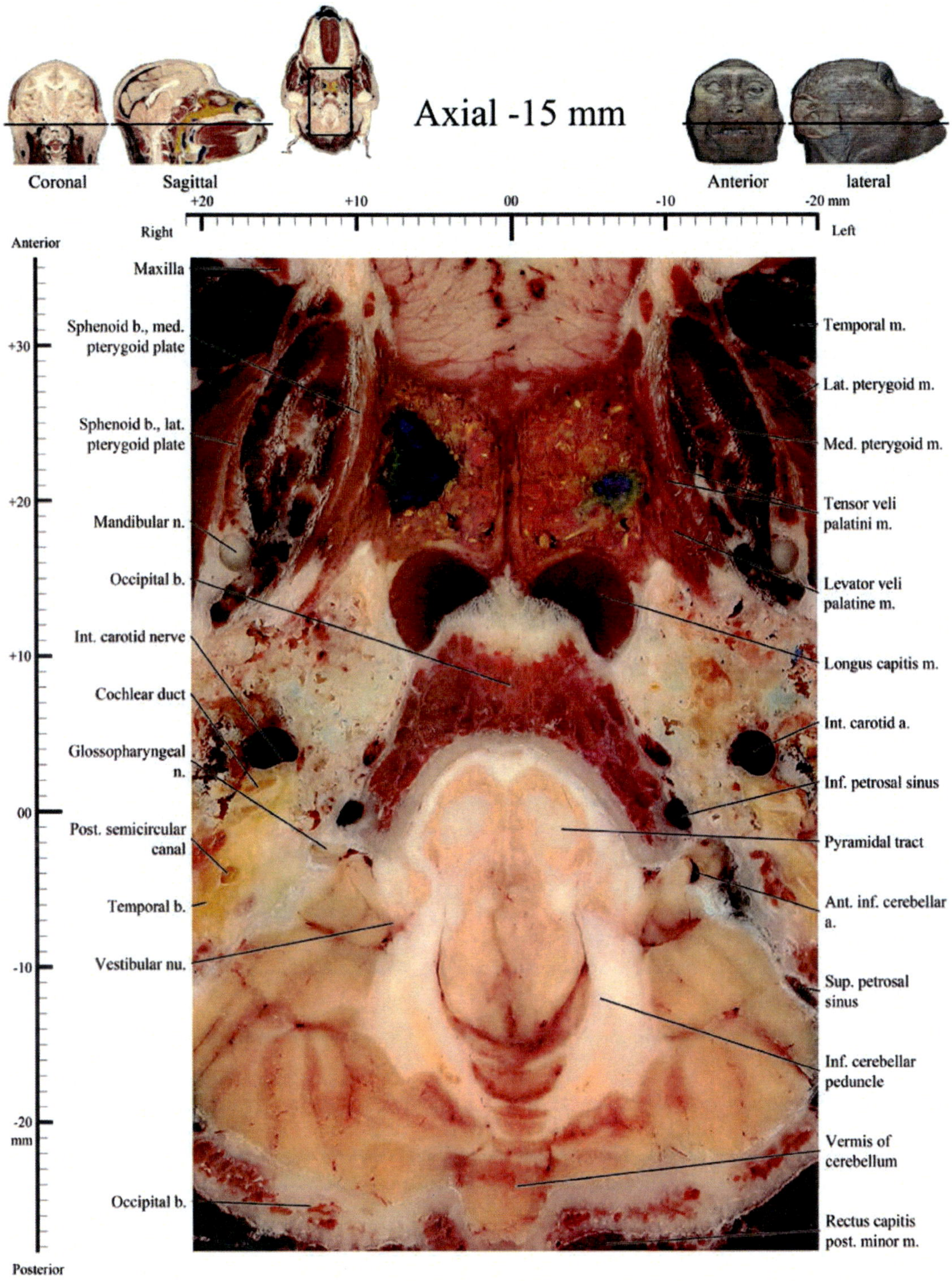

Coronal
Sagittal
Axial -15 mm
Anterior
lateral
+20
+10
00
-10
-20 mm
Right
Left
Anterior
Posterior
+30
+20
+10
00
-10
-20
mm
Maxilla
Sphenoid b., med. pterygoid plate
Sphenoid b., lat. pterygoid plate
Mandibular n.
Occipital b.
Int. carotid nerve
Cochlear duct
Glossopharyngeal n.
Post. semicircular canal
Temporal b.
Vestibular nu.
Occipital b.
Temporal m.
Lat. pterygoid m.
Med. pterygoid m.
Tensor veli palatini m.
Levator veli palatine m.
Longus capitis m.
Int. carotid a.
Inf. petrosal sinus
Pyramidal tract
Ant. inf. cerebellar a.
Sup. petrosal sinus
Inf. cerebellar peduncle
Vermis of cerebellum
Rectus capitis post. minor m.

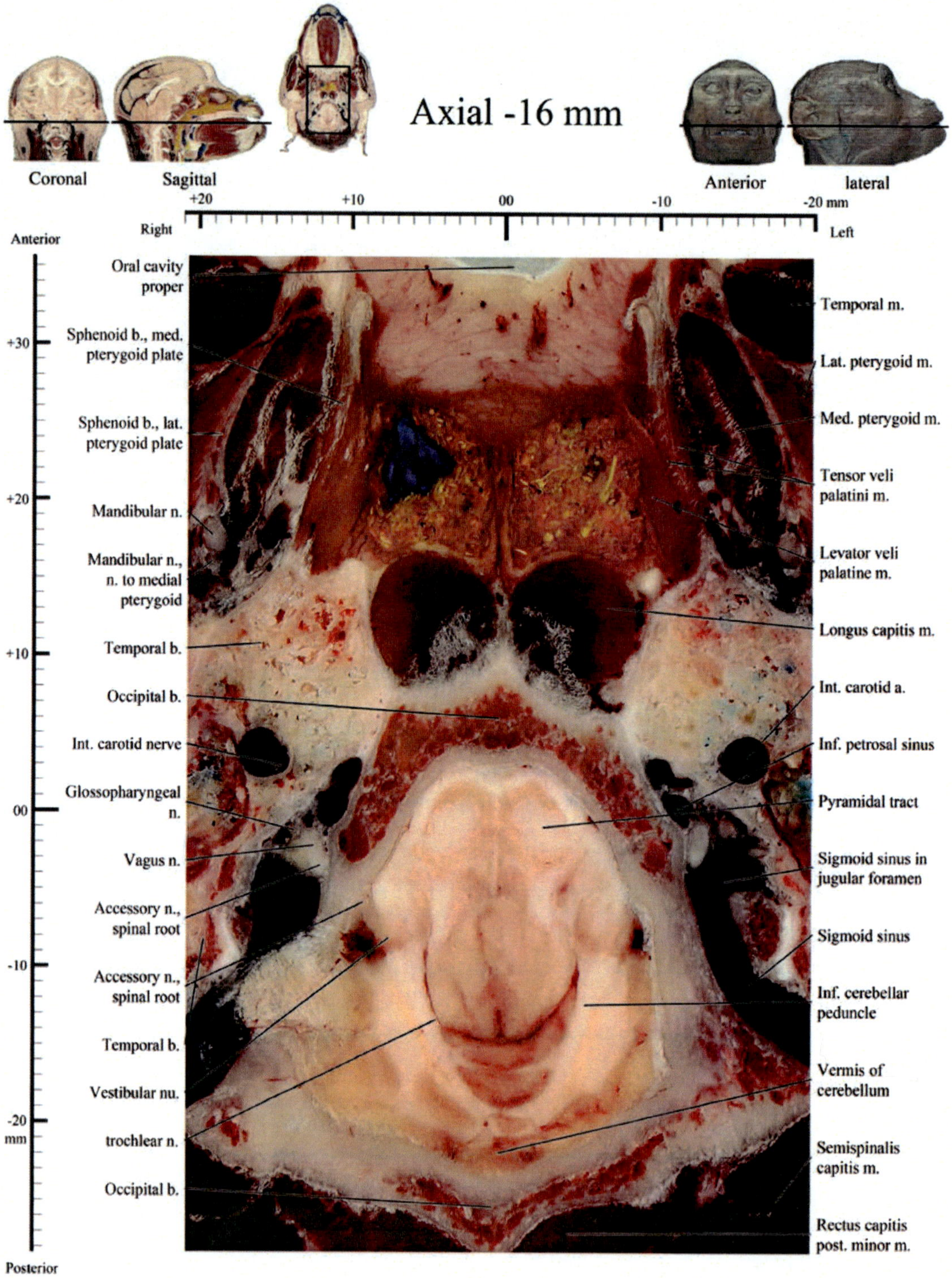
Coronal
Sagittal
Axial -16 mm
Anterior
lateral
+20
+10
00
-10
-20 mm
Right
Left
Anterior
Posterior
+30
+20
+10
00
-10
-20
mm
Oral cavity proper
Sphenoid b., med. pterygoid plate
Sphenoid b., lat. pterygoid plate
Mandibular n.
Mandibular n., n. to medial pterygoid
Temporal b.
Occipital b.
Int. carotid nerve
Glossopharyngeal n.
Vagus n.
Accessory n., spinal root
Accessory n., spinal root
Temporal b.
Vestibular nu.
trochlear n.
Occipital b.
Temporal m.
Lat. pterygoid m.
Med. pterygoid m.
Tensor veli palatini m.
Levator veli palatine m.
Longus capitis m.
Int. carotid a.
Inf. petrosal sinus
Pyramidal tract
Sigmoid sinus in jugular foramen
Sigmoid sinus
Inf. cerebellar peduncle
Vermis of cerebellum
Semispinalis capitis m.
Rectus capitis post. minor m.

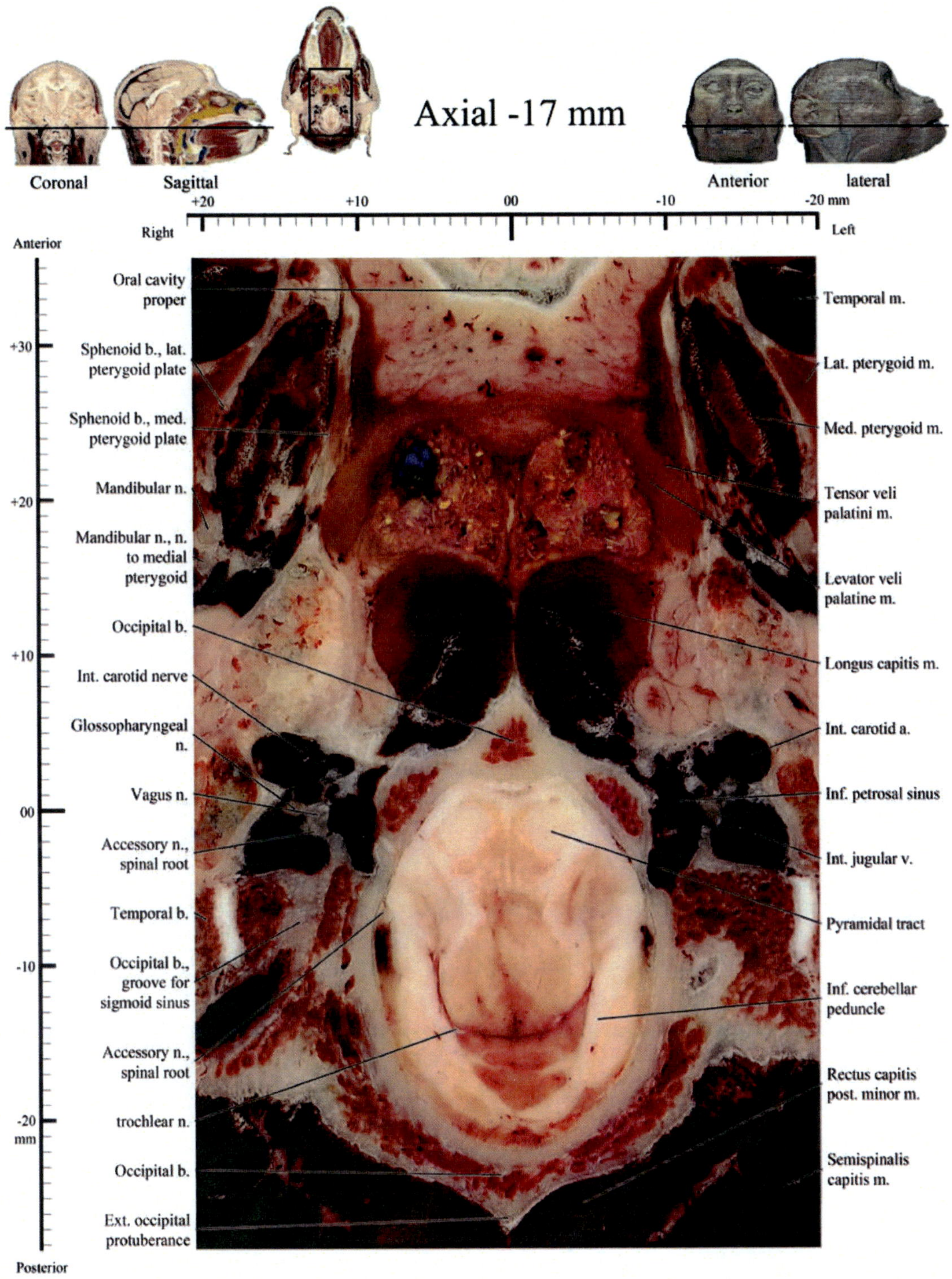
Coronal
Sagittal
Axial -17 mm
Anterior
lateral
+20
+10
00
-10
-20 mm
Right
Left
Anterior
Posterior
+30
+20
+10
00
-10
-20
mm
Oral cavity proper
Sphenoid b., lat. pterygoid plate
Sphenoid b., med. pterygoid plate
Mandibular n.
Mandibular n., n. to medial pterygoid
Occipital b.
Int. carotid nerve
Glossopharyngeal n.
Vagus n.
Accessory n., spinal root
Temporal b.
Occipital b., groove for sigmoid sinus
Accessory n., spinal root
trochlear n.
Occipital b.
Ext. occipital protuberance
Temporal m.
Lat. pterygoid m.
Med. pterygoid m.
Tensor veli palatini m.
Levator veli palatine m.
Longus capitis m.
Int. carotid a.
Inf. petrosal sinus
Int. jugular v.
Pyramidal tract
Inf. cerebellar peduncle
Rectus capitis post. minor m.
Semispinalis capitis m.

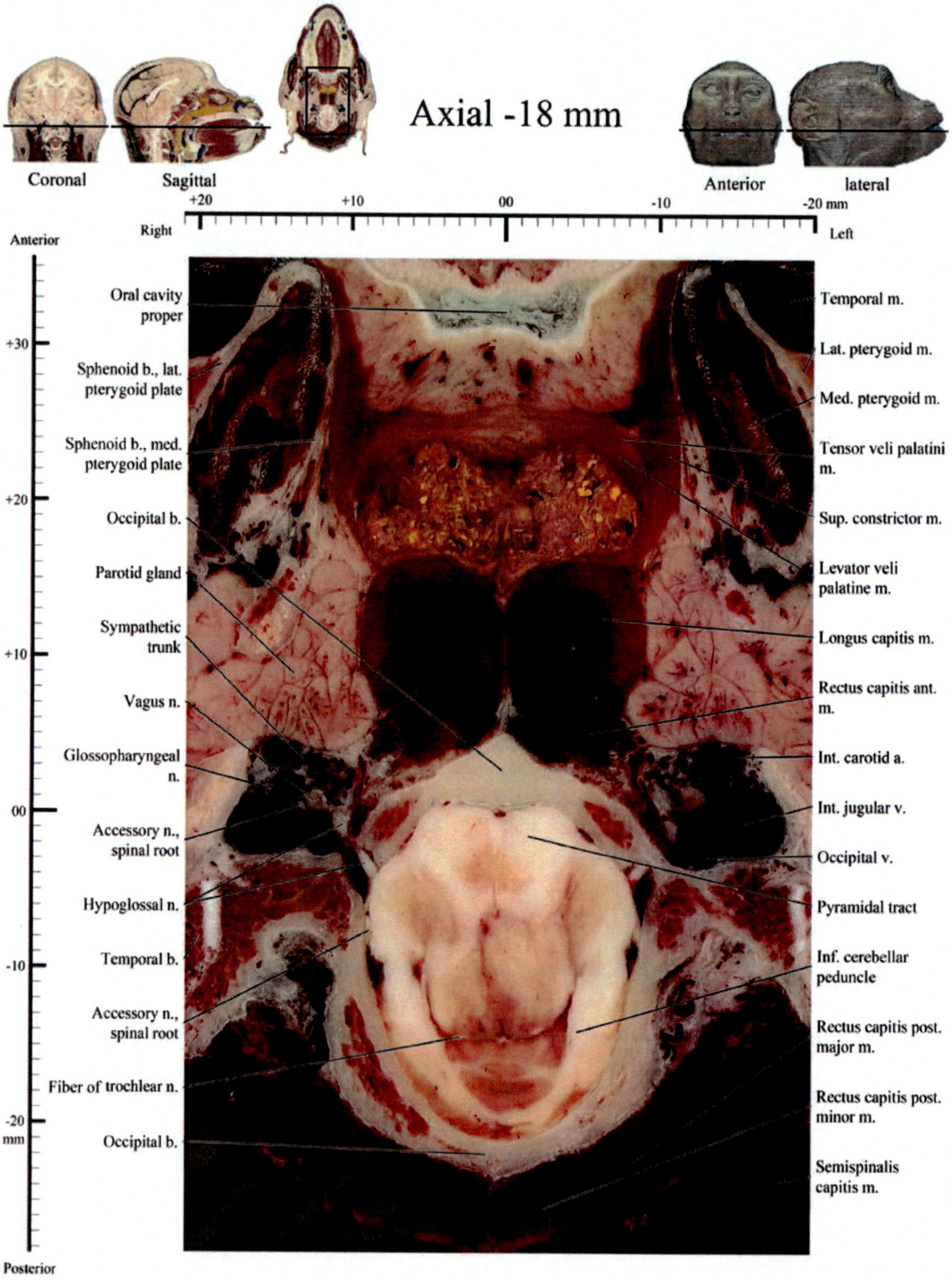
Axial -18 mm
Coronal
Sagittal
Anterior
lateral
+20
+10
00
-10
-20 mm
Right
Left
Anterior
Posterior
+30
+20
+10
00
-10
-20
mm
Oral cavity proper
Sphenoid b., lat. pterygoid plate
Sphenoid b., med. pterygoid plate
Occipital b.
Parotid gland
Sympathetic trunk
Vagus n.
Glossopharyngeal n.
Accessory n., spinal root
Hypoglossal n.
Temporal b.
Accessory n., spinal root
Fiber of trochlear n.
Occipital b.
Temporal m.
Lat. pterygoid m.
Med. pterygoid m.
Tensor veli palatini m.
Sup. constrictor m.
Levator veli palatine m.
Longus capitis m.
Rectus capitis ant. m.
Int. carotid a.
Int. jugular v.
Occipital v.
Pyramidal tract
Inf. cerebellar peduncle
Rectus capitis post. major m.
Rectus capitis post. minor m.
Semispinalis capitis m.

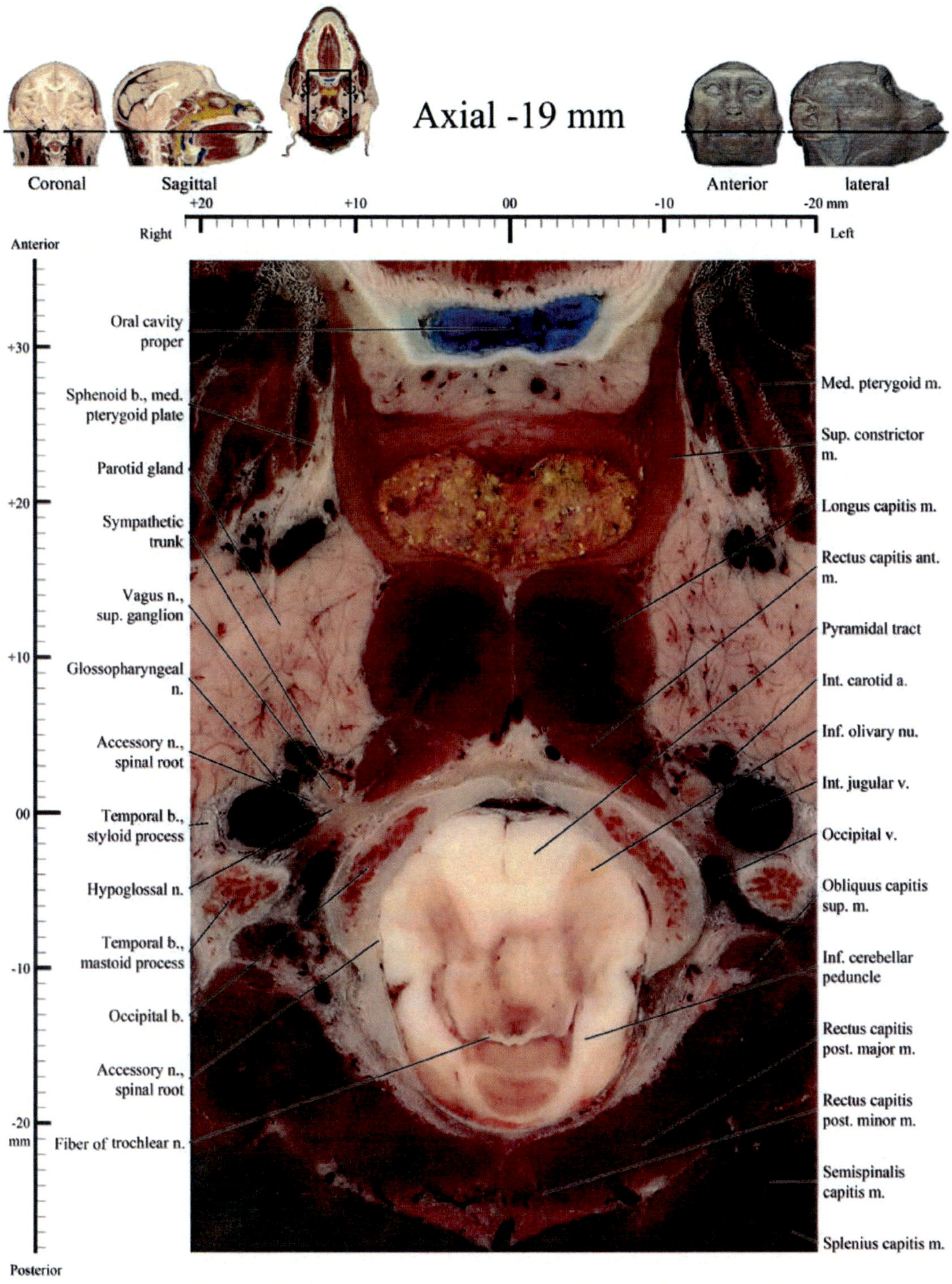
Coronal
Sagittal
Axial -19 mm
Anterior
lateral
+20
+10
00
-10
-20 mm
Right
Left
Anterior
+30
+20
+10
00
-10
-20
mm
Posterior
Oral cavity proper
Sphenoid b., med. pterygoid plate
Parotid gland
Sympathetic trunk
Vagus n., sup. ganglion
Glossopharyngeal n.
Accessory n., spinal root
Temporal b., styloid process
Hypoglossal n.
Temporal b., mastoid process
Occipital b.
Accessory n., spinal root
Fiber of trochlear n.
Med. pterygoid m.
Sup. constrictor m.
Longus capitis m.
Rectus capitis ant. m.
Pyramidal tract
Int. carotid a.
Inf. olivary nu.
Int. jugular v.
Occipital v.
Obliquus capitis sup. m.
Inf. cerebellar peduncle
Rectus capitis post. major m.
Rectus capitis post. minor m.
Semispinalis capitis m.
Splenius capitis m.

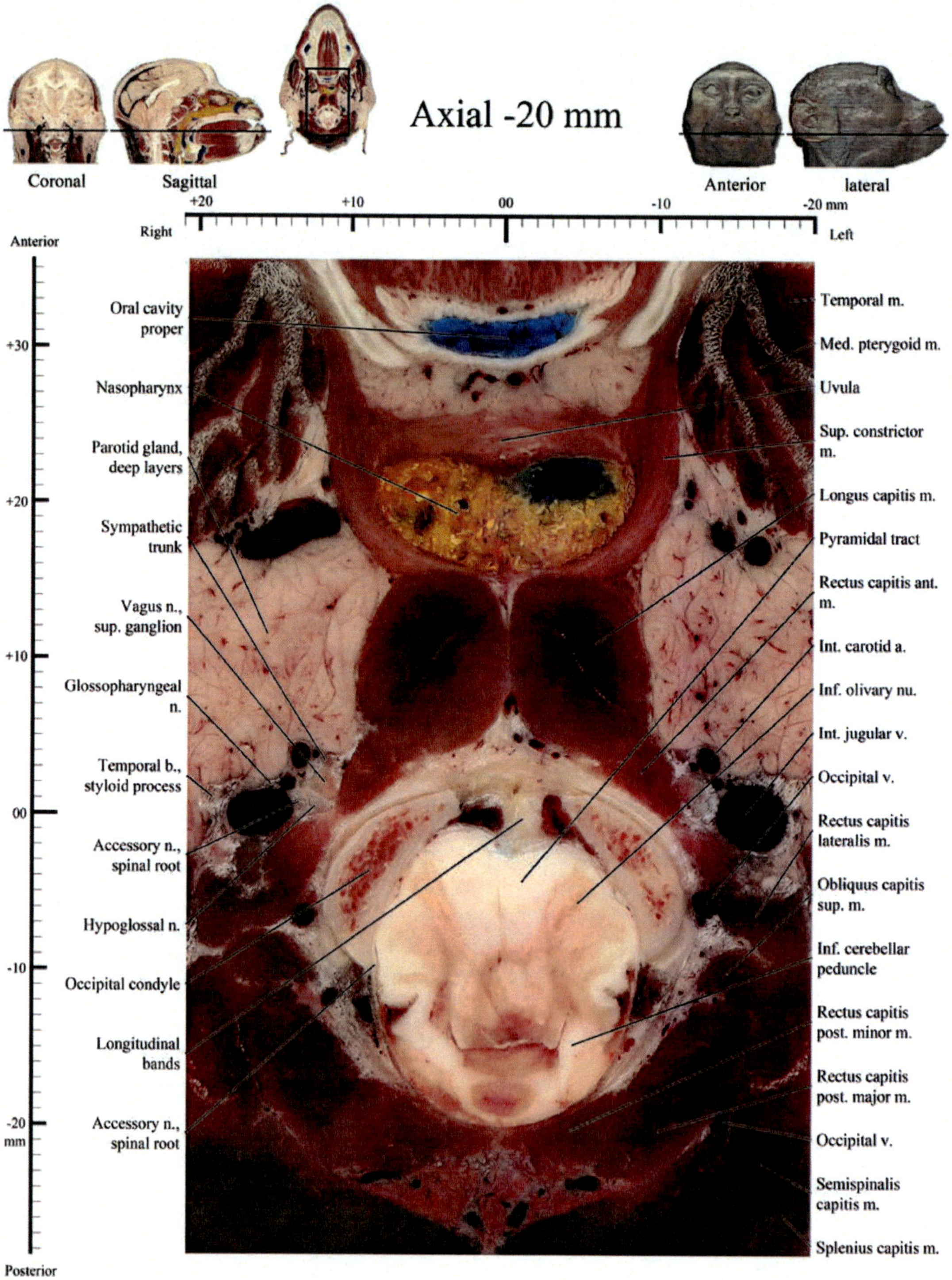
Axial -20 mm
Coronal
Sagittal
Anterior
lateral
+20
+10
00
-10
-20 mm
Right
Left
Anterior
Posterior
+30
+20
+10
00
-10
-20
mm
Oral cavity proper
Nasopharynx
Parotid gland, deep layers
Sympathetic trunk
Vagus n., sup. ganglion
Glossopharyngeal n.
Temporal b., styloid process
Accessory n., spinal root
Hypoglossal n.
Occipital condyle
Longitudinal bands
Accessory n., spinal root
Temporal m.
Med. pterygoid m.
Uvula
Sup. constrictor m.
Longus capitis m.
Pyramidal tract
Rectus capitis ant. m.
Int. carotid a.
Inf. olivary nu.
Int. jugular v.
Occipital v.
Rectus capitis lateralis m.
Obliquus capitis sup. m.
Inf. cerebellar peduncle
Rectus capitis post. minor m.
Rectus capitis post. major m.
Occipital v.
Semispinalis capitis m.
Splenius capitis m.

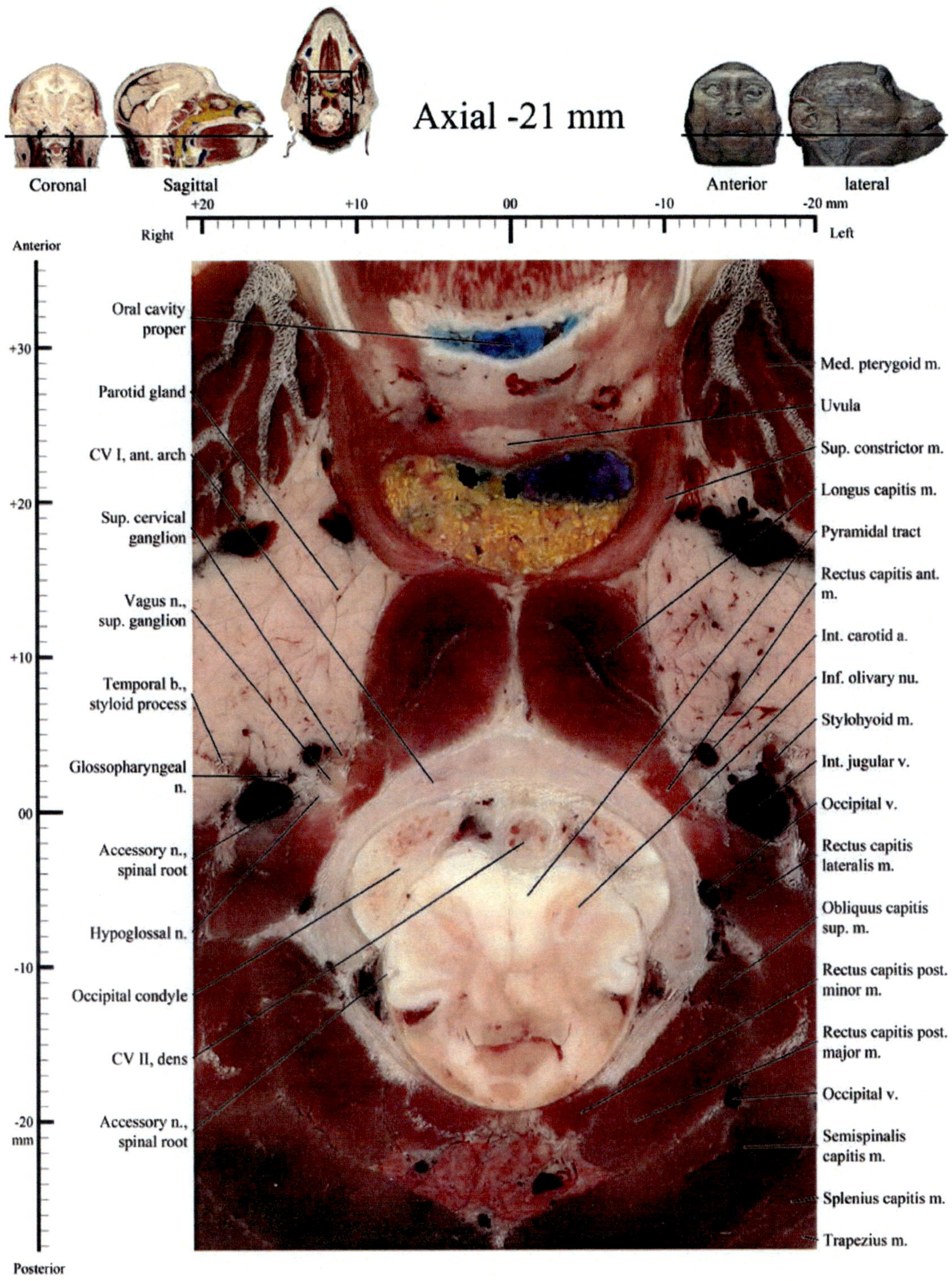

Coronal
Sagittal
Axial -21 mm
Anterior
lateral
+20
+10
00
-10
-20 mm
Right
Left
Anterior
Posterior
+30
+20
+10
00
-10
-20
mm
Oral cavity proper
Parotid gland
CV I, ant. arch
Sup. cervical ganglion
Vagus n., sup. ganglion
Temporal b., styloid process
Glossopharyngeal n.
Accessory n., spinal root
Hypoglossal n.
Occipital condyle
CV II, dens
Accessory n., spinal root
Med. pterygoid m.
Uvula
Sup. constrictor m.
Longus capitis m.
Pyramidal tract
Rectus capitis ant. m.
Int. carotid a.
Inf. olivary nu.
Stylohyoid m.
Int. jugular v.
Occipital v.
Rectus capitis lateralis m.
Obliquus capitis sup. m.
Rectus capitis post. minor m.
Rectus capitis post. major m.
Occipital v.
Semispinalis capitis m.
Splenius capitis m.
Trapezius m.

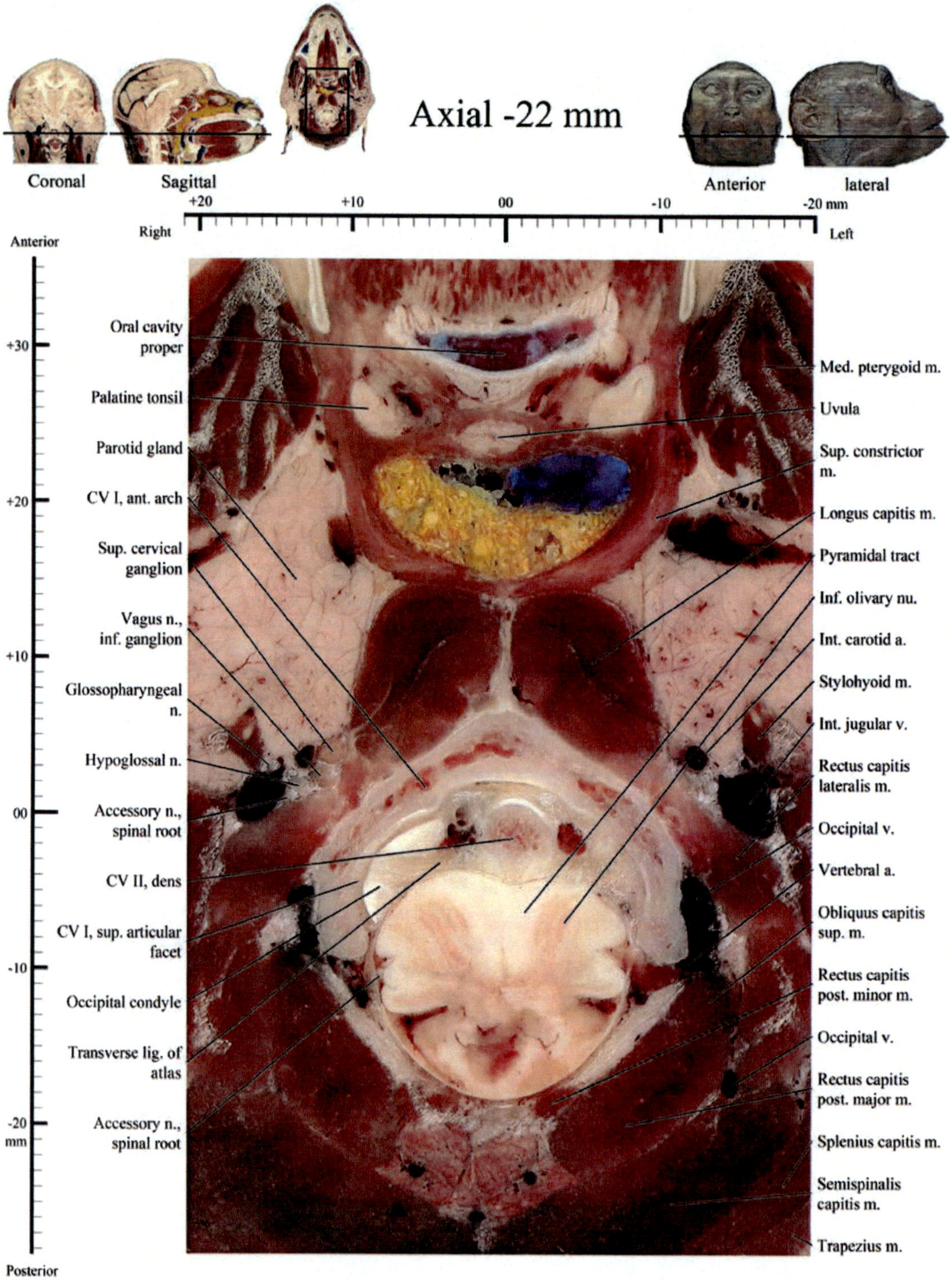
Axial -22 mm
Coronal
Sagittal
Anterior
lateral
+20
+10
00
-10
-20 mm
Right
Left
Anterior
Posterior
+30
+20
+10
00
-10
-20
mm
Oral cavity proper
Palatine tonsil
Parotid gland
CV I, ant. arch
Sup. cervical ganglion
Vagus n., inf. ganglion
Glossopharyngeal n.
Hypoglossal n.
Accessory n., spinal root
CV II, dens
CV I, sup. articular facet
Occipital condyle
Transverse lig. of atlas
Accessory n., spinal root
Med. pterygoid m.
Uvula
Sup. constrictor m.
Longus capitis m.
Pyramidal tract
Inf. olivary nu.
Int. carotid a.
Stylohyoid m.
Int. jugular v.
Rectus capitis lateralis m.
Occipital v.
Vertebral a.
Obliquus capitis sup. m.
Rectus capitis post. minor m.
Occipital v.
Rectus capitis post. major m.
Splenius capitis m.
Semispinalis capitis m.
Trapezius m.

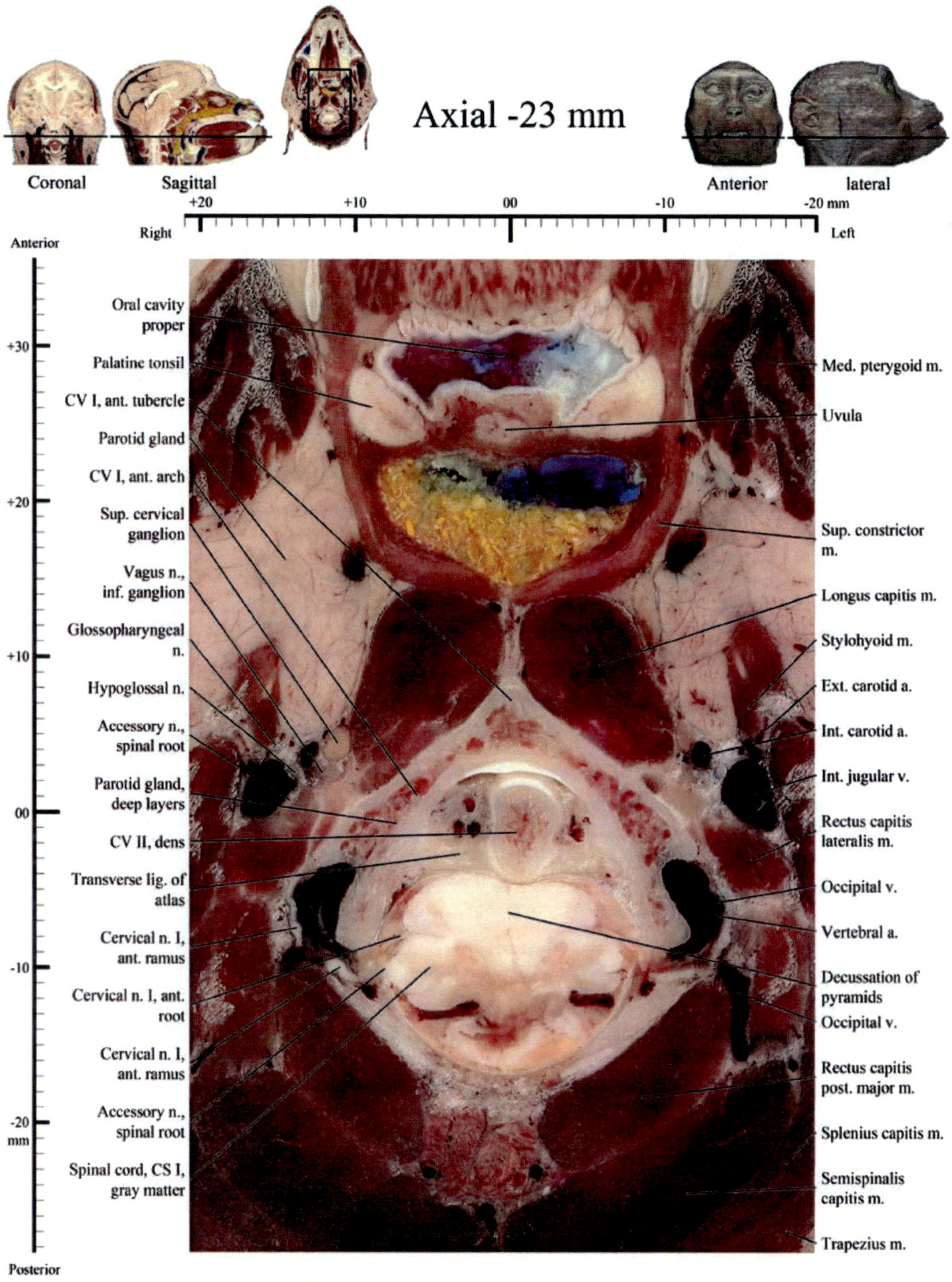

Axial -23 mm
Coronal
Sagittal
Anterior
lateral
+20
+10
00
-10
-20 mm
Right
Left
Anterior
Posterior
+30
+20
+10
00
-10
-20
mm
Oral cavity proper
Palatine tonsil
CV I, ant. tubercle
Parotid gland
CV I, ant. arch
Sup. cervical ganglion
Vagus n., inf. ganglion
Glossopharyngeal n.
Hypoglossal n.
Accessory n., spinal root
Parotid gland, deep layers
CV II, dens
Transverse lig. of atlas
Cervical n. I, ant. ramus
Cervical n. I, ant. root
Cervical n. I, ant. ramus
Accessory n., spinal root
Spinal cord, CS I, gray matter
Med. pterygoid m.
Uvula
Sup. constrictor m.
Longus capitis m.
Stylohyoid m.
Ext. carotid a.
Int. carotid a.
Int. jugular v.
Rectus capitis lateralis m.
Occipital v.
Vertebral a.
Decussation of pyramids
Occipital v.
Rectus capitis post. major m.
Splenius capitis m.
Semispinalis capitis m.
Trapezius m.

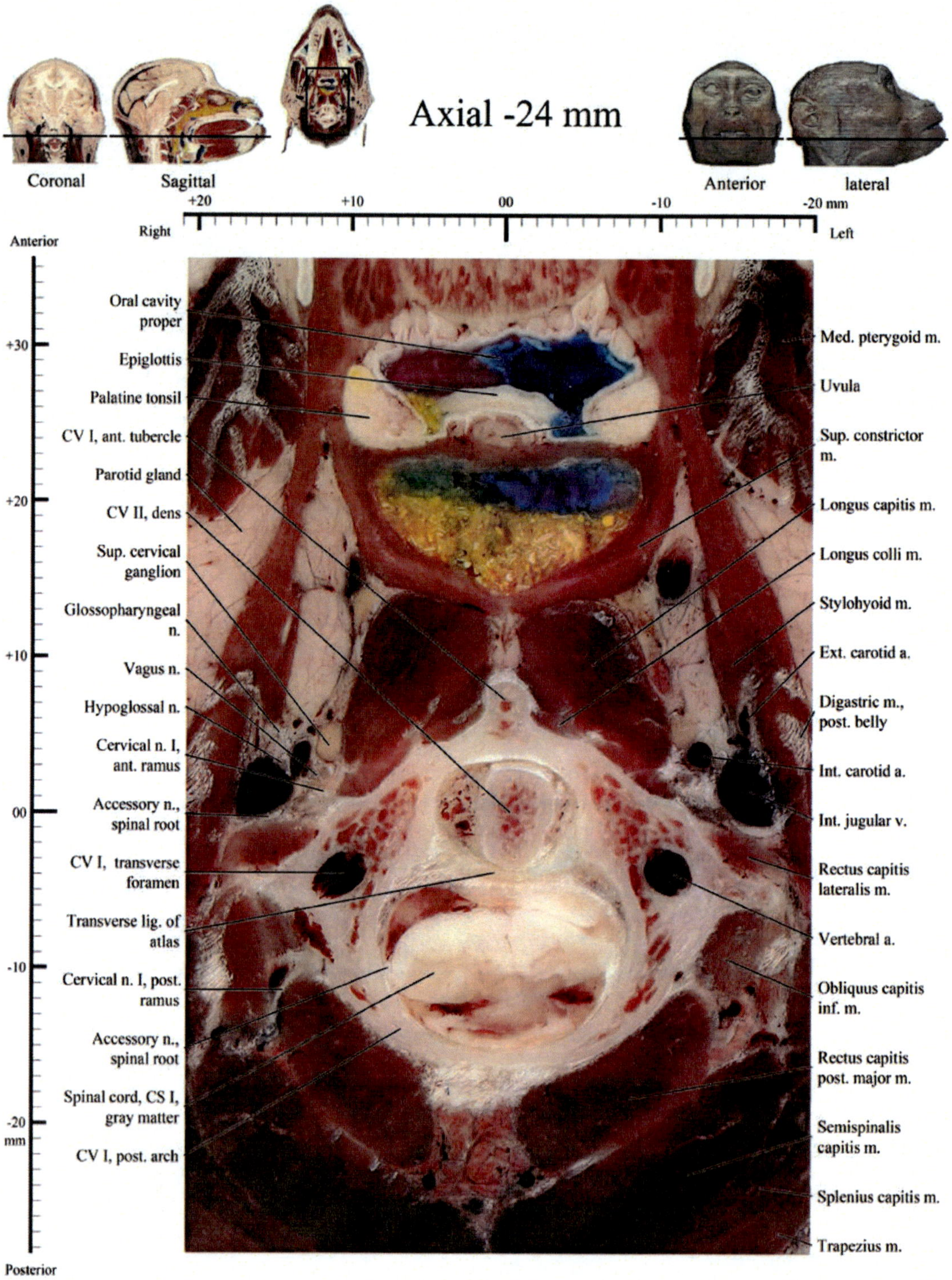
Axial -24 mm
Coronal
Sagittal
Anterior
lateral
+20
+10
00
-10
-20 mm
Right
Left
Anterior
+30
+20
+10
00
-10
-20
mm
Posterior
Oral cavity proper
Epiglottis
Palatine tonsil
CV I, ant. tubercle
Parotid gland
CV II, dens
Sup. cervical ganglion
Glossopharyngeal n.
Vagus n.
Hypoglossal n.
Cervical n. I, ant. ramus
Accessory n., spinal root
CV I, transverse foramen
Transverse lig. of atlas
Cervical n. I, post. ramus
Accessory n., spinal root
Spinal cord, CS I, gray matter
CV I, post. arch
Med. pterygoid m.
Uvula
Sup. constrictor m.
Longus capitis m.
Longus colli m.
Stylohyoid m.
Ext. carotid a.
Digastric m., post. belly
Int. carotid a.
Int. jugular v.
Rectus capitis lateralis m.
Vertebral a.
Obliquus capitis inf. m.
Rectus capitis post. major m.
Semispinalis capitis m.
Splenius capitis m.
Trapezius m.

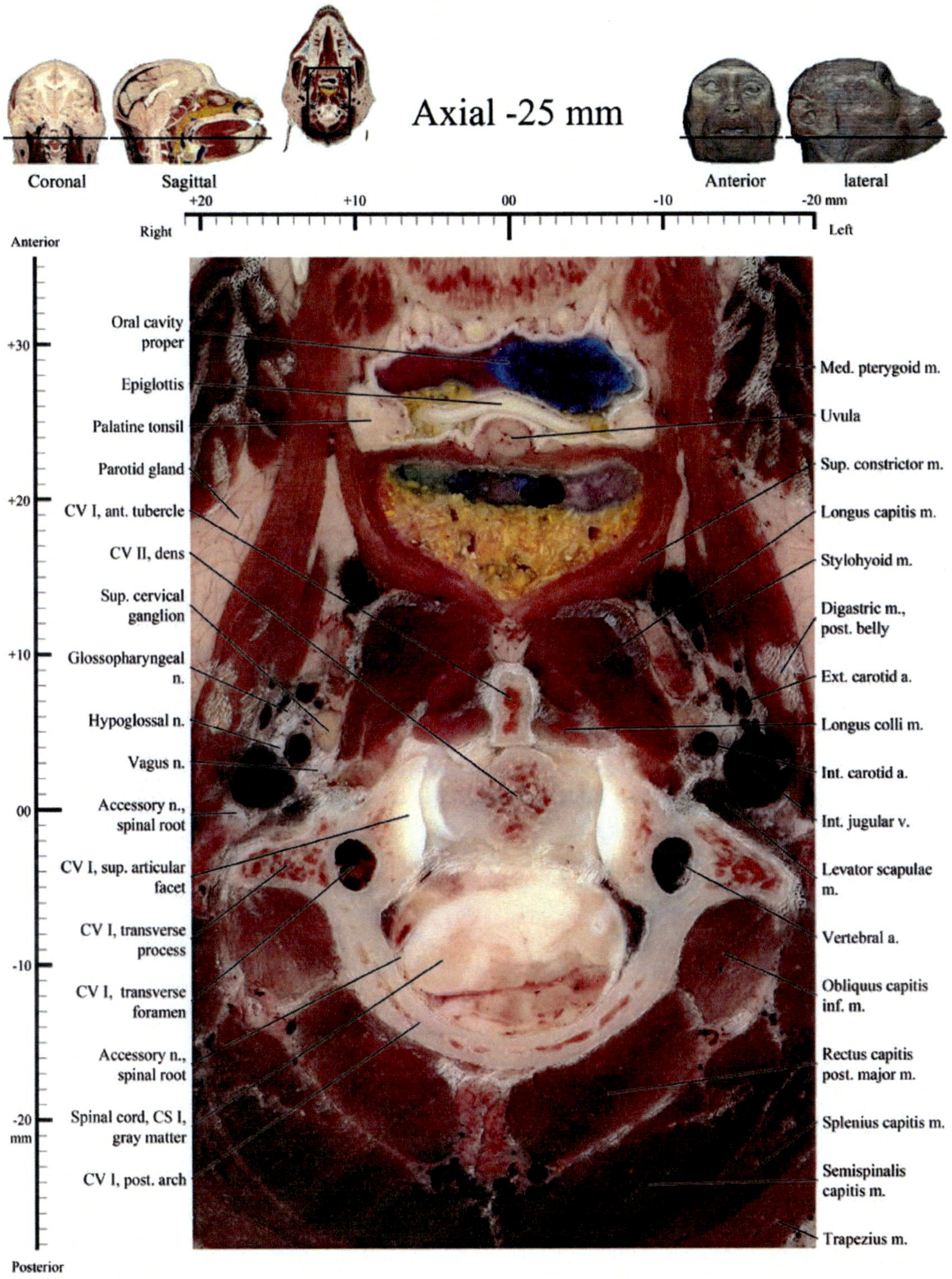
Coronal
Sagittal
Axial -25 mm
Anterior
lateral
+20
+10
00
-10
-20 mm
Right
Left
Anterior
+30
+20
+10
00
-10
-20
mm
Posterior
Oral cavity proper
Epiglottis
Palatine tonsil
Parotid gland
CV I, ant. tubercle
CV II, dens
Sup. cervical ganglion
Glossopharyngeal n.
Hypoglossal n.
Vagus n.
Accessory n., spinal root
CV I, sup. articular facet
CV I, transverse process
CV I, transverse foramen
Accessory n., spinal root
Spinal cord, CS I, gray matter
CV I, post. arch
Med. pterygoid m.
Uvula
Sup. constrictor m.
Longus capitis m.
Stylohyoid m.
Digastric m., post. belly
Ext. carotid a.
Longus colli m.
Int. carotid a.
Int. jugular v.
Levator scapulae m.
Vertebral a.
Obliquus capitis inf. m.
Rectus capitis post. major m.
Splenius capitis m.
Semispinalis capitis m.
Trapezius m.

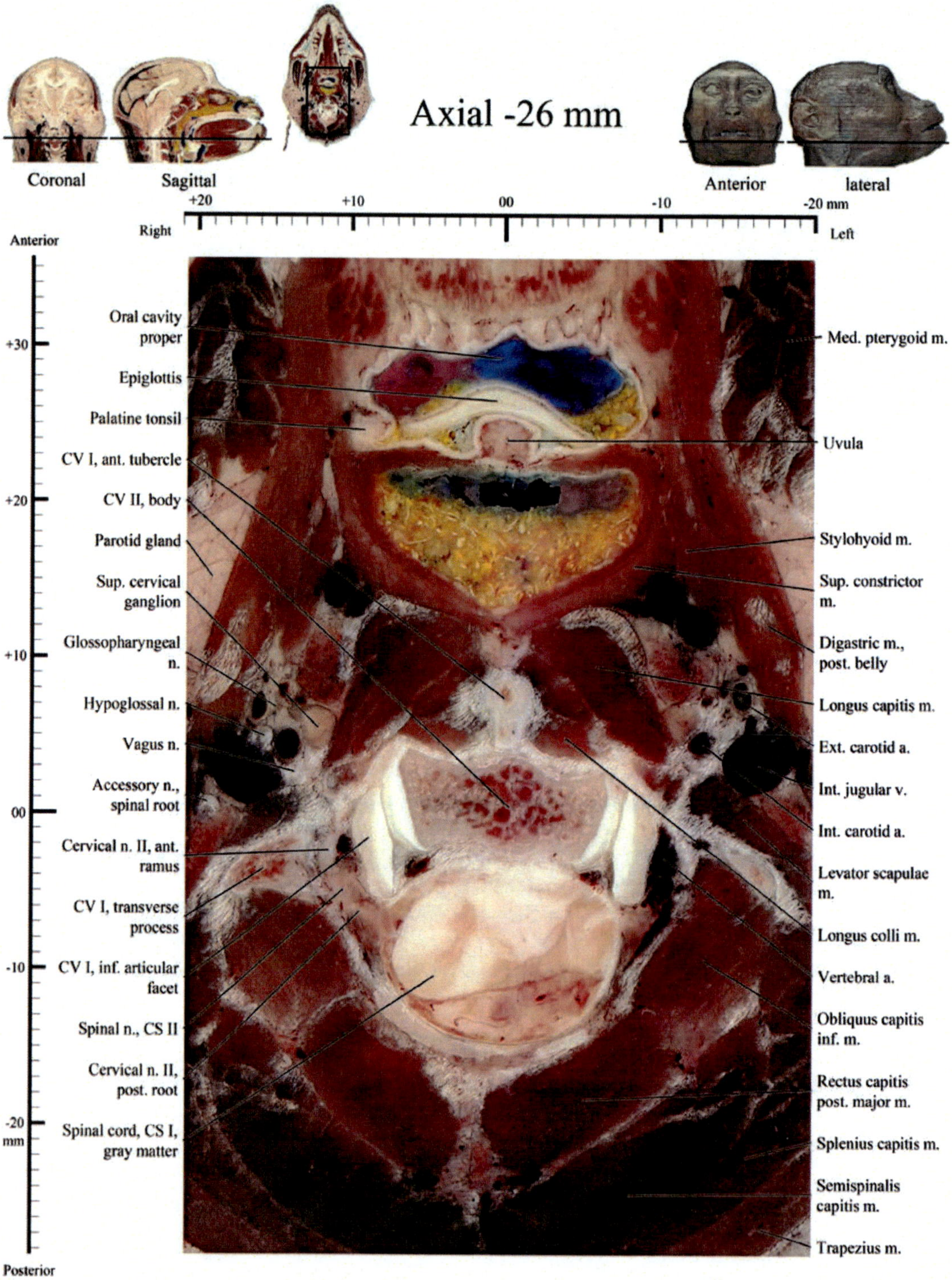
Coronal
Sagittal
Axial -26 mm
Anterior
lateral
+20
+10
00
-10
-20 mm
Right
Left
Anterior
Posterior
+30
+20
+10
00
-10
-20 mm
Oral cavity proper
Epiglottis
Palatine tonsil
CV I, ant. tubercle
CV II, body
Parotid gland
Sup. cervical ganglion
Glossopharyngeal n.
Hypoglossal n.
Vagus n.
Accessory n., spinal root
Cervical n. II, ant. ramus
CV I, transverse process
CV I, inf. articular facet
Spinal n., CS II
Cervical n. II, post. root
Spinal cord, CS I, gray matter
Med. pterygoid m.
Uvula
Stylohyoid m.
Sup. constrictor m.
Digastric m., post. belly
Longus capitis m.
Ext. carotid a.
Int. jugular v.
Int. carotid a.
Levator scapulae m.
Longus colli m.
Vertebral a.
Obliquus capitis inf. m.
Rectus capitis post. major m.
Splenius capitis m.
Semispinalis capitis m.
Trapezius m.

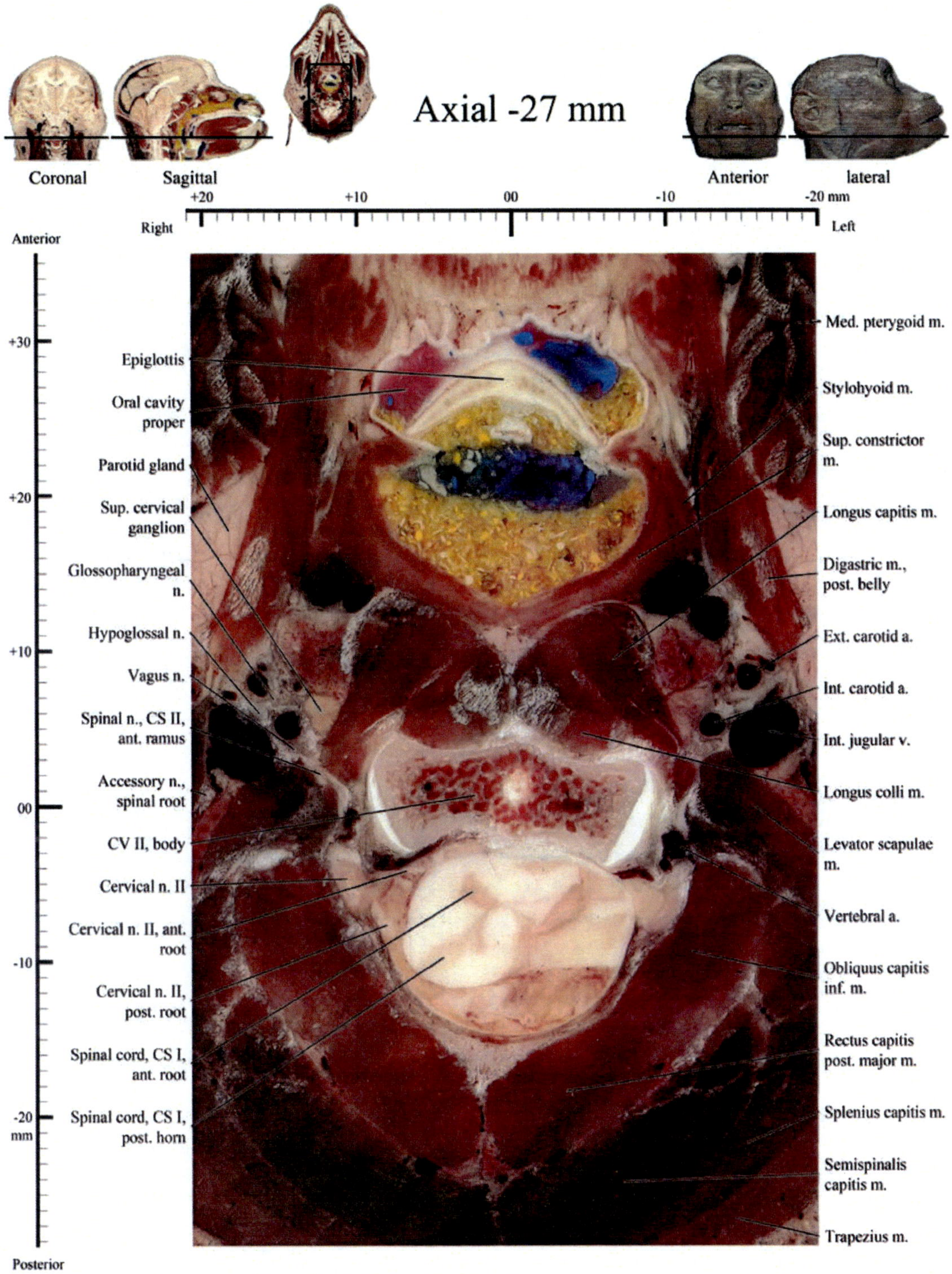

Axial -27 mm
Coronal
Sagittal
Anterior
lateral
+20
+10
00
-10
-20 mm
Right
Left
Anterior
+30
+20
+10
00
-10
-20
mm
Posterior
Epiglottis
Oral cavity proper
Parotid gland
Sup. cervical ganglion
Glossopharyngeal n.
Hypoglossal n.
Vagus n.
Spinal n., CS II, ant. ramus
Accessory n., spinal root
CV II, body
Cervical n. II
Cervical n. II, ant. root
Cervical n. II, post. root
Spinal cord, CS I, ant. root
Spinal cord, CS I, post. horn
Med. pterygoid m.
Stylohyoid m.
Sup. constrictor m.
Longus capitis m.
Digastric m., post. belly
Ext. carotid a.
Int. carotid a.
Int. jugular v.
Longus colli m.
Levator scapulae m.
Vertebral a.
Obliquus capitis inf. m.
Rectus capitis post. major m.
Splenius capitis m.
Semispinalis capitis m.
Trapezius m.

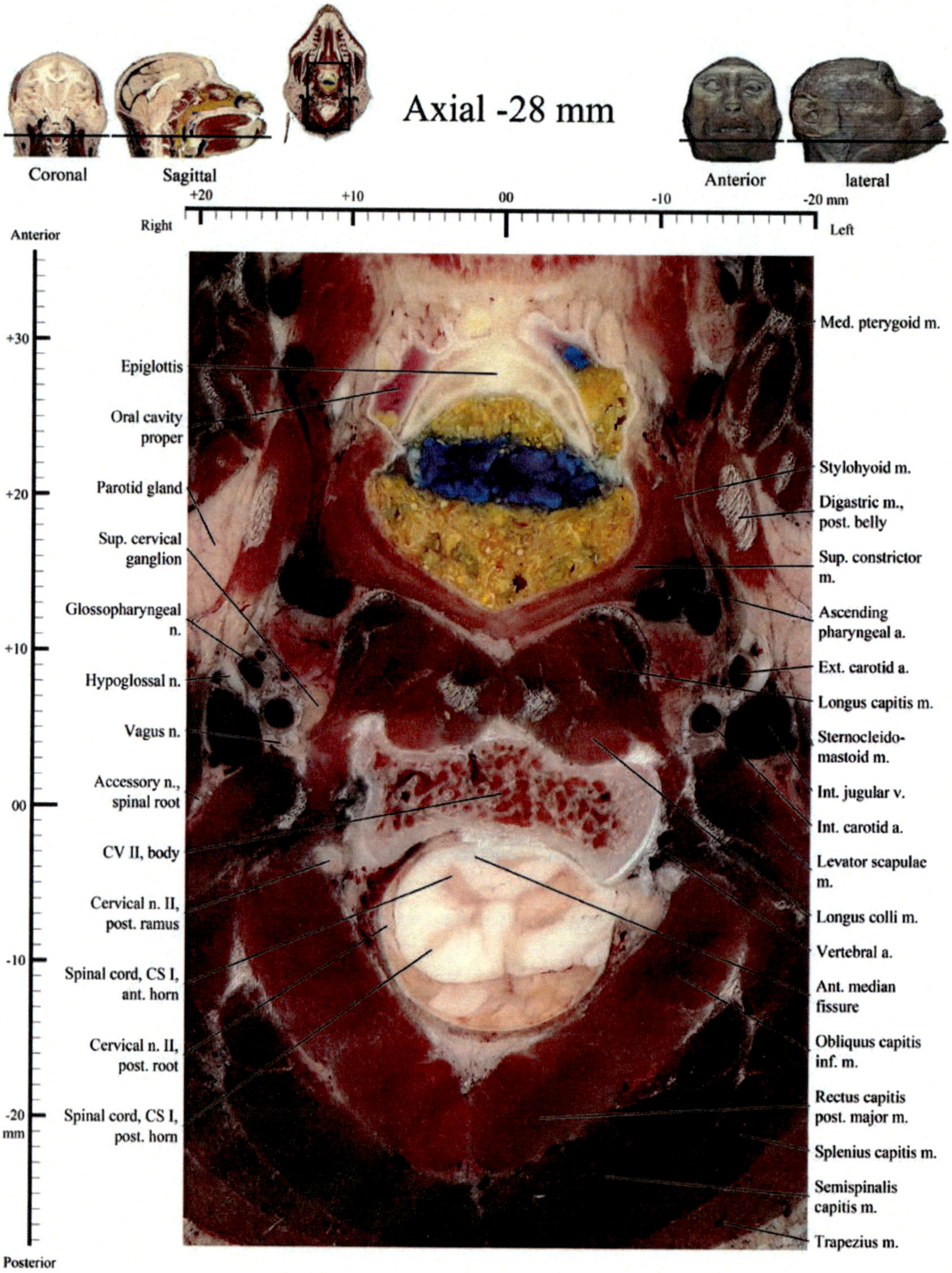
Coronal
Sagittal
Axial -28 mm
Anterior
lateral
+20
+10
00
-10
-20 mm
Right
Left
Anterior
+30
+20
+10
00
-10
-20
mm
Posterior
Epiglottis
Oral cavity proper
Parotid gland
Sup. cervical ganglion
Glossopharyngeal n.
Hypoglossal n.
Vagus n.
Accessory n., spinal root
CV II, body
Cervical n. II, post. ramus
Spinal cord, CS I, ant. horn
Cervical n. II, post. root
Spinal cord, CS I, post. horn
Med. pterygoid m.
Stylohyoid m.
Digastric m., post. belly
Sup. constrictor m.
Ascending pharyngeal a.
Ext. carotid a.
Longus capitis m.
Sternocleido-mastoid m.
Int. jugular v.
Int. carotid a.
Levator scapulae m.
Longus colli m.
Vertebral a.
Ant. median fissure
Obliquus capitis inf. m.
Rectus capitis post. major m.
Splenius capitis m.
Semispinalis capitis m.
Trapezius m.

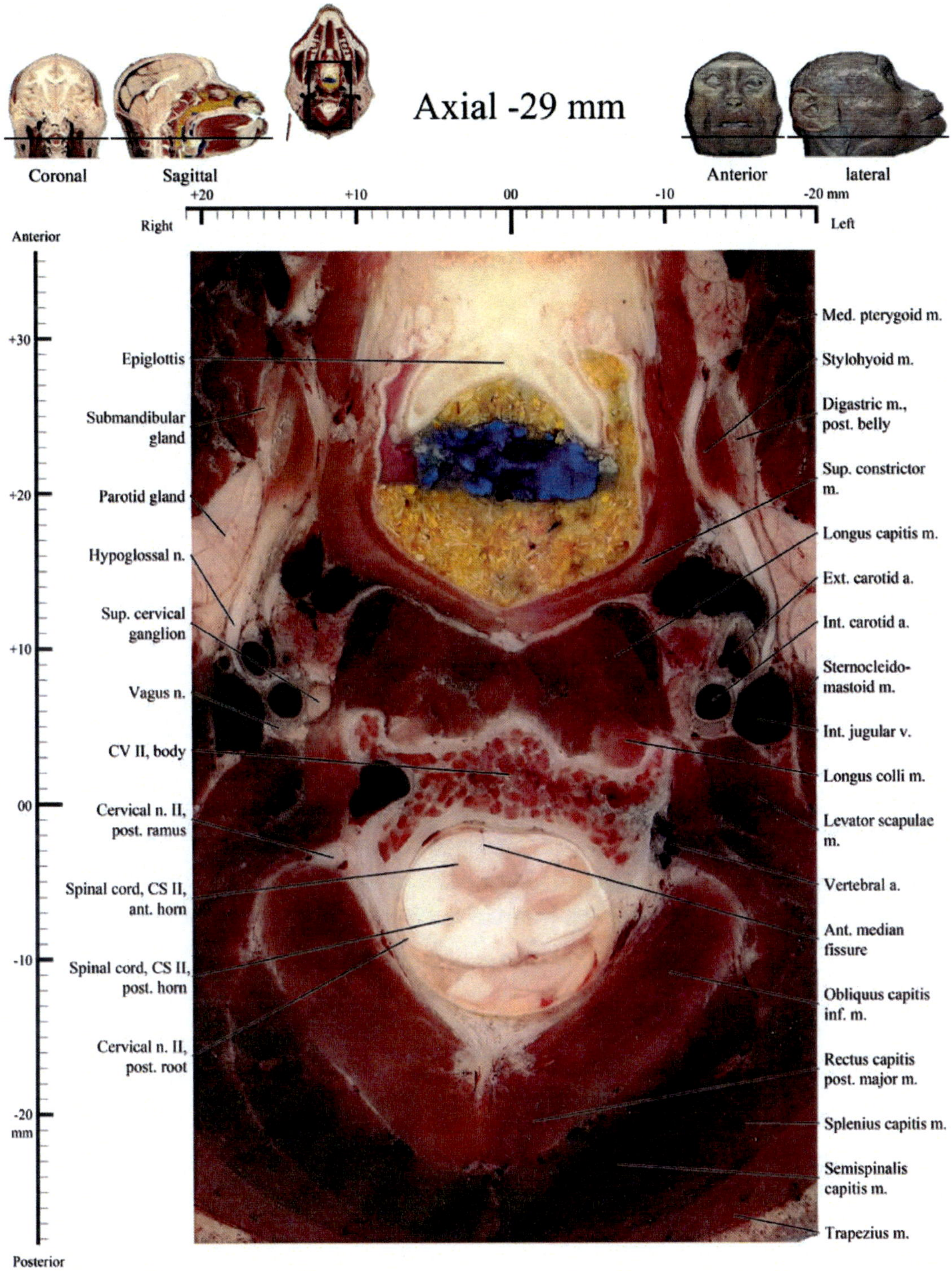

Coronal
Sagittal
Axial -29 mm
Anterior
lateral
+20
+10
00
-10
-20 mm
Right
Left
Anterior
+30
+20
+10
00
-10
-20
mm
Posterior
Epiglottis
Submandibular gland
Parotid gland
Hypoglossal n.
Sup. cervical ganglion
Vagus n.
CV II, body
Cervical n. II, post. ramus
Spinal cord, CS II, ant. horn
Spinal cord, CS II, post. horn
Cervical n. II, post. root
Med. pterygoid m.
Stylohyoid m.
Digastric m., post. belly
Sup. constrictor m.
Longus capitis m.
Ext. carotid a.
Int. carotid a.
Sternocleido-mastoid m.
Int. jugular v.
Longus colli m.
Levator scapulae m.
Vertebral a.
Ant. median fissure
Obliquus capitis inf. m.
Rectus capitis post. major m.
Splenius capitis m.
Semispinalis capitis m.
Trapezius m.

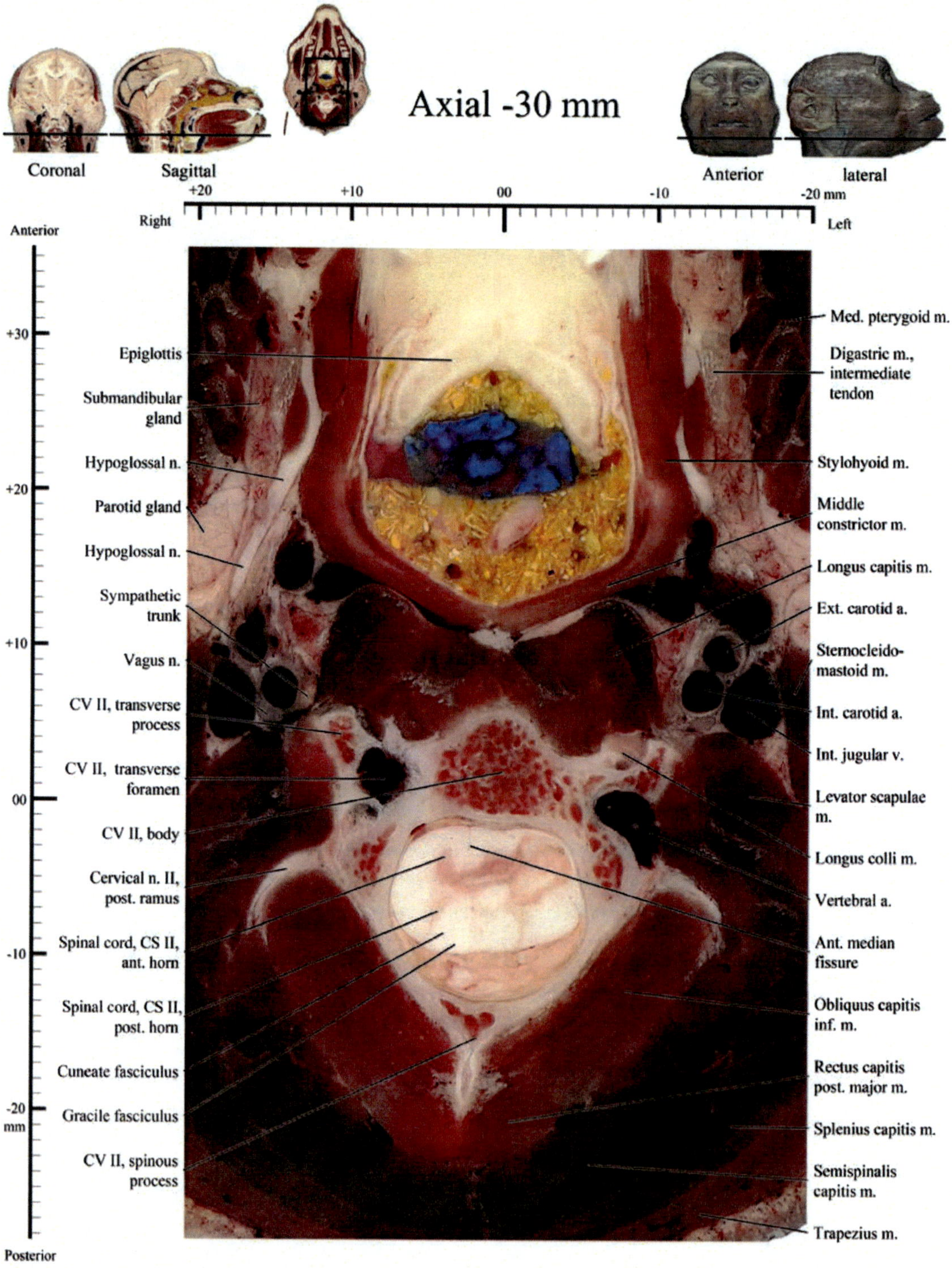
Coronal
Sagittal
Axial -30 mm
Anterior
lateral
+20
+10
00
-10
-20 mm
Right
Left
Anterior
+30
+20
+10
00
-10
-20
mm
Posterior
Epiglottis
Submandibular gland
Hypoglossal n.
Parotid gland
Hypoglossal n.
Sympathetic trunk
Vagus n.
CV II, transverse process
CV II, transverse foramen
CV II, body
Cervical n. II, post. ramus
Spinal cord, CS II, ant. horn
Spinal cord, CS II, post. horn
Cuneate fasciculus
Gracile fasciculus
CV II, spinous process
Med. pterygoid m.
Digastric m., intermediate tendon
Stylohyoid m.
Middle constrictor m.
Longus capitis m.
Ext. carotid a.
Sternocleido-mastoid m.
Int. carotid a.
Int. jugular v.
Levator scapulae m.
Longus colli m.
Vertebral a.
Ant. median fissure
Obliquus capitis inf. m.
Rectus capitis post. major m.
Splenius capitis m.
Semispinalis capitis m.
Trapezius m.

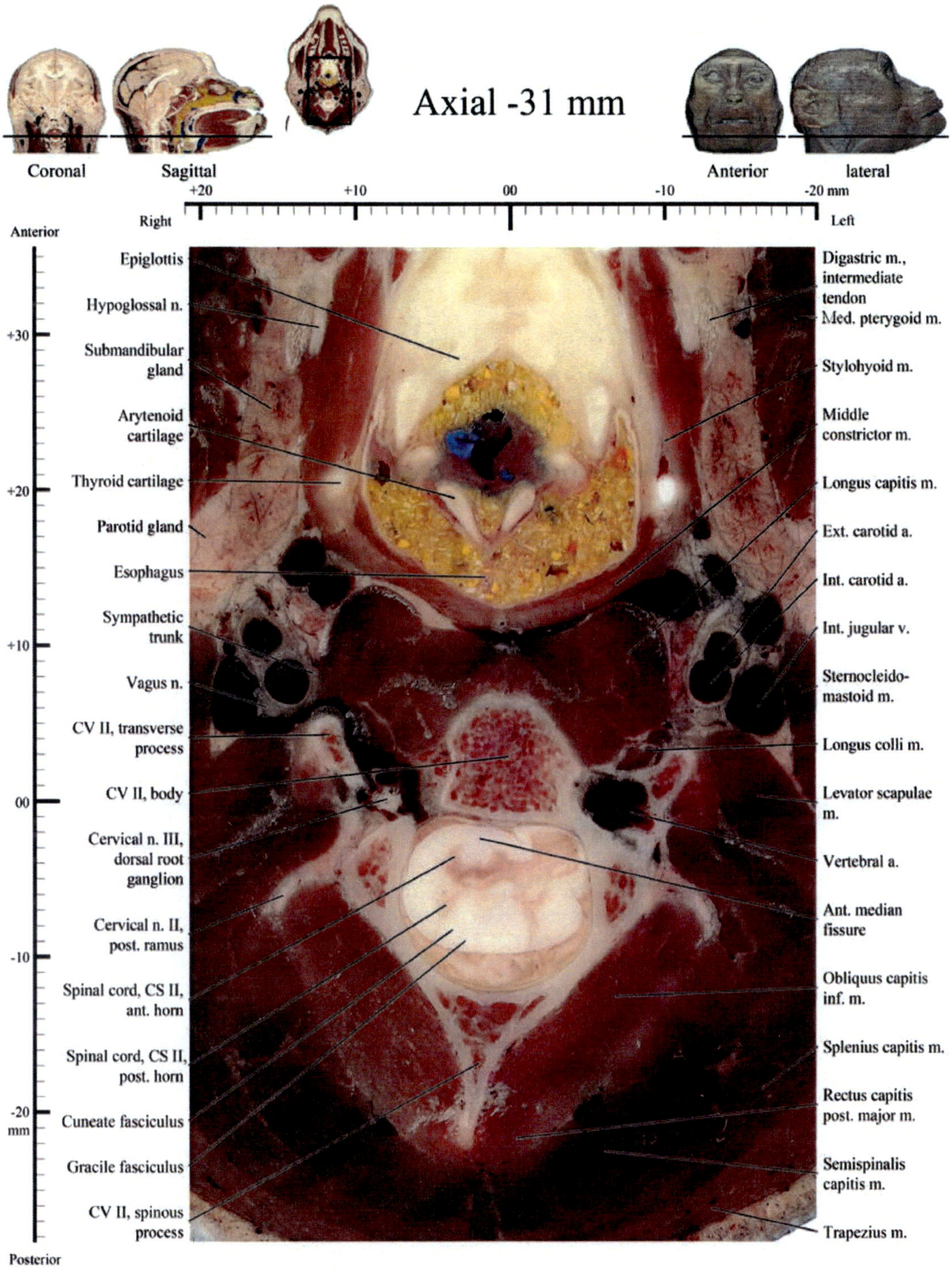

Coronal
Sagittal
Axial -31 mm
Anterior
lateral
+20
+10
00
-10
-20 mm
Right
Left
Anterior
Posterior
+30
+20
+10
00
-10
-20
mm
Epiglottis
Hypoglossal n.
Submandibular gland
Arytenoid cartilage
Thyroid cartilage
Parotid gland
Esophagus
Sympathetic trunk
Vagus n.
CV II, transverse process
CV II, body
Cervical n. III, dorsal root ganglion
Cervical n. II, post. ramus
Spinal cord, CS II, ant. horn
Spinal cord, CS II, post. horn
Cuneate fasciculus
Gracile fasciculus
CV II, spinous process
Digastric m., intermediate tendon
Med. pterygoid m.
Stylohyoid m.
Middle constrictor m.
Longus capitis m.
Ext. carotid a.
Int. carotid a.
Int. jugular v.
Sternocleido-mastoid m.
Longus colli m.
Levator scapulae m.
Vertebral a.
Ant. median fissure
Obliquus capitis inf. m.
Splenius capitis m.
Rectus capitis post. major m.
Semispinalis capitis m.
Trapezius m.

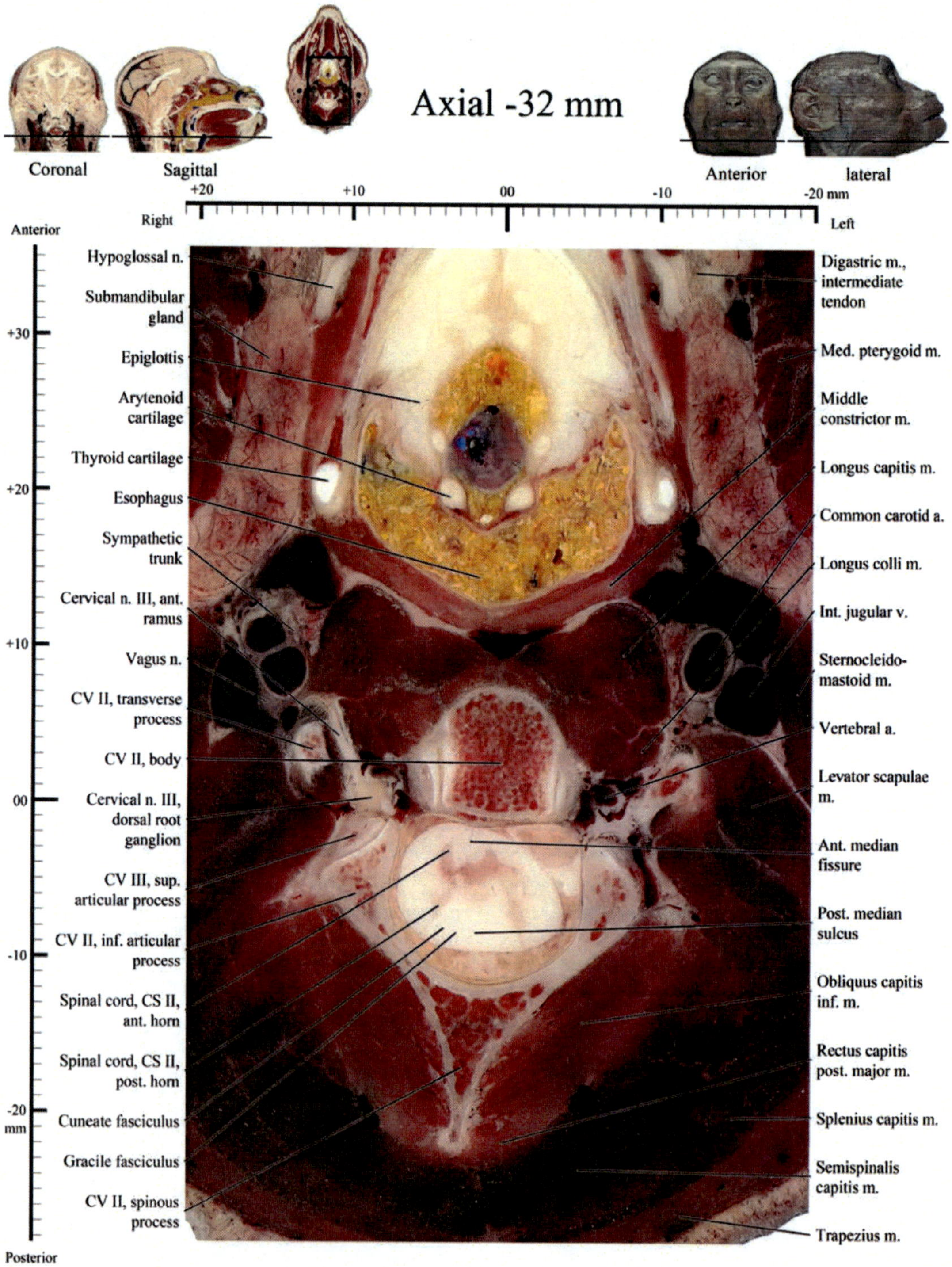
Coronal
Sagittal
Axial -32 mm
Anterior
lateral
+20
+10
00
-10
-20 mm
Right
Left
Anterior
Posterior
+30
+20
+10
00
-10
-20
mm
Hypoglossal n.
Submandibular gland
Epiglottis
Arytenoid cartilage
Thyroid cartilage
Esophagus
Sympathetic trunk
Cervical n. III, ant. ramus
Vagus n.
CV II, transverse process
CV II, body
Cervical n. III, dorsal root ganglion
CV III, sup. articular process
CV II, inf. articular process
Spinal cord, CS II, ant. horn
Spinal cord, CS II, post. horn
Cuneate fasciculus
Gracile fasciculus
CV II, spinous process
Digastric m., intermediate tendon
Med. pterygoid m.
Middle constrictor m.
Longus capitis m.
Common carotid a.
Longus colli m.
Int. jugular v.
Sternocleido-mastoid m.
Vertebral a.
Levator scapulae m.
Ant. median fissure
Post. median sulcus
Obliquus capitis inf. m.
Rectus capitis post. major m.
Splenius capitis m.
Semispinalis capitis m.
Trapezius m.

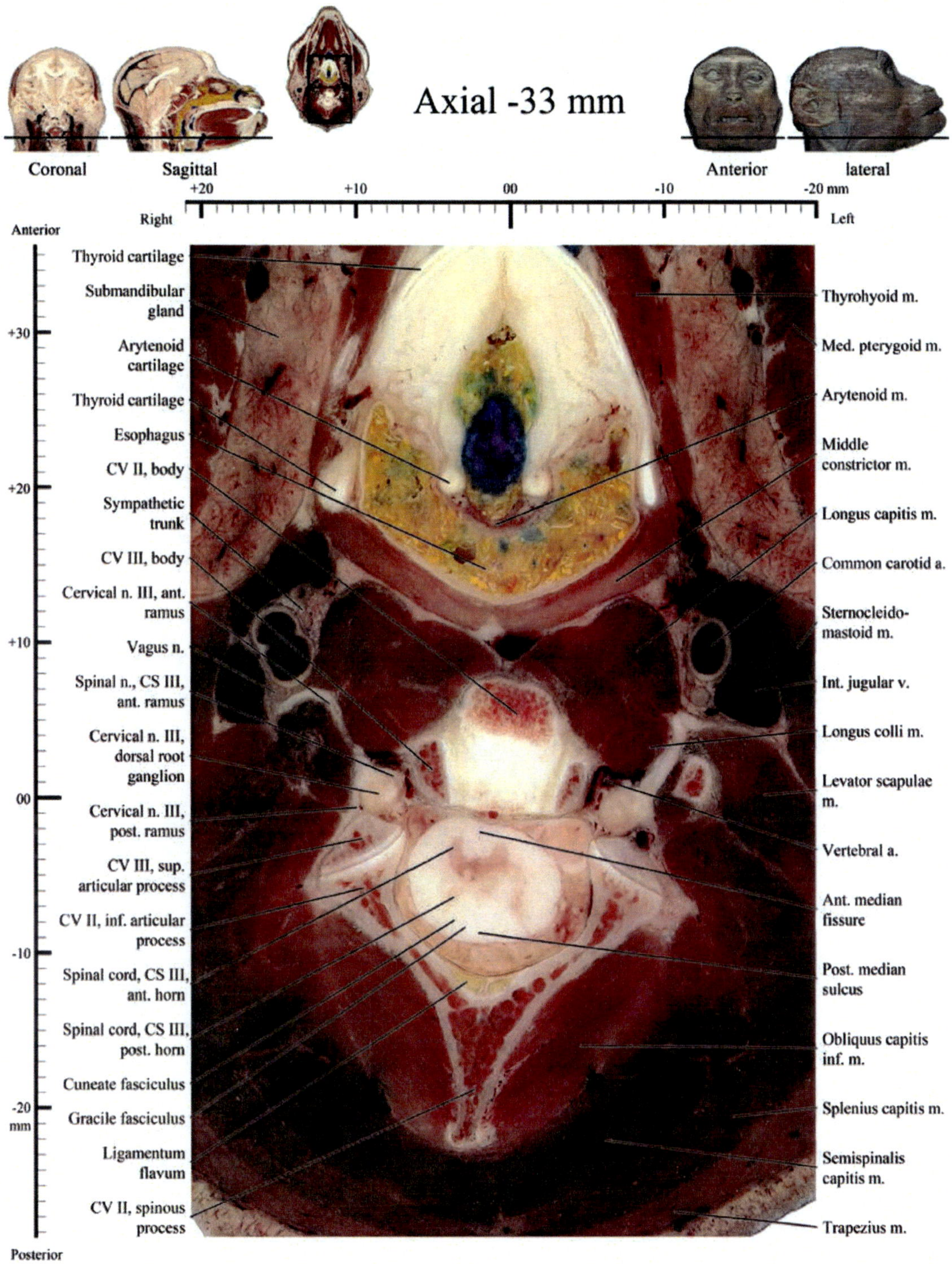
Axial -33 mm
Coronal
Sagittal
Anterior
lateral
+20
+10
00
-10
-20 mm
Right
Left
Anterior
Posterior
+30
+20
+10
00
-10
-20
mm
Thyroid cartilage
Submandibular gland
Arytenoid cartilage
Thyroid cartilage
Esophagus
CV II, body
Sympathetic trunk
CV III, body
Cervical n. III, ant. ramus
Vagus n.
Spinal n., CS III, ant. ramus
Cervical n. III, dorsal root ganglion
Cervical n. III, post. ramus
CV III, sup. articular process
CV II, inf. articular process
Spinal cord, CS III, ant. horn
Spinal cord, CS III, post. horn
Cuneate fasciculus
Gracile fasciculus
Ligamentum flavum
CV II, spinous process
Thyrohyoid m.
Med. pterygoid m.
Arytenoid m.
Middle constrictor m.
Longus capitis m.
Common carotid a.
Sternocleido-mastoid m.
Int. jugular v.
Longus colli m.
Levator scapulae m.
Vertebral a.
Ant. median fissure
Post. median sulcus
Obliquus capitis inf. m.
Splenius capitis m.
Semispinalis capitis m.
Trapezius m.

Axial -34 mm

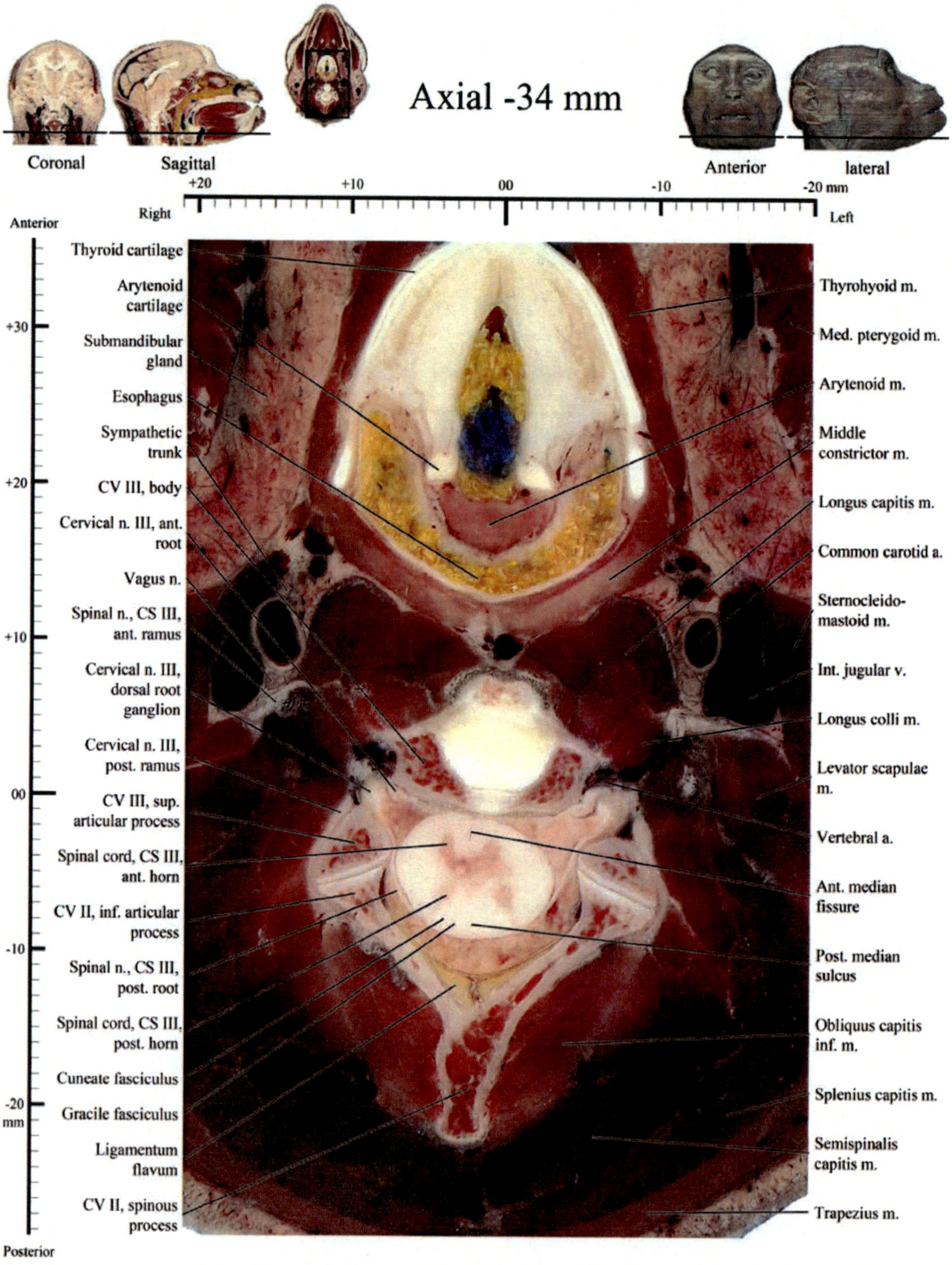
Coronal
Sagittal
Anterior
lateral
+20
+10
00
-10
-20 mm
Right
Left
Anterior
Posterior
+30
+20
+10
00
-10
-20
mm
Thyroid cartilage
Arytenoid cartilage
Submandibular gland
Esophagus
Sympathetic trunk
CV III, body
Cervical n. III, ant. root
Vagus n.
Spinal n., CS III, ant. ramus
Cervical n. III, dorsal root ganglion
Cervical n. III, post. ramus
CV III, sup. articular process
Spinal cord, CS III, ant. horn
CV II, inf. articular process
Spinal n., CS III, post. root
Spinal cord, CS III, post. horn
Cuneate fasciculus
Gracile fasciculus
Ligamentum flavum
CV II, spinous process
Thyrohyoid m.
Med. pterygoid m.
Arytenoid m.
Middle constrictor m.
Longus capitis m.
Common carotid a.
Sternocleido-mastoid m.
Int. jugular v.
Longus colli m.
Levator scapulae m.
Vertebral a.
Ant. median fissure
Post. median sulcus
Obliquus capitis inf. m.
Splenius capitis m.
Semispinalis capitis m.
Trapezius m.

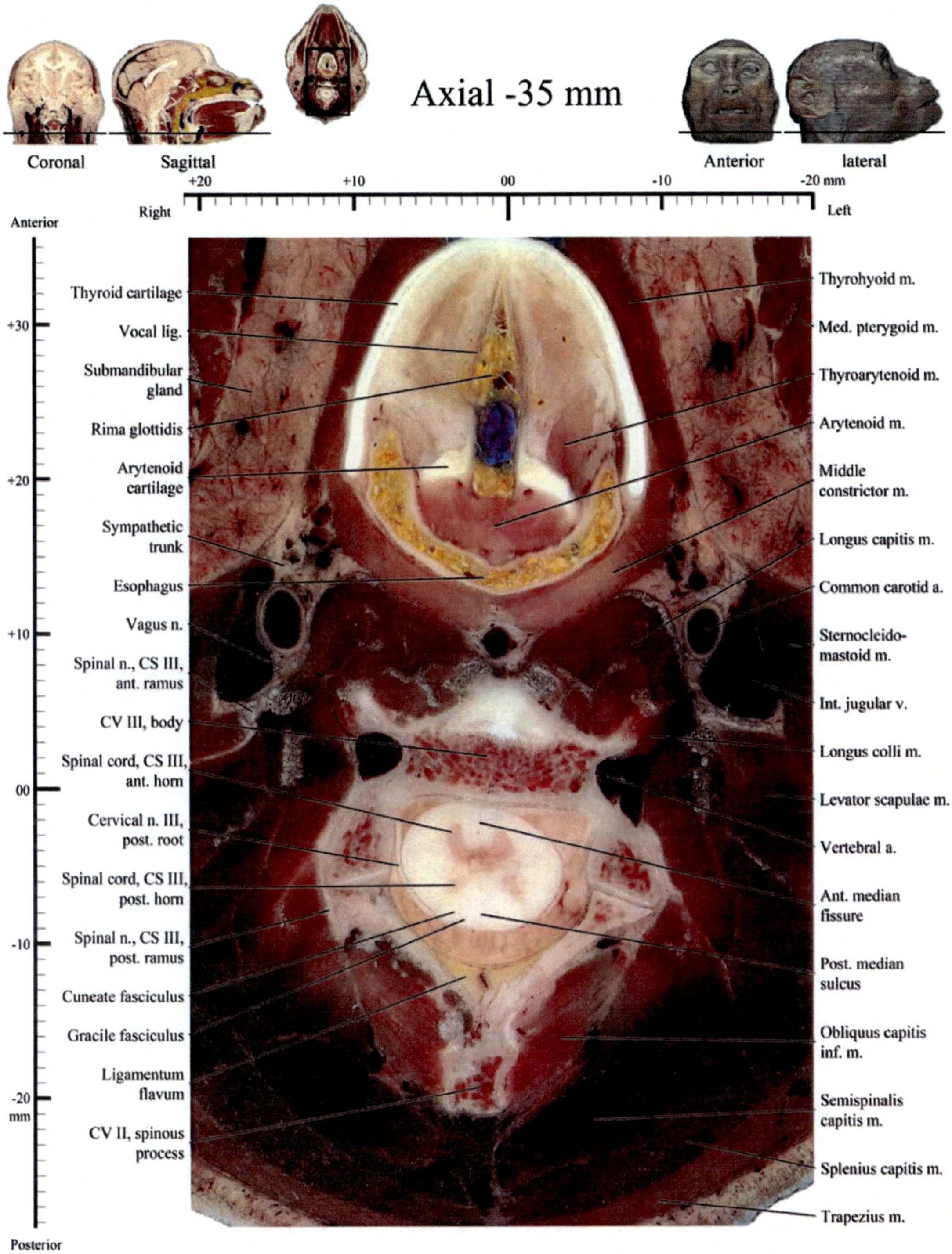
Axial -35 mm
Coronal
Sagittal
Anterior
lateral
+20
+10
00
-10
-20 mm
Right
Left
Anterior
Posterior
+30
+20
+10
00
-10
-20
mm
Thyroid cartilage
Vocal lig.
Submandibular gland
Rima glottidis
Arytenoid cartilage
Sympathetic trunk
Esophagus
Vagus n.
Spinal n., CS III, ant. ramus
CV III, body
Spinal cord, CS III, ant. horn
Cervical n. III, post. root
Spinal cord, CS III, post. horn
Spinal n., CS III, post. ramus
Cuneate fasciculus
Gracile fasciculus
Ligamentum flavum
CV II, spinous process
Thyrohyoid m.
Med. pterygoid m.
Thyroarytenoid m.
Arytenoid m.
Middle constrictor m.
Longus capitis m.
Common carotid a.
Sternocleido-mastoid m.
Int. jugular v.
Longus colli m.
Levator scapulae m.
Vertebral a.
Ant. median fissure
Post. median sulcus
Obliquus capitis inf. m.
Semispinalis capitis m.
Splenius capitis m.
Trapezius m.

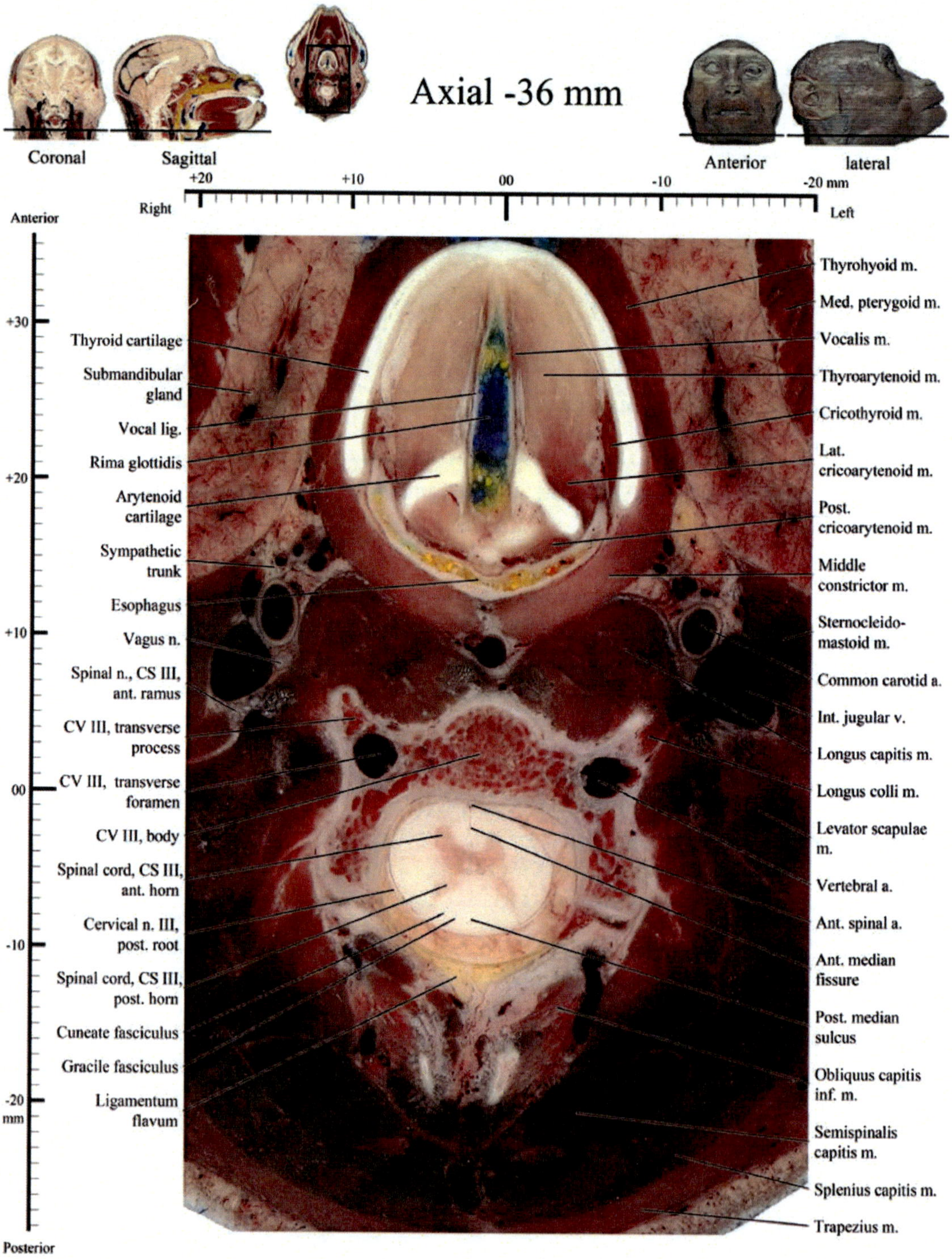
Axial -36 mm
Coronal
Sagittal
Anterior
lateral
+20
+10
00
-10
-20 mm
Right
Left
Anterior
Posterior
+30
+20
+10
00
-10
-20
mm
Thyroid cartilage
Submandibular gland
Vocal lig.
Rima glottidis
Arytenoid cartilage
Sympathetic trunk
Esophagus
Vagus n.
Spinal n., CS III, ant. ramus
CV III, transverse process
CV III, transverse foramen
CV III, body
Spinal cord, CS III, ant. horn
Cervical n. III, post. root
Spinal cord, CS III, post. horn
Cuneate fasciculus
Gracile fasciculus
Ligamentum flavum
Thyrohyoid m.
Med. pterygoid m.
Vocalis m.
Thyroarytenoid m.
Cricothyroid m.
Lat. cricoarytenoid m.
Post. cricoarytenoid m.
Middle constrictor m.
Sternocleido-mastoid m.
Common carotid a.
Int. jugular v.
Longus capitis m.
Longus colli m.
Levator scapulae m.
Vertebral a.
Ant. spinal a.
Ant. median fissure
Post. median sulcus
Obliquus capitis inf. m.
Semispinalis capitis m.
Splenius capitis m.
Trapezius m.

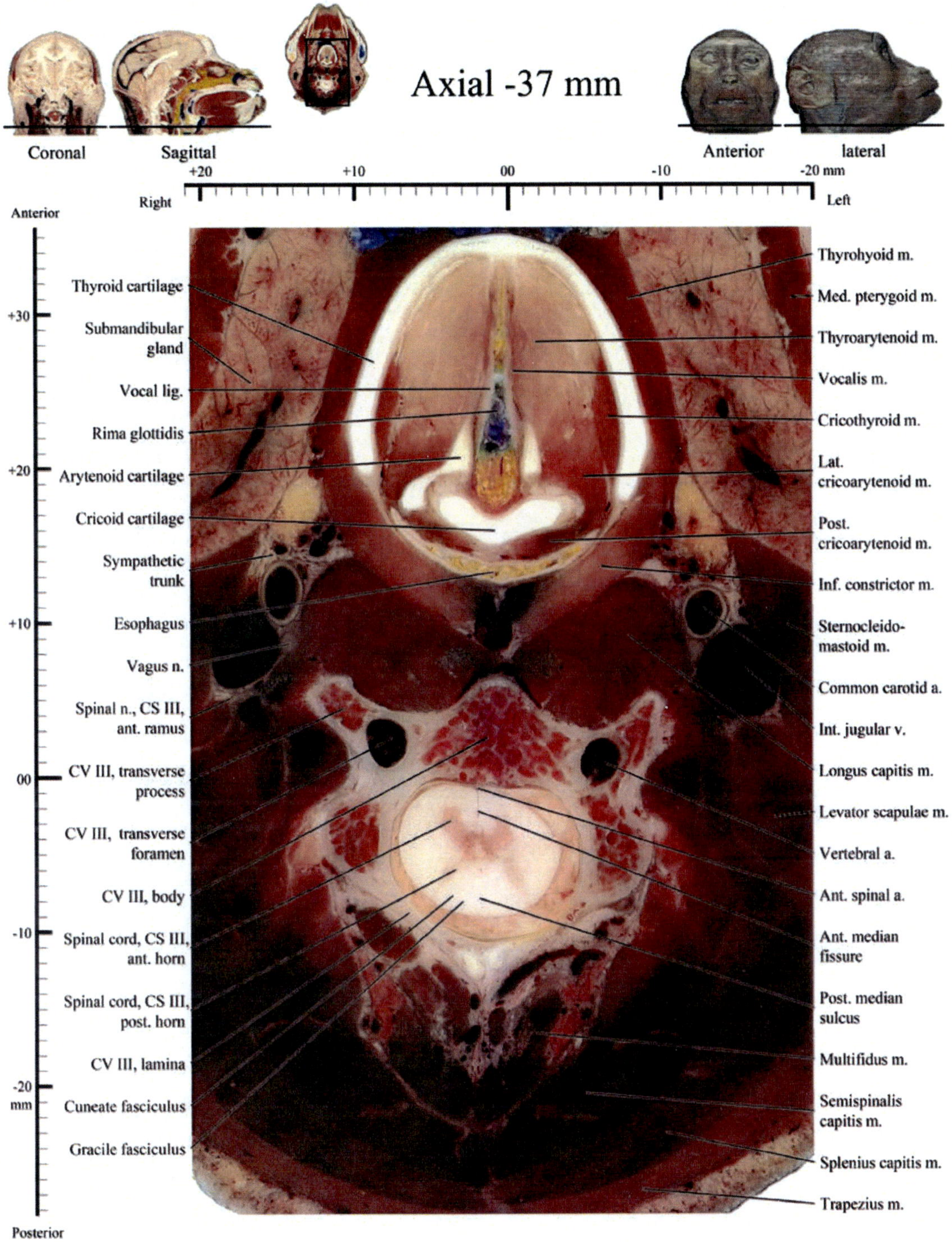
Axial -37 mm
Coronal
Sagittal
Anterior
lateral
+20
+10
00
-10
-20 mm
Right
Left
Anterior
Posterior
+30
+20
+10
00
-10
-20
mm
Thyroid cartilage
Submandibular gland
Vocal lig.
Rima glottidis
Arytenoid cartilage
Cricoid cartilage
Sympathetic trunk
Esophagus
Vagus n.
Spinal n., CS III, ant. ramus
CV III, transverse process
CV III, transverse foramen
CV III, body
Spinal cord, CS III, ant. horn
Spinal cord, CS III, post. horn
CV III, lamina
Cuneate fasciculus
Gracile fasciculus
Thyrohyoid m.
Med. pterygoid m.
Thyroarytenoid m.
Vocalis m.
Cricothyroid m.
Lat. cricoarytenoid m.
Post. cricoarytenoid m.
Inf. constrictor m.
Sternocleido-mastoid m.
Common carotid a.
Int. jugular v.
Longus capitis m.
Levator scapulae m.
Vertebral a.
Ant. spinal a.
Ant. median fissure
Post. median sulcus
Multifidus m.
Semispinalis capitis m.
Splenius capitis m.
Trapezius m.

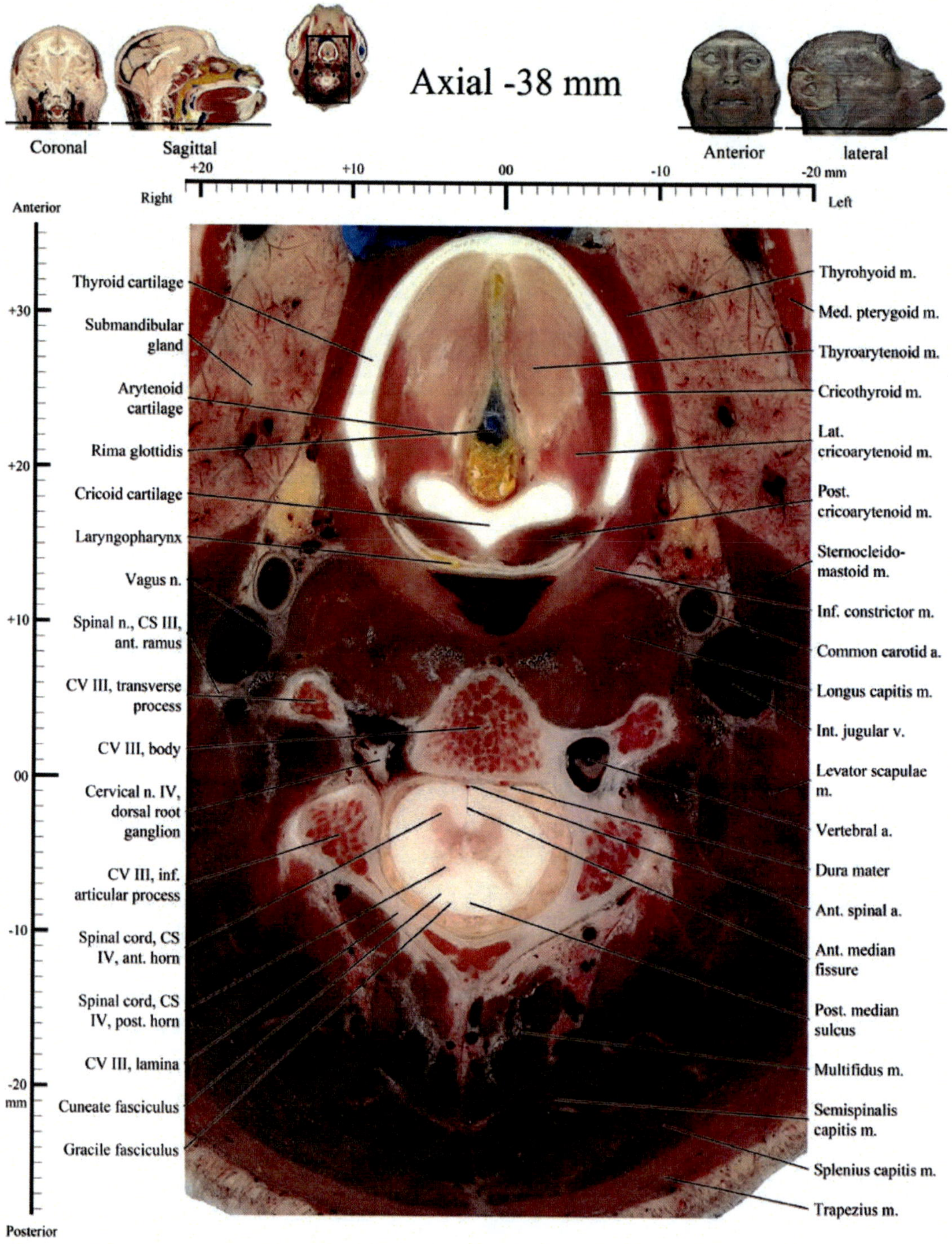
Coronal
Sagittal
Axial -38 mm
Anterior
lateral
+20
+10
00
-10
-20 mm
Right
Left
Anterior
Posterior
+30
+20
+10
00
-10
-20
mm
Thyroid cartilage
Submandibular gland
Arytenoid cartilage
Rima glottidis
Cricoid cartilage
Laryngopharynx
Vagus n.
Spinal n., CS III, ant. ramus
CV III, transverse process
CV III, body
Cervical n. IV, dorsal root ganglion
CV III, inf. articular process
Spinal cord, CS IV, ant. horn
Spinal cord, CS IV, post. horn
CV III, lamina
Cuneate fasciculus
Gracile fasciculus
Thyrohyoid m.
Med. pterygoid m.
Thyroarytenoid m.
Cricothyroid m.
Lat. cricoarytenoid m.
Post. cricoarytenoid m.
Sternocleido-mastoid m.
Inf. constrictor m.
Common carotid a.
Longus capitis m.
Int. jugular v.
Levator scapulae m.
Vertebral a.
Dura mater
Ant. spinal a.
Ant. median fissure
Post. median sulcus
Multifidus m.
Semispinalis capitis m.
Splenius capitis m.
Trapezius m.

Axial -39 mm

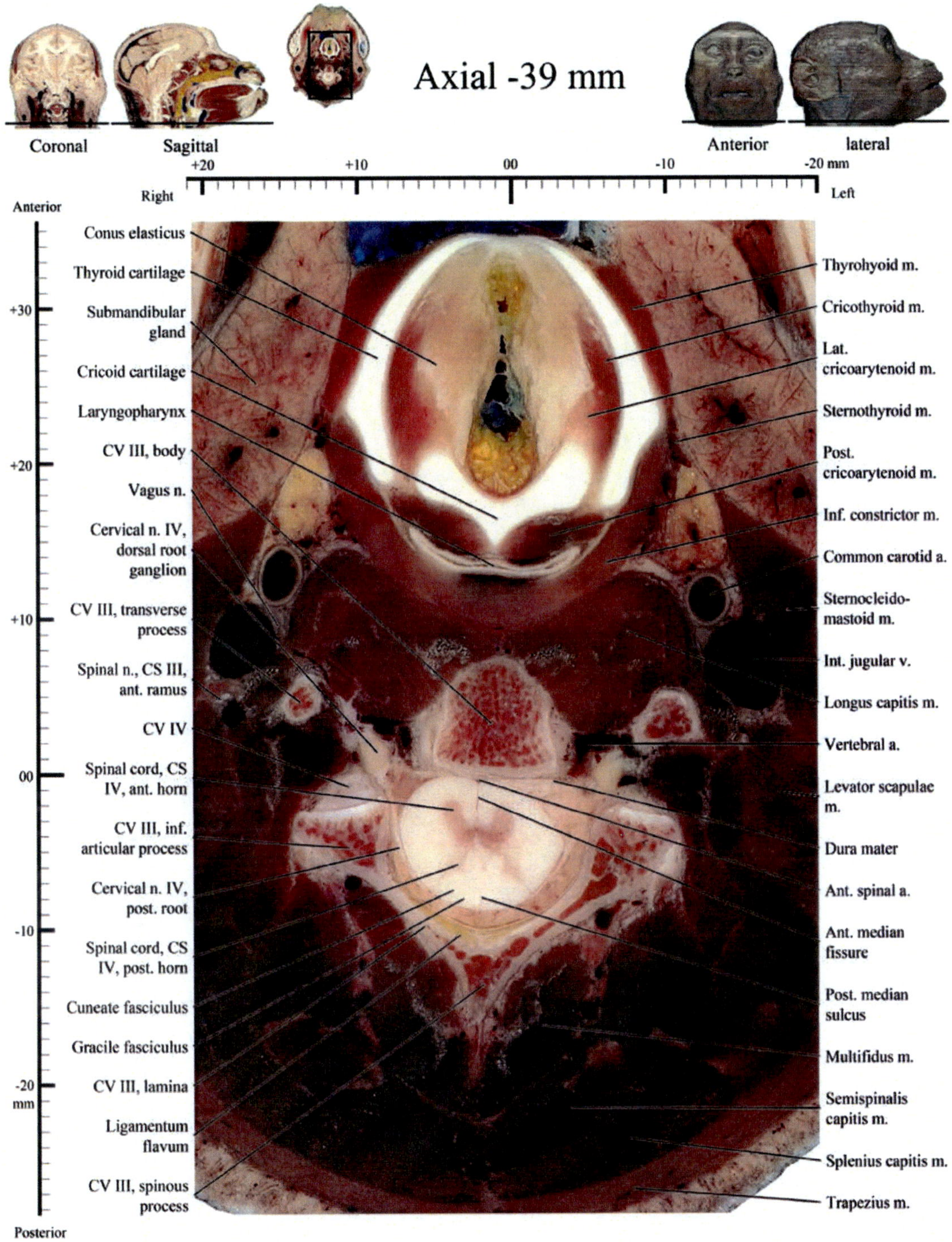
Coronal
Sagittal
Anterior
lateral
+20
+10
00
-10
-20 mm
Right
Left
Anterior
Posterior
+30
+20
+10
00
-10
-20
mm
Conus elasticus
Thyroid cartilage
Submandibular gland
Cricoid cartilage
Laryngopharynx
CV III, body
Vagus n.
Cervical n. IV, dorsal root ganglion
CV III, transverse process
Spinal n., CS III, ant. ramus
CV IV
Spinal cord, CS IV, ant. horn
CV III, inf. articular process
Cervical n. IV, post. root
Spinal cord, CS IV, post. horn
Cuneate fasciculus
Gracile fasciculus
CV III, lamina
Ligamentum flavum
CV III, spinous process
Thyrohyoid m.
Cricothyroid m.
Lat. cricoarytenoid m.
Sternothyroid m.
Post. cricoarytenoid m.
Inf. constrictor m.
Common carotid a.
Sternocleido-mastoid m.
Int. jugular v.
Longus capitis m.
Vertebral a.
Levator scapulae m.
Dura mater
Ant. spinal a.
Ant. median fissure
Post. median sulcus
Multifidus m.
Semispinalis capitis m.
Splenius capitis m.
Trapezius m.

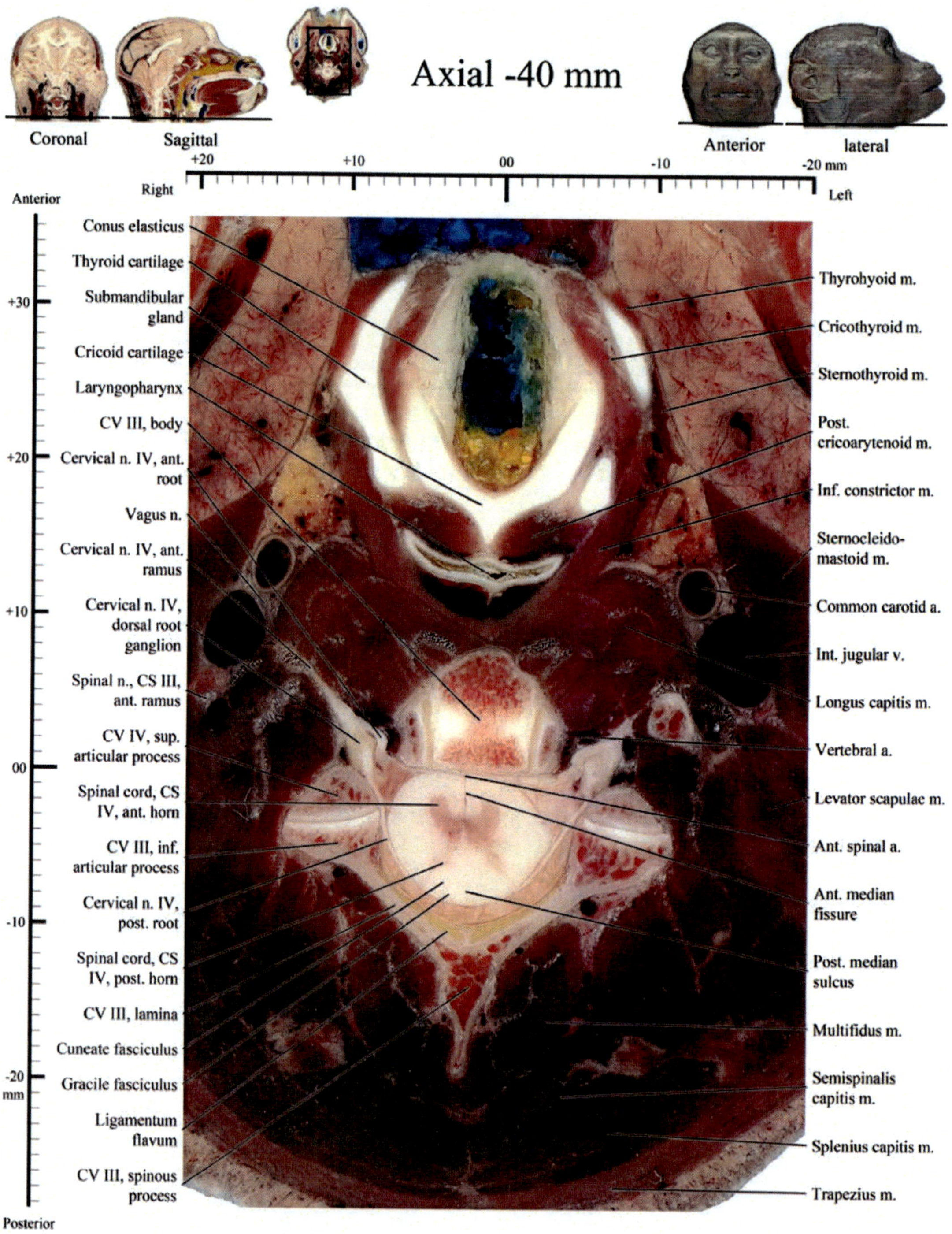

Coronal
Sagittal
Axial -40 mm
Anterior
lateral
+20
+10
00
-10
-20 mm
Right
Left
Anterior
Posterior
+30
+20
+10
00
-10
-20 mm
Conus elasticus
Thyroid cartilage
Submandibular gland
Cricoid cartilage
Laryngopharynx
CV III, body
Cervical n. IV, ant. root
Vagus n.
Cervical n. IV, ant. ramus
Cervical n. IV, dorsal root ganglion
Spinal n., CS III, ant. ramus
CV IV, sup. articular process
Spinal cord, CS IV, ant. horn
CV III, inf. articular process
Cervical n. IV, post. root
Spinal cord, CS IV, post. horn
CV III, lamina
Cuneate fasciculus
Gracile fasciculus
Ligamentum flavum
CV III, spinous process
Thyrohyoid m.
Cricothyroid m.
Sternothyroid m.
Post. cricoarytenoid m.
Inf. constrictor m.
Sternocleido-mastoid m.
Common carotid a.
Int. jugular v.
Longus capitis m.
Vertebral a.
Levator scapulae m.
Ant. spinal a.
Ant. median fissure
Post. median sulcus
Multifidus m.
Semispinalis capitis m.
Splenius capitis m.
Trapezius m.

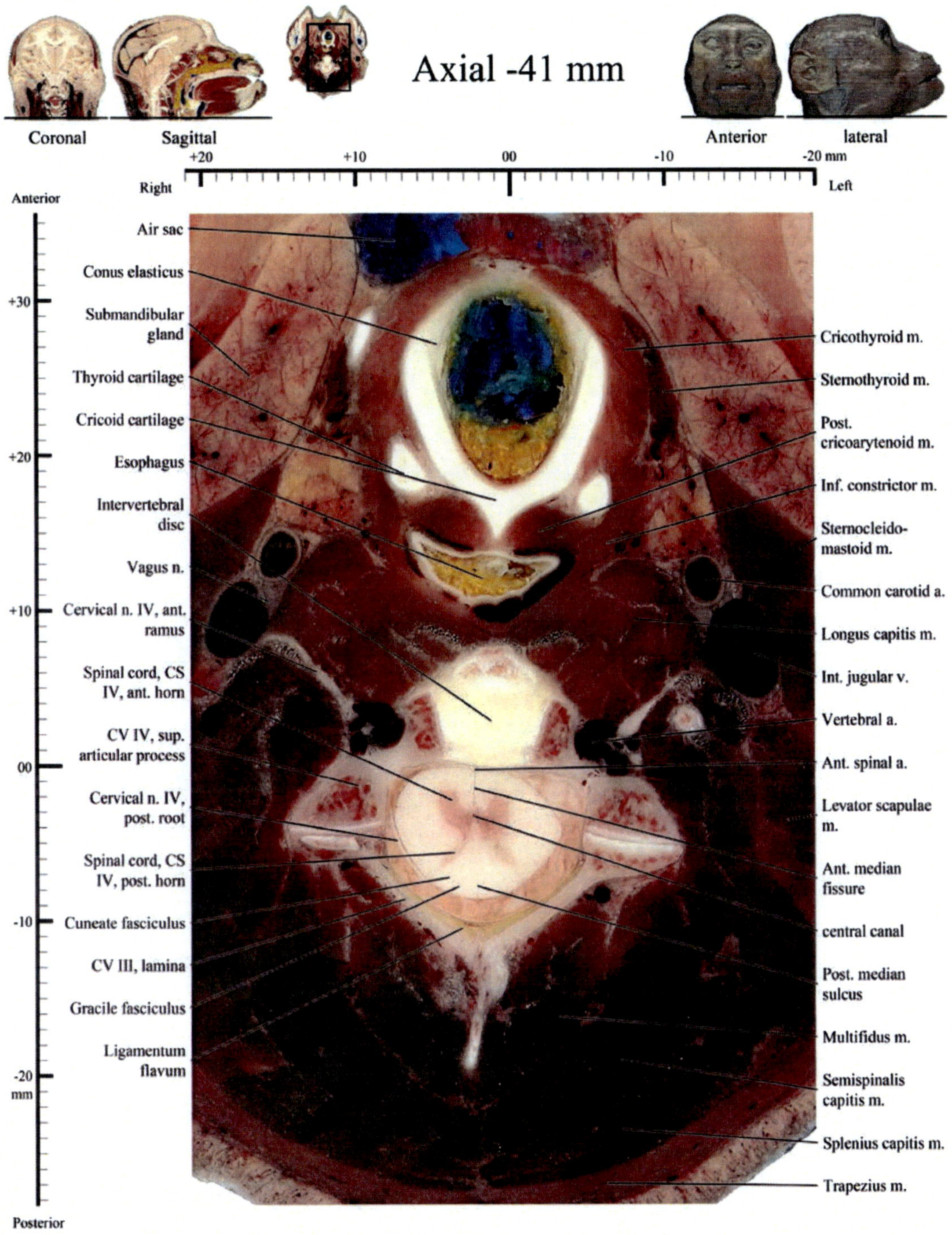
Coronal
Sagittal
Axial -41 mm
Anterior
lateral
+20
+10
00
-10
-20 mm
Right
Left
Anterior
Posterior
+30
+20
+10
00
-10
-20
mm
Air sac
Conus elasticus
Submandibular gland
Thyroid cartilage
Cricoid cartilage
Esophagus
Intervertebral disc
Vagus n.
Cervical n. IV, ant. ramus
Spinal cord, CS IV, ant. horn
CV IV, sup. articular process
Cervical n. IV, post. root
Spinal cord, CS IV, post. horn
Cuneate fasciculus
CV III, lamina
Gracile fasciculus
Ligamentum flavum
Cricothyroid m.
Sternothyroid m.
Post. cricoarytenoid m.
Inf. constrictor m.
Sternocleido-mastoid m.
Common carotid a.
Longus capitis m.
Int. jugular v.
Vertebral a.
Ant. spinal a.
Levator scapulae m.
Ant. median fissure
central canal
Post. median sulcus
Multifidus m.
Semispinalis capitis m.
Splenius capitis m.
Trapezius m.

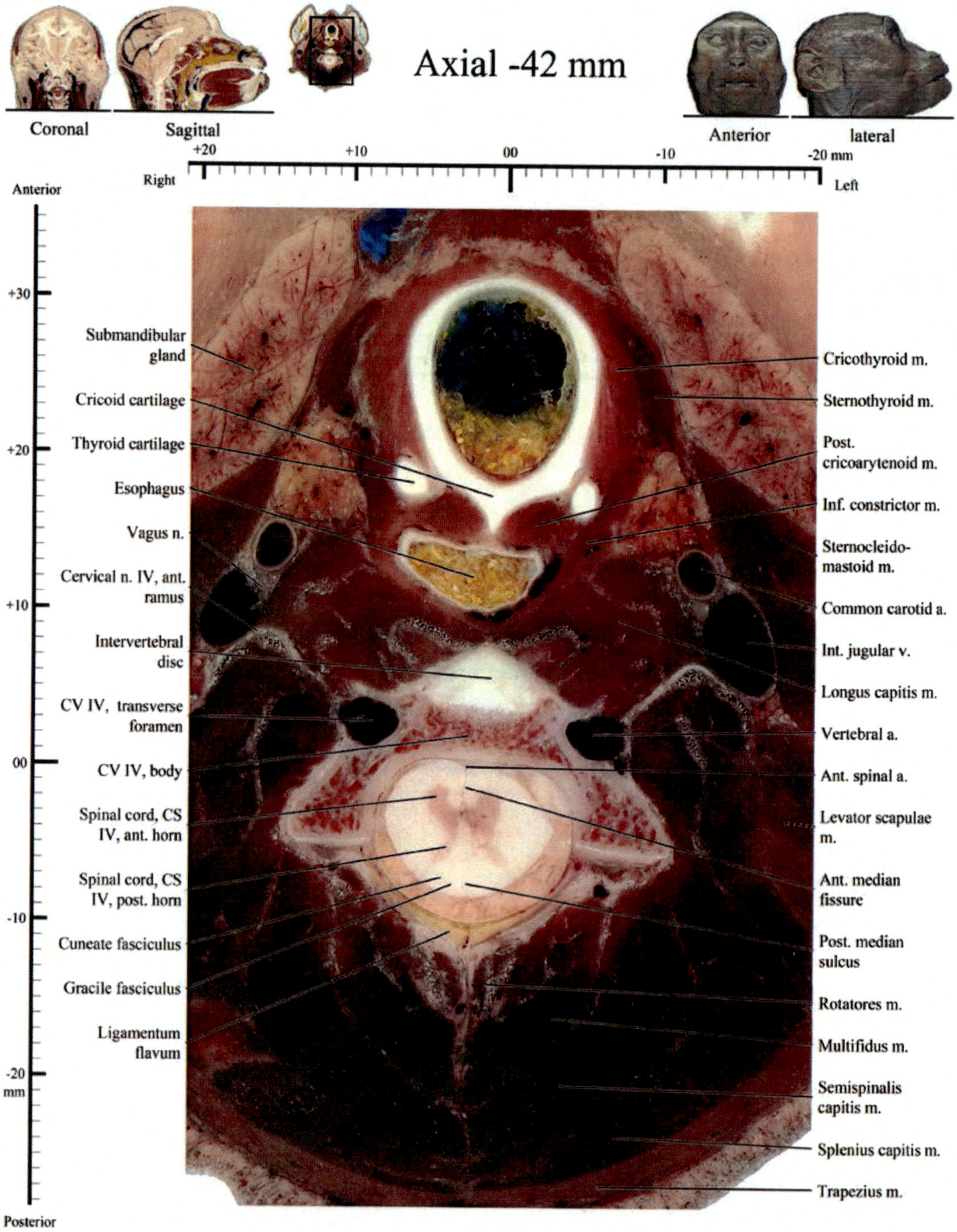
Coronal
Sagittal
Axial -42 mm
Anterior
lateral
+20
+10
00
-10
-20 mm
Right
Left
Anterior
+30
+20
+10
00
-10
-20 mm
Posterior
Submandibular gland
Cricoid cartilage
Thyroid cartilage
Esophagus
Vagus n.
Cervical n. IV, ant. ramus
Intervertebral disc
CV IV, transverse foramen
CV IV, body
Spinal cord, CS IV, ant. horn
Spinal cord, CS IV, post. horn
Cuneate fasciculus
Gracile fasciculus
Ligamentum flavum
Cricothyroid m.
Sternothyroid m.
Post. cricoarytenoid m.
Inf. constrictor m.
Sternocleido-mastoid m.
Common carotid a.
Int. jugular v.
Longus capitis m.
Vertebral a.
Ant. spinal a.
Levator scapulae m.
Ant. median fissure
Post. median sulcus
Rotatores m.
Multifidus m.
Semispinalis capitis m.
Splenius capitis m.
Trapezius m.

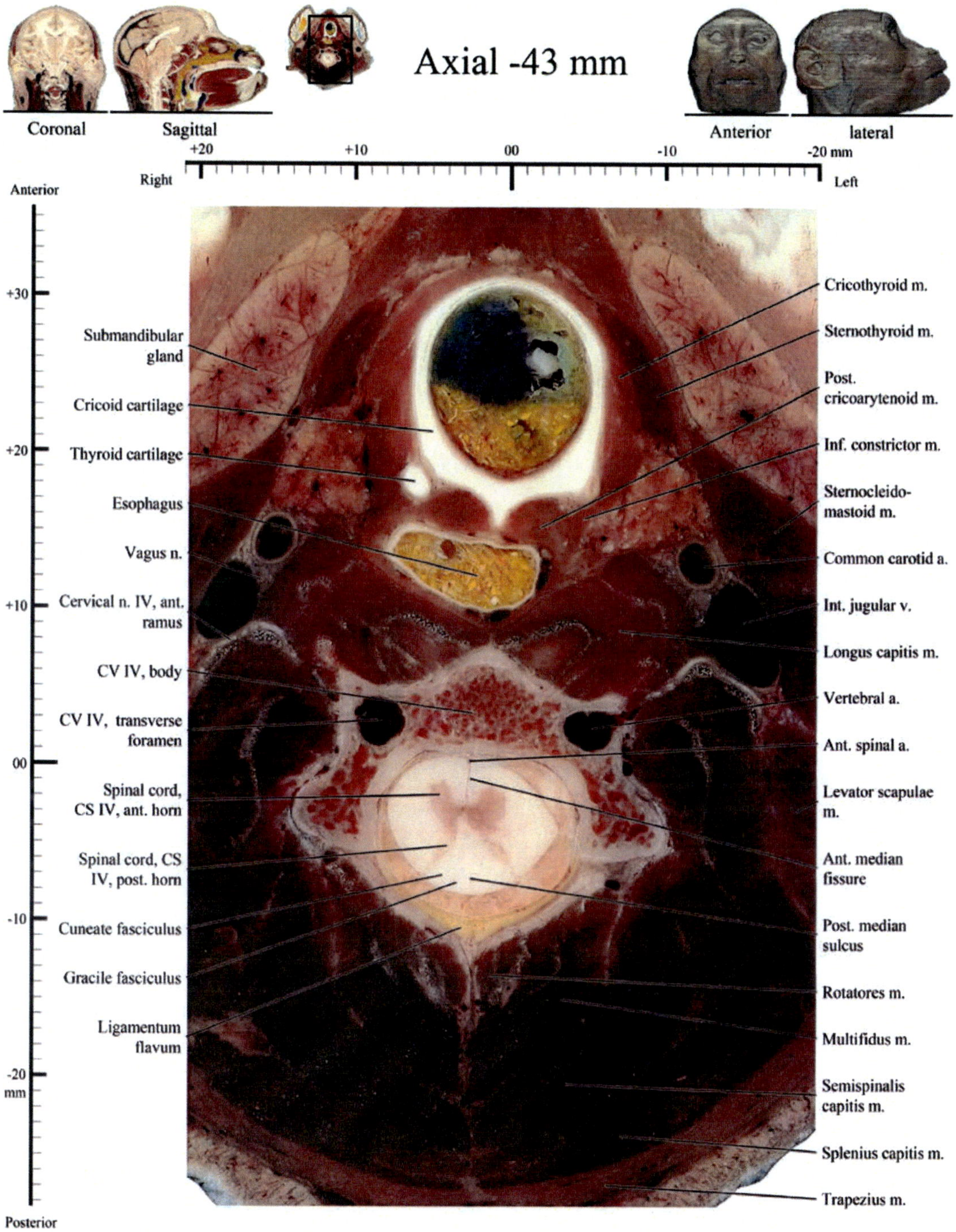
Coronal
Sagittal
Axial -43 mm
Anterior
lateral
+20
+10
00
-10
-20 mm
Right
Left
Anterior
Posterior
+30
+20
+10
00
-10
-20 mm
Submandibular gland
Cricoid cartilage
Thyroid cartilage
Esophagus
Vagus n.
Cervical n. IV, ant. ramus
CV IV, body
CV IV, transverse foramen
Spinal cord, CS IV, ant. horn
Spinal cord, CS IV, post. horn
Cuneate fasciculus
Gracile fasciculus
Ligamentum flavum
Cricothyroid m.
Sternothyroid m.
Post. cricoarytenoid m.
Inf. constrictor m.
Sternocleido-mastoid m.
Common carotid a.
Int. jugular v.
Longus capitis m.
Vertebral a.
Ant. spinal a.
Levator scapulae m.
Ant. median fissure
Post. median sulcus
Rotatores m.
Multifidus m.
Semispinalis capitis m.
Splenius capitis m.
Trapezius m.

6 Coronal (Expanded)

All images, 0.024 mm × 0.024 mm sized pixel

+36 mm ~ +31 mm, 5 mm intervals
+31 mm ~ +27 mm, 4 mm intervals
+27 mm ~ 0 mm, 1 mm intervals
0 mm ~ −2 mm, 0.5 mm intervals
−2 mm ~ −13 mm, 1 mm intervals
−13 mm ~ 38 mm, 25 mm intervals

J. S. Park, *Cross-Sectional Atlas of Rhesus Monkey Head*, https://doi.org/10.1007/978-981-19-3254-0_6

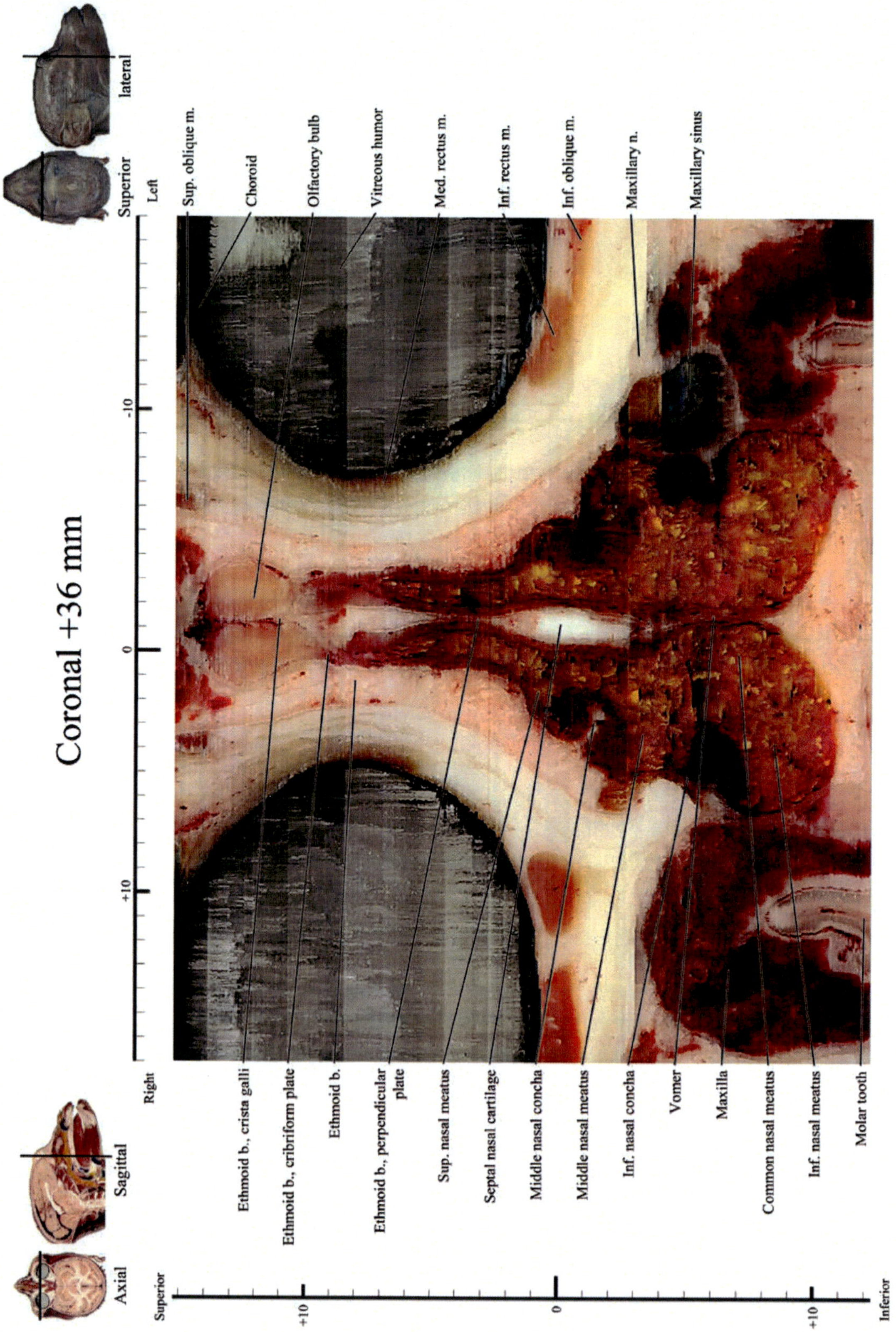
Coronal +36 mm
Superior
lateral
Axial
Sagittal
Left
Right
Sup. oblique m.
Choroid
Olfactory bulb
Vitreous humor
Med. rectus m.
Inf. rectus m.
Inf. oblique m.
Maxillary n.
Maxillary sinus
Ethmoid b., crista galli
Ethmoid b., cribriform plate
Ethmoid b.
Ethmoid b., perpendicular plate
Sup. nasal meatus
Septal nasal cartilage
Middle nasal concha
Middle nasal meatus
Inf. nasal concha
Vomer
Maxilla
Common nasal meatus
Inf. nasal meatus
Molar tooth
-10
0
+10
Superior
+10
0
+10
Inferior

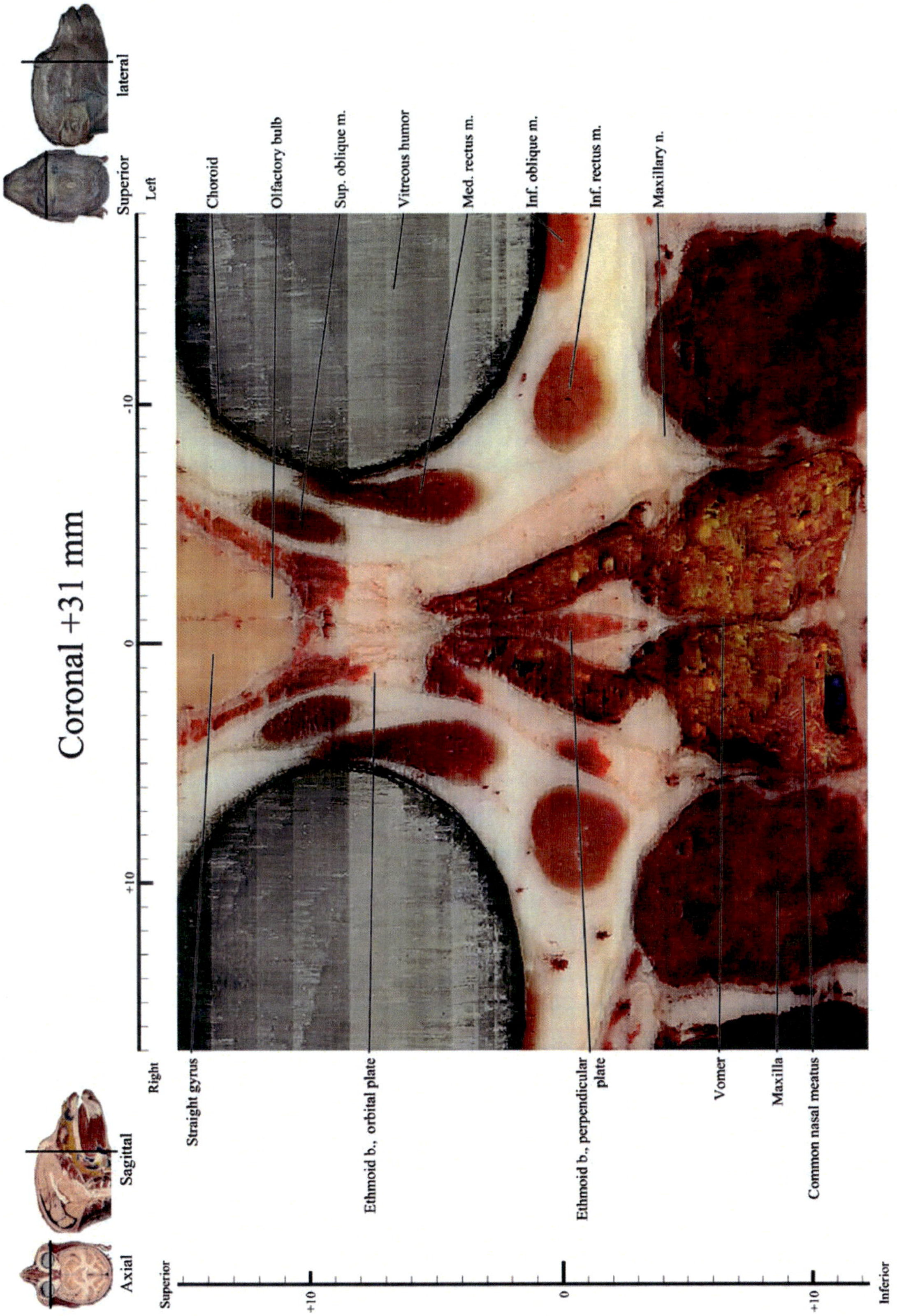
Coronal +31 mm
Superior
lateral
Left
Right
Sagittal
Axial
Choroid
Olfactory bulb
Sup. oblique m.
Vitreous humor
Med. rectus m.
Inf. oblique m.
Inf. rectus m.
Maxillary n.
Straight gyrus
Ethmoid b., orbital plate
Ethmoid b., perpendicular plate
Vomer
Maxilla
Common nasal meatus
-10
0
+10
Superior
+10
0
+10
Inferior

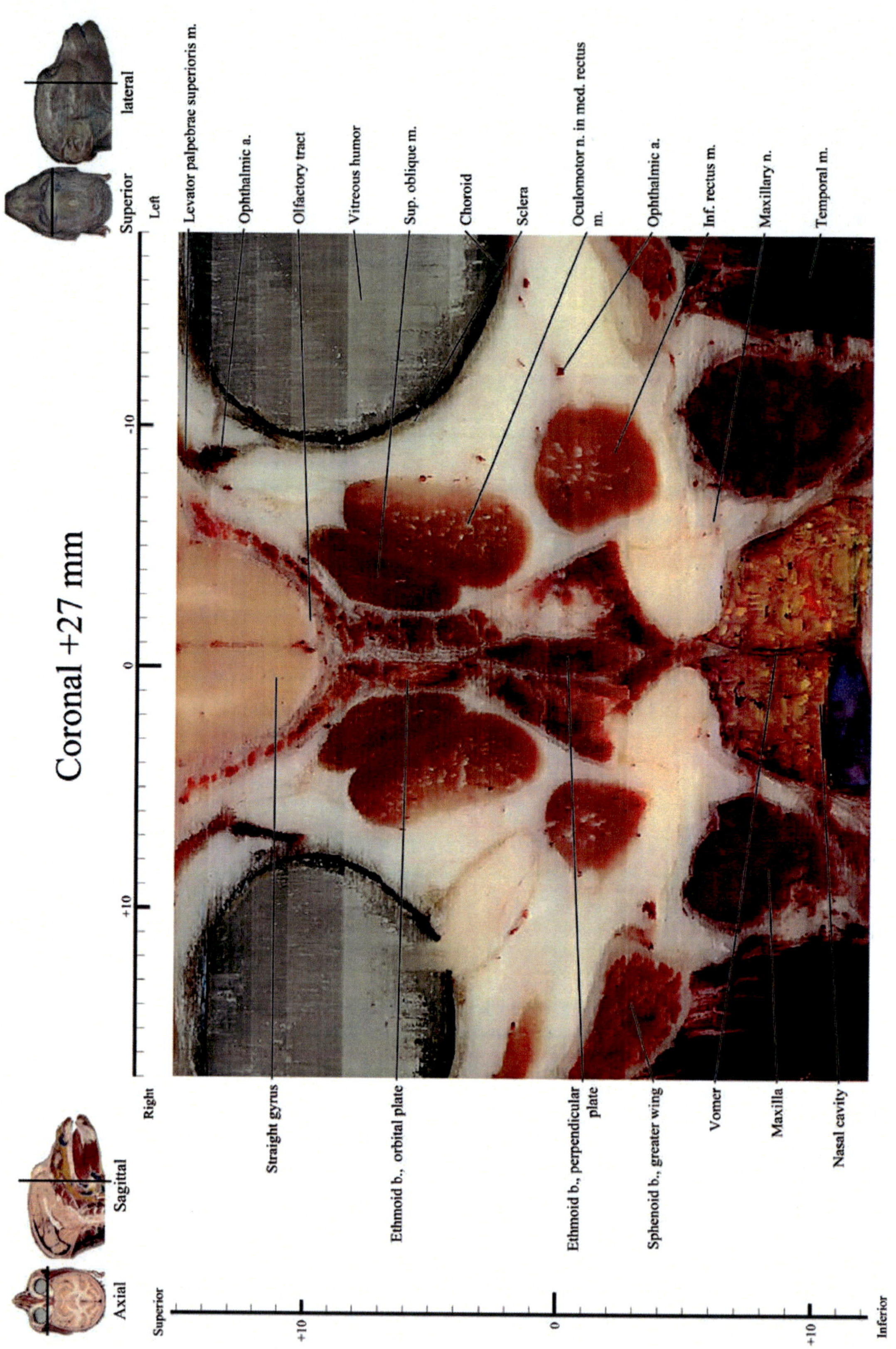
Coronal +27 mm
Superior
lateral
Sagittal
Axial
Left
Right
Superior
Inferior
-10
0
+10
+10
0
+10
Levator palpebrae superioris m.
Ophthalmic a.
Olfactory tract
Vitreous humor
Sup. oblique m.
Choroid
Sclera
Oculomotor n. in med. rectus m.
Ophthalmic a.
Inf. rectus m.
Maxillary n.
Temporal m.
Straight gyrus
Ethmoid b., orbital plate
Ethmoid b., perpendicular plate
Sphenoid b., greater wing
Vomer
Maxilla
Nasal cavity

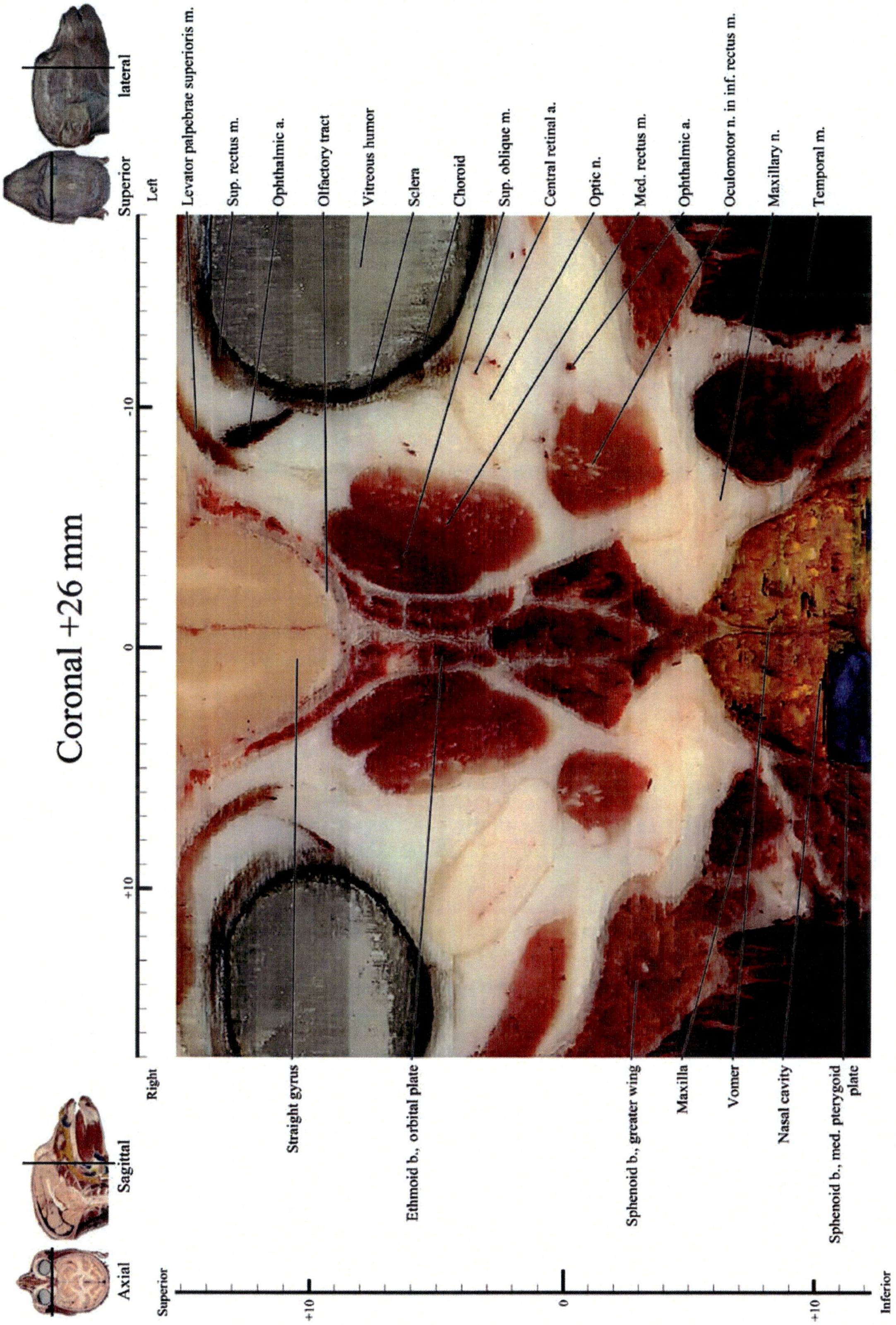
Coronal +26 mm
Superior
lateral
Axial
Sagital
Left
Right
Superior
Inferior
-10
0
+10
Levator palpebrae superioris m.
Sup. rectus m.
Ophthalmic a.
Olfactory tract
Vitreous humor
Sclera
Choroid
Sup. oblique m.
Central retinal a.
Optic n.
Med. rectus m.
Ophthalmic a.
Oculomotor n. in inf. rectus m.
Maxillary n.
Temporal m.
Straight gyrus
Ethmoid b., orbital plate
Sphenoid b., greater wing
Maxilla
Vomer
Nasal cavity
Sphenoid b., med. pterygoid plate

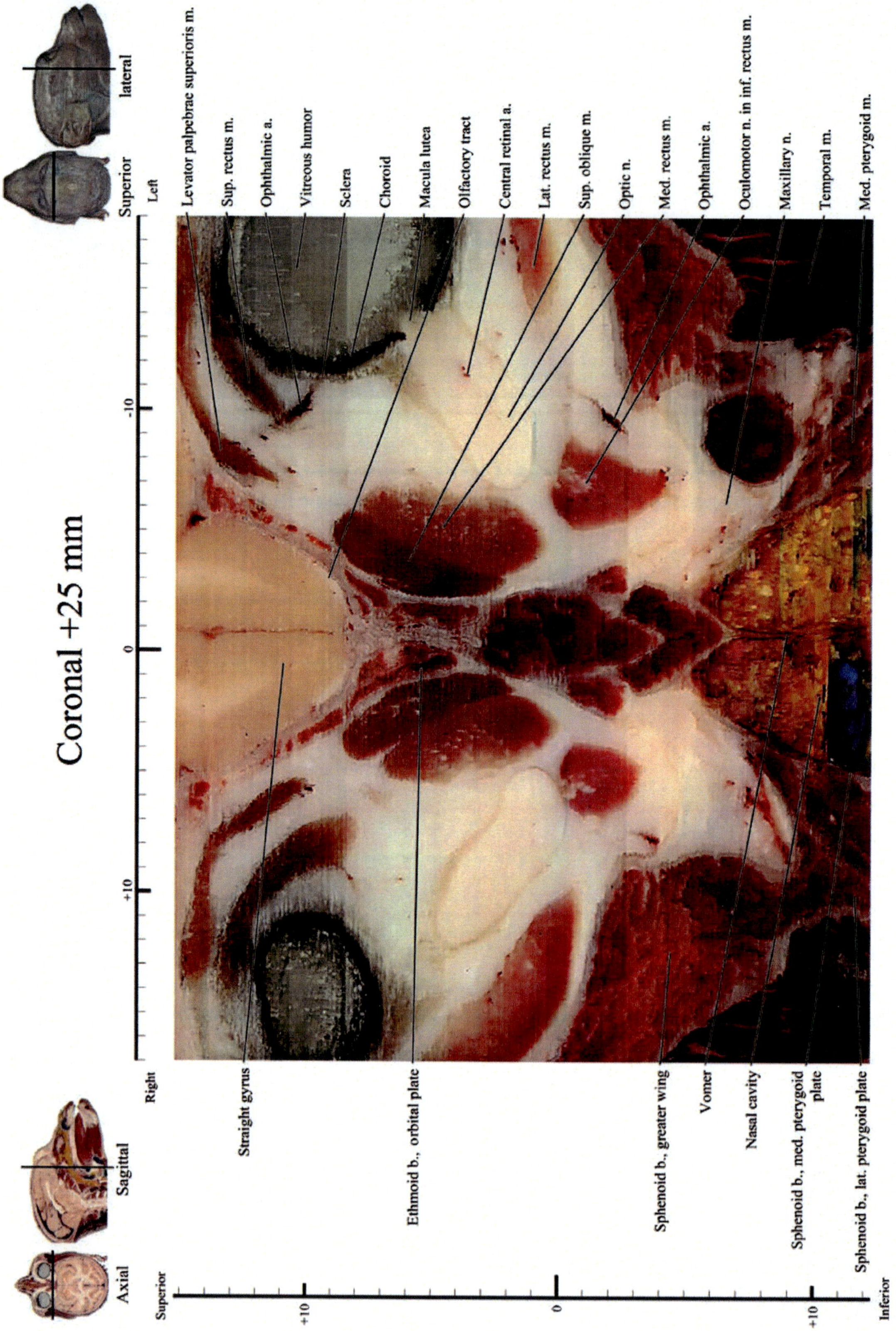
Coronal +25 mm
Superior
lateral
Axial
Sagittal
Left
Right
Superior
Inferior
-10
0
+10
+10
0
+10
Levator palpebrae superioris m.
Sup. rectus m.
Ophthalmic a.
Vitreous humor
Sclera
Choroid
Macula lutea
Olfactory tract
Central retinal a.
Lat. rectus m.
Sup. oblique m.
Optic n.
Med. rectus m.
Ophthalmic a.
Oculomotor n. in inf. rectus m.
Maxillary n.
Temporal m.
Med. pterygoid m.
Straight gyrus
Ethmoid b., orbital plate
Sphenoid b., greater wing
Vomer
Nasal cavity
Sphenoid b., med. pterygoid plate
Sphenoid b., lat. pterygoid plate

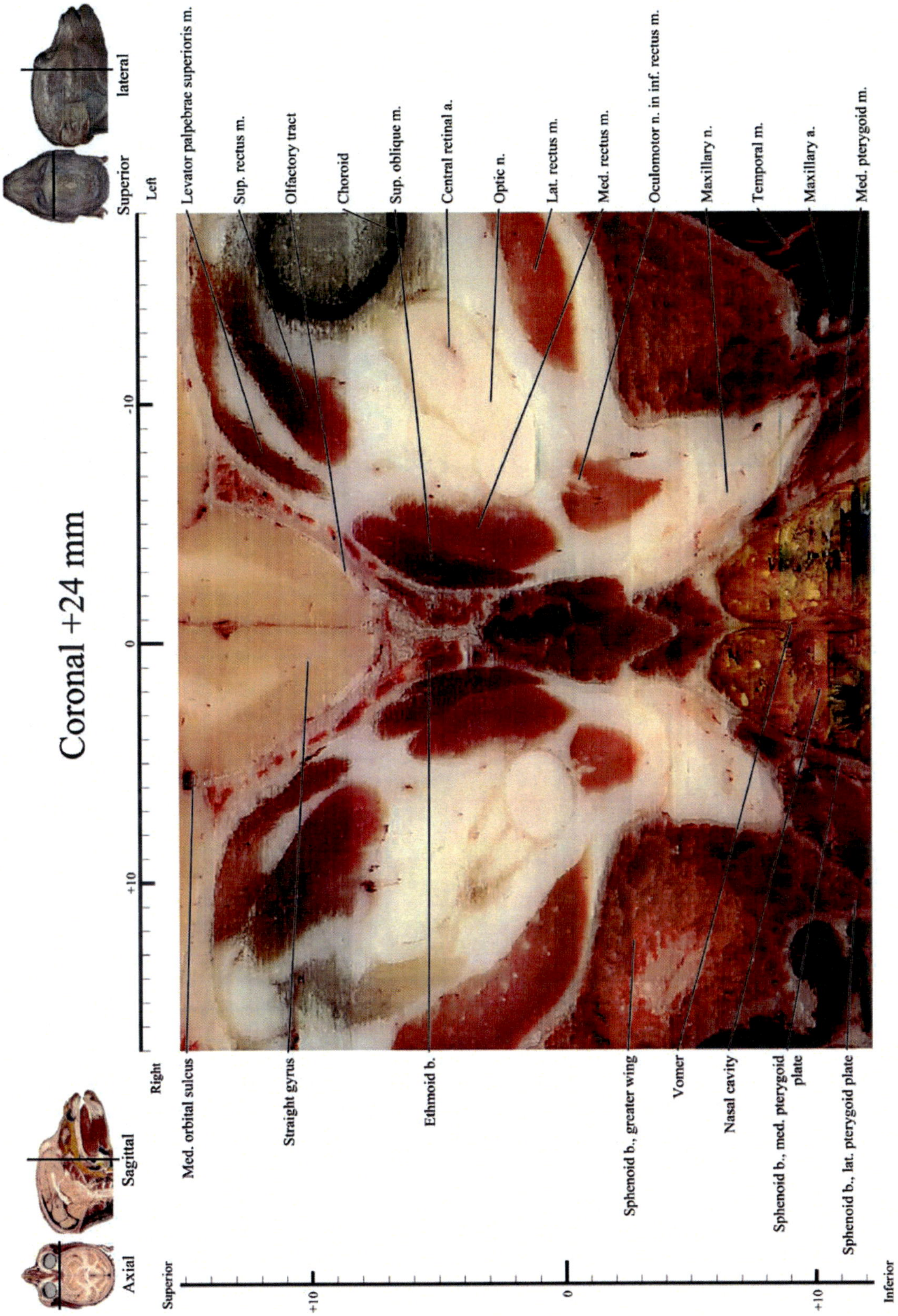
Coronal +24 mm
Superior
lateral
Left
Right
Axial
Sagittal
Superior
Inferior
-10
0
+10
Levator palpebrae superioris m.
Sup. rectus m.
Olfactory tract
Choroid
Sup. oblique m.
Central retinal a.
Optic n.
Lat. rectus m.
Med. rectus m.
Oculomotor n. in inf. rectus m.
Maxillary n.
Temporal m.
Maxillary a.
Med. pterygoid m.
Med. orbital sulcus
Straight gyrus
Ethmoid b.
Sphenoid b., greater wing
Vomer
Nasal cavity
Sphenoid b., med. pterygoid plate
Sphenoid b., lat. pterygoid plate

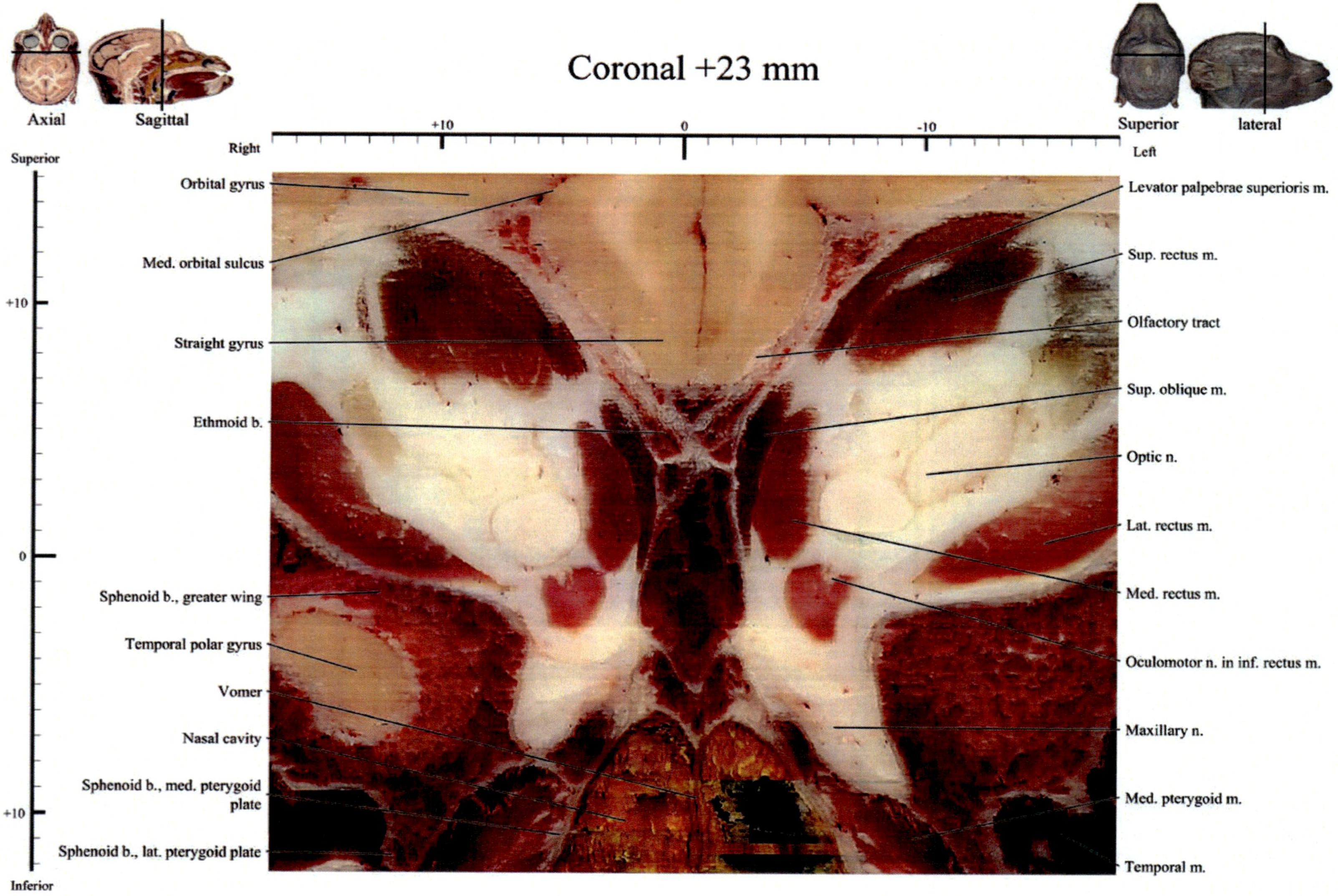
Coronal +23 mm
Axial
Sagittal
Superior
lateral
Right
Left
Superior
Inferior
+10
0
-10
Orbital gyrus
Med. orbital sulcus
Straight gyrus
Ethmoid b.
Sphenoid b., greater wing
Temporal polar gyrus
Vomer
Nasal cavity
Sphenoid b., med. pterygoid plate
Sphenoid b., lat. pterygoid plate
Levator palpebrae superioris m.
Sup. rectus m.
Olfactory tract
Sup. oblique m.
Optic n.
Lat. rectus m.
Med. rectus m.
Oculomotor n. in inf. rectus m.
Maxillary n.
Med. pterygoid m.
Temporal m.

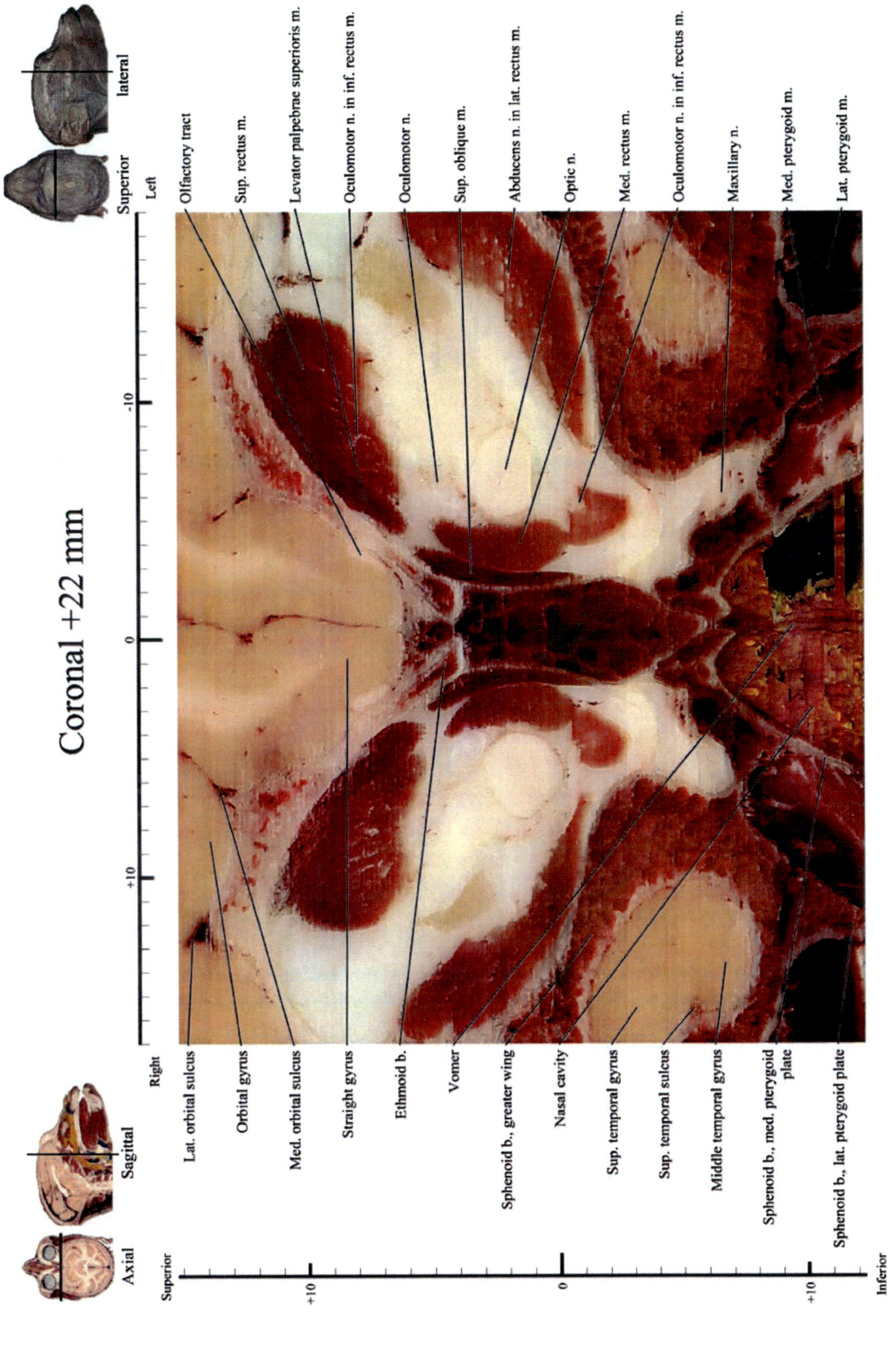
Coronal +22 mm
Superior
lateral
Sagittal
Axial
Left
Right
Olfactory tract
Sup. rectus m.
Levator palpebrae superioris m.
Oculomotor n. in inf. rectus m.
Oculomotor n.
Sup. oblique m.
Abducens n. in lat. rectus m.
Optic n.
Med. rectus m.
Oculomotor n. in inf. rectus m.
Maxillary n.
Med. pterygoid m.
Lat. pterygoid m.
Lat. orbital sulcus
Orbital gyrus
Med. orbital sulcus
Straight gyrus
Ethmoid b.
Vomer
Sphenoid b., greater wing
Nasal cavity
Sup. temporal gyrus
Sup. temporal sulcus
Middle temporal gyrus
Sphenoid b., med. pterygoid plate
Sphenoid b., lat. pterygoid plate
-10
0
+10
Superior
+10
0
+10
Inferior

Coronal +21 mm

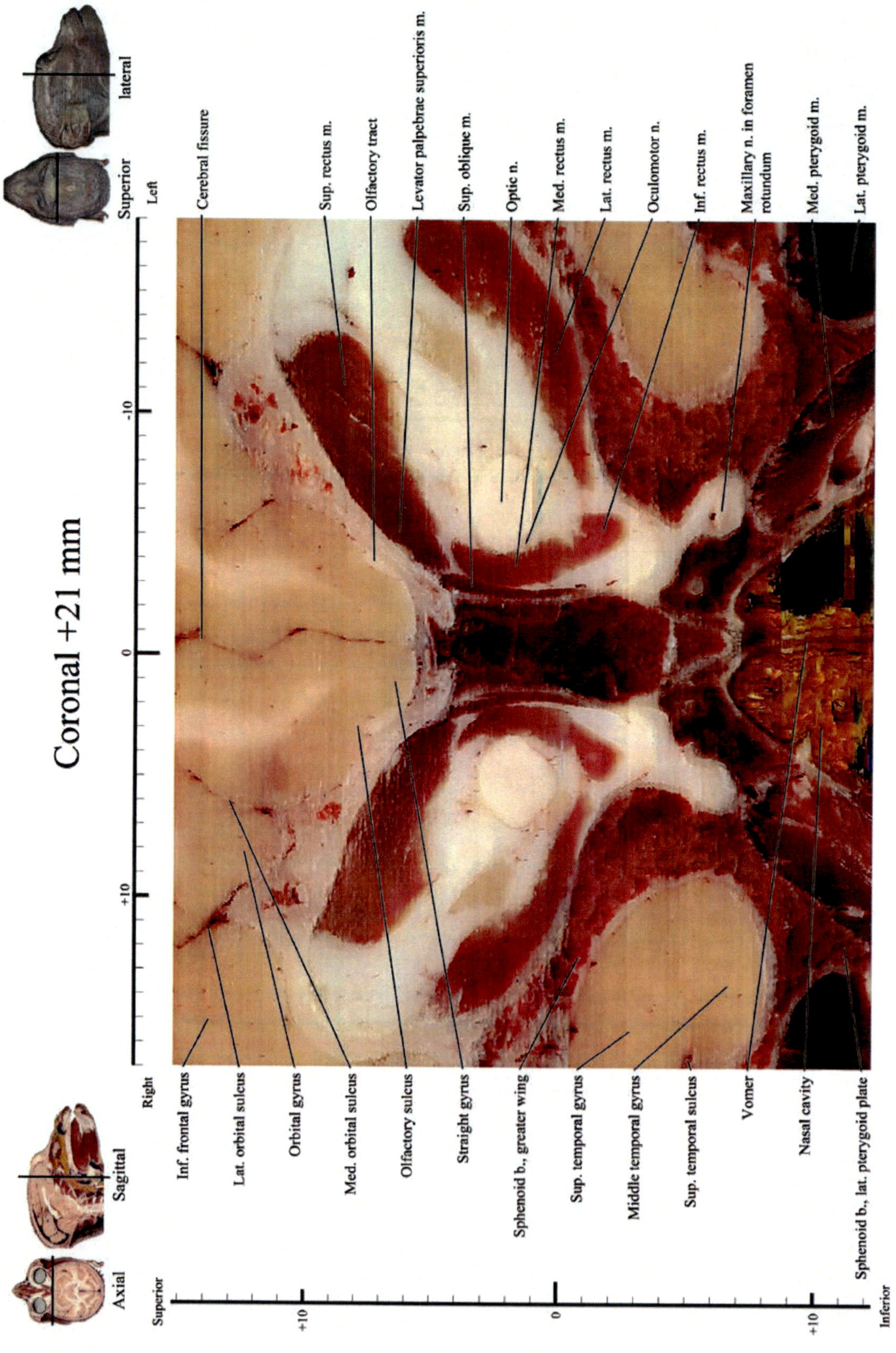

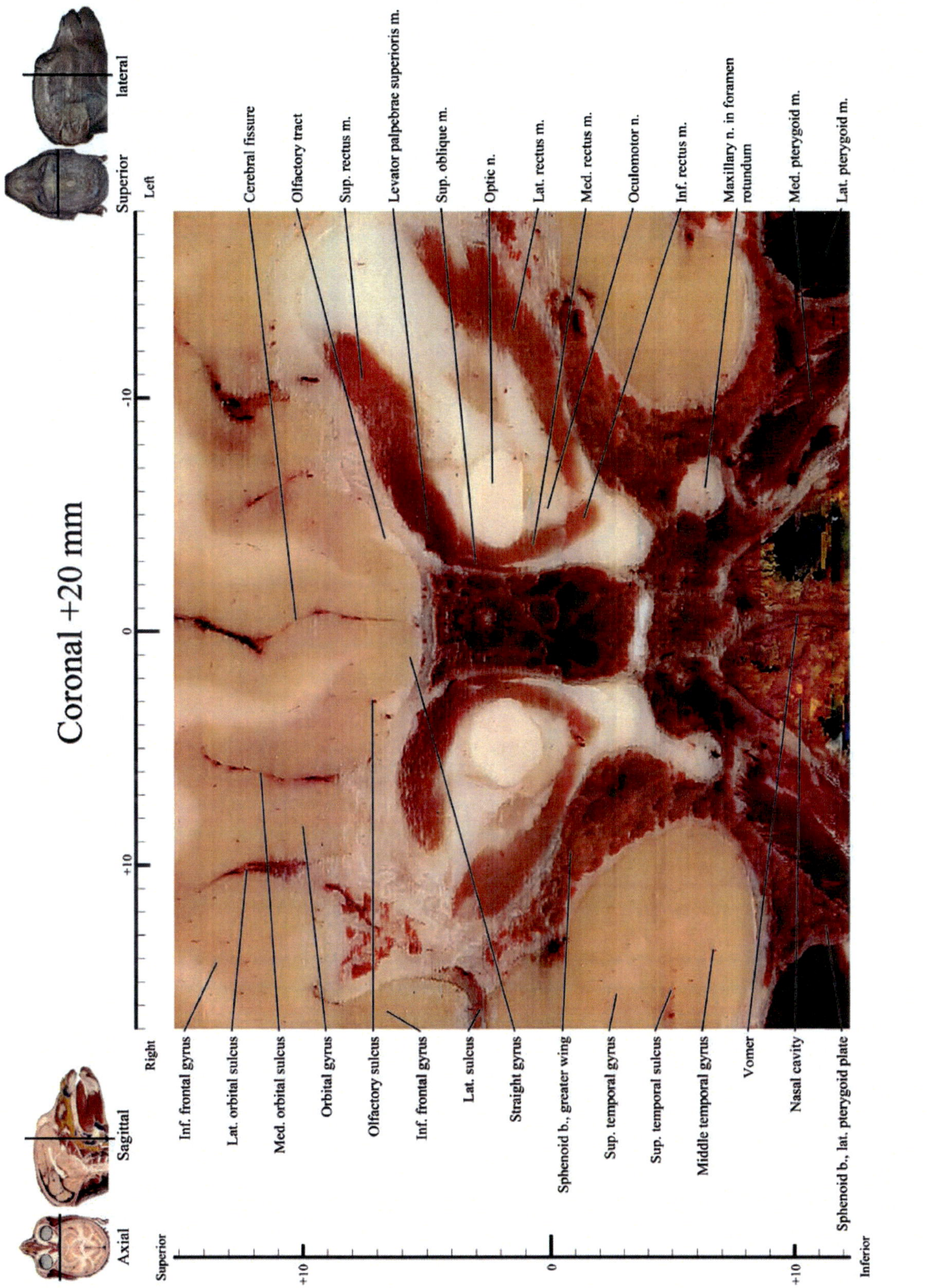
Coronal +20 mm
Superior
Left
lateral
Axial
Sagittal
Right
Superior
Inferior
-10
0
+10
+10
0
+10
Cerebral fissure
Olfactory tract
Sup. rectus m.
Levator palpebrae superioris m.
Sup. oblique m.
Optic n.
Lat. rectus m.
Med. rectus m.
Oculomotor n.
Inf. rectus m.
Maxillary n. in foramen rotundum
Med. pterygoid m.
Lat. pterygoid m.
Inf. frontal gyrus
Lat. orbital sulcus
Med. orbital sulcus
Orbital gyrus
Olfactory sulcus
Inf. frontal gyrus
Lat. sulcus
Straight gyrus
Sphenoid b., greater wing
Sup. temporal gyrus
Sup. temporal sulcus
Middle temporal gyrus
Vomer
Nasal cavity
Sphenoid b., lat. pterygoid plate

Coronal +19 mm

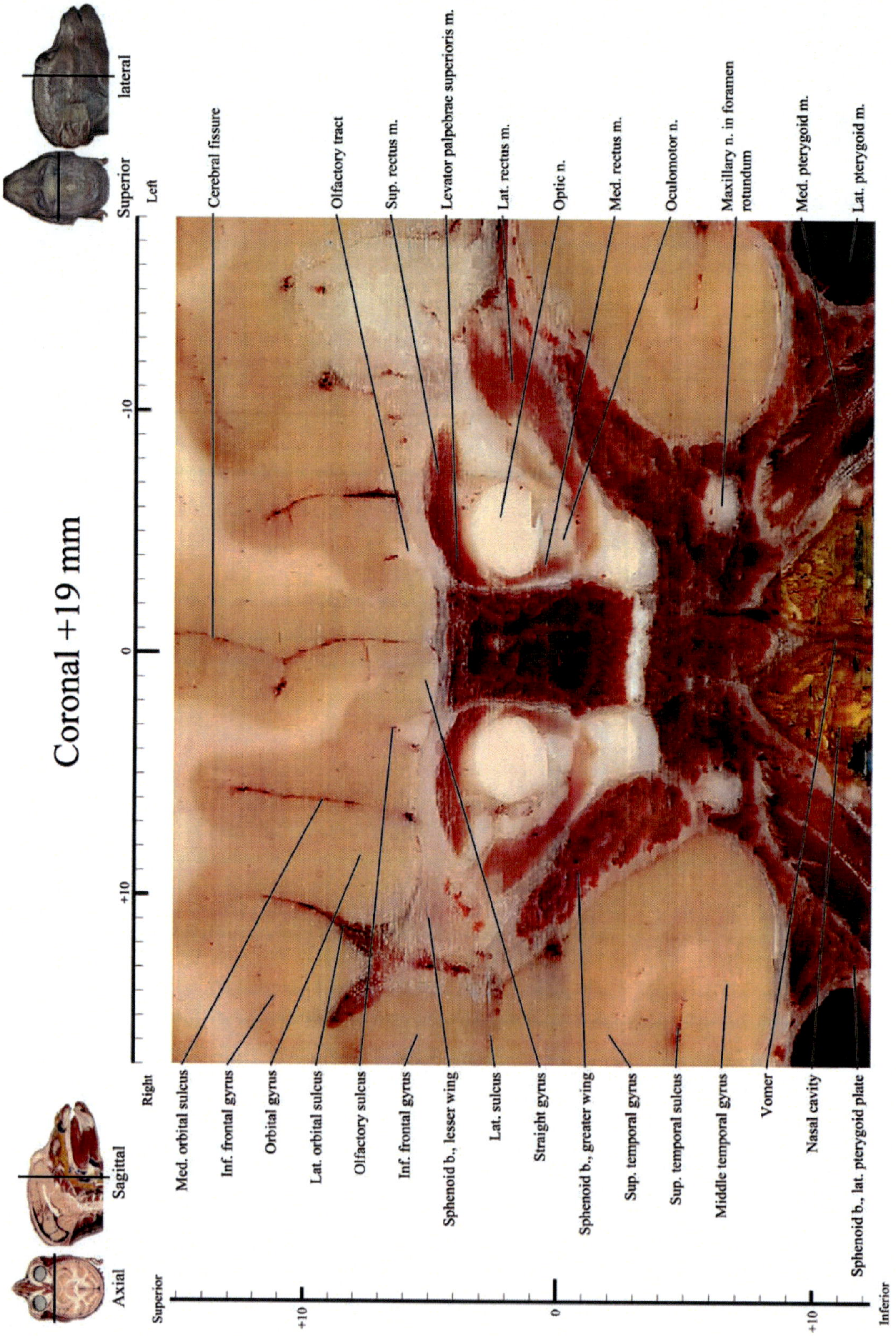

Coronal +18 mm

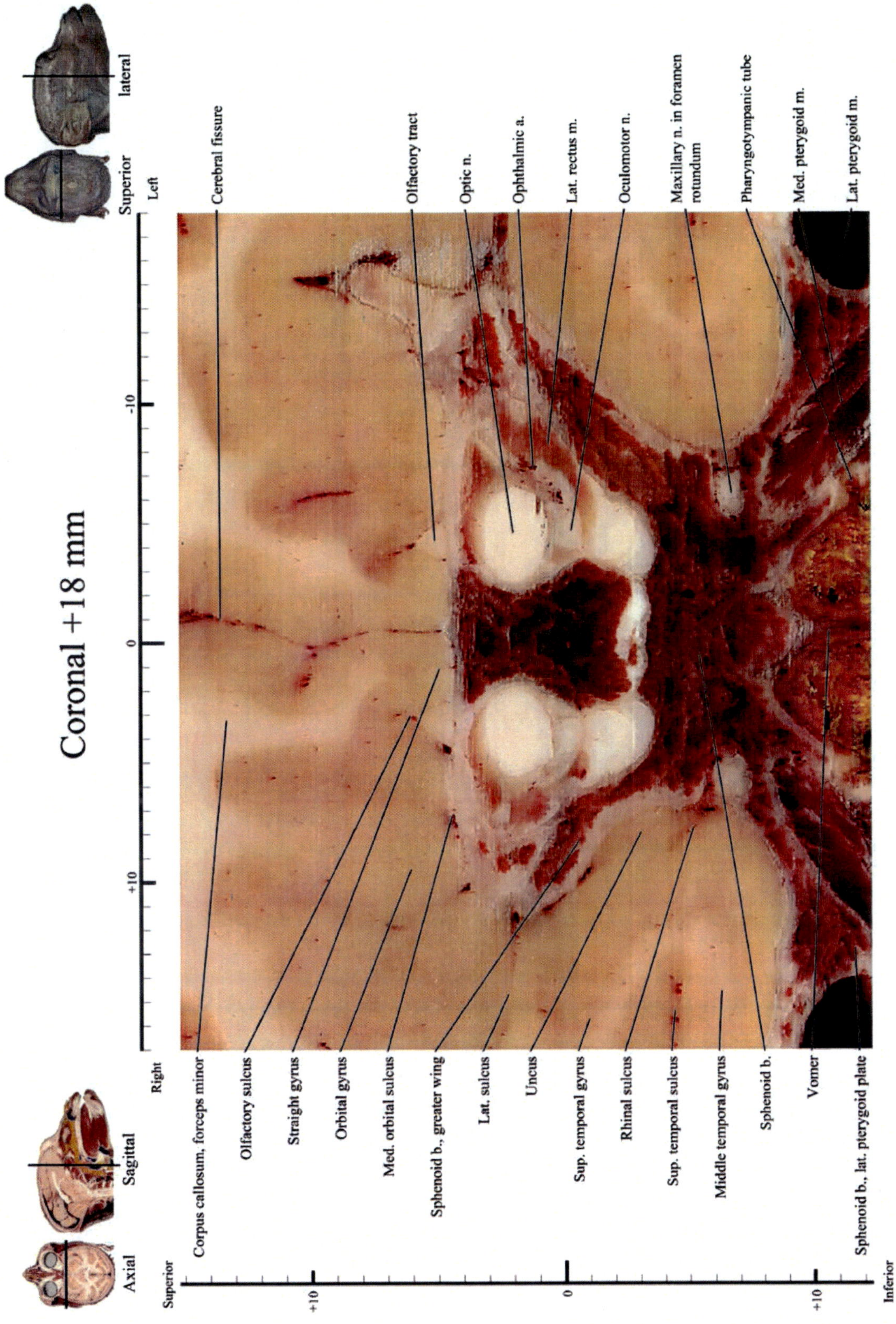

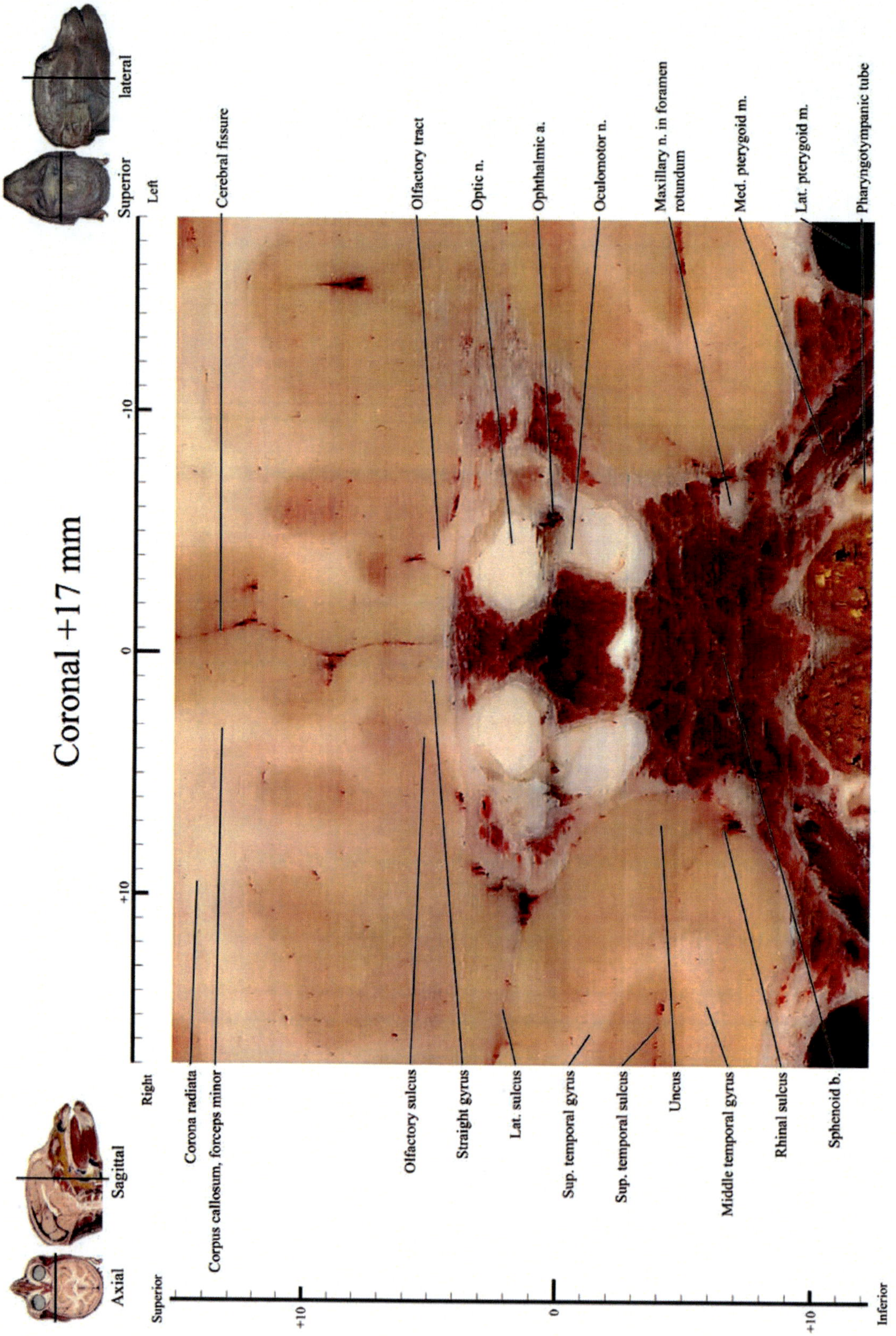
Coronal +17 mm
Superior
lateral
Left
Right
Axial
Sagittal
Cerebral fissure
Olfactory tract
Optic n.
Ophthalmic a.
Oculomotor n.
Maxillary n. in foramen rotundum
Med. pterygoid m.
Lat. pterygoid m.
Pharyngotympanic tube
Corona radiata
Corpus callosum, forceps minor
Olfactory sulcus
Straight gyrus
Lat. sulcus
Sup. temporal gyrus
Sup. temporal sulcus
Uncus
Middle temporal gyrus
Rhinal sulcus
Sphenoid b.
-10
0
+10
Superior
+10
0
+10
Inferior

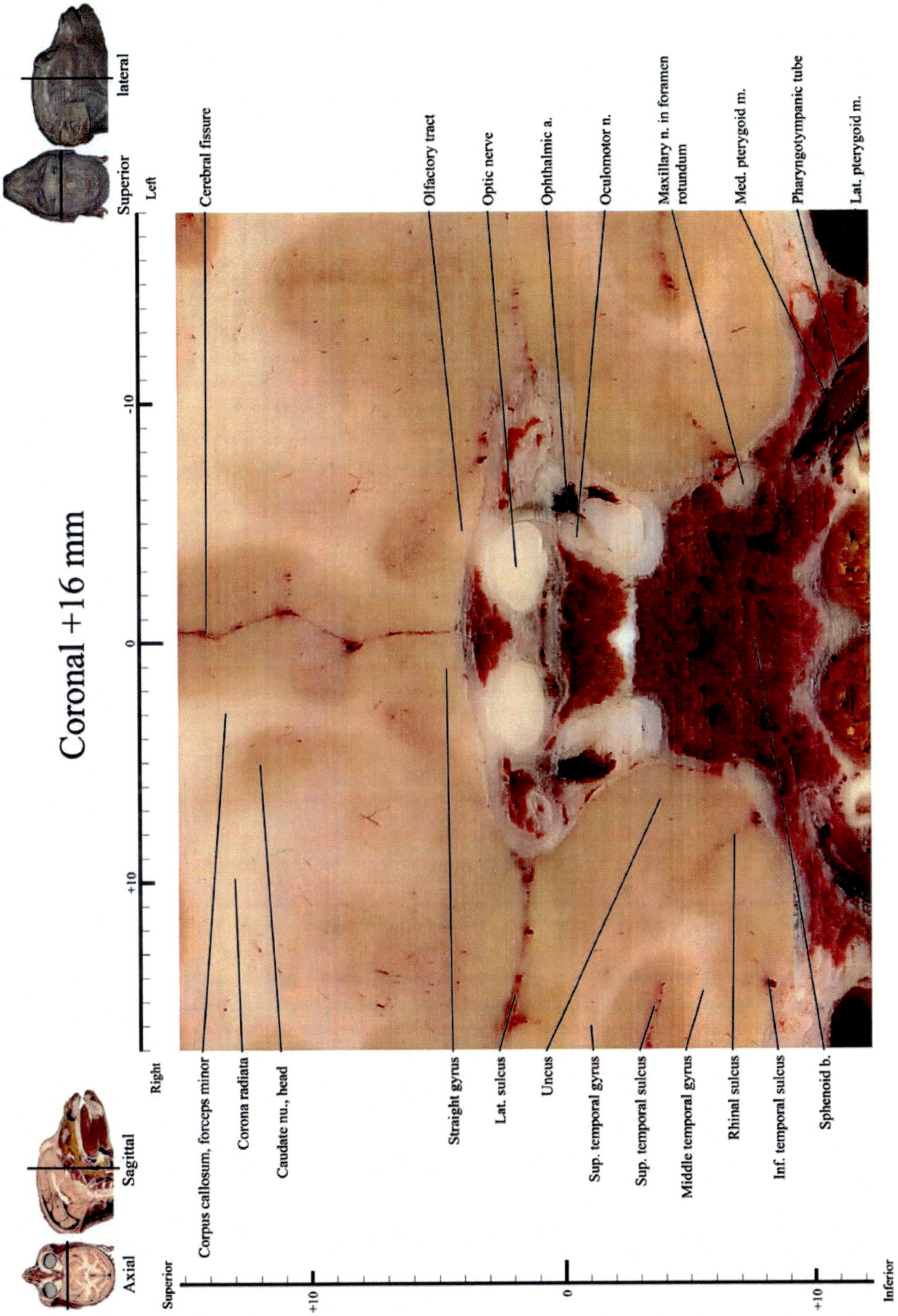
Coronal +16 mm
Superior
lateral
Left
Cerebral fissure
Olfactory tract
Optic nerve
Ophthalmic a.
Oculomotor n.
Maxillary n. in foramen rotundum
Med. pterygoid m.
Pharyngotympanic tube
Lat. pterygoid m.
-10
0
+10
Right
Corpus callosum, forceps minor
Corona radiata
Caudate nu., head
Straight gyrus
Lat. sulcus
Uncus
Sup. temporal gyrus
Sup. temporal sulcus
Middle temporal gyrus
Rhinal sulcus
Inf. temporal sulcus
Sphenoid b.
Sagittal
Axial
Superior
+10
0
+10
Inferior

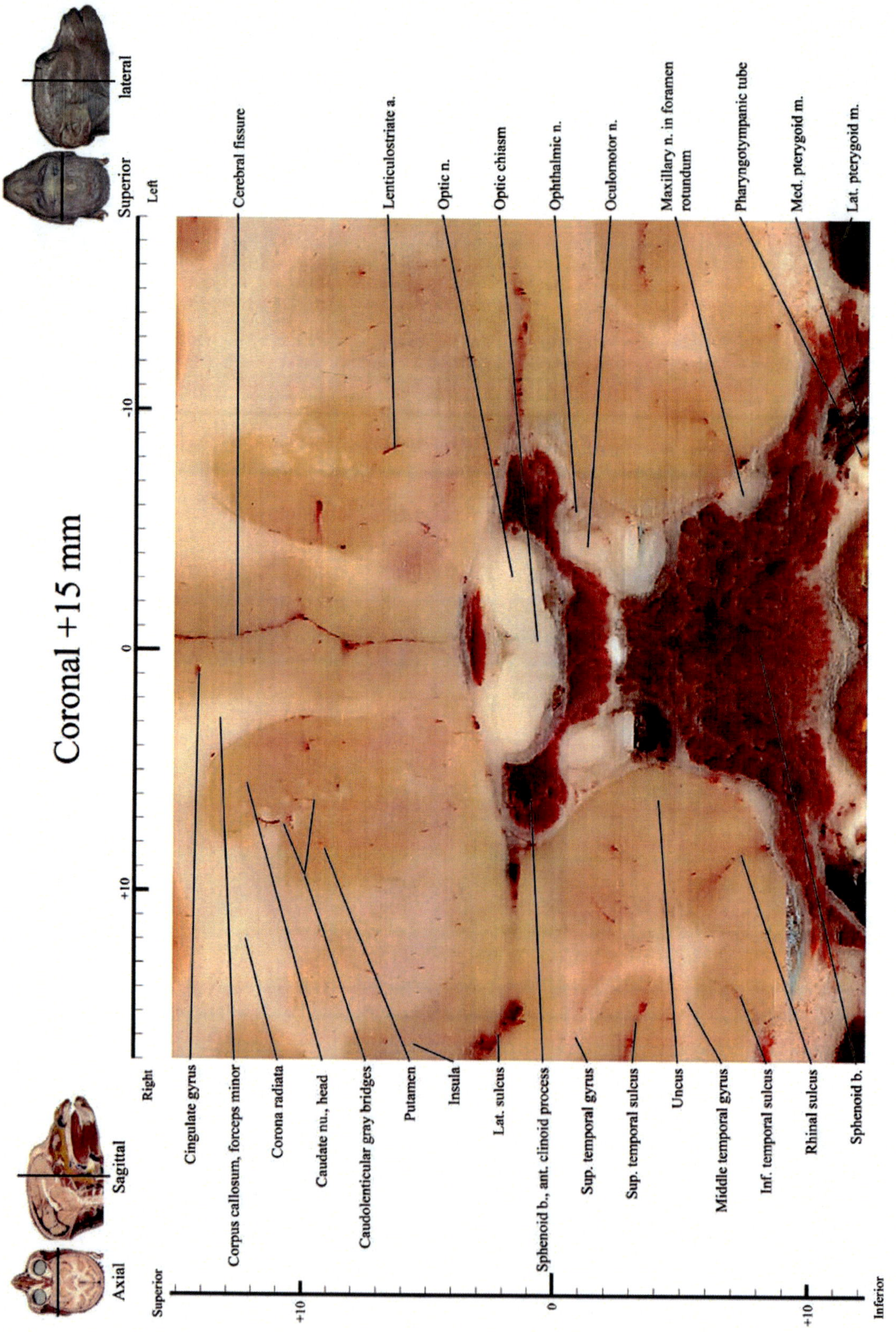
Coronal +15 mm
Superior
lateral
Left
Right
-10
0
+10
Cerebral fissure
Lenticulostriate a.
Optic n.
Optic chiasm
Ophthalmic n.
Oculomotor n.
Maxillary n. in foramen rotundum
Pharyngotympanic tube
Med. pterygoid m.
Lat. pterygoid m.
Cingulate gyrus
Corpus callosum, forceps minor
Corona radiata
Caudate nu., head
Caudolenticular gray bridges
Putamen
Insula
Lat. sulcus
Sphenoid b., ant. clinoid process
Sup. temporal gyrus
Sup. temporal sulcus
Uncus
Middle temporal gyrus
Inf. temporal sulcus
Rhinal sulcus
Sphenoid b.
Axial
Sagittal
Superior
+10
0
+10
Inferior

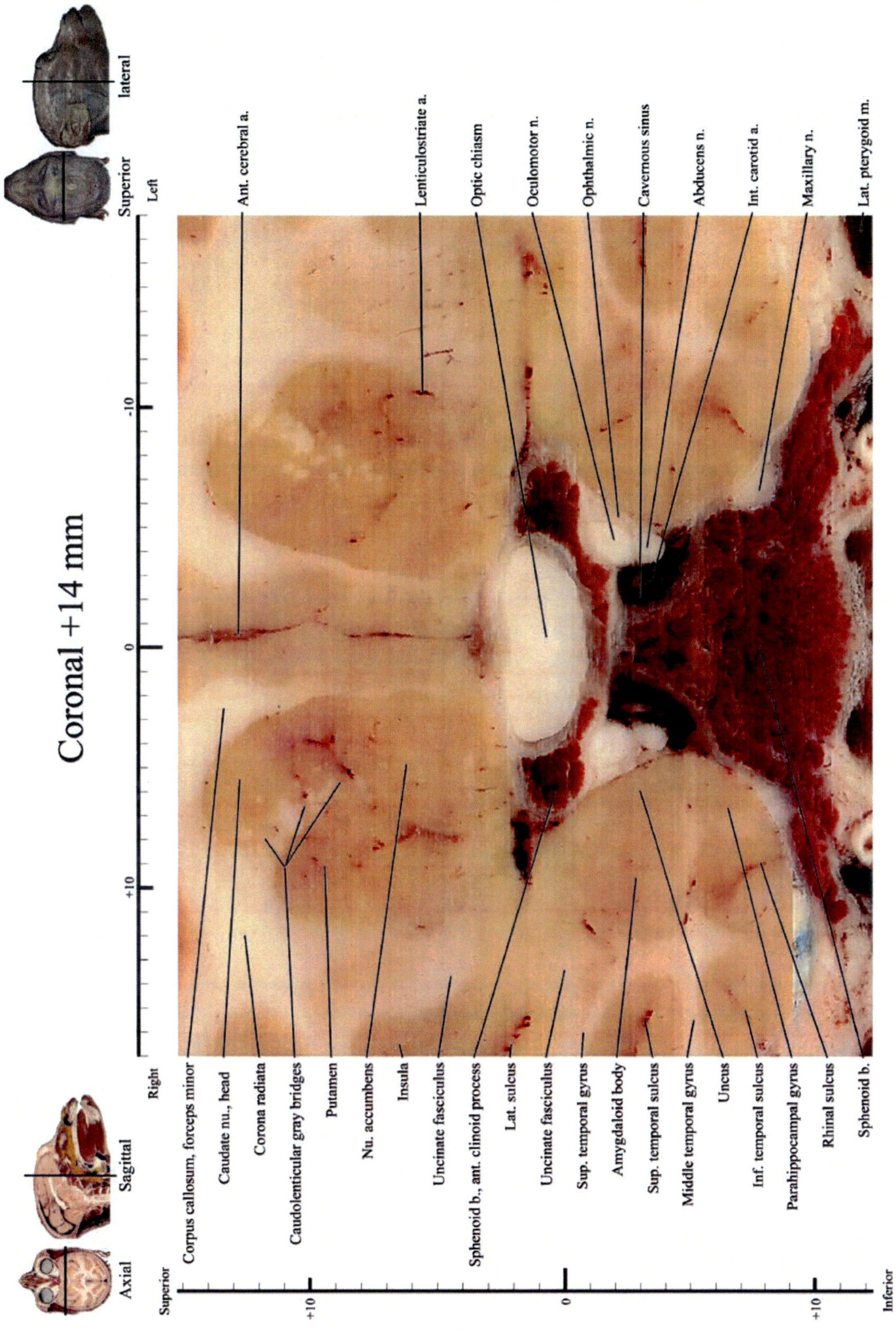
Coronal +14 mm
Superior
Left
lateral
Axial
Sagittal
Right
Superior
Inferior
-10
0
+10
Ant. cerebral a.
Lenticulostriate a.
Optic chiasm
Oculomotor n.
Ophthalmic n.
Cavernous sinus
Abducens n.
Int. carotid a.
Maxillary n.
Lat. pterygoid m.
Corpus callosum, forceps minor
Caudate nu., head
Corona radiata
Caudolenticular gray bridges
Putamen
Nu. accumbens
Insula
Uncinate fasciculus
Sphenoid b., ant. clinoid process
Lat. sulcus
Uncinate fasciculus
Sup. temporal gyrus
Amygdaloid body
Sup. temporal sulcus
Middle temporal gyrus
Uncus
Inf. temporal sulcus
Parahippocampal gyrus
Rhinal sulcus
Sphenoid b.

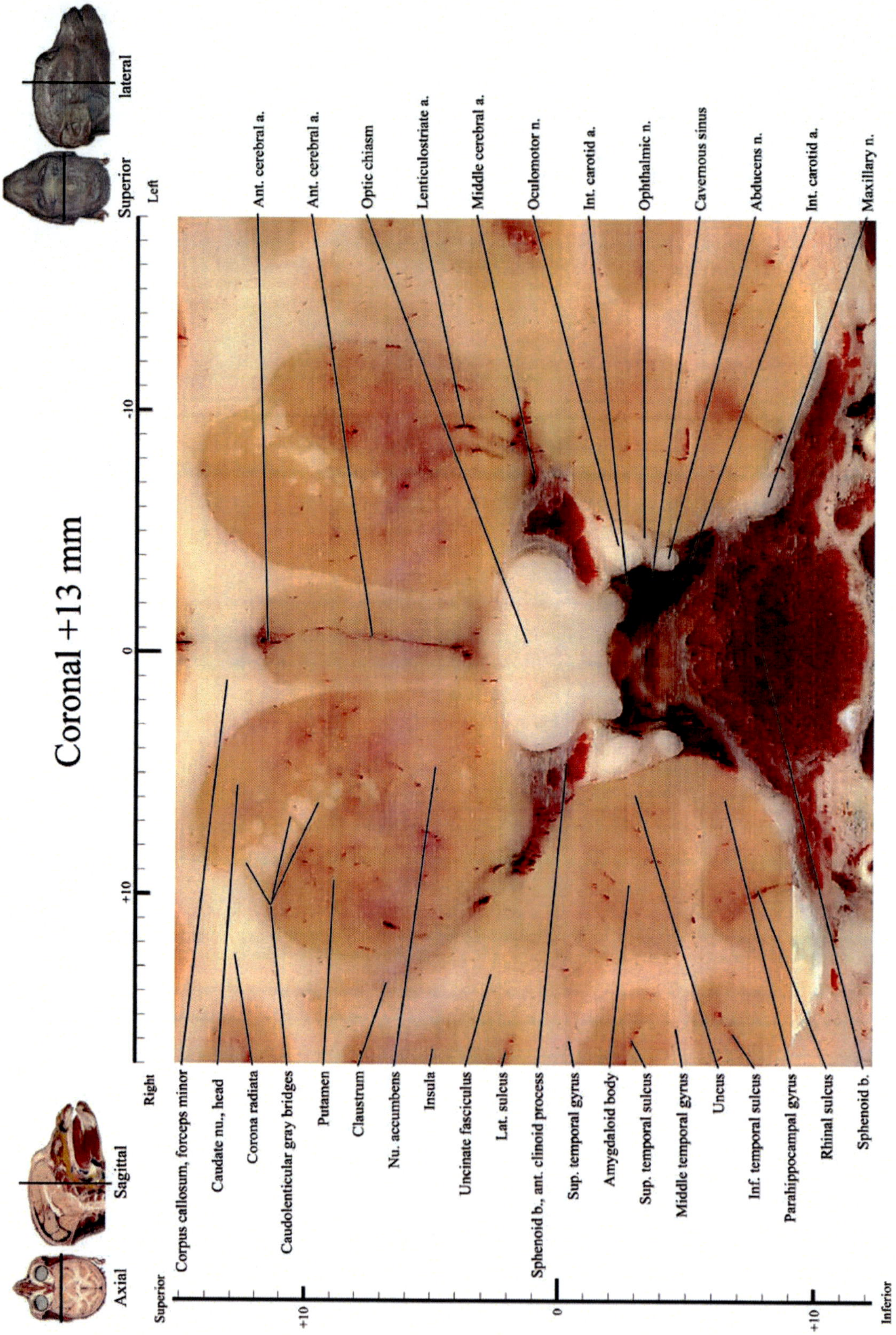
Coronal +13 mm
Axial
Sagittal
Superior
lateral
Superior
Inferior
+10
0
+10
Right
Left
-10
0
+10
Ant. cerebral a.
Ant. cerebral a.
Optic chiasm
Lenticulostriate a.
Middle cerebral a.
Oculomotor n.
Int. carotid a.
Ophthalmic n.
Cavernous sinus
Abducens n.
Int. carotid a.
Maxillary n.
Corpus callosum, forceps minor
Caudate nu., head
Corona radiata
Caudolenticular gray bridges
Putamen
Claustrum
Nu. accumbens
Insula
Uncinate fasciculus
Lat. sulcus
Sphenoid b., ant. clinoid process
Sup. temporal gyrus
Amygdaloid body
Sup. temporal sulcus
Middle temporal gyrus
Uncus
Inf. temporal sulcus
Parahippocampal gyrus
Rhinal sulcus
Sphenoid b.

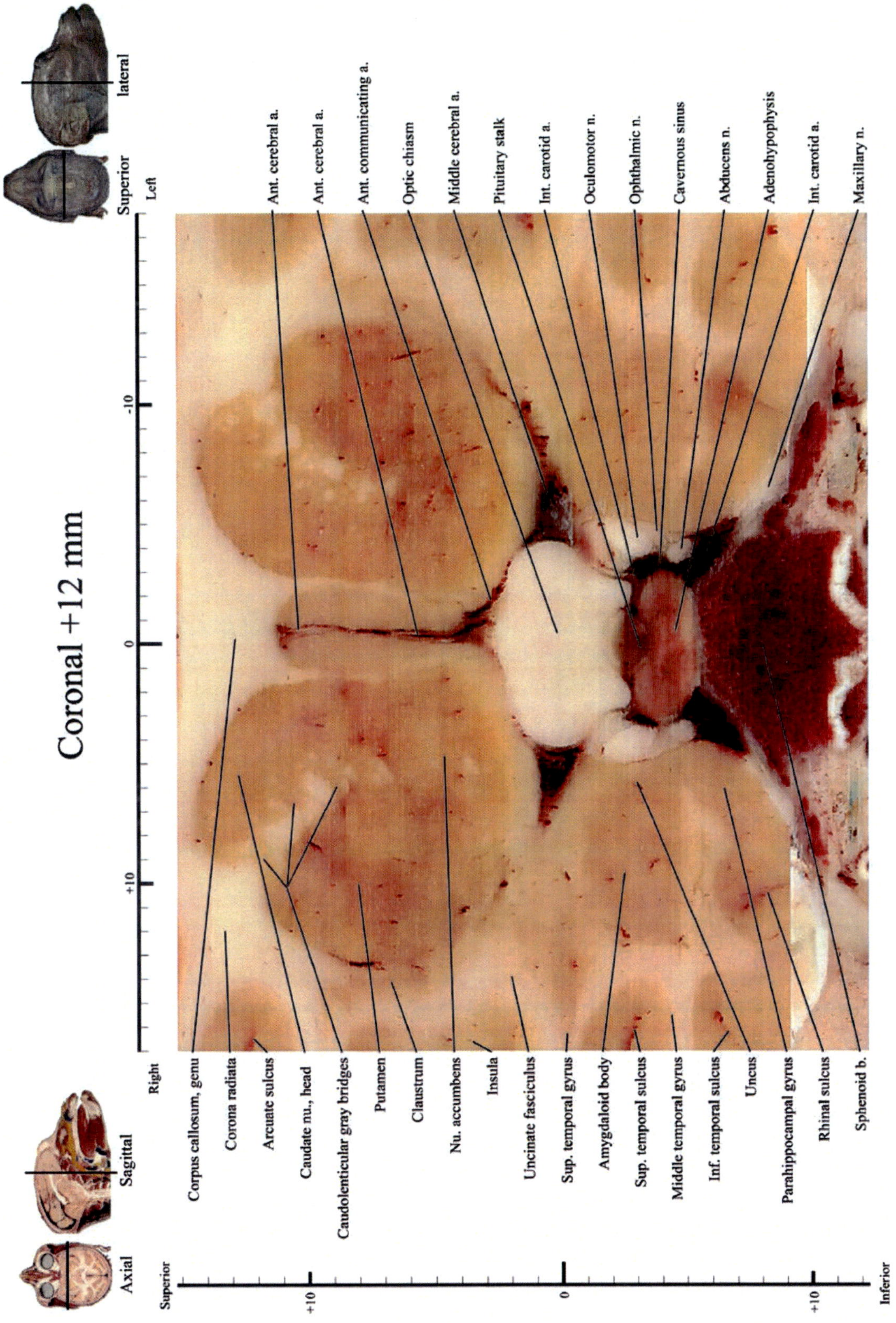
Coronal +12 mm
Superior
lateral
Left
Right
Sagittal
Axial
-10
0
+10
Superior
+10
0
+10
Inferior
Ant. cerebral a.
Ant. cerebral a.
Ant. communicating a.
Optic chiasm
Middle cerebral a.
Pituitary stalk
Int. carotid a.
Oculomotor n.
Ophthalmic n.
Cavernous sinus
Abducens n.
Adenohypophysis
Int. carotid a.
Maxillary n.
Corpus callosum, genu
Corona radiata
Arcuate sulcus
Caudate nu., head
Caudolenticular gray bridges
Putamen
Claustrum
Nu. accumbens
Insula
Uncinate fasciculus
Sup. temporal gyrus
Amygdaloid body
Sup. temporal sulcus
Middle temporal gyrus
Inf. temporal sulcus
Uncus
Parahippocampal gyrus
Rhinal sulcus
Sphenoid b.

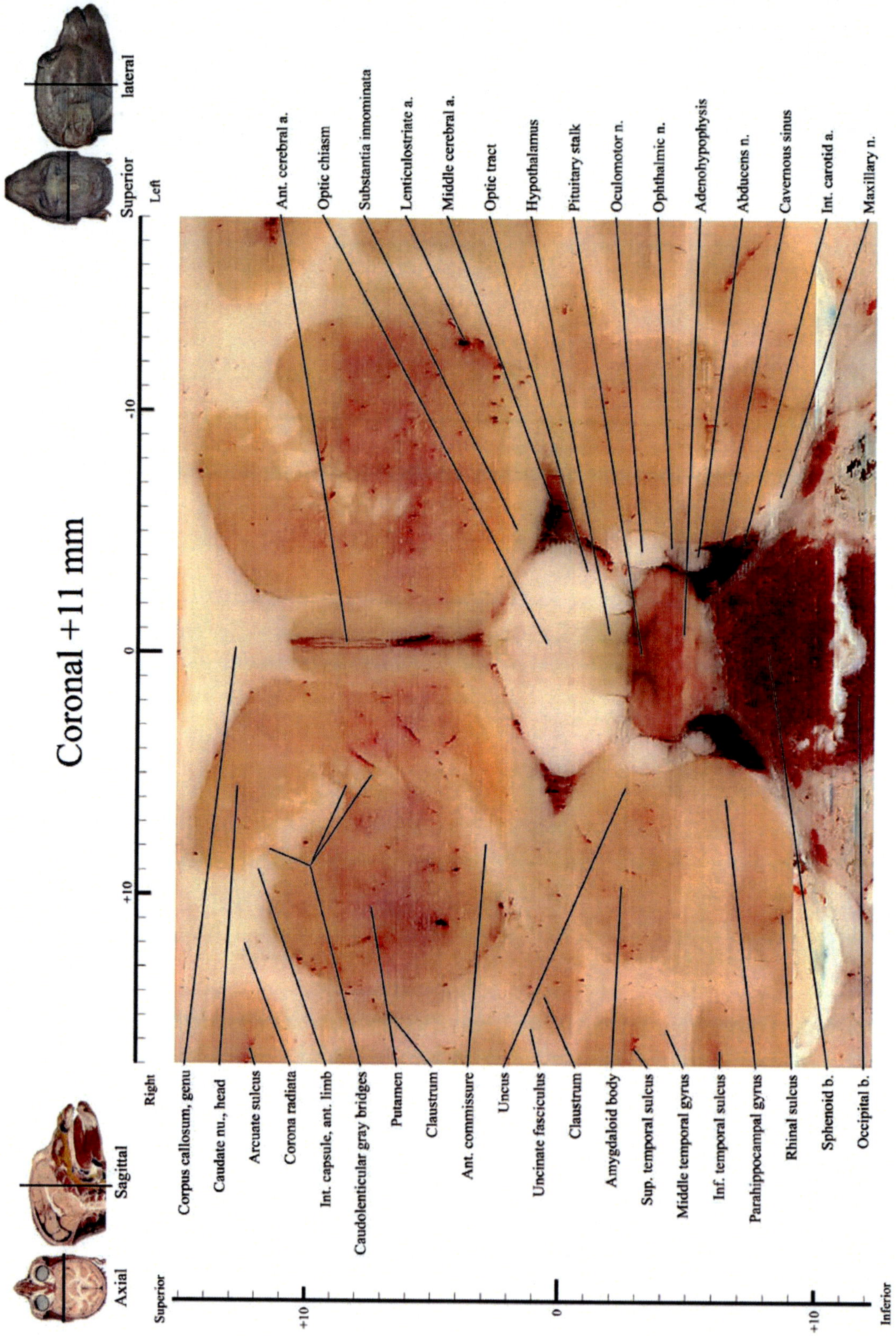
Coronal +11 mm
Superior
lateral
Axial
Sagittal
Left
Right
Ant. cerebral a.
Optic chiasm
Substantia innominata
Lenticulostriate a.
Middle cerebral a.
Optic tract
Hypothalamus
Pituitary stalk
Oculomotor n.
Ophthalmic n.
Adenohypophysis
Abducens n.
Cavernous sinus
Int. carotid a.
Maxillary n.
Corpus callosum, genu
Caudate nu., head
Arcuate sulcus
Corona radiata
Int. capsule, ant. limb
Caudolenticular gray bridges
Putamen
Claustrum
Ant. commissure
Uncus
Uncinate fasciculus
Claustrum
Amygdaloid body
Sup. temporal sulcus
Middle temporal gyrus
Inf. temporal sulcus
Parahippocampal gyrus
Rhinal sulcus
Sphenoid b.
Occipital b.
-10
0
+10
Superior
+10
0
+10
Inferior

Coronal +10 mm

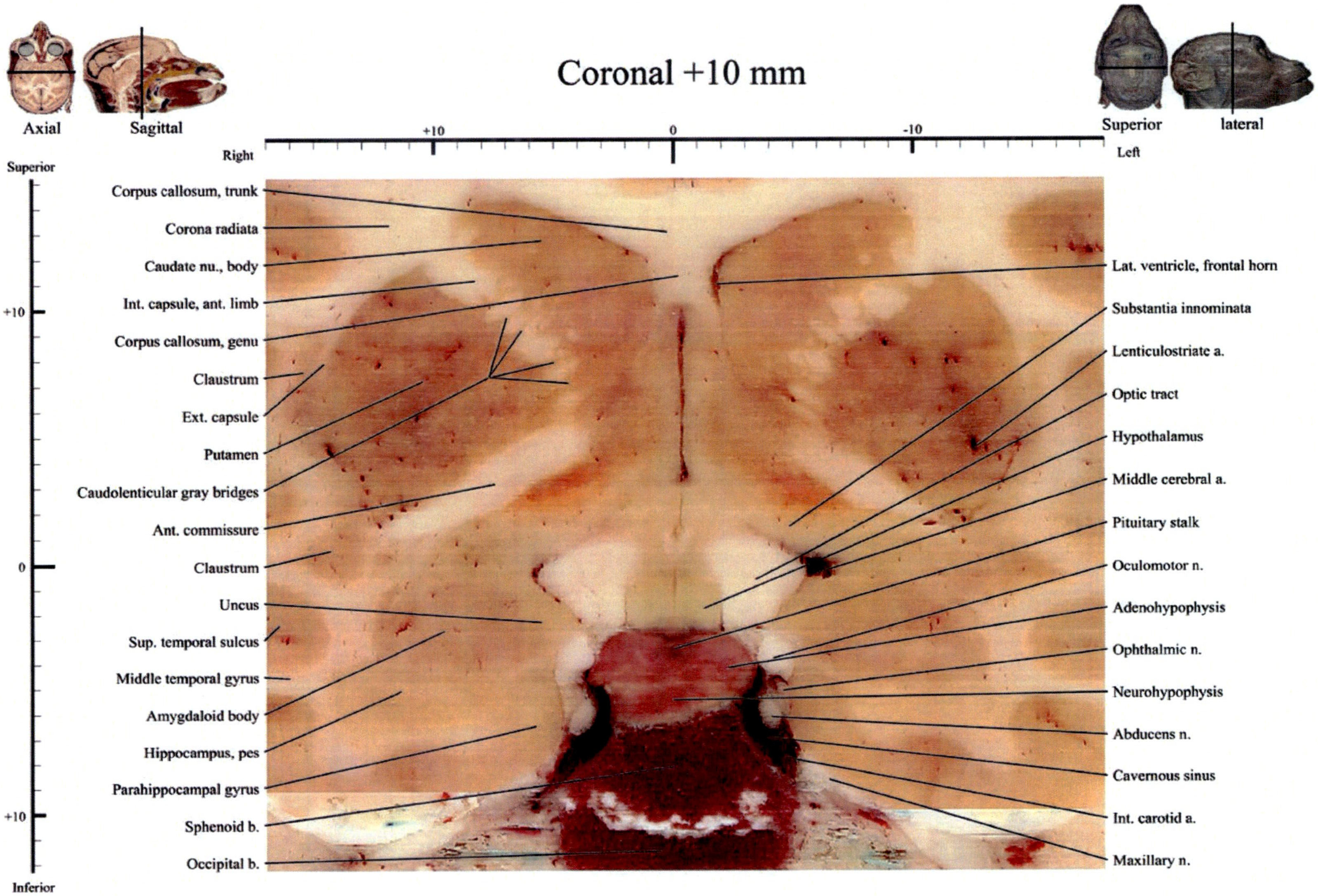

Coronal +9 mm

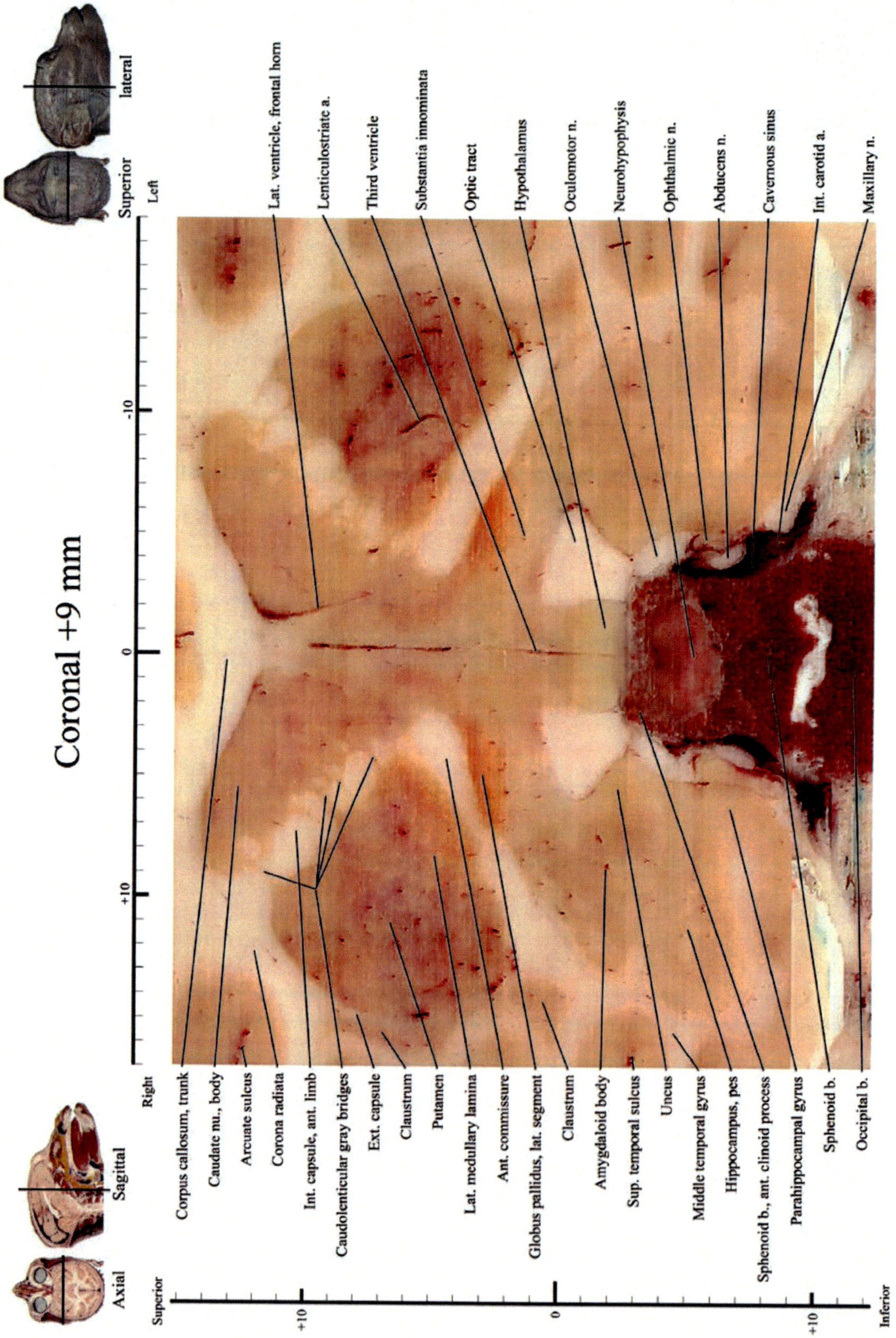
Superior
lateral
Left
Lat. ventricle, frontal horn
Lenticulostriate a.
Third ventricle
Substantia innominata
Optic tract
Hypothalamus
Oculomotor n.
Neurohypophysis
Ophthalmic n.
Abducens n.
Cavernous sinus
Int. carotid a.
Maxillary n.
-10
0
+10
Right
Corpus callosum, trunk
Caudate nu., body
Arcuate sulcus
Corona radiata
Int. capsule, ant. limb
Caudolenticular gray bridges
Ext. capsule
Claustrum
Putamen
Lat. medullary lamina
Ant. commissure
Globus pallidus, lat. segment
Claustrum
Amygdaloid body
Sup. temporal sulcus
Uncus
Middle temporal gyrus
Hippocampus, pes
Sphenoid b., ant. clinoid process
Parahippocampal gyrus
Sphenoid b.
Occipital b.
Axial
Sagittal
Superior
+10
0
+10
Inferior

Coronal +8 mm

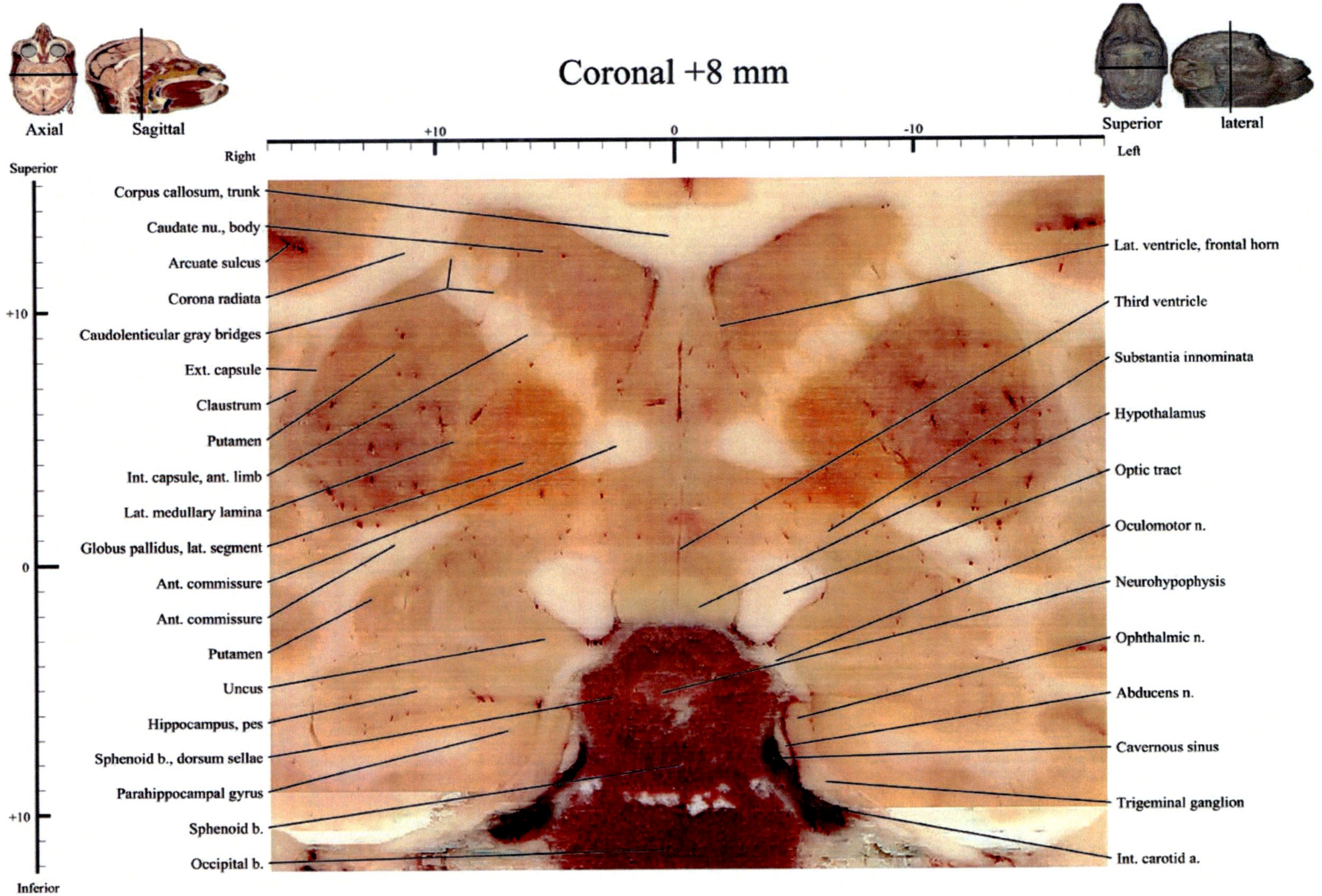
Axial
Sagittal
Superior
lateral
Right
Left
+10
0
-10
Superior
Inferior
+10
0
+10
Corpus callosum, trunk
Caudate nu., body
Arcuate sulcus
Corona radiata
Caudolenticular gray bridges
Ext. capsule
Claustrum
Putamen
Int. capsule, ant. limb
Lat. medullary lamina
Globus pallidus, lat. segment
Ant. commissure
Ant. commissure
Putamen
Uncus
Hippocampus, pes
Sphenoid b., dorsum sellae
Parahippocampal gyrus
Sphenoid b.
Occipital b.
Lat. ventricle, frontal horn
Third ventricle
Substantia innominata
Hypothalamus
Optic tract
Oculomotor n.
Neurohypophysis
Ophthalmic n.
Abducens n.
Cavernous sinus
Trigeminal ganglion
Int. carotid a.

Coronal +7 mm

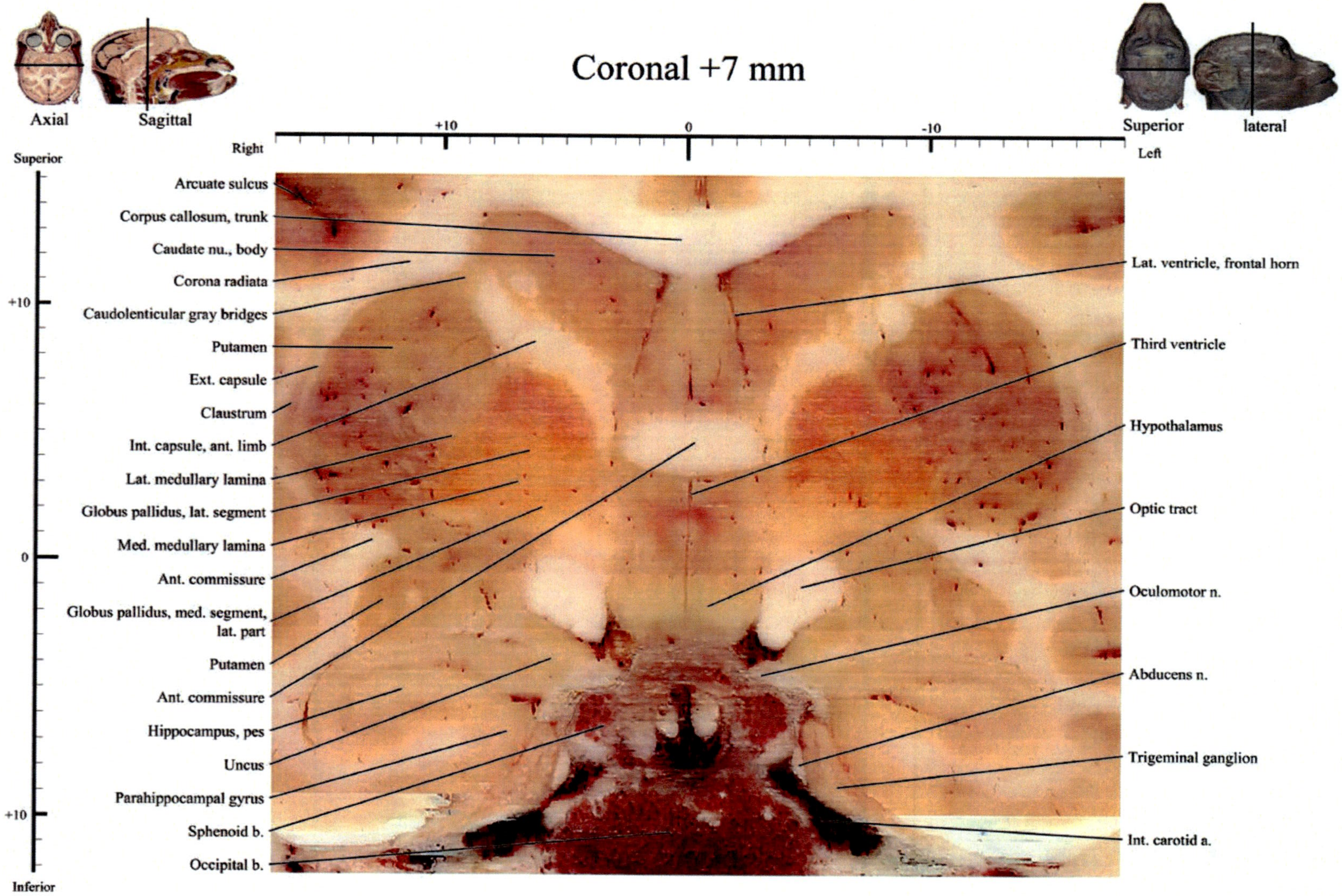
Axial
Sagittal
Superior
lateral
Right
Left
Superior
Inferior
+10
0
-10
+10
0
+10
Arcuate sulcus
Corpus callosum, trunk
Caudate nu., body
Corona radiata
Caudolenticular gray bridges
Putamen
Ext. capsule
Claustrum
Int. capsule, ant. limb
Lat. medullary lamina
Globus pallidus, lat. segment
Med. medullary lamina
Ant. commissure
Globus pallidus, med. segment, lat. part
Putamen
Ant. commissure
Hippocampus, pes
Uncus
Parahippocampal gyrus
Sphenoid b.
Occipital b.
Lat. ventricle, frontal horn
Third ventricle
Hypothalamus
Optic tract
Oculomotor n.
Abducens n.
Trigeminal ganglion
Int. carotid a.

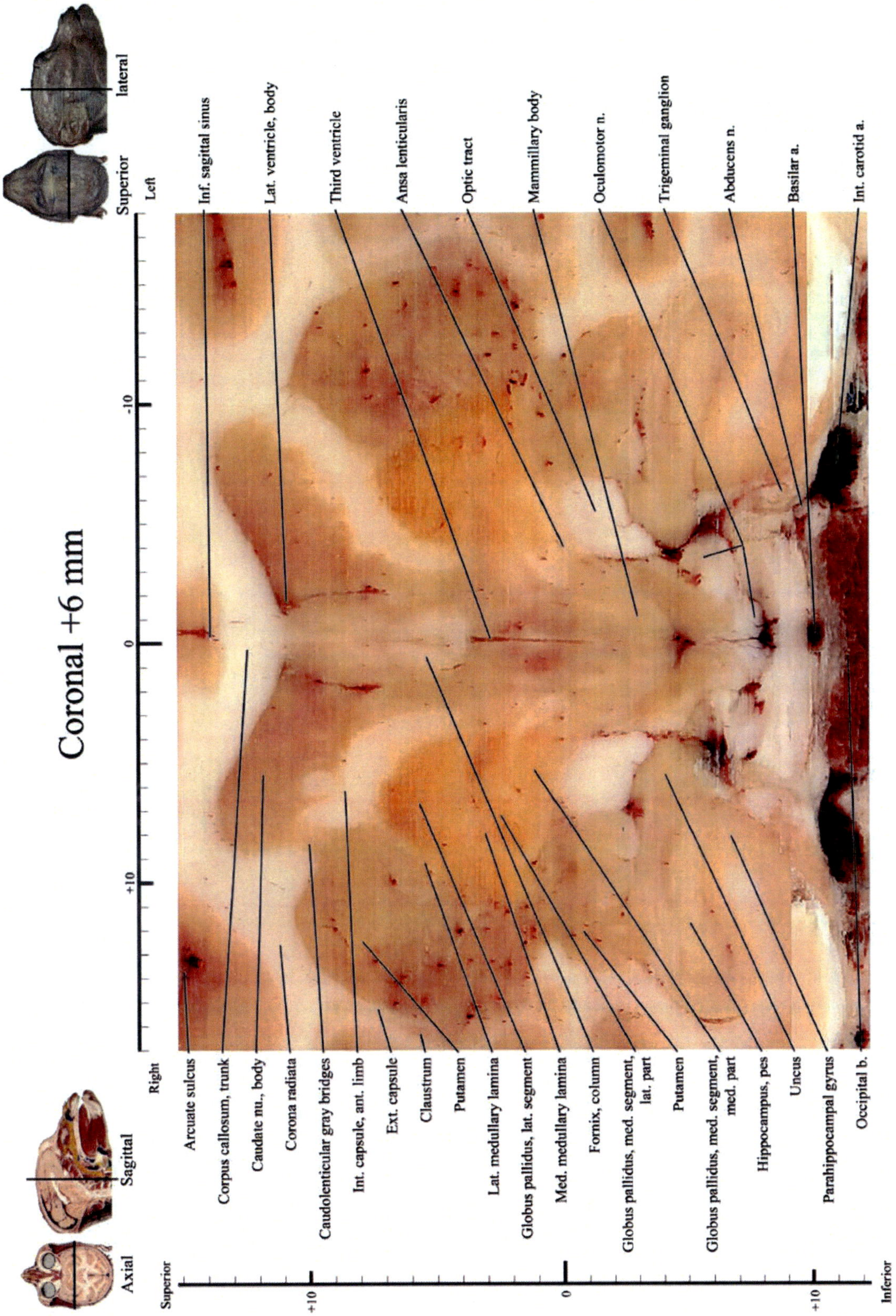

Coronal +6 mm
Superior
Left
lateral
Right
Axial
Sagittal
Inferior
-10
0
+10
Inf. sagittal sinus
Lat. ventricle, body
Third ventricle
Ansa lenticularis
Optic tract
Mammillary body
Oculomotor n.
Trigeminal ganglion
Abducens n.
Basilar a.
Int. carotid a.
Arcuate sulcus
Corpus callosum, trunk
Caudate nu., body
Corona radiata
Caudolenticular gray bridges
Int. capsule, ant. limb
Ext. capsule
Claustrum
Putamen
Lat. medullary lamina
Globus pallidus, lat. segment
Med. medullary lamina
Fornix, column
Globus pallidus, med. segment, lat. part
Putamen
Globus pallidus, med. segment, med. part
Hippocampus, pes
Uncus
Parahippocampal gyrus
Occipital b.

Coronal +5 mm

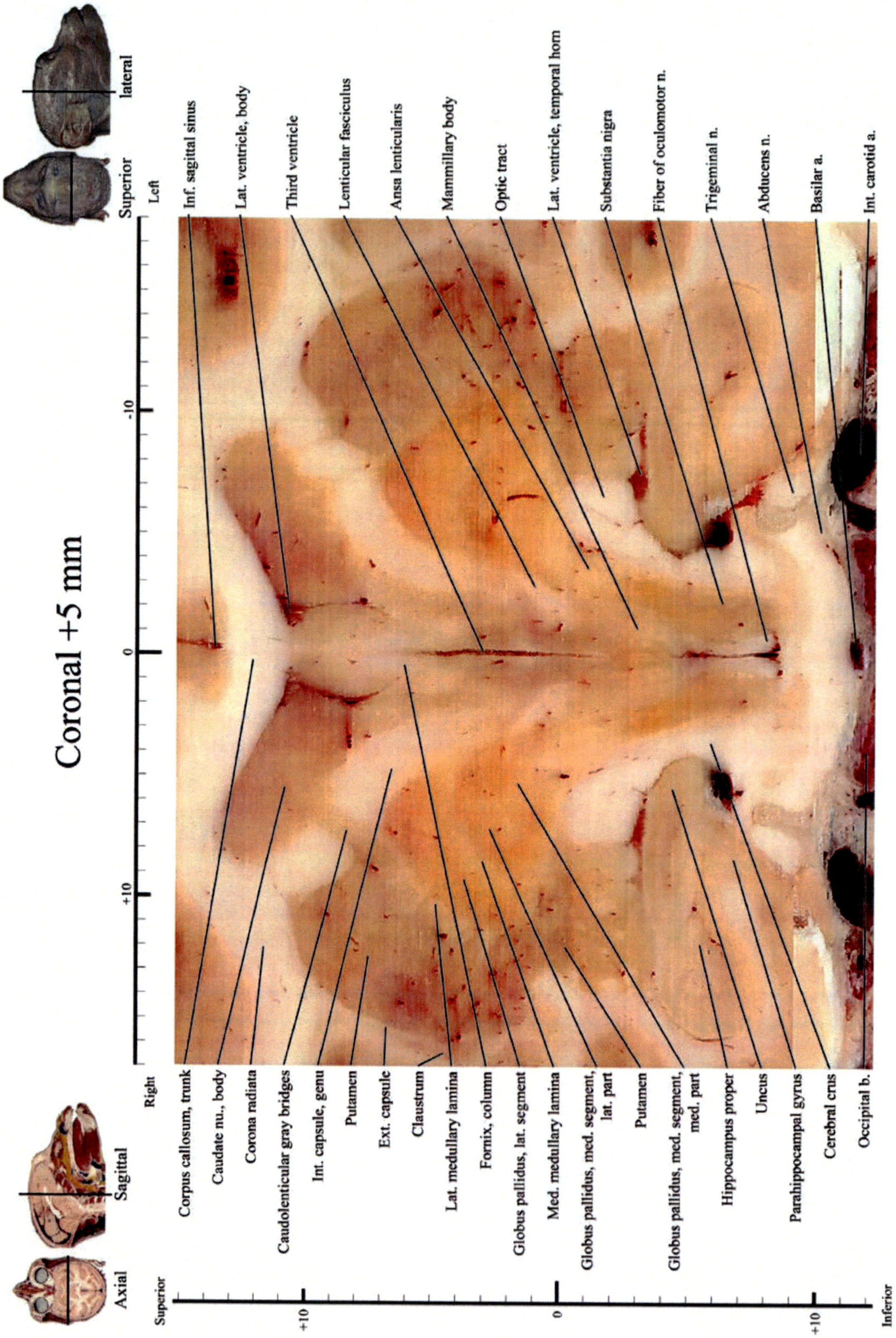
Superior
lateral
Left
Axial
Sagittal
Right
Inf. sagittal sinus
Lat. ventricle, body
Third ventricle
Lenticular fasciculus
Ansa lenticularis
Mammillary body
Optic tract
Lat. ventricle, temporal horn
Substantia nigra
Fiber of oculomotor n.
Trigeminal n.
Abducens n.
Basilar a.
Int. carotid a.
Corpus callosum, trunk
Caudate nu., body
Corona radiata
Caudolenticular gray bridges
Int. capsule, genu
Putamen
Ext. capsule
Claustrum
Lat. medullary lamina
Fornix, column
Globus pallidus, lat. segment
Med. medullary lamina
Globus pallidus, med. segment, lat. part
Putamen
Globus pallidus, med. segment, med. part
Hippocampus proper
Uncus
Parahippocampal gyrus
Cerebral crus
Occipital b.
-10
0
+10
Superior
+10
0
+10
Inferior

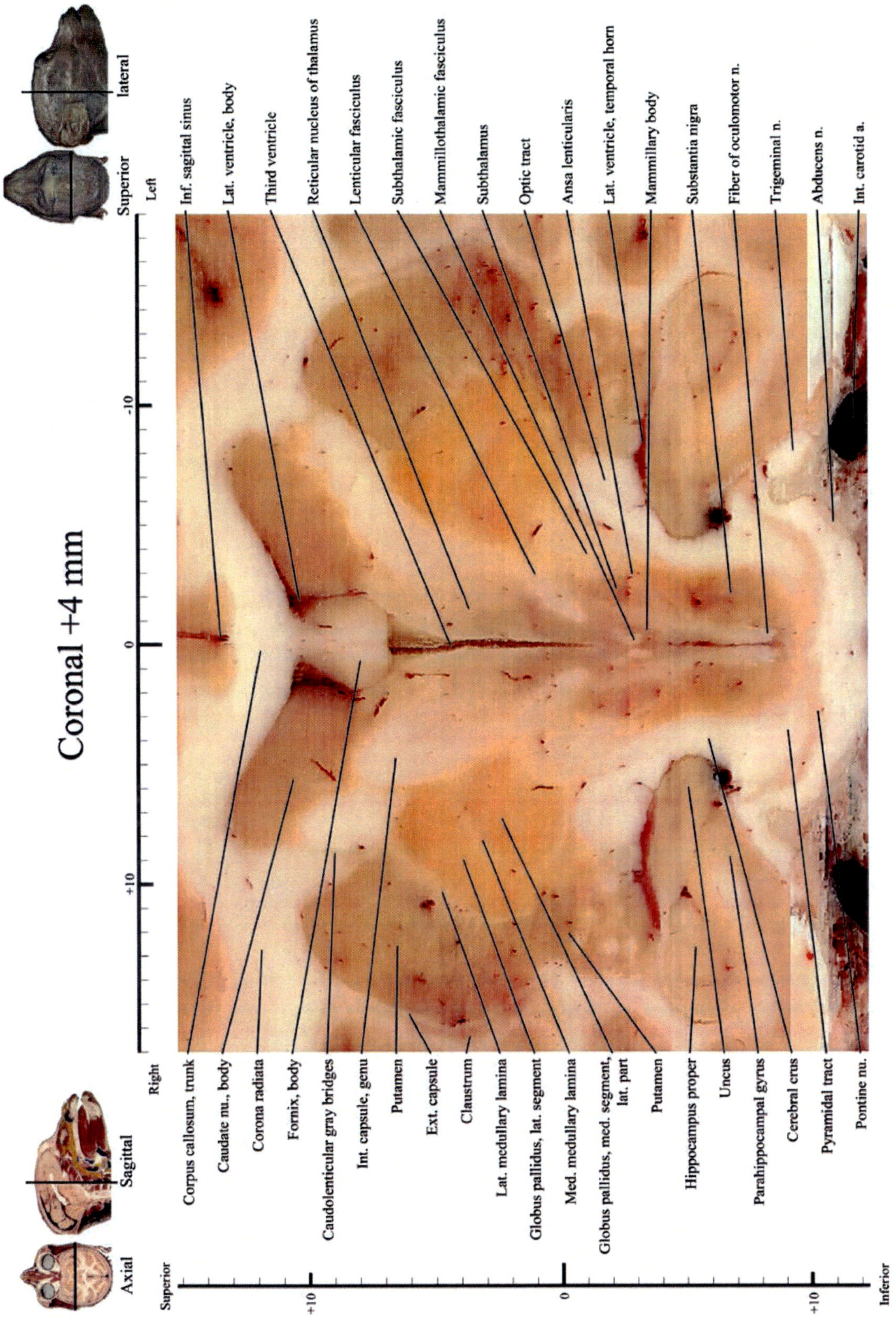
Coronal +4 mm
Superior
lateral
Axial
Sagittal
Left
Right
Superior
Inferior
-10
0
+10
Inf. sagittal sinus
Lat. ventricle, body
Third ventricle
Reticular nucleus of thalamus
Lenticular fasciculus
Subthalamic fasciculus
Mammillothalamic fasciculus
Subthalamus
Optic tract
Ansa lenticularis
Lat. ventricle, temporal horn
Mammillary body
Substantia nigra
Fiber of oculomotor n.
Trigeminal n.
Abducens n.
Int. carotid a.
Corpus callosum, trunk
Caudate nu., body
Corona radiata
Fornix, body
Caudolenticular gray bridges
Int. capsule, genu
Putamen
Ext. capsule
Claustrum
Lat. medullary lamina
Globus pallidus, lat. segment
Med. medullary lamina
Globus pallidus, med. segment, lat. part
Putamen
Hippocampus proper
Uncus
Parahippocampal gyrus
Cerebral crus
Pyramidal tract
Pontine nu.

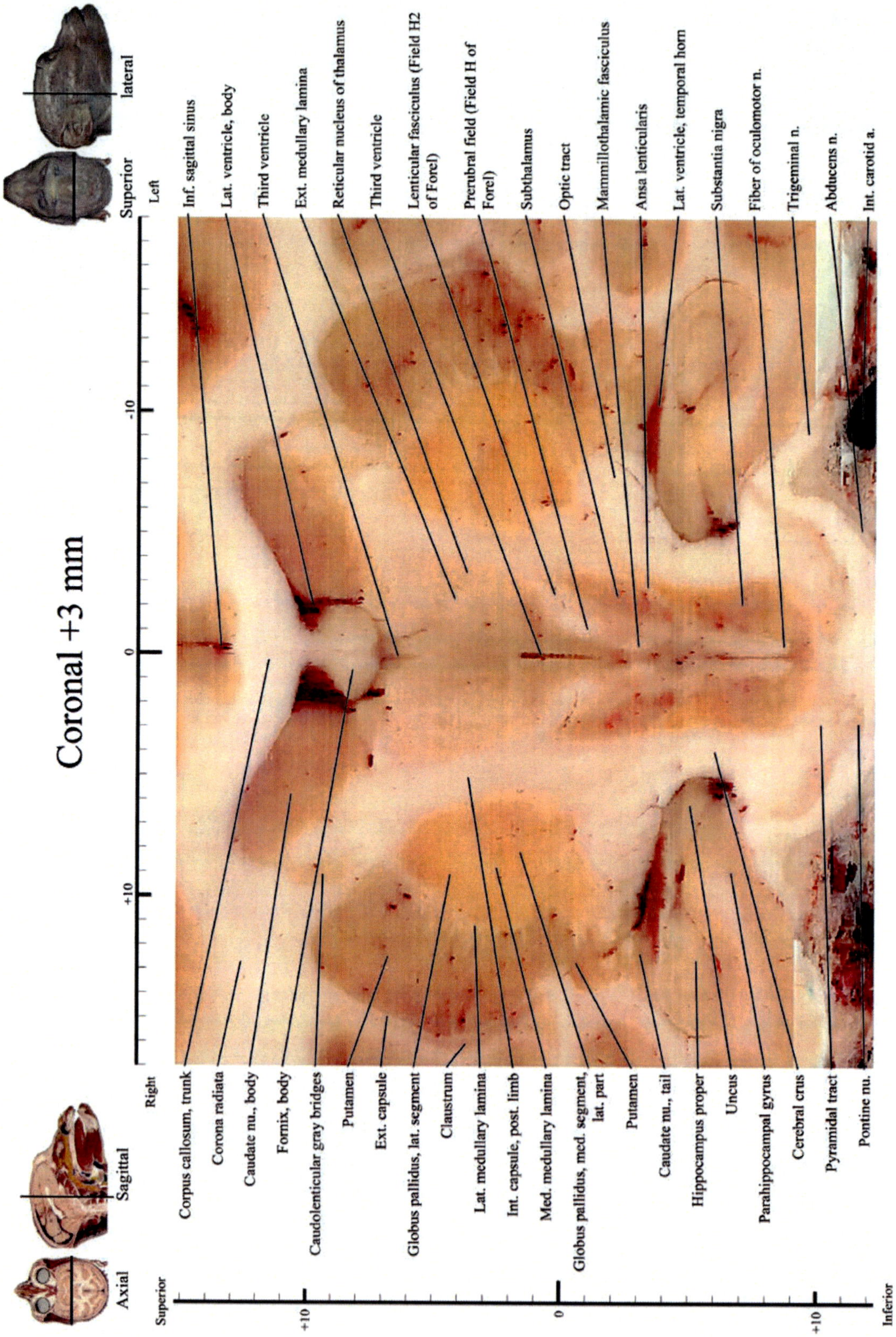
Coronal +3 mm
Superior
lateral
Left
Right
Axial
Sagittal
Superior
Inferior
-10
0
+10
+10
0
+10
Inf. sagittal sinus
Lat. ventricle, body
Third ventricle
Ext. medullary lamina
Reticular nucleus of thalamus
Third ventricle
Lenticular fasciculus (Field H2 of Forel)
Prerubral field (Field H of Forel)
Subthalamus
Optic tract
Mammillothalamic fasciculus
Ansa lenticularis
Lat. ventricle, temporal horn
Substantia nigra
Fiber of oculomotor n.
Trigeminal n.
Abducens n.
Int. carotid a.
Corpus callosum, trunk
Corona radiata
Caudate nu., body
Fornix, body
Caudolenticular gray bridges
Putamen
Ext. capsule
Globus pallidus, lat. segment
Claustrum
Lat. medullary lamina
Int. capsule, post. limb
Med. medullary lamina
Globus pallidus, med. segment, lat. part
Putamen
Caudate nu., tail
Hippocampus proper
Uncus
Parahippocampal gyrus
Cerebral crus
Pyramidal tract
Pontine nu.

Coronal +2 mm

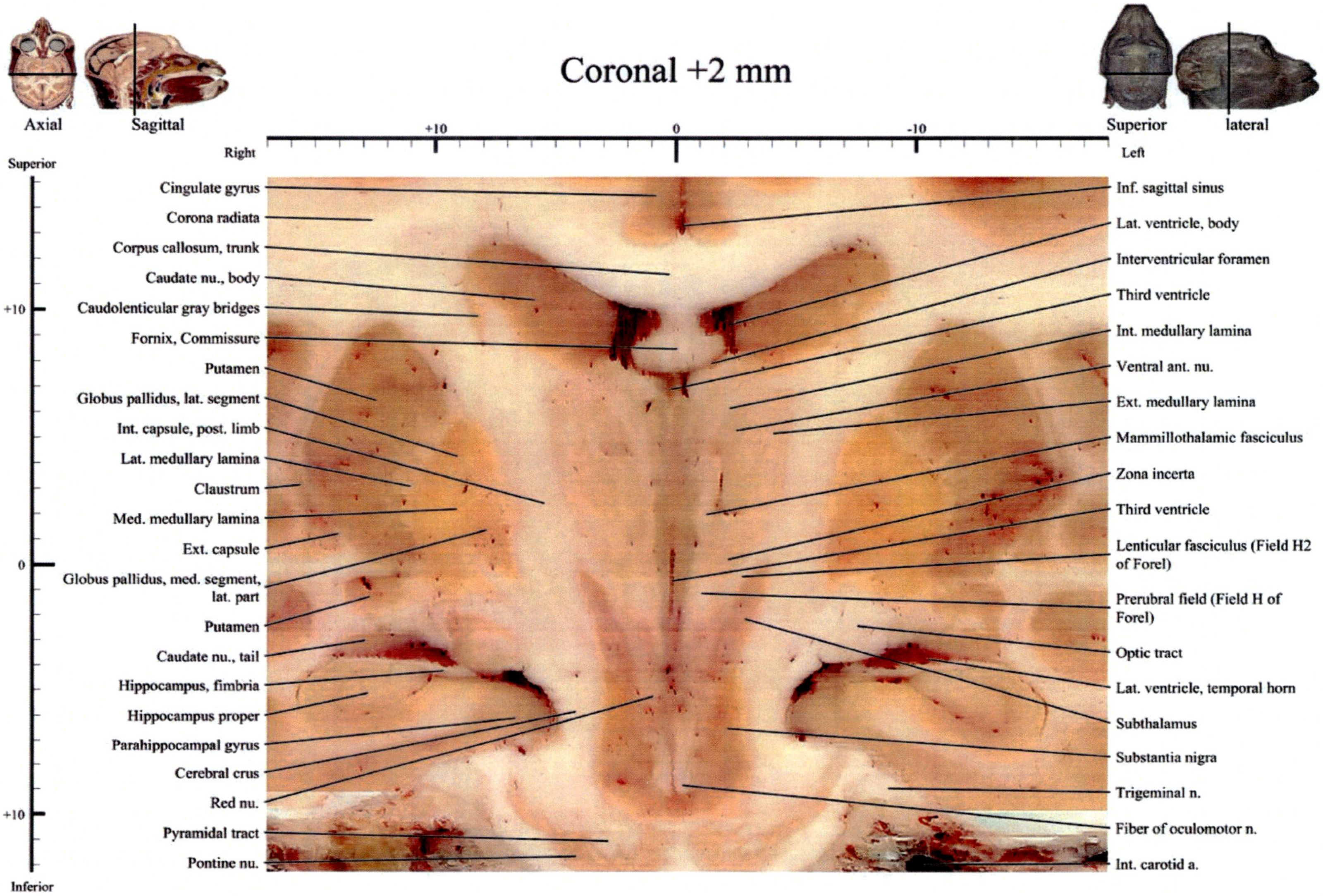
Axial
Sagittal
Superior
lateral
Right
Left
+10
0
-10
Superior
Inferior
+10
0
+10
Cingulate gyrus
Corona radiata
Corpus callosum, trunk
Caudate nu., body
Caudolenticular gray bridges
Fornix, Commissure
Putamen
Globus pallidus, lat. segment
Int. capsule, post. limb
Lat. medullary lamina
Claustrum
Med. medullary lamina
Ext. capsule
Globus pallidus, med. segment, lat. part
Putamen
Caudate nu., tail
Hippocampus, fimbria
Hippocampus proper
Parahippocampal gyrus
Cerebral crus
Red nu.
Pyramidal tract
Pontine nu.
Inf. sagittal sinus
Lat. ventricle, body
Interventricular foramen
Third ventricle
Int. medullary lamina
Ventral ant. nu.
Ext. medullary lamina
Mammillothalamic fasciculus
Zona incerta
Third ventricle
Lenticular fasciculus (Field H2 of Forel)
Prerubral field (Field H of Forel)
Optic tract
Lat. ventricle, temporal horn
Subthalamus
Substantia nigra
Trigeminal n.
Fiber of oculomotor n.
Int. carotid a.

Coronal +1 mm

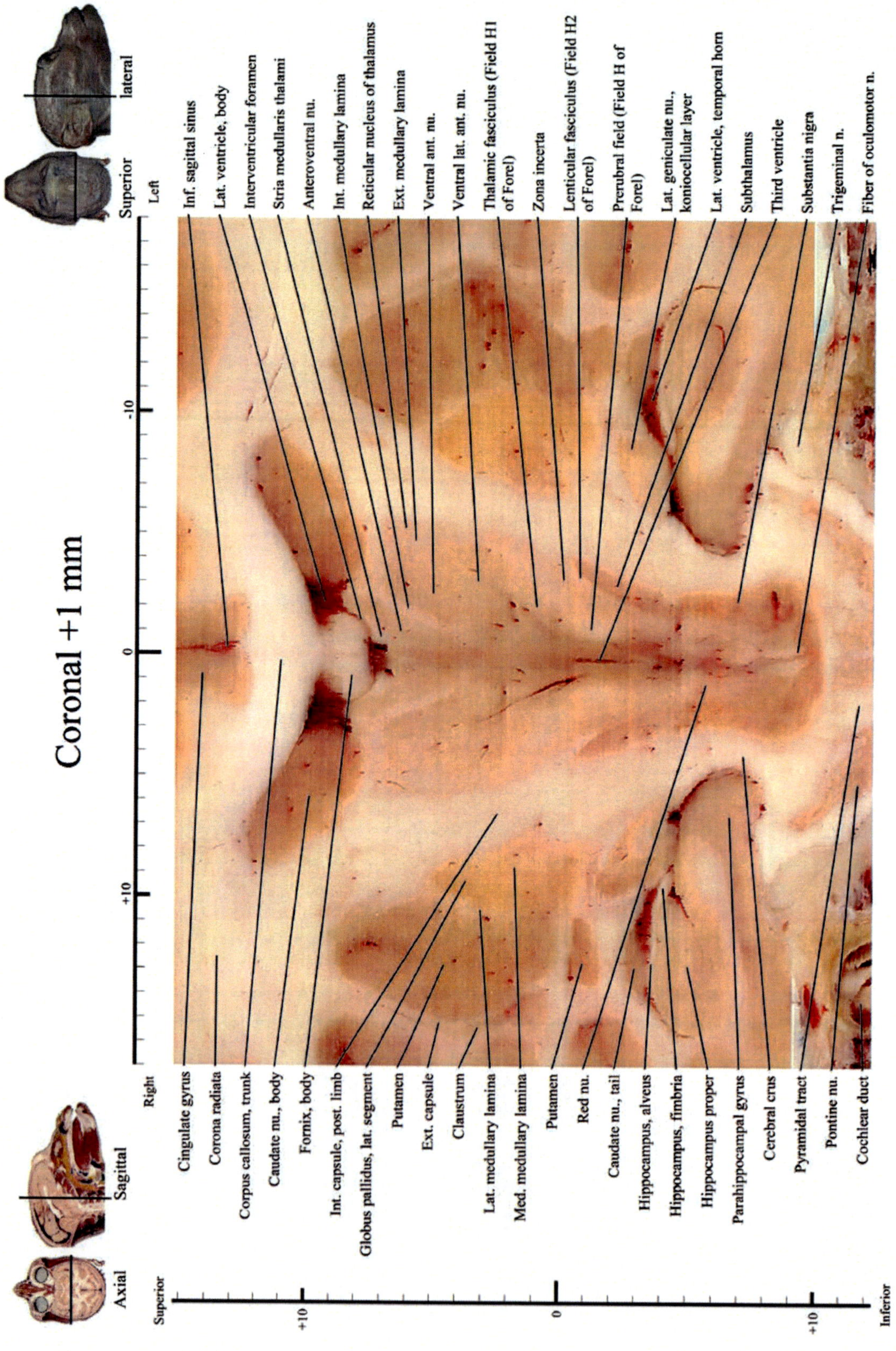
Axial
Sagittal
Superior
lateral
Right
Left
Superior
Inferior
+10
0
-10
+10
Inf. sagital sinus
Lat. ventricle, body
Interventricular foramen
Stria medullaris thalami
Anteroventral nu.
Int. medullary lamina
Reticular nucleus of thalamus
Ext. medullary lamina
Ventral ant. nu.
Ventral lat. ant. nu.
Thalamic fasciculus (Field H1 of Forel)
Zona incerta
Lenticular fasciculus (Field H2 of Forel)
Prerubral field (Field H of Forel)
Lat. geniculate nu., koniocellular layer
Lat. ventricle, temporal horn
Subthalamus
Third ventricle
Substantia nigra
Trigeminal n.
Fiber of oculomotor n.
Cingulate gyrus
Corona radiata
Corpus callosum, trunk
Caudate nu., body
Fornix, body
Int. capsule, post. limb
Globus pallidus, lat. segment
Putamen
Ext. capsule
Claustrum
Lat. medullary lamina
Med. medullary lamina
Putamen
Red nu.
Caudate nu., tail
Hippocampus, alveus
Hippocampus, fimbria
Hippocampus proper
Parahippocampal gyrus
Cerebral crus
Pyramidal tract
Pontine nu.
Cochlear duct

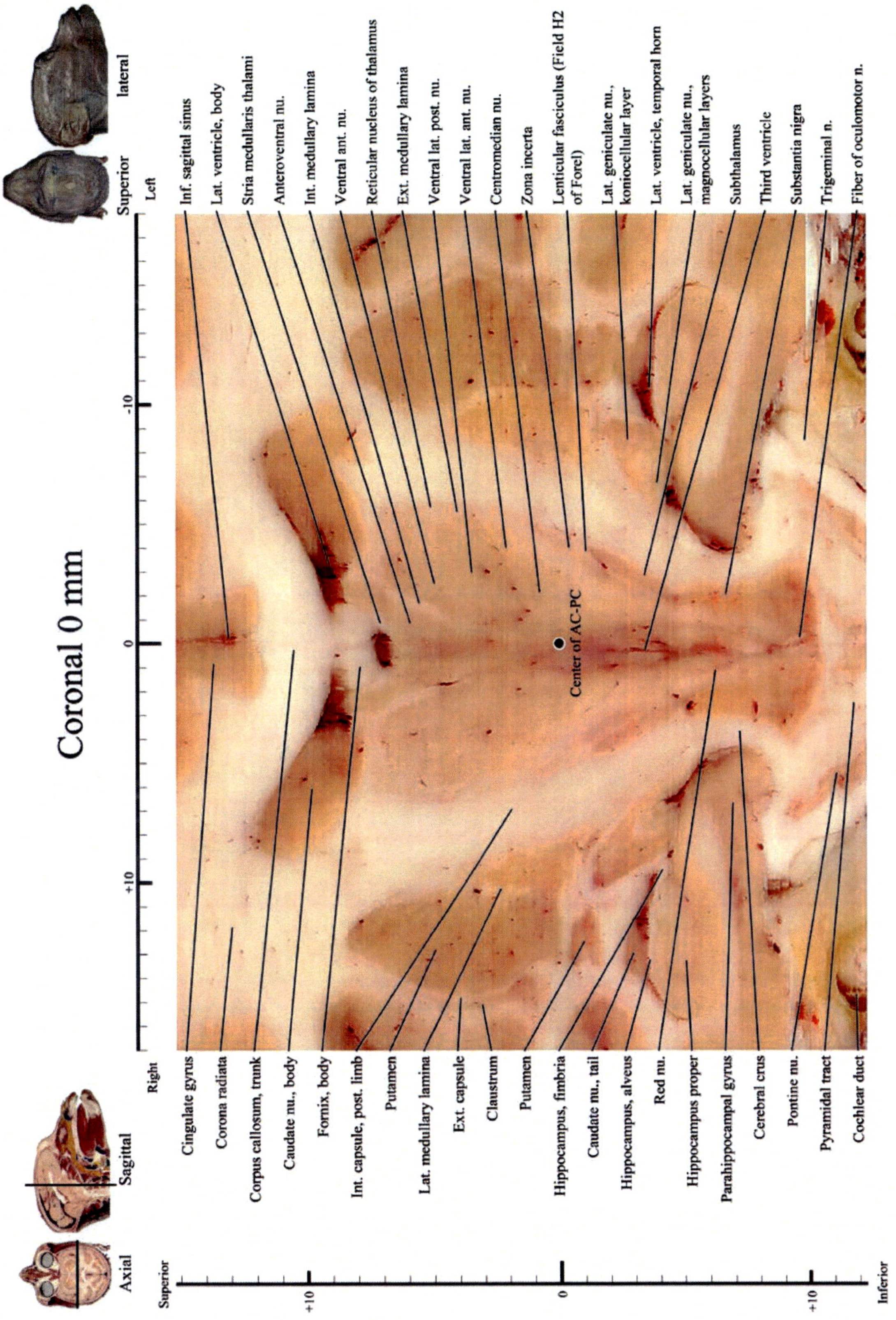
Coronal 0 mm
Superior
lateral
Left
Right
Sagittal
Axial
Inferior
-10
0
+10
Center of AC-PC
Inf. sagittal sinus
Lat. ventricle, body
Stria medullaris thalami
Anteroventral nu.
Int. medullary lamina
Ventral ant. nu.
Reticular nucleus of thalamus
Ext. medullary lamina
Ventral lat. post. nu.
Ventral lat. ant. nu.
Centromedian nu.
Zona incerta
Lenticular fasciculus (Field H2 of Forel)
Lat. geniculate nu., koniocellular layer
Lat. ventricle, temporal horn
Lat. geniculate nu., magnocellular layers
Subthalamus
Third ventricle
Substantia nigra
Trigeminal n.
Fiber of oculomotor n.
Cingulate gyrus
Corona radiata
Corpus callosum, trunk
Caudate nu., body
Fornix, body
Int. capsule, post. limb
Putamen
Lat. medullary lamina
Ext. capsule
Claustrum
Putamen
Hippocampus, fimbria
Caudate nu., tail
Hippocampus, alveus
Red nu.
Hippocampus proper
Parahippocampal gyrus
Cerebral crus
Pontine nu.
Pyramidal tract
Cochlear duct

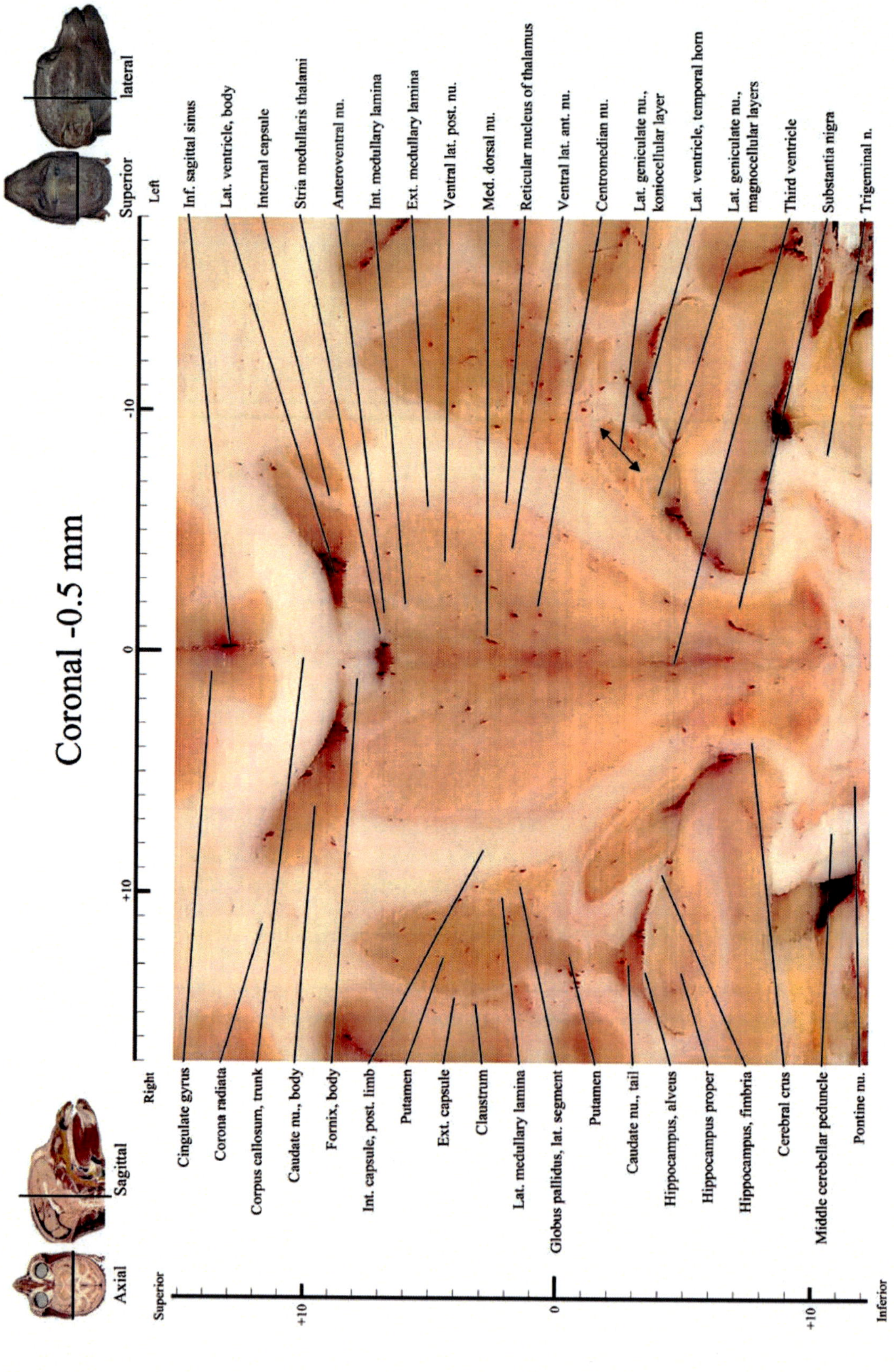
Coronal -0.5 mm
Superior
lateral
Left
Inf. sagittal sinus
Lat. ventricle, body
Internal capsule
Stria medullaris thalami
Anteroventral nu.
Int. medullary lamina
Ext. medullary lamina
Ventral lat. post. nu.
Med. dorsal nu.
Reticular nucleus of thalamus
Ventral lat. ant. nu.
Centromedian nu.
Lat. geniculate nu., koniocellular layer
Lat. ventricle, temporal horn
Lat. geniculate nu., magnocellular layers
Third ventricle
Substantia nigra
Trigeminal n.
-10
0
+10
Axial
Sagittal
Right
Cingulate gyrus
Corona radiata
Corpus callosum, trunk
Caudate nu., body
Fornix, body
Int. capsule, post. limb
Putamen
Ext. capsule
Claustrum
Lat. medullary lamina
Globus pallidus, lat. segment
Putamen
Caudate nu., tail
Hippocampus, alveus
Hippocampus proper
Hippocampus, fimbria
Cerebral crus
Middle cerebellar peduncle
Pontine nu.
Superior
+10
0
+10
Inferior

Coronal -1 mm

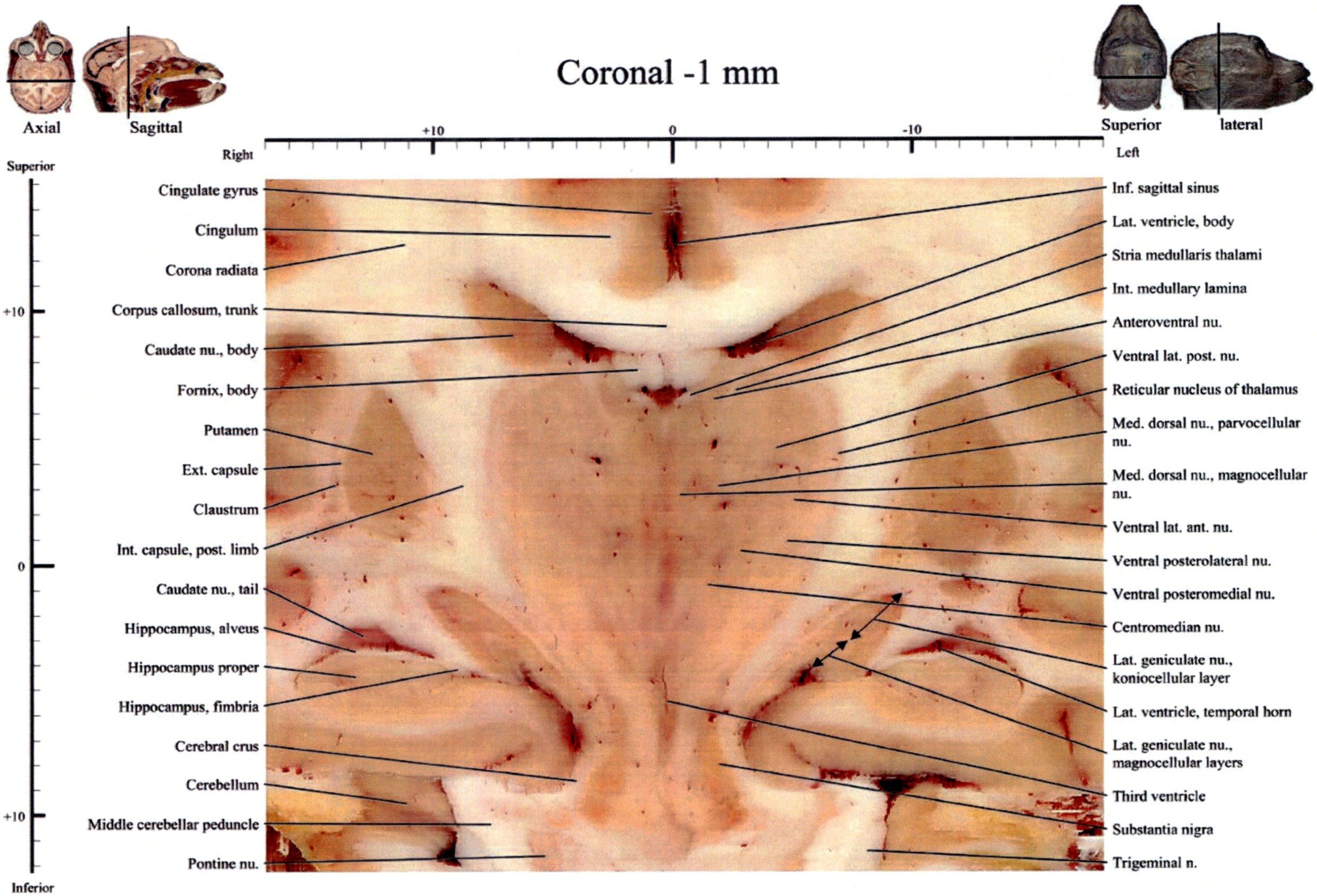
Axial
Sagittal
Superior
lateral
Right
Left
+10
0
-10
Superior
Inferior
+10
0
+10
Cingulate gyrus
Cingulum
Corona radiata
Corpus callosum, trunk
Caudate nu., body
Fornix, body
Putamen
Ext. capsule
Claustrum
Int. capsule, post. limb
Caudate nu., tail
Hippocampus, alveus
Hippocampus proper
Hippocampus, fimbria
Cerebral crus
Cerebellum
Middle cerebellar peduncle
Pontine nu.
Inf. sagittal sinus
Lat. ventricle, body
Stria medullaris thalami
Int. medullary lamina
Anteroventral nu.
Ventral lat. post. nu.
Reticular nucleus of thalamus
Med. dorsal nu., parvocellular nu.
Med. dorsal nu., magnocellular nu.
Ventral lat. ant. nu.
Ventral posterolateral nu.
Ventral posteromedial nu.
Centromedian nu.
Lat. geniculate nu., koniocellular layer
Lat. ventricle, temporal horn
Lat. geniculate nu., magnocellular layers
Third ventricle
Substantia nigra
Trigeminal n.

Coronal -1.5 mm

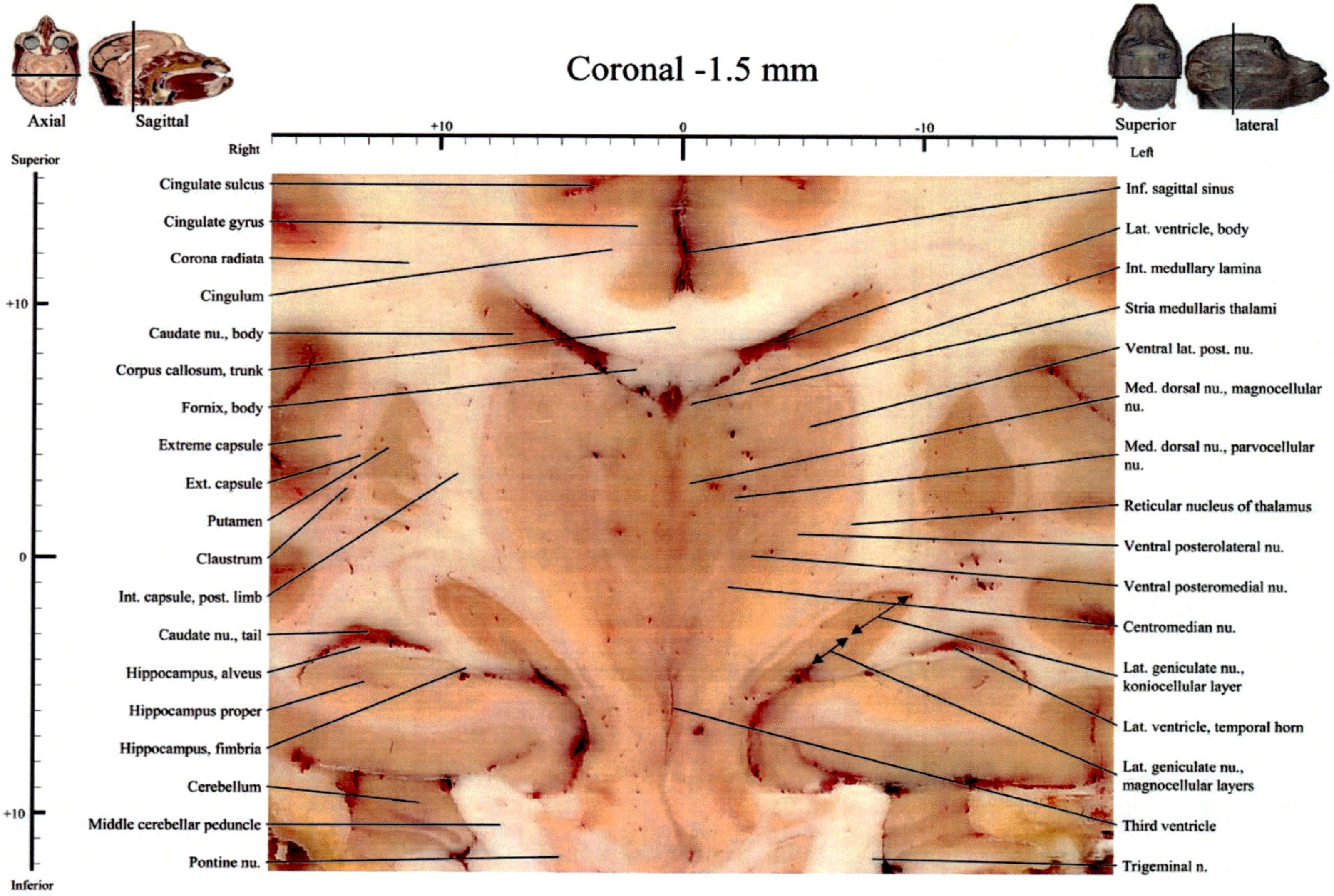

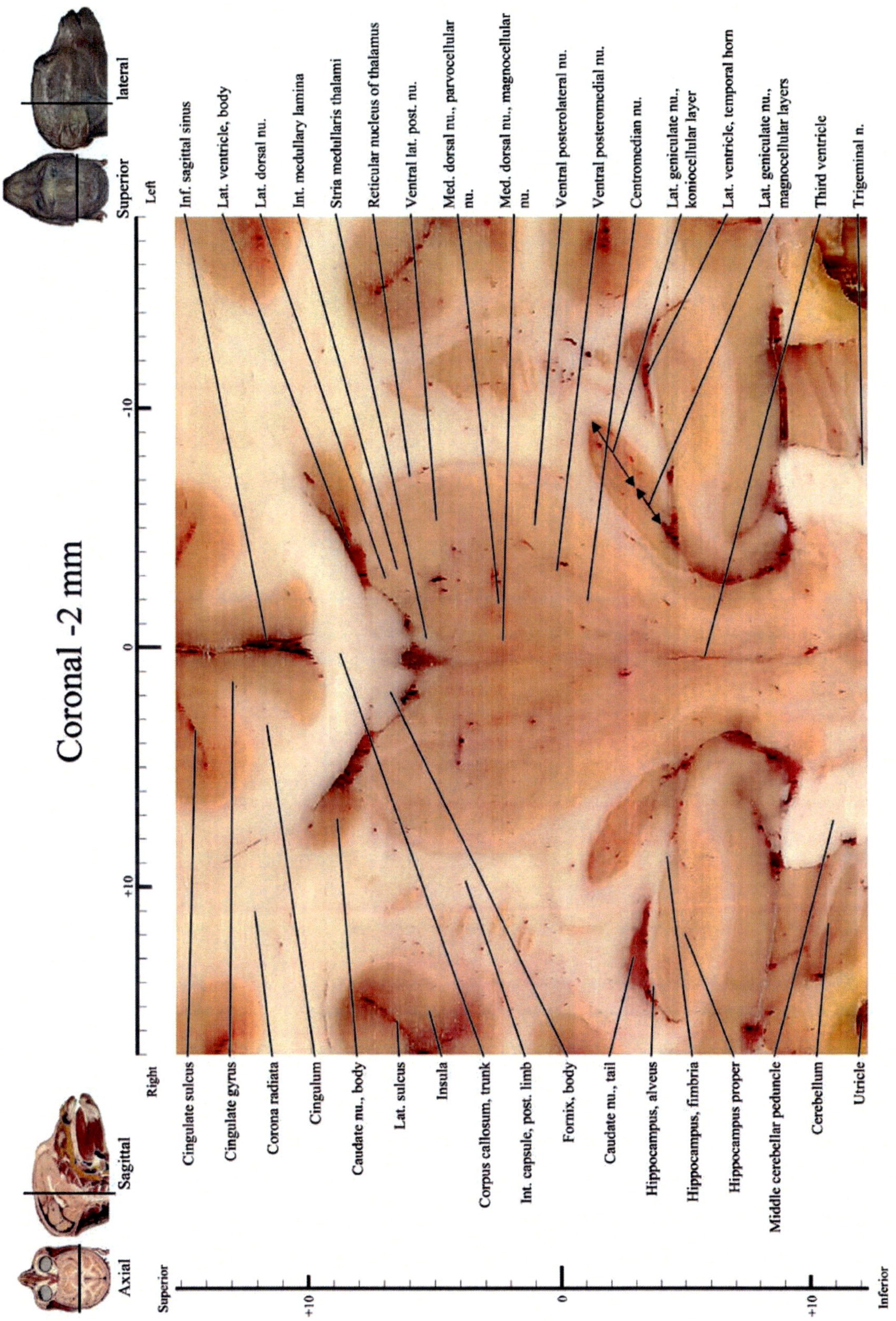
Coronal -2 mm
Superior
lateral
Left
Inf. sagittal sinus
Lat. ventricle, body
Lat. dorsal nu.
Int. medullary lamina
Stria medullaris thalami
Reticular nucleus of thalamus
Ventral lat. post. nu.
Med. dorsal nu., parvocellular nu.
Med. dorsal nu., magnocellular nu.
Ventral posterolateral nu.
Ventral posteromedial nu.
Centromedian nu.
Lat. geniculate nu., koniocellular layer
Lat. ventricle, temporal horn
Lat. geniculate nu., magnocellular layers
Third ventricle
Trigeminal n.
-10
0
+10
Right
Cingulate sulcus
Cingulate gyrus
Corona radiata
Cingulum
Caudate nu., body
Lat. sulcus
Insula
Corpus callosum, trunk
Int. capsule, post. limb
Fornix, body
Caudate nu., tail
Hippocampus, alveus
Hippocampus, fimbria
Hippocampus proper
Middle cerebellar peduncle
Cerebellum
Utricle
Sagittal
Axial
Superior
+10
0
+10
Inferior

Coronal -3 mm

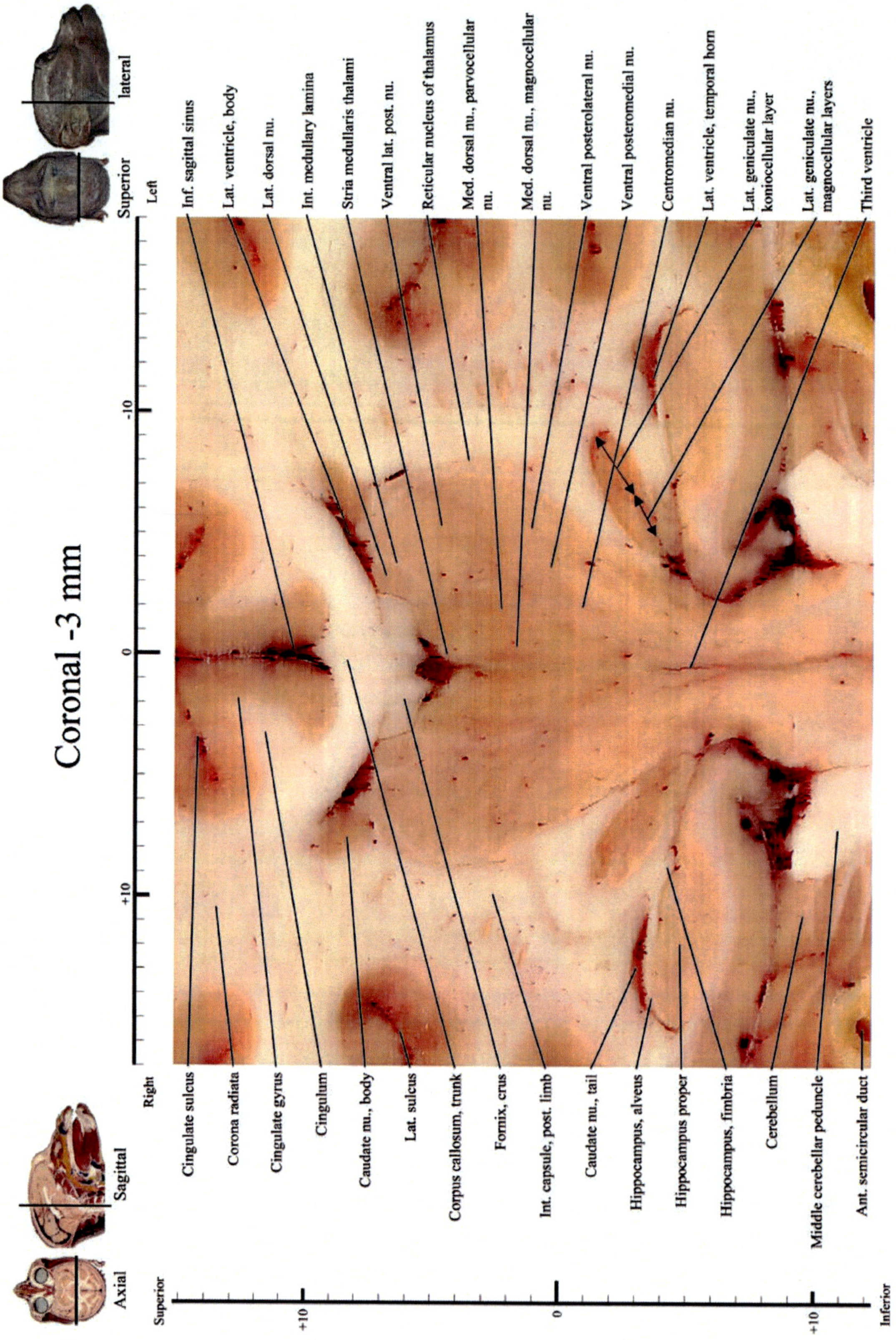
Superior
lateral
Left
Inf. sagittal sinus
Lat. ventricle, body
Lat. dorsal nu.
Int. medullary lamina
Stria medullaris thalami
Ventral lat. post. nu.
Reticular nucleus of thalamus
Med. dorsal nu., parvocellular nu.
Med. dorsal nu., magnocellular nu.
Ventral posterolateral nu.
Ventral posteromedial nu.
Centromedian nu.
Lat. ventricle, temporal horn
Lat. geniculate nu., koniocellular layer
Lat. geniculate nu., magnocellular layers
Third ventricle
-10
0
+10
Right
Cingulate sulcus
Corona radiata
Cingulate gyrus
Cingulum
Caudate nu., body
Lat. sulcus
Corpus callosum, trunk
Fornix, crus
Int. capsule, post. limb
Caudate nu., tail
Hippocampus, alveus
Hippocampus proper
Hippocampus, fimbria
Cerebellum
Middle cerebellar peduncle
Ant. semicircular duct
Axial
Sagittal
Superior
+10
0
+10
Inferior

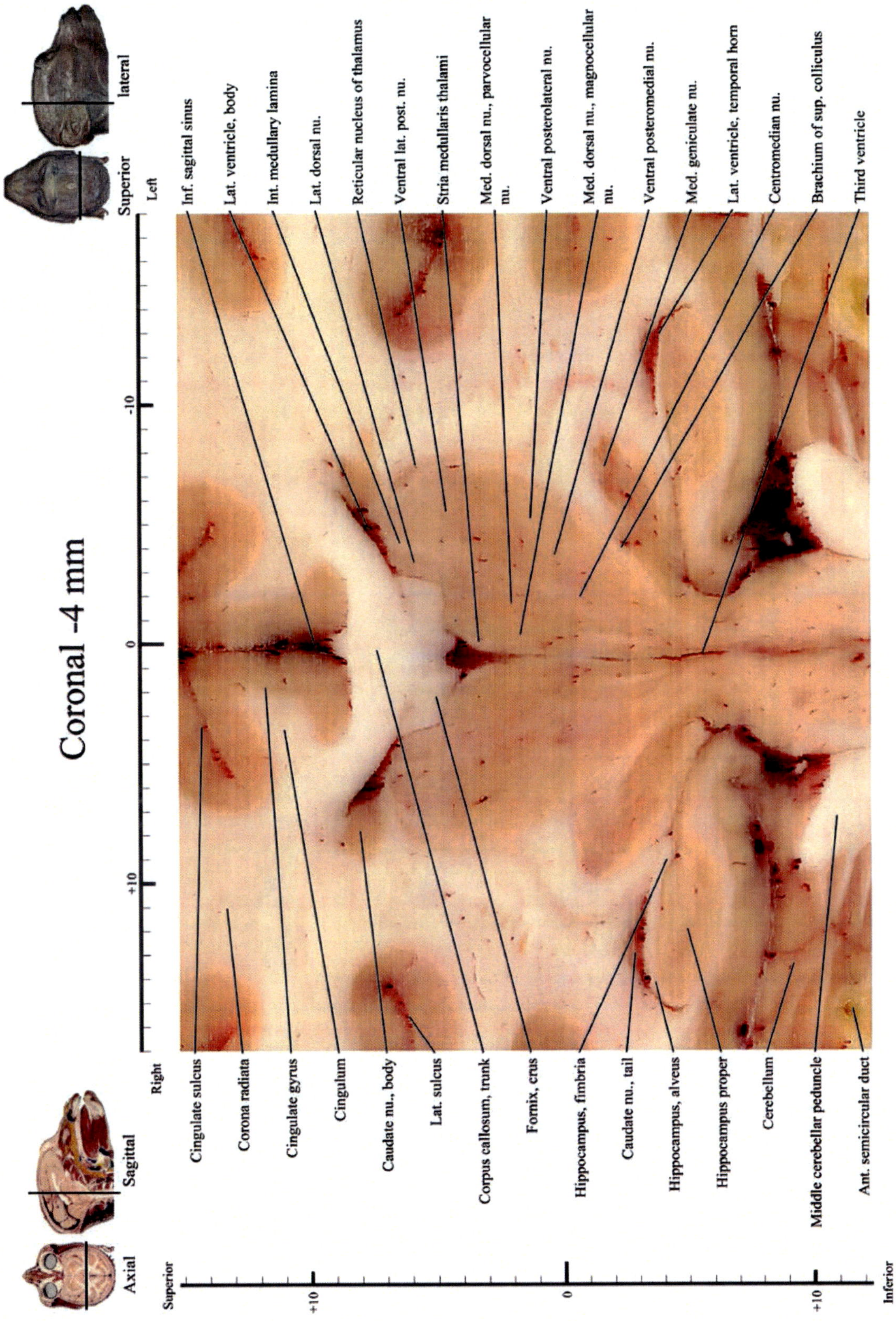
Coronal -4 mm
Superior
lateral
Left
Right
Sagittal
Axial
Superior
Inferior
-10
0
+10
+10
0
+10
Inf. sagittal sinus
Lat. ventricle, body
Int. medullary lamina
Lat. dorsal nu.
Reticular nucleus of thalamus
Ventral lat. post. nu.
Stria medullaris thalami
Med. dorsal nu., parvocellular nu.
Ventral posterolateral nu.
Med. dorsal nu., magnocellular nu.
Ventral posteromedial nu.
Med. geniculate nu.
Lat. ventricle, temporal horn
Centromedian nu.
Brachium of sup. colliculus
Third ventricle
Cingulate sulcus
Corona radiata
Cingulate gyrus
Cingulum
Caudate nu., body
Lat. sulcus
Corpus callosum, trunk
Fornix, crus
Hippocampus, fimbria
Caudate nu., tail
Hippocampus, alveus
Hippocampus proper
Cerebellum
Middle cerebellar peduncle
Ant. semicircular duct

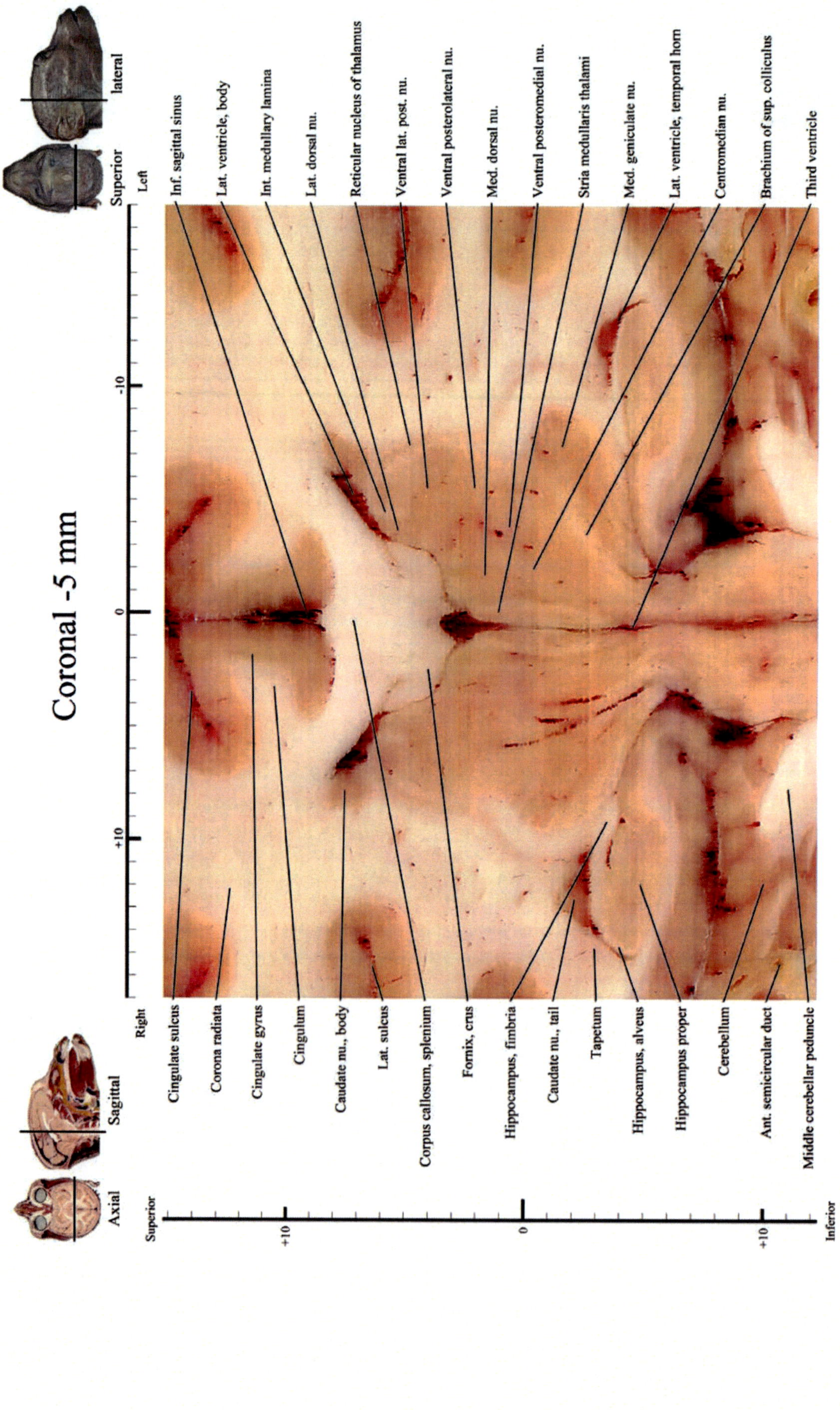
Coronal -5 mm
Superior
lateral
Sagittal
Axial
Left
Right
Superior
Inferior
-10
0
+10
Inf. sagittal sinus
Lat. ventricle, body
Int. medullary lamina
Lat. dorsal nu.
Reticular nucleus of thalamus
Ventral lat. post. nu.
Ventral posterolateral nu.
Med. dorsal nu.
Ventral posteromedial nu.
Stria medullaris thalami
Med. geniculate nu.
Lat. ventricle, temporal horn
Centromedian nu.
Brachium of sup. colliculus
Third ventricle
Cingulate sulcus
Corona radiata
Cingulate gyrus
Cingulum
Caudate nu., body
Lat. sulcus
Corpus callosum, splenium
Fornix, crus
Hippocampus, fimbria
Caudate nu., tail
Tapetum
Hippocampus, alveus
Hippocampus proper
Cerebellum
Ant. semicircular duct
Middle cerebellar peduncle

Coronal -6 mm

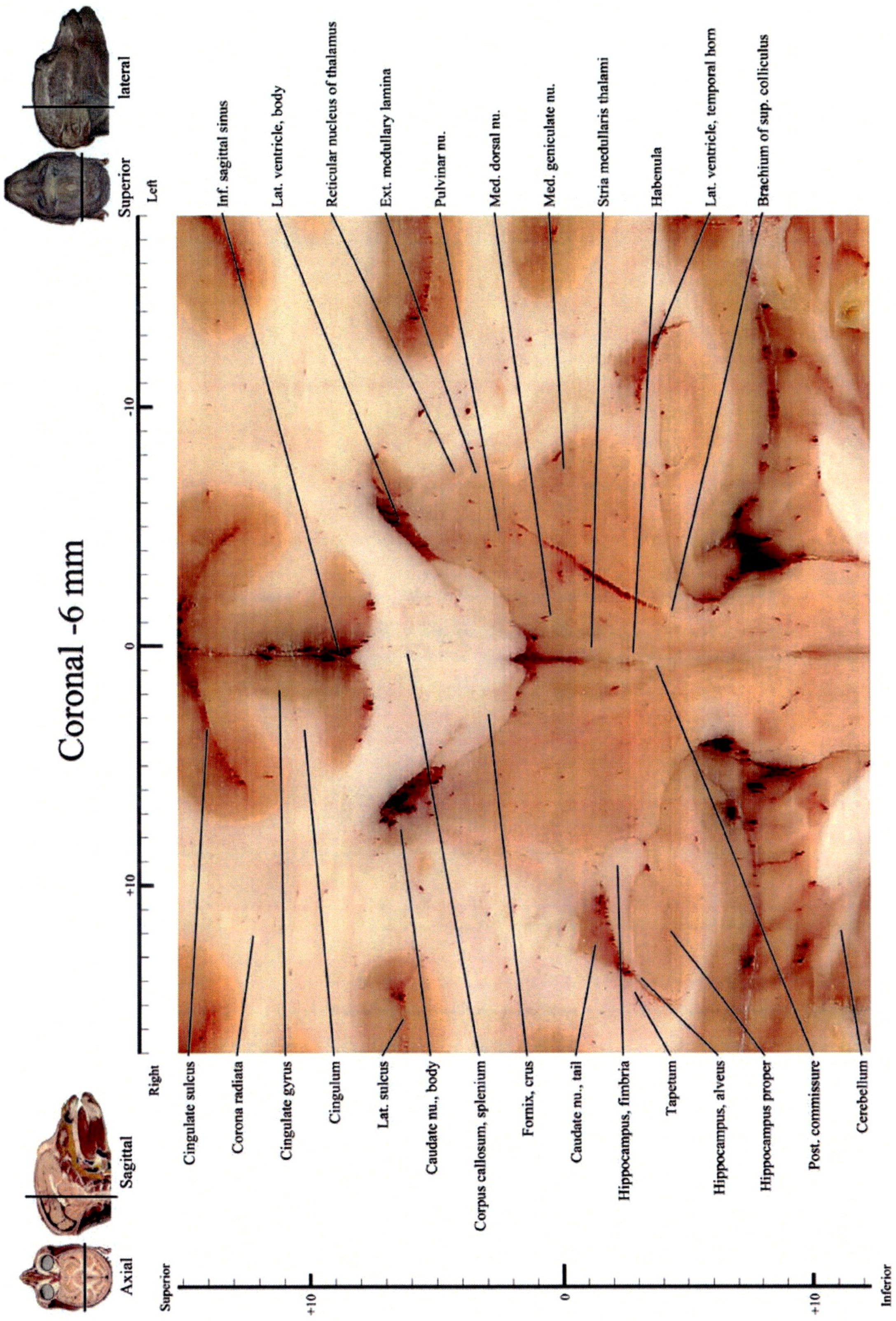
Superior
lateral
Sagittal
Axial
Left
Right
Superior
Inferior
-10
0
+10
+10
0
+10
Inf. sagittal sinus
Lat. ventricle, body
Reticular nucleus of thalamus
Ext. medullary lamina
Pulvinar nu.
Med. dorsal nu.
Med. geniculate nu.
Stria medullaris thalami
Habenula
Lat. ventricle, temporal horn
Brachium of sup. colliculus
Cingulate sulcus
Corona radiata
Cingulate gyrus
Cingulum
Lat. sulcus
Caudate nu., body
Corpus callosum, splenium
Fornix, crus
Caudate nu., tail
Hippocampus, fimbria
Tapetum
Hippocampus, alveus
Hippocampus proper
Post. commissure
Cerebellum

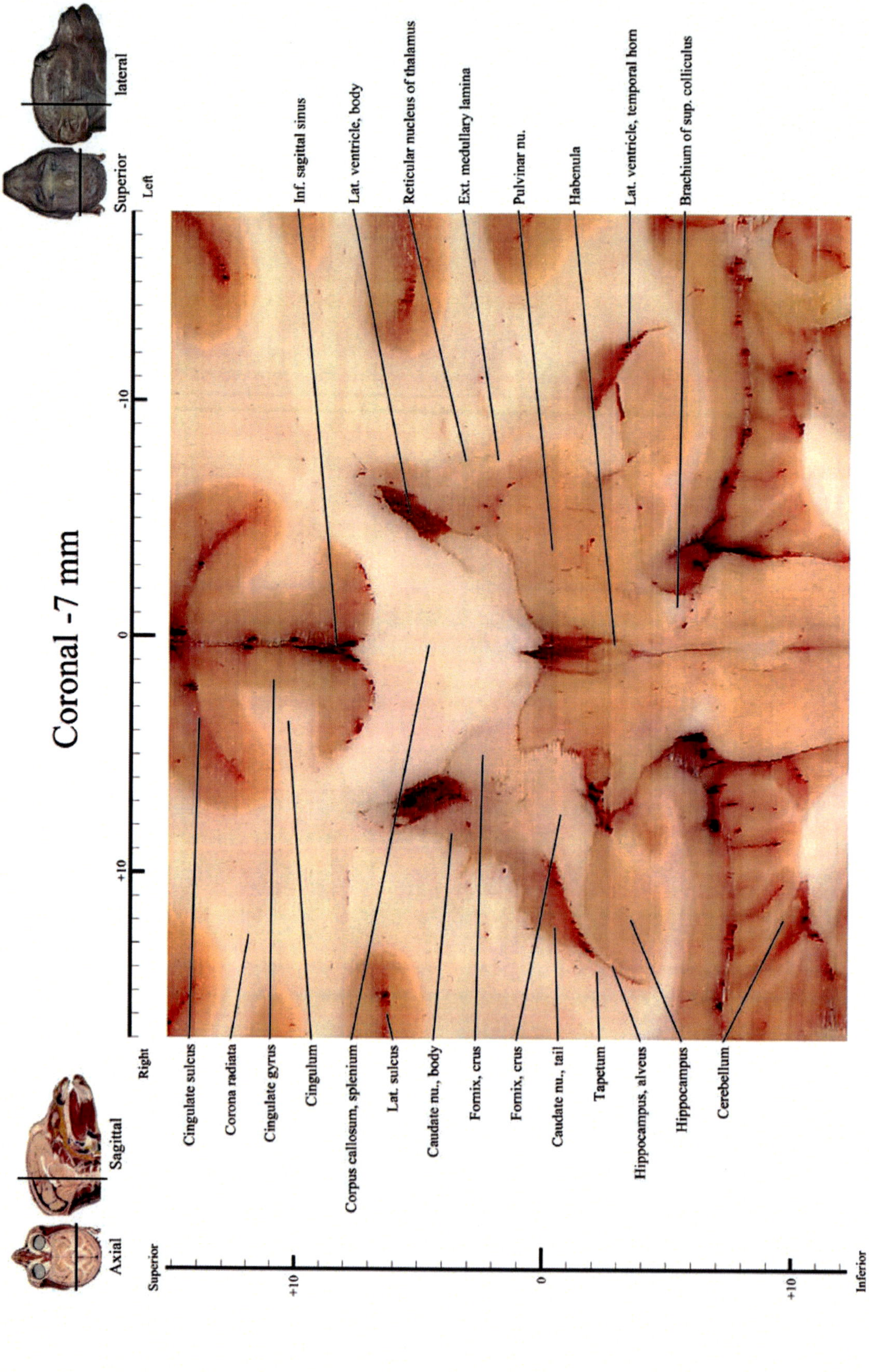
Coronal -7 mm
Superior
lateral
Axial
Sagittal
Left
Right
Inf. sagital sinus
Lat. ventricle, body
Reticular nucleus of thalamus
Ext. medullary lamina
Pulvinar nu.
Habenula
Lat. ventricle, temporal horn
Brachium of sup. colliculus
Cingulate sulcus
Corona radiata
Cingulate gyrus
Cingulum
Corpus callosum, splenium
Lat. sulcus
Caudate nu., body
Fornix, crus
Fornix, crus
Caudate nu., tail
Tapetum
Hippocampus, alveus
Hippocampus
Cerebellum
-10
0
+10
Superior
+10
0
+10
Inferior

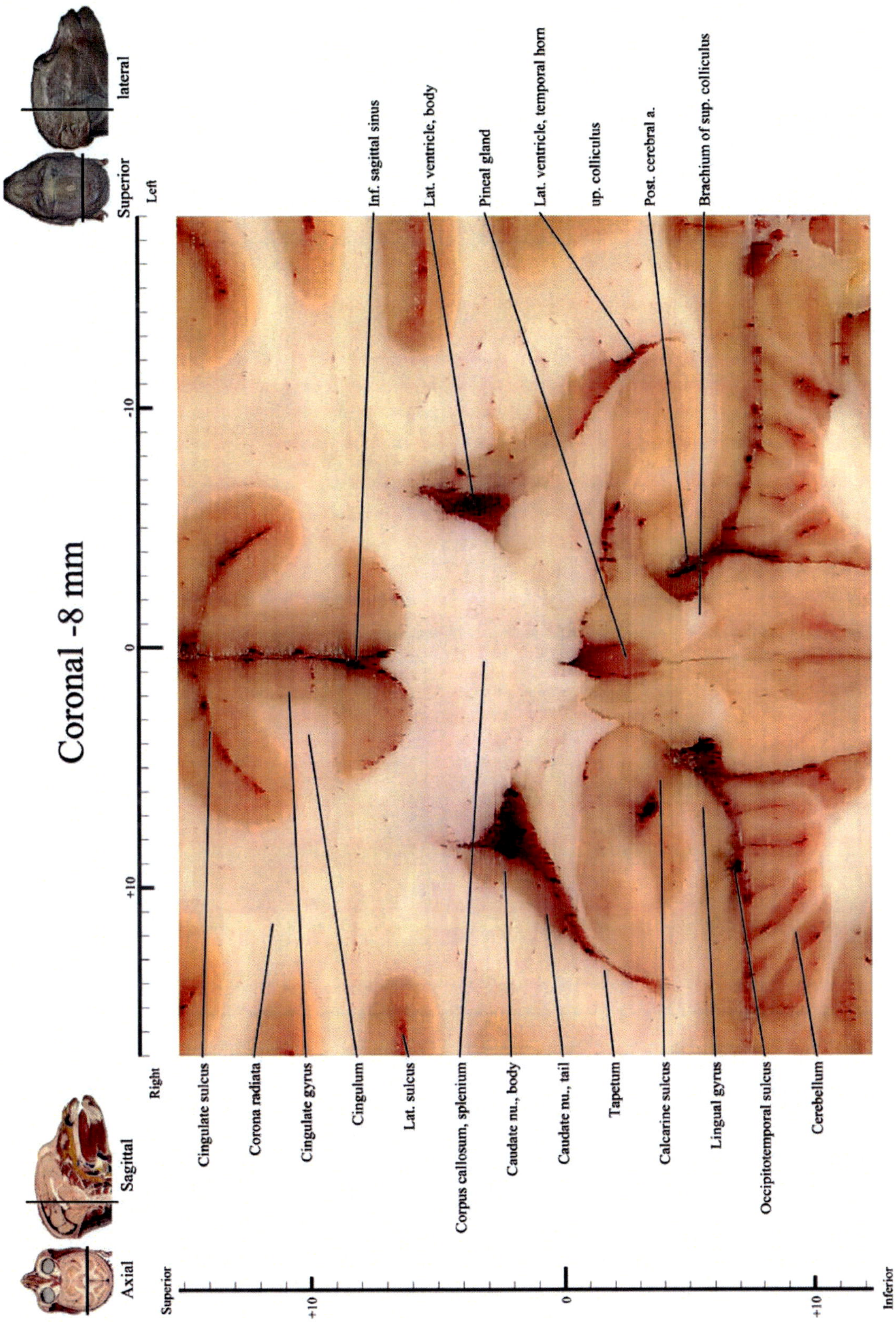
Coronal -8 mm
lateral
Superior
Left
Sagittal
Axial
Right
Superior
Inferior
-10
0
+10
Inf. sagittal sinus
Lat. ventricle, body
Pineal gland
Lat. ventricle, temporal horn
up. colliculus
Post. cerebral a.
Brachium of sup. colliculus
Cingulate sulcus
Corona radiata
Cingulate gyrus
Cingulum
Lat. sulcus
Corpus callosum, splenium
Caudate nu., body
Caudate nu., tail
Tapetum
Calcarine sulcus
Lingual gyrus
Occipitotemporal sulcus
Cerebellum

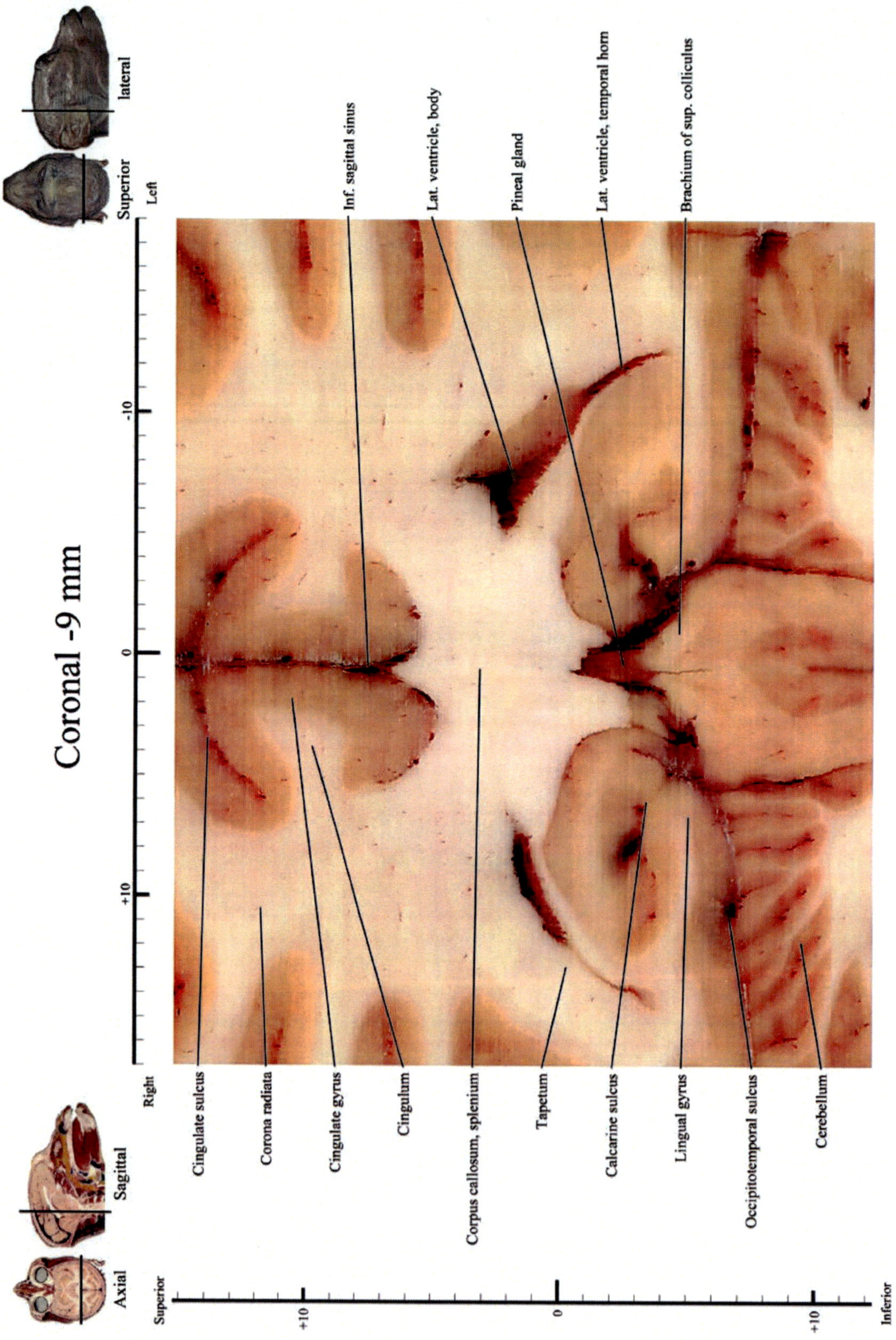
Coronal -9 mm
lateral
Superior
Left
Right
Sagittal
Axial
Superior
Inferior
-10
0
+10
Inf. sagittal sinus
Lat. ventricle, body
Pineal gland
Lat. ventricle, temporal horn
Brachium of sup. colliculus
Cingulate sulcus
Corona radiata
Cingulate gyrus
Cingulum
Corpus callosum, splenium
Tapetum
Calcarine sulcus
Lingual gyrus
Occipitotemporal sulcus
Cerebellum

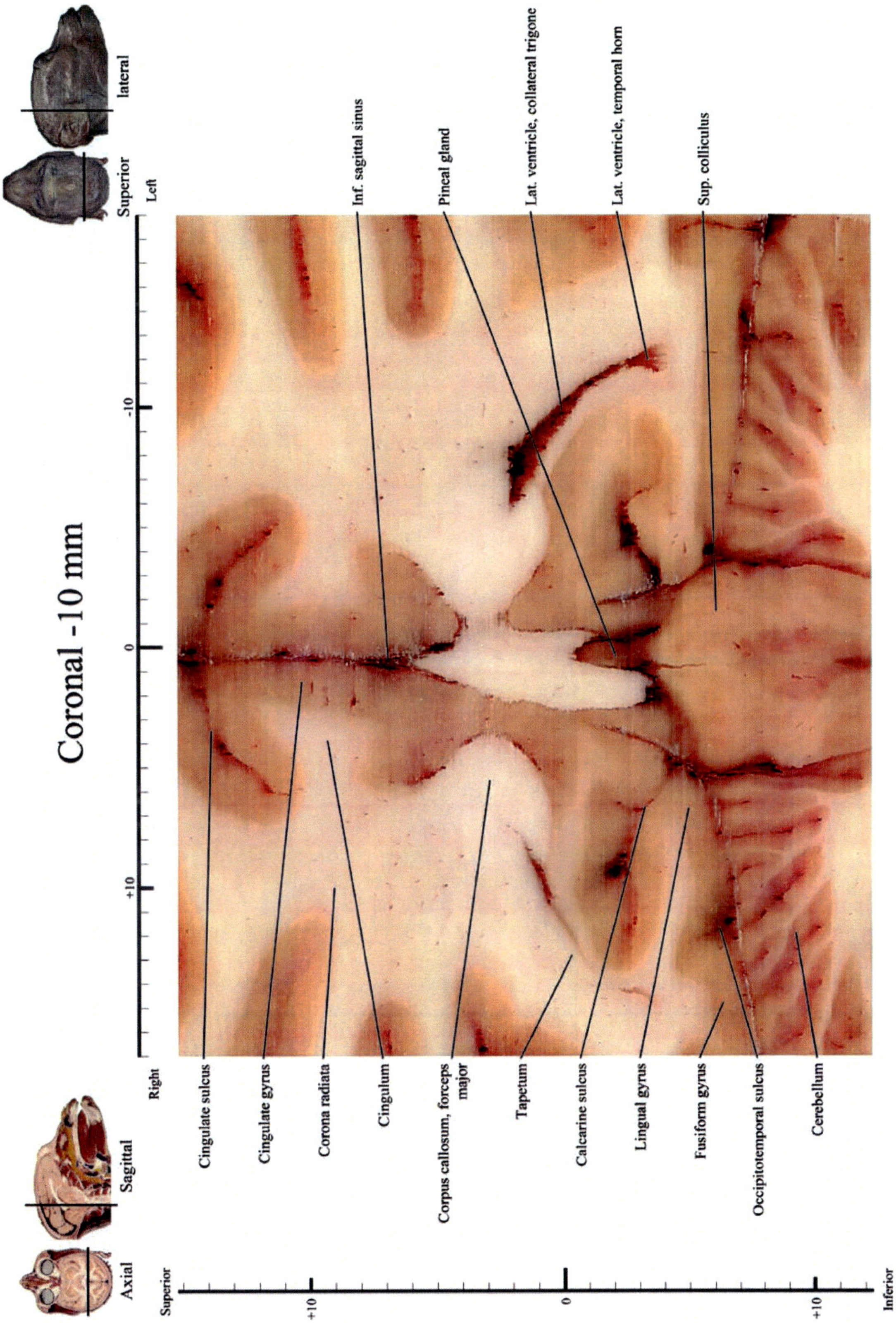
Coronal -10 mm
Superior
lateral
Sagittal
Axial
Left
Right
Inf. sagittal sinus
Pineal gland
Lat. ventricle, collateral trigone
Lat. ventricle, temporal horn
Sup. colliculus
Cingulate sulcus
Cingulate gyrus
Corona radiata
Cingulum
Corpus callosum, forceps major
Tapetum
Calcarine sulcus
Lingual gyrus
Fusiform gyrus
Occipitotemporal sulcus
Cerebellum
-10
0
+10
Superior
+10
0
+10
Inferior

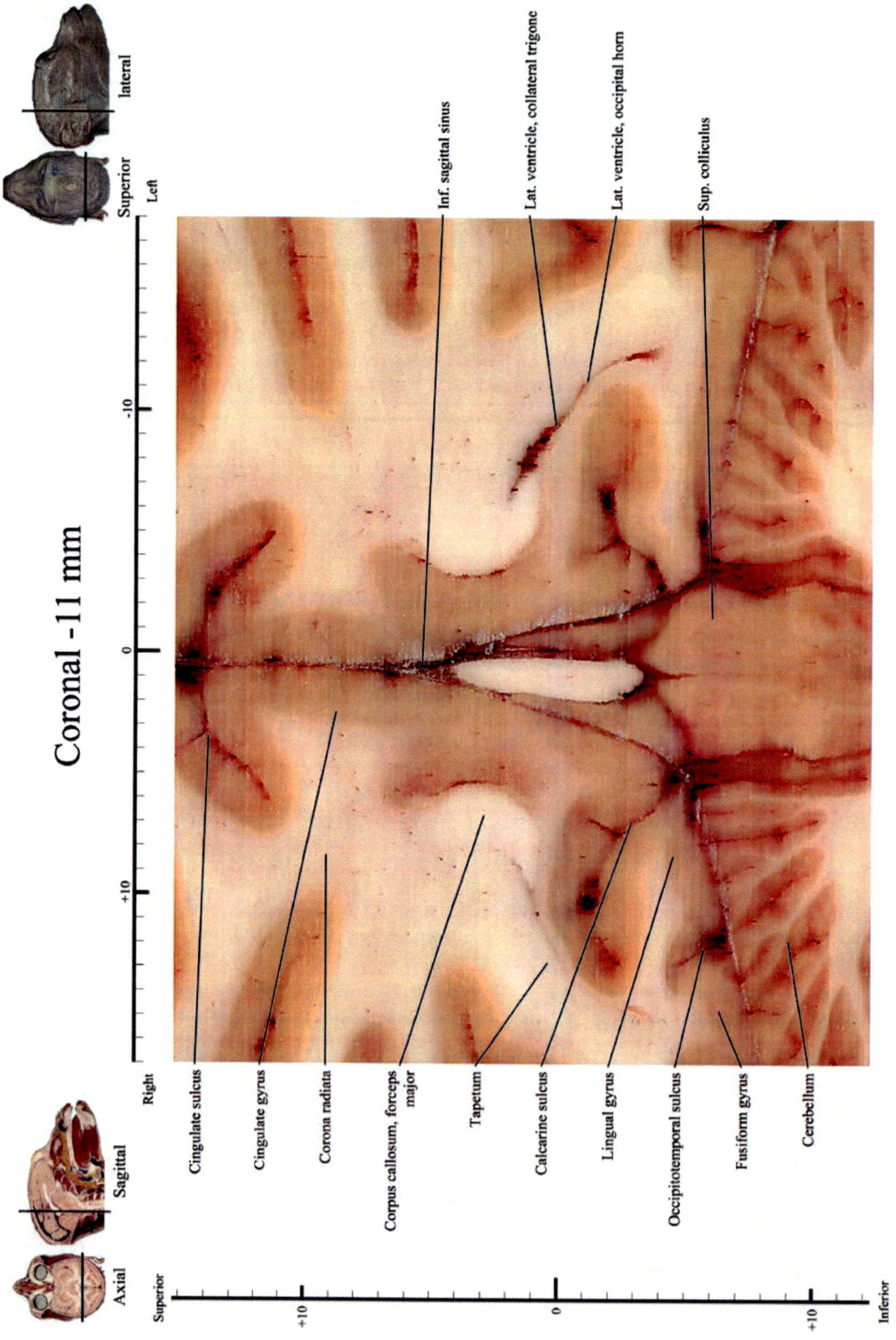
Coronal -11 mm
Superior
lateral
Left
Right
Axial
Sagittal
-10
0
+10
Superior
Inferior
+10
0
+10
Inf. sagittal sinus
Lat. ventricle, collateral trigone
Lat. ventricle, occipital horn
Sup. colliculus
Cingulate sulcus
Cingulate gyrus
Corona radiata
Corpus callosum, forceps major
Tapetum
Calcarine sulcus
Lingual gyrus
Occipitotemporal sulcus
Fusiform gyrus
Cerebellum

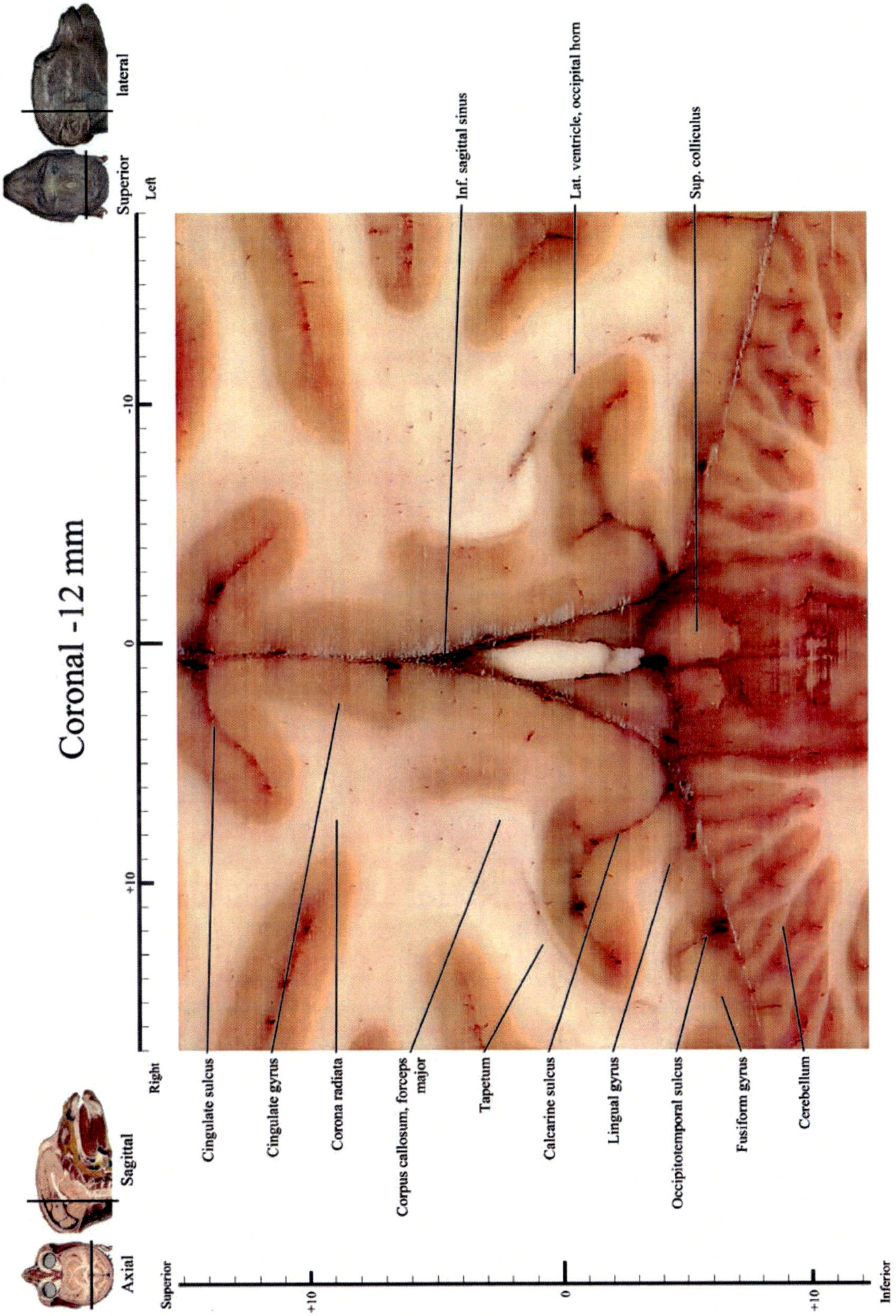
Coronal -12 mm
Superior
lateral
Left
Right
Sagittal
Axial
Superior
Inferior
-10
0
+10
Inf. sagittal sinus
Lat. ventricle, occipital horn
Sup. colliculus
Cingulate sulcus
Cingulate gyrus
Corona radiata
Corpus callosum, forceps major
Tapetum
Calcarine sulcus
Lingual gyrus
Occipitotemporal sulcus
Fusiform gyrus
Cerebellum

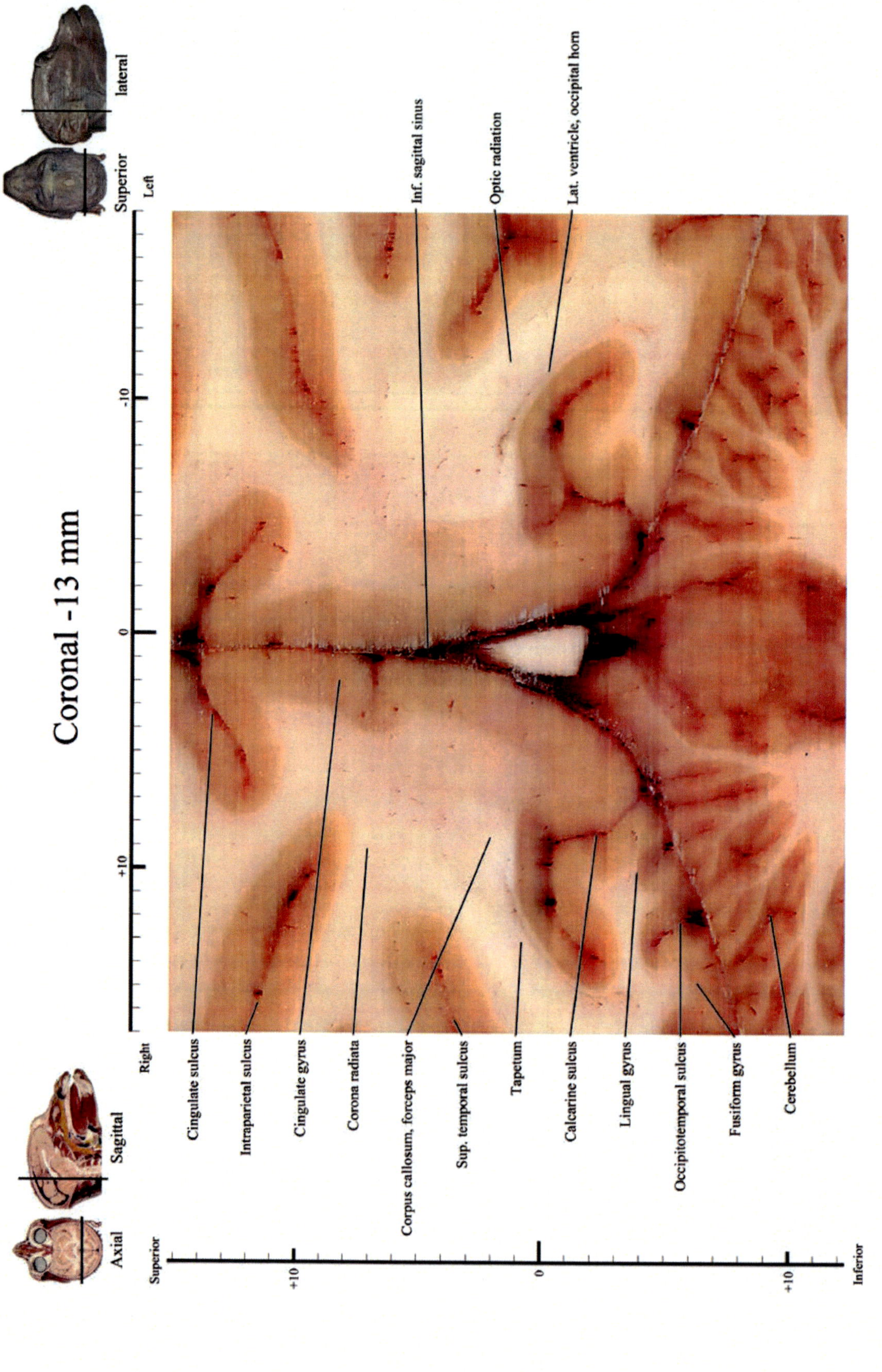
Coronal -13 mm
Axial
Sagittal
Superior
lateral
Right
Left
Superior
Inferior
+10
0
-10
Inf. sagittal sinus
Optic radiation
Lat. ventricle, occipital horn
Cingulate sulcus
Intraparietal sulcus
Cingulate gyrus
Corona radiata
Corpus callosum, forceps major
Sup. temporal sulcus
Tapetum
Calcarine sulcus
Lingual gyrus
Occipitotemporal sulcus
Fusiform gyrus
Cerebellum

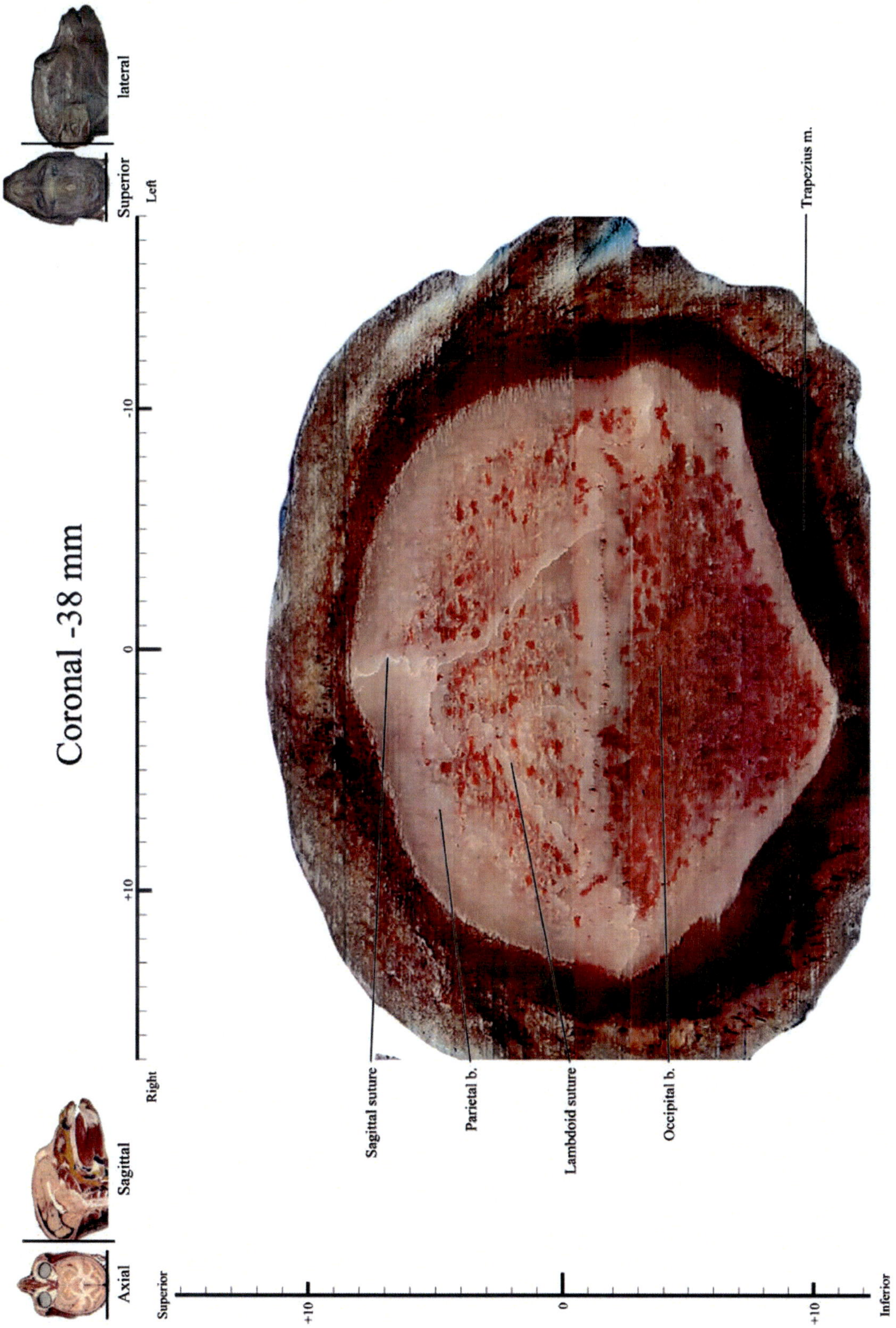
Coronal -38 mm
lateral
Superior
Left
Right
Sagittal
Axial
Superior
Inferior
-10
0
+10
Trapezius m.
Sagittal suture
Parietal b.
Lambdoid suture
Occipital b.

7 Sagittal (Expanded)

All images, 0.024 mm × 0.024 mm sized pixel

0 mm ~ −16 mm, 0.5 mm intervals
−16 mm ~ −18 mm, 1 mm intervals

J. S. Park, *Cross-Sectional Atlas of Rhesus Monkey Head*, https://doi.org/10.1007/978-981-19-3254-0_7

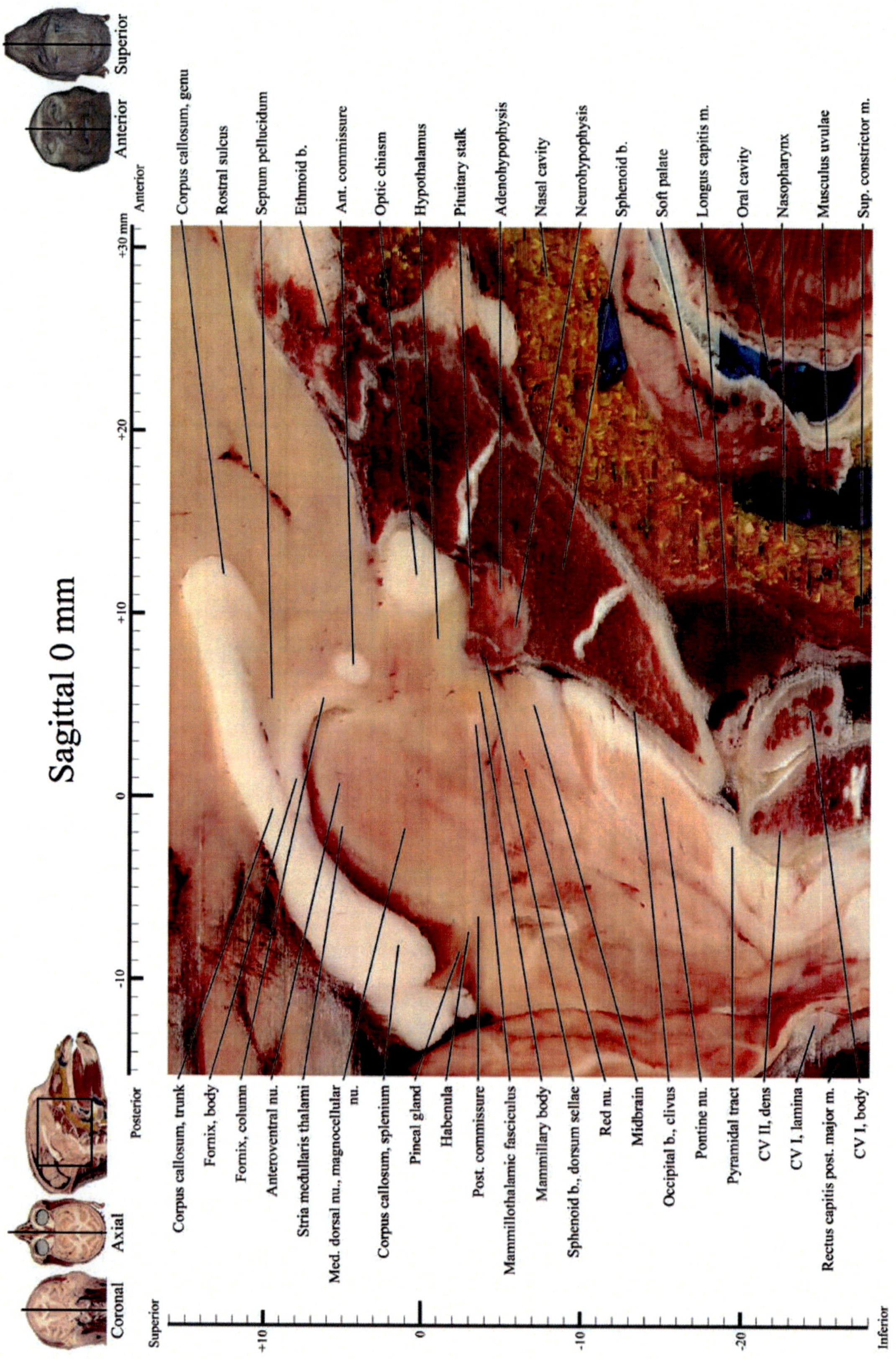
Sagittal 0 mm
Anterior
Superior
Coronal
Axial
Posterior
Inferior
+30 mm
+20
+10
0
-10
-20
Corpus callosum, genu
Rostral sulcus
Septum pellucidum
Ethmoid b.
Ant. commissure
Optic chiasm
Hypothalamus
Pituitary stalk
Adenohypophysis
Nasal cavity
Neurohypophysis
Sphenoid b.
Soft palate
Longus capitis m.
Oral cavity
Nasopharynx
Musculus uvulae
Sup. constrictor m.
Corpus callosum, trunk
Fornix, body
Fornix, column
Anteroventral nu.
Stria medullaris thalami
Med. dorsal nu., magnocellular nu.
Corpus callosum, splenium
Pineal gland
Habenula
Post. commissure
Mammillothalamic fasciculus
Mammillary body
Sphenoid b., dorsum sellae
Red nu.
Midbrain
Occipital b., clivus
Pontine nu.
Pyramidal tract
CV II, dens
CV I, lamina
Rectus capitis post. major m.
CV I, body

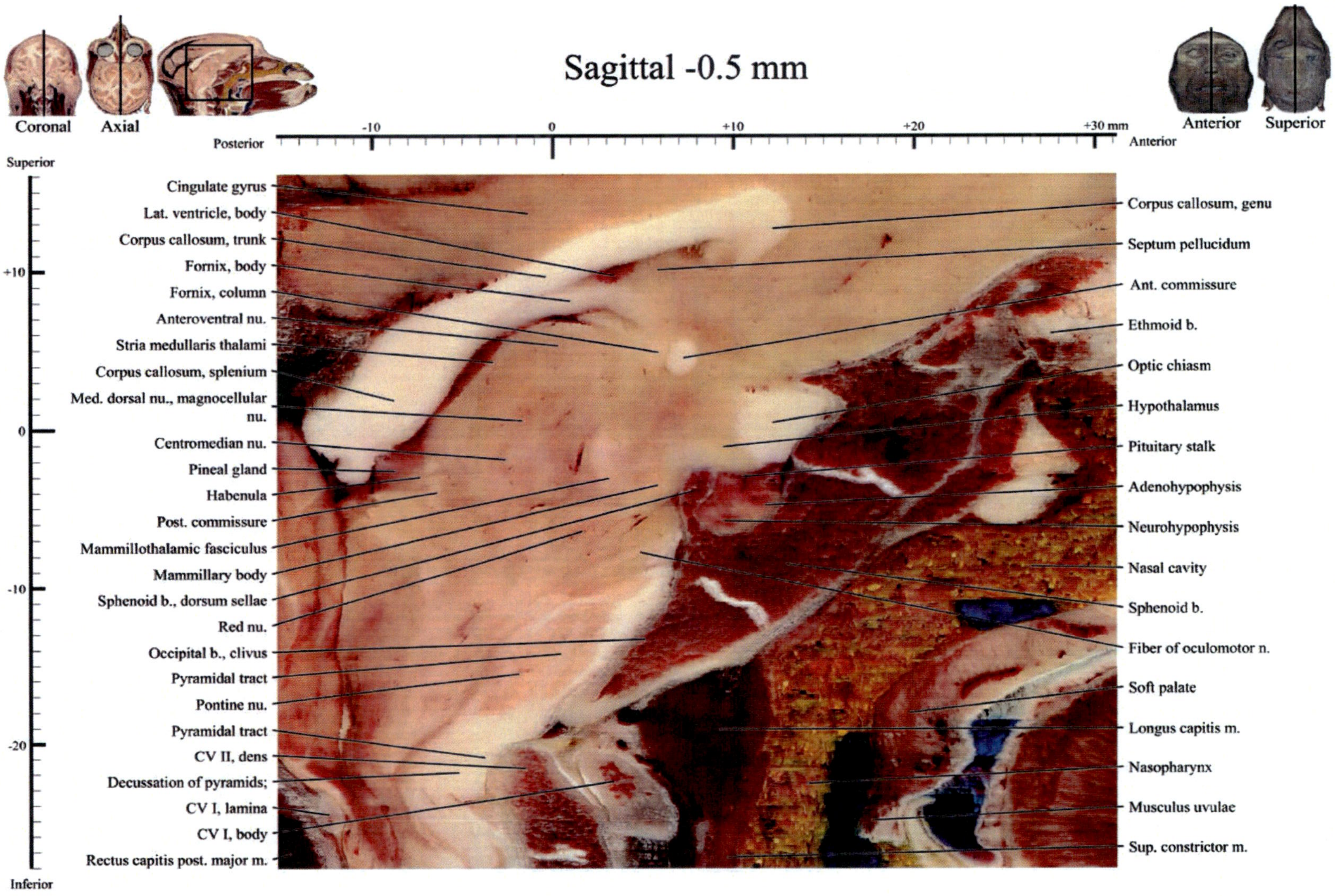
Sagittal -0.5 mm
Coronal
Axial
Anterior
Superior
Posterior
Anterior
-10
0
+10
+20
+30 mm
Superior
Inferior
+10
0
-10
-20
Cingulate gyrus
Lat. ventricle, body
Corpus callosum, trunk
Fornix, body
Fornix, column
Anteroventral nu.
Stria medullaris thalami
Corpus callosum, splenium
Med. dorsal nu., magnocellular nu.
Centromedian nu.
Pineal gland
Habenula
Post. commissure
Mammillothalamic fasciculus
Mammillary body
Sphenoid b., dorsum sellae
Red nu.
Occipital b., clivus
Pyramidal tract
Pontine nu.
Pyramidal tract
CV II, dens
Decussation of pyramids;
CV I, lamina
CV I, body
Rectus capitis post. major m.
Corpus callosum, genu
Septum pellucidum
Ant. commissure
Ethmoid b.
Optic chiasm
Hypothalamus
Pituitary stalk
Adenohypophysis
Neurohypophysis
Nasal cavity
Sphenoid b.
Fiber of oculomotor n.
Soft palate
Longus capitis m.
Nasopharynx
Musculus uvulae
Sup. constrictor m.

Sagittal -1 mm

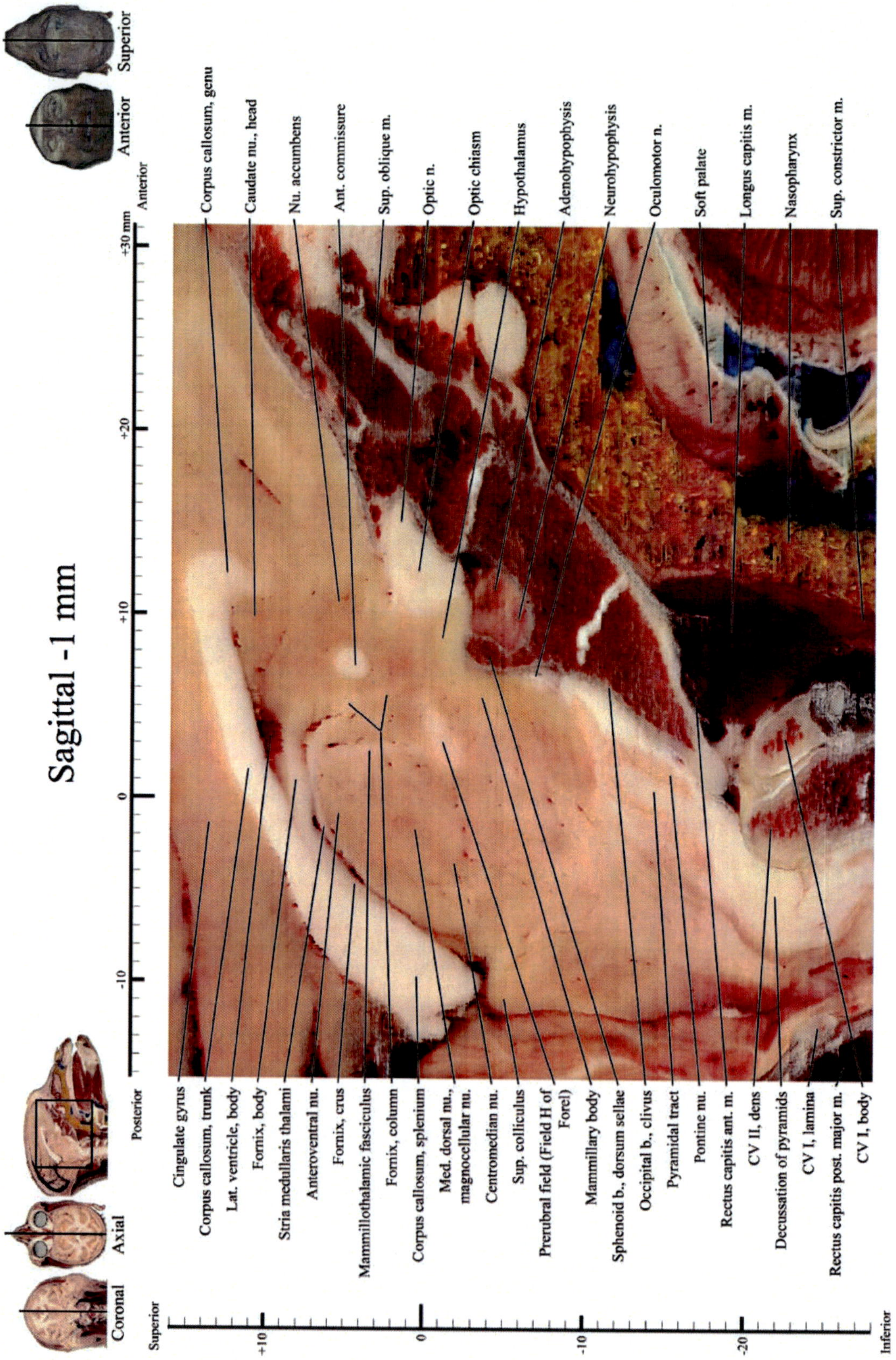

Sagittal -1.5 mm

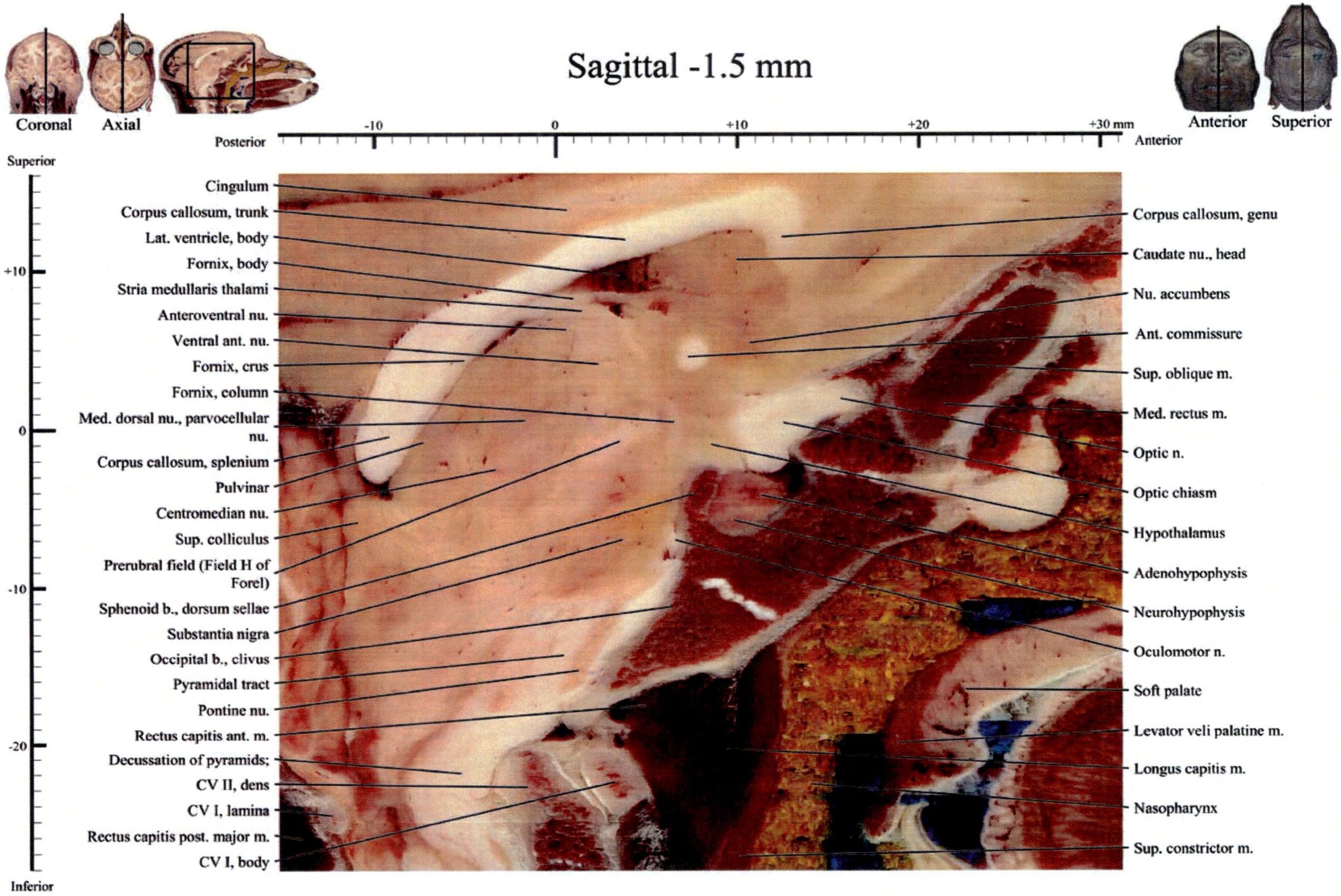

Sagittal -2 mm

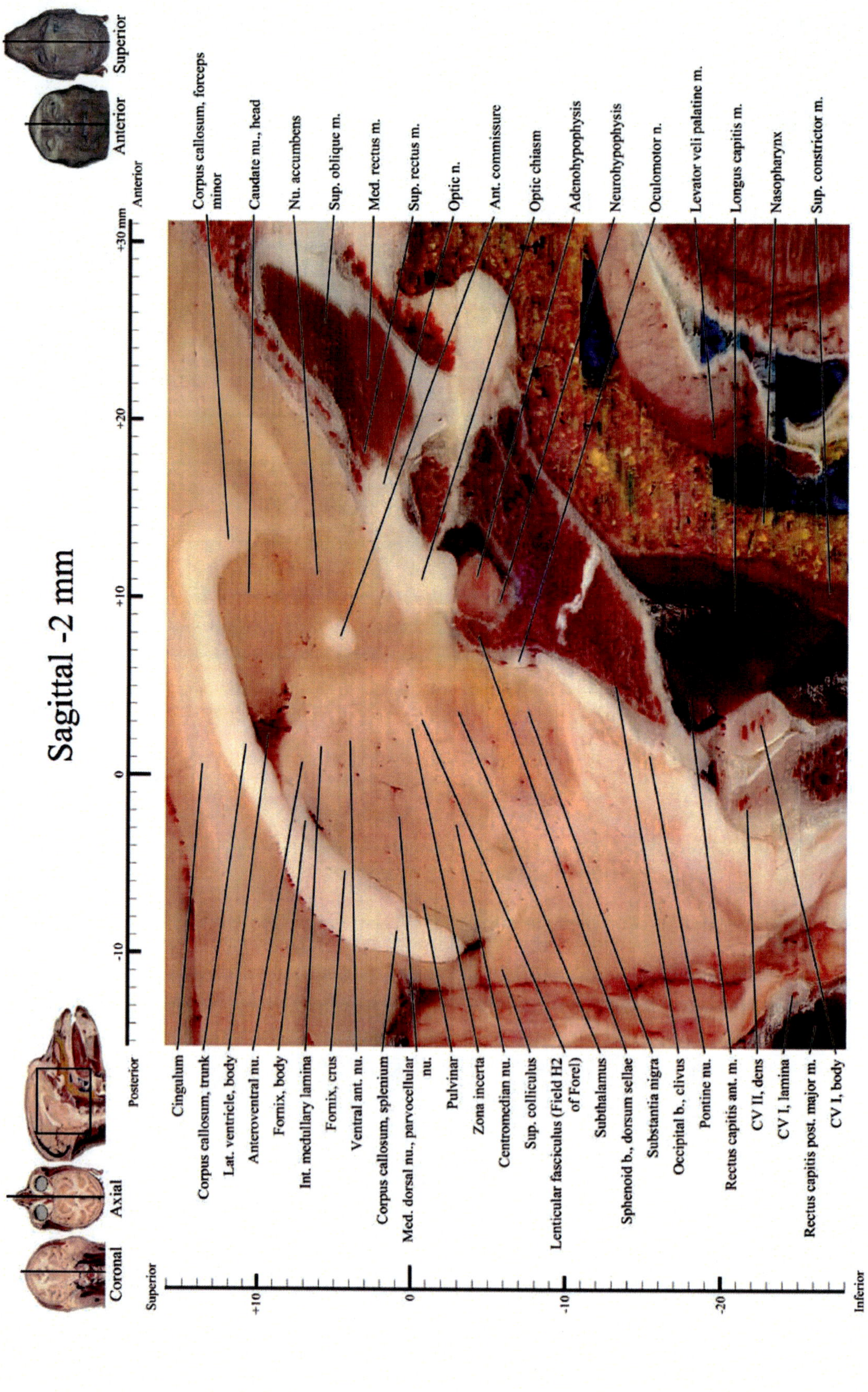

Sagittal -2.5 mm

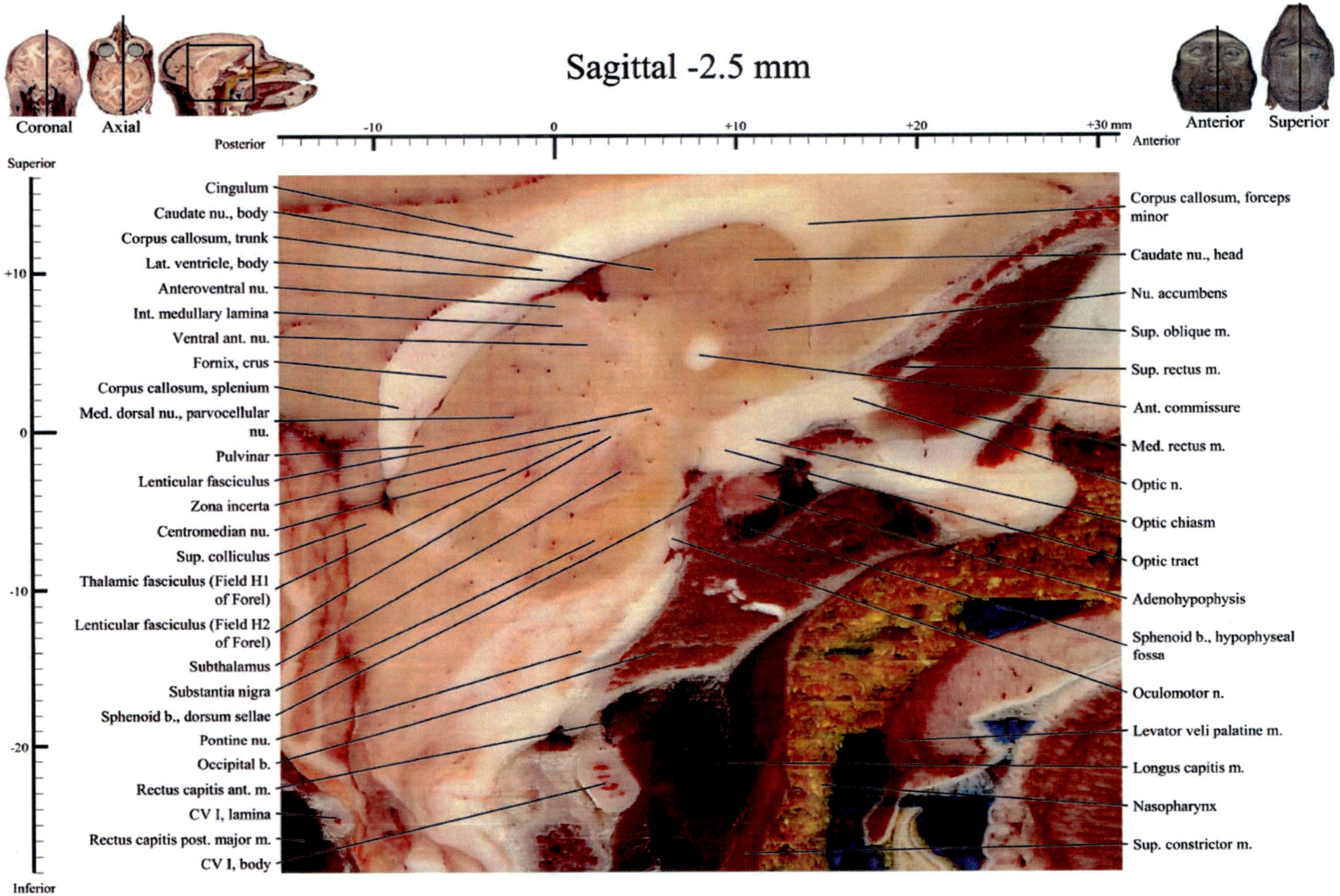

Sagittal -3 mm

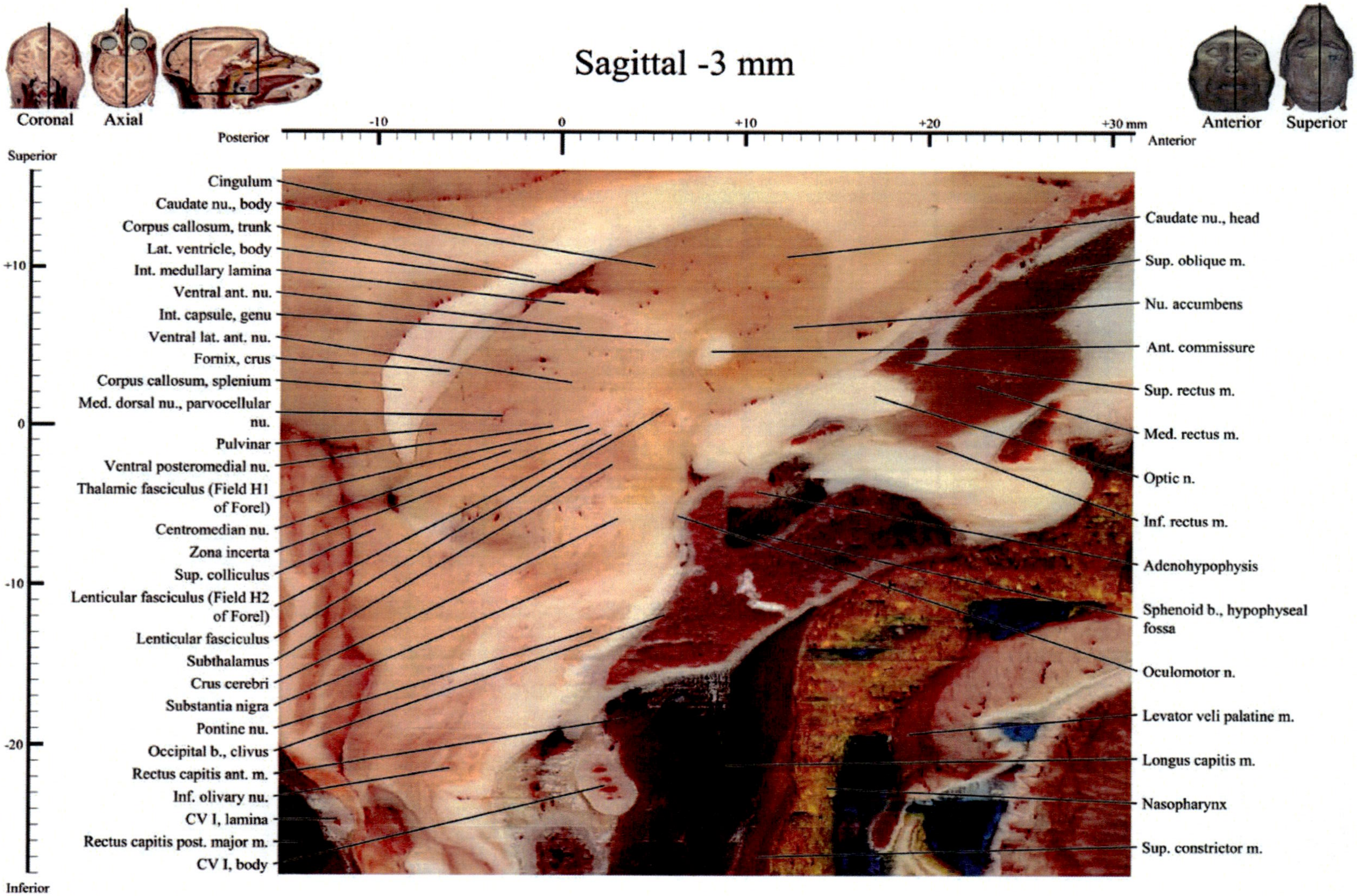

Sagittal -3.5 mm

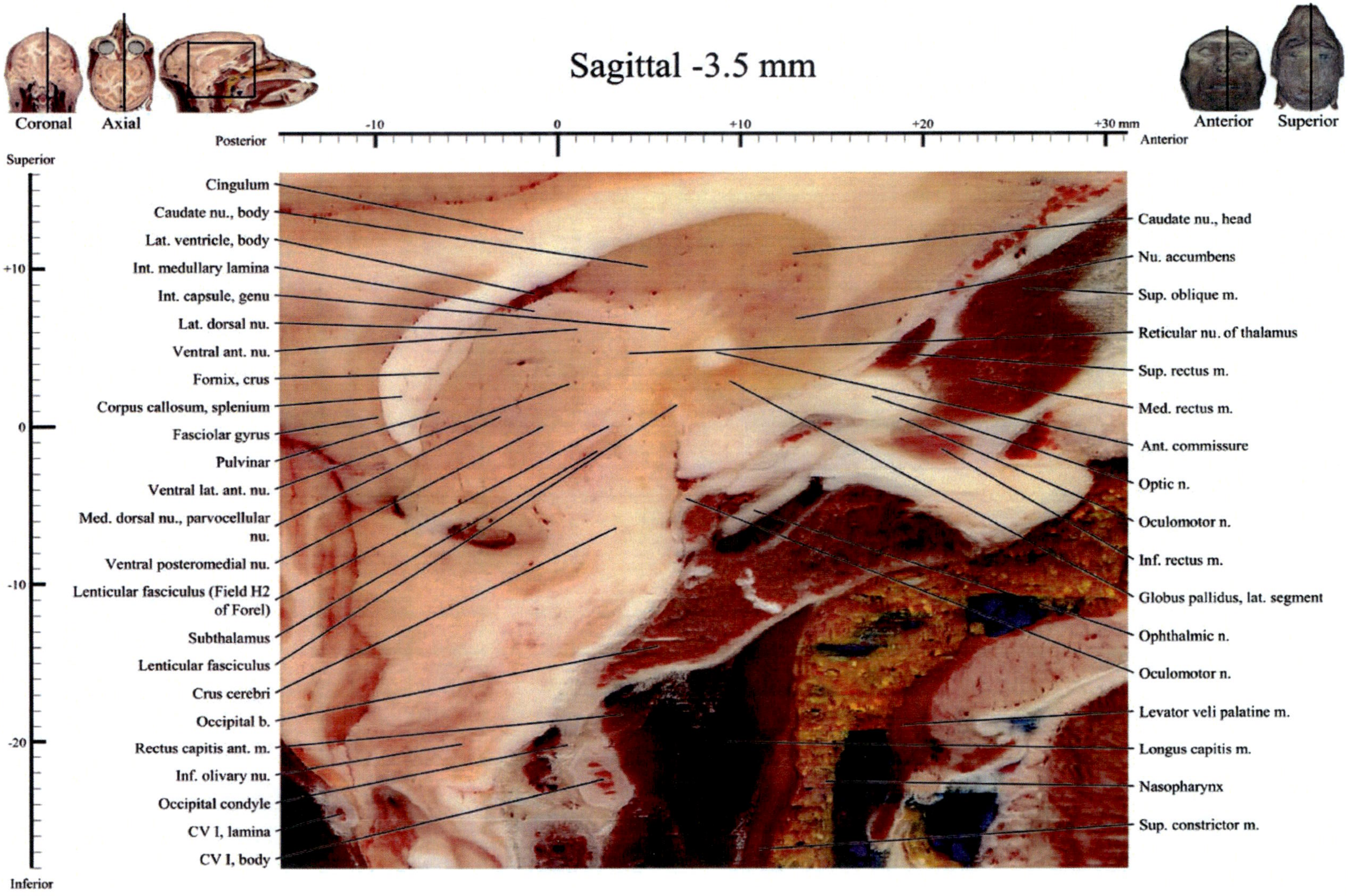

Sagittal -4 mm

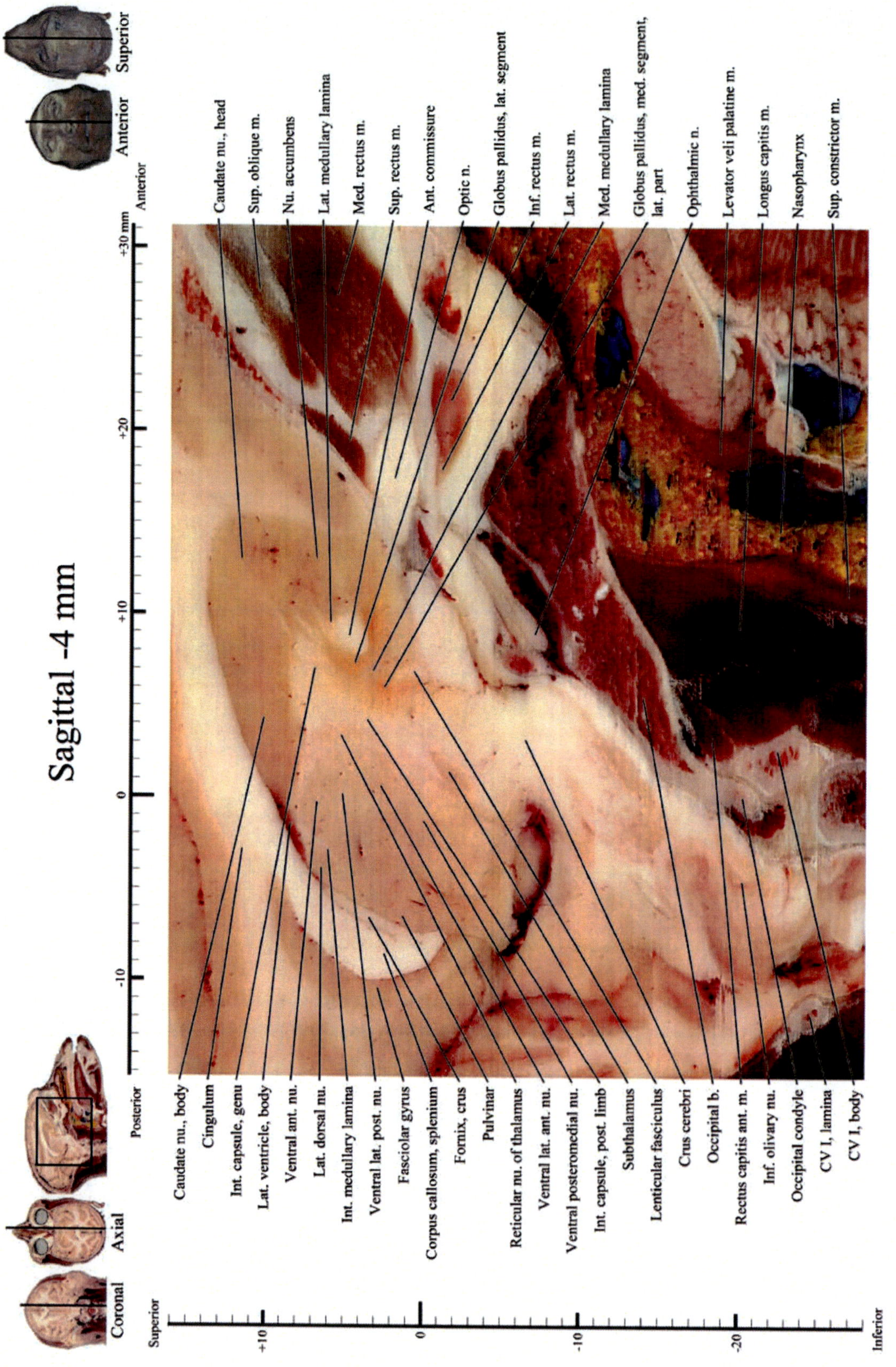

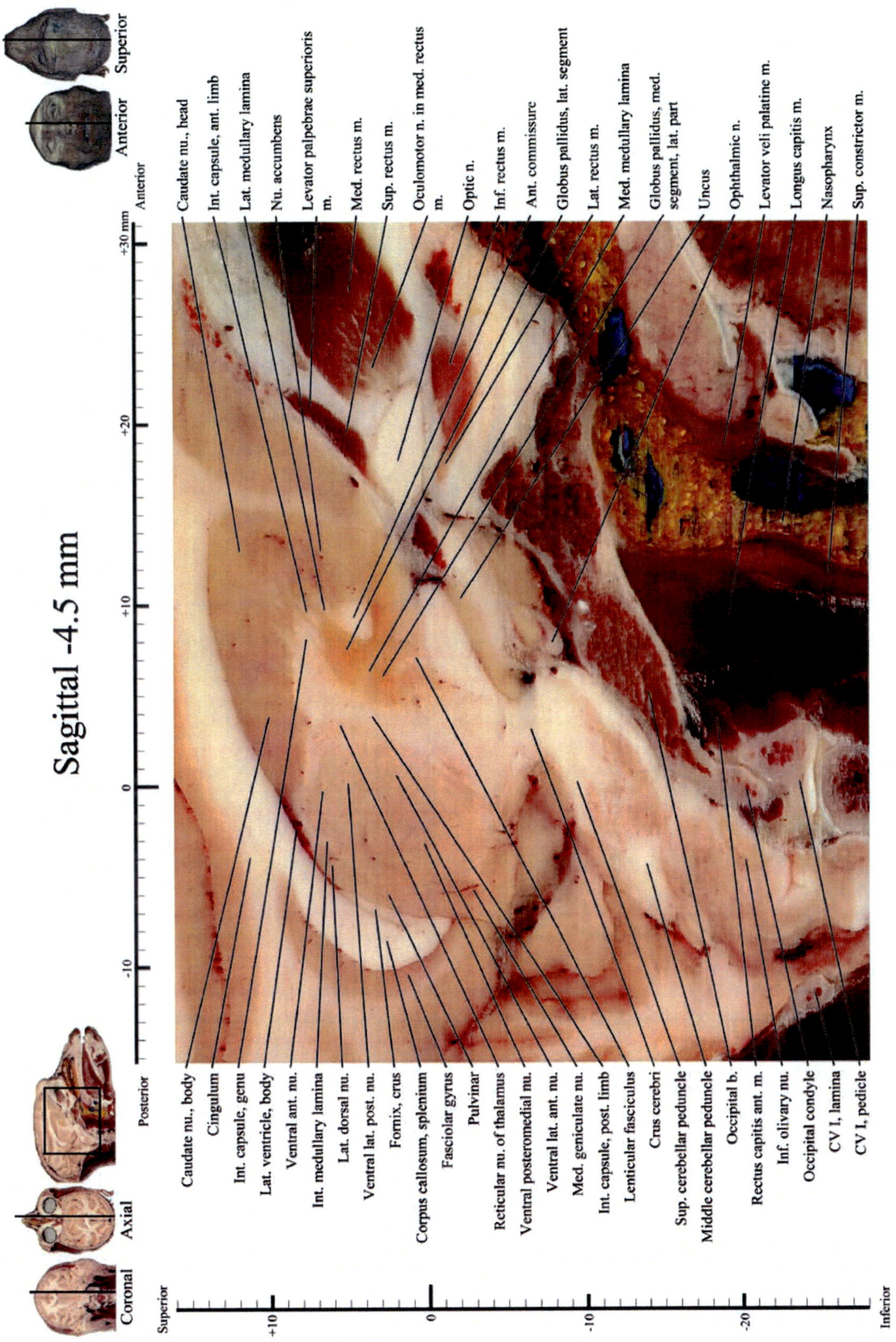
Sagittal -4.5 mm
Anterior
Superior
Coronal
Axial
Anterior
Posterior
Superior
Inferior
+30 mm
+20
+10
0
-10
-20
Caudate nu., head
Int. capsule, ant. limb
Lat. medullary lamina
Nu. accumbens
Levator palpebrae superioris m.
Med. rectus m.
Sup. rectus m.
Oculomotor n. in med. rectus m.
Optic n.
Inf. rectus m.
Ant. commissure
Globus pallidus, lat. segment
Lat. rectus m.
Med. medullary lamina
Globus pallidus, med. segment, lat. part
Uncus
Ophthalmic n.
Levator veli palatine m.
Longus capitis m.
Nasopharynx
Sup. constrictor m.
Caudate nu., body
Cingulum
Int. capsule, genu
Lat. ventricle, body
Ventral ant. nu.
Int. medullary lamina
Lat. dorsal nu.
Ventral lat. post. nu.
Fornix, crus
Corpus callosum, splenium
Fasciolar gyrus
Pulvinar
Reticular nu. of thalamus
Ventral posteromedial nu.
Ventral lat. ant. nu.
Med. geniculate nu.
Int. capsule, post. limb
Lenticular fasciculus
Crus cerebri
Sup. cerebellar peduncle
Middle cerebellar peduncle
Occipital b.
Rectus capitis ant. m.
Inf. olivary nu.
Occipital condyle
CV I, lamina
CV I, pedicle

Sagittal -5 mm

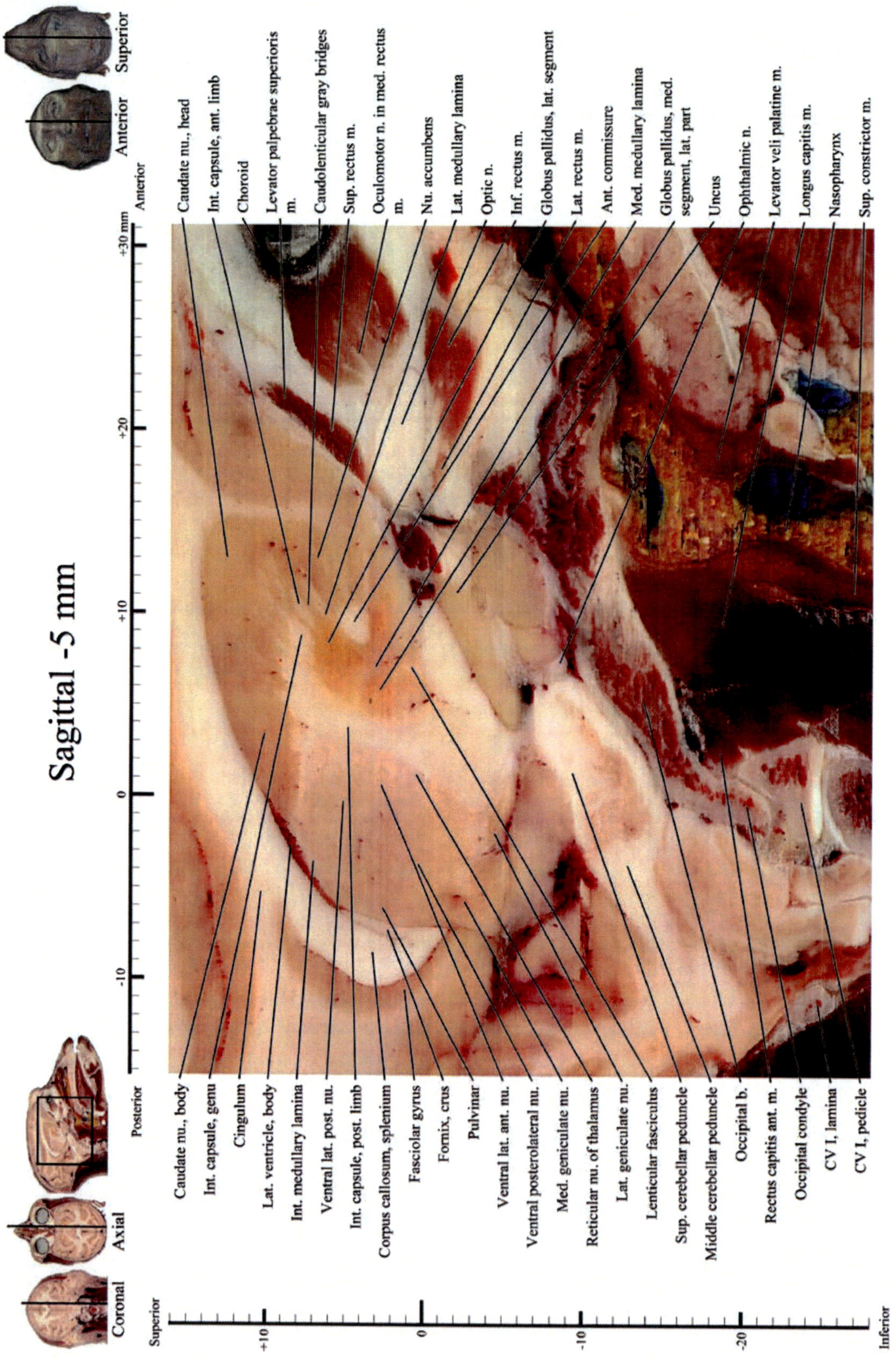

Sagittal -5.5 mm

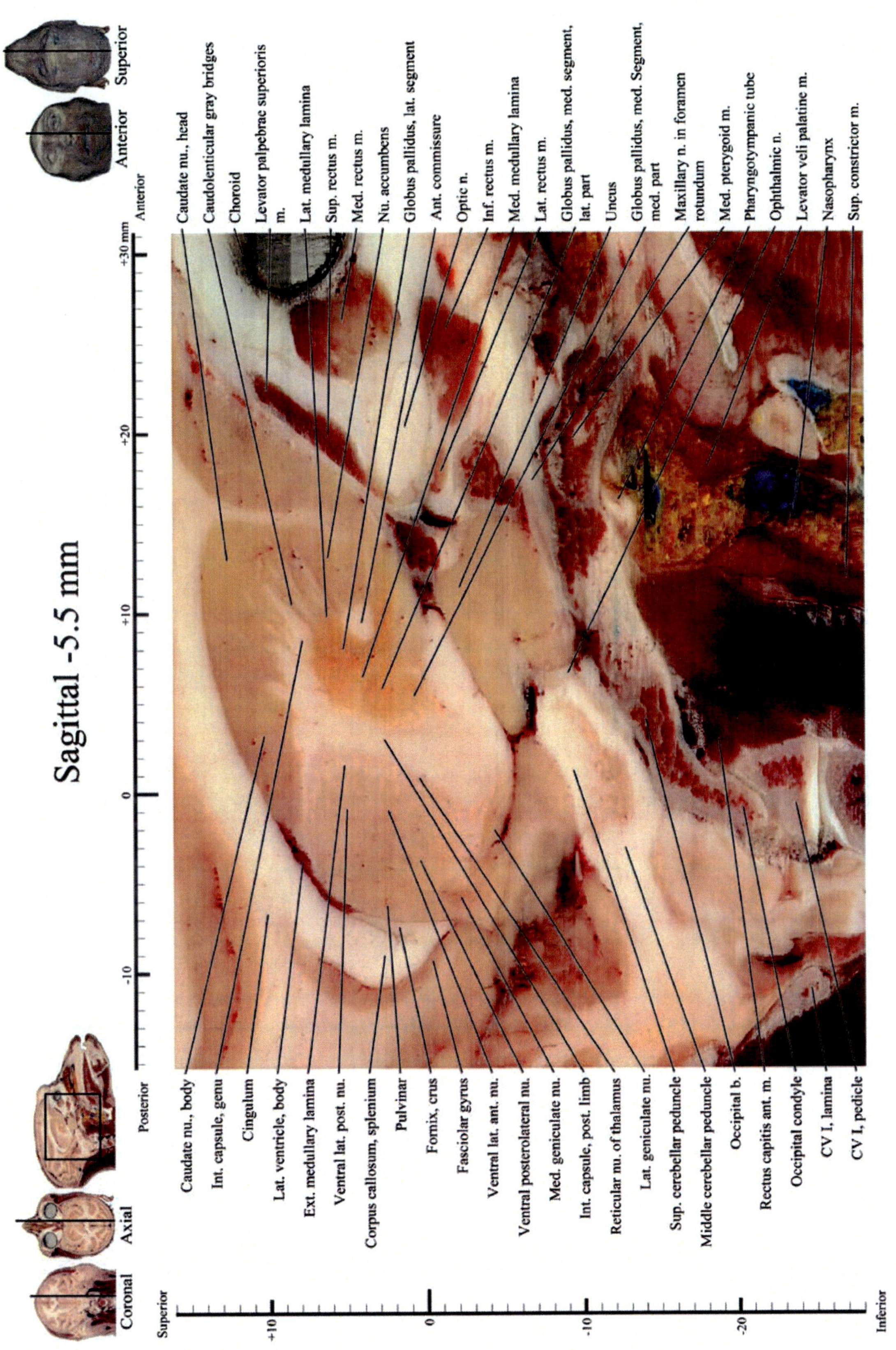

Sagittal -6 mm

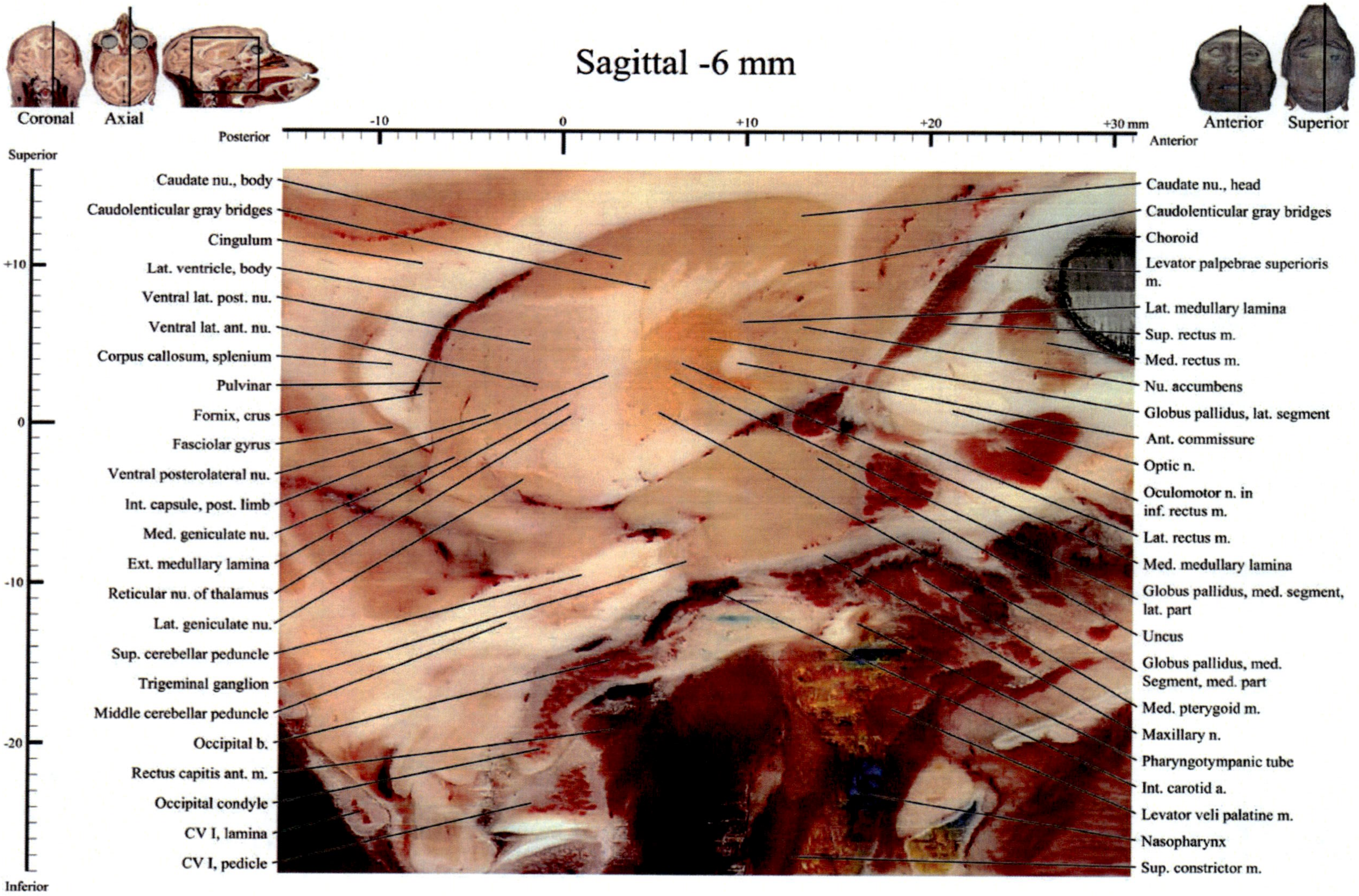

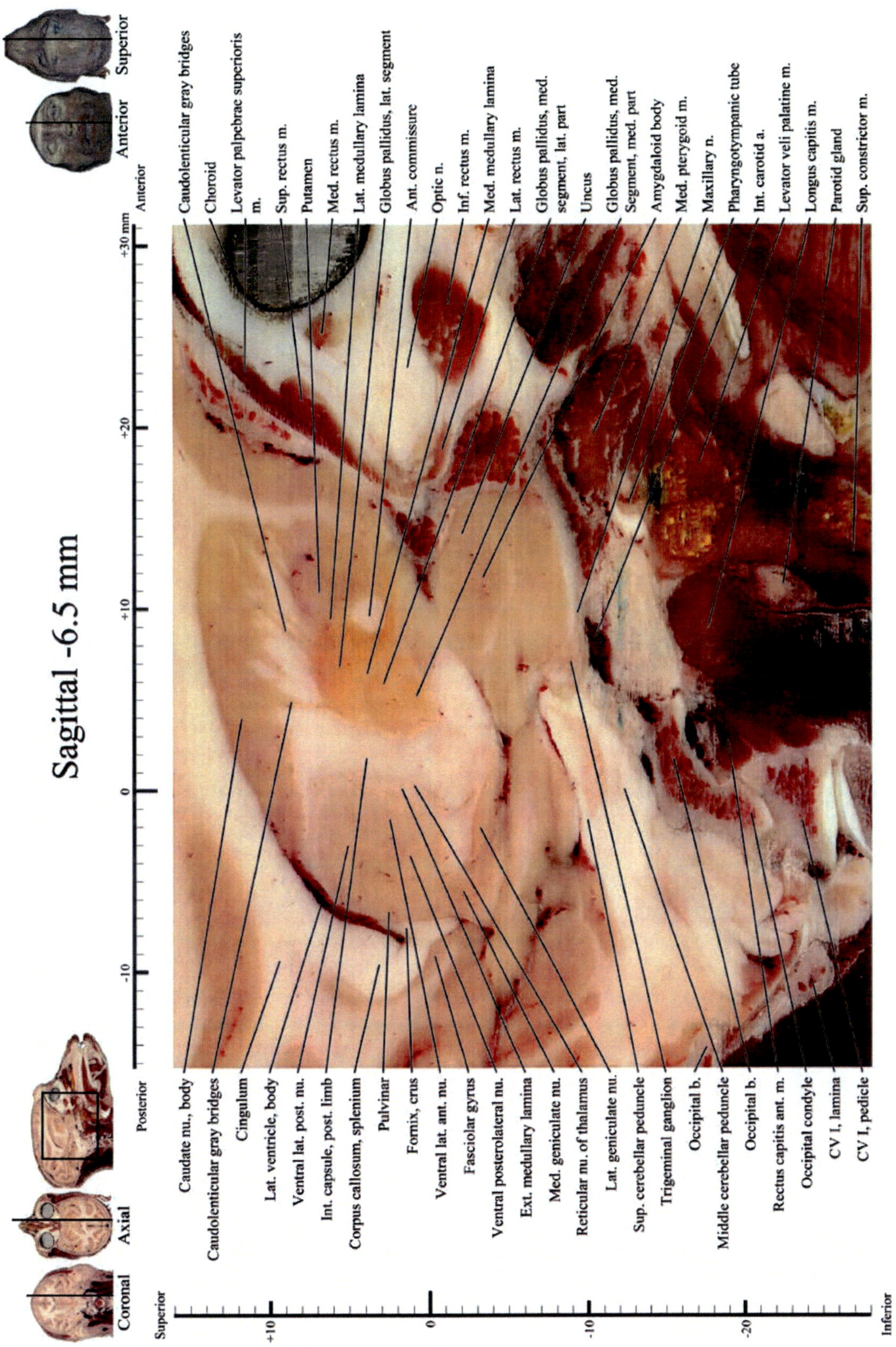
Sagittal -6.5 mm
Anterior
Superior
Coronal
Axial
Anterior
Posterior
Superior
Inferior
+30 mm
+20
+10
0
-10
+10
0
-10
-20
Caudolenticular gray bridges
Choroid
Levator palpebrae superioris m.
Sup. rectus m.
Putamen
Med. rectus m.
Lat. medullary lamina
Globus pallidus, lat. segment
Ant. commissure
Optic n.
Inf. rectus m.
Med. medullary lamina
Lat. rectus m.
Globus pallidus, med. segment, lat. part
Uncus
Globus pallidus, med. Segment, med. part
Amygdaloid body
Med. pterygoid m.
Maxillary n.
Pharyngotympanic tube
Int. carotid a.
Levator veli palatine m.
Longus capitis m.
Parotid gland
Sup. constrictor m.
Caudate nu., body
Caudolenticular gray bridges
Cingulum
Lat. ventricle, body
Ventral lat. post. nu.
Int. capsule, post. limb
Corpus callosum, splenium
Pulvinar
Fornix, crus
Ventral lat. ant. nu.
Fasciolar gyrus
Ventral posterolateral nu.
Ext. medullary lamina
Med. geniculate nu.
Reticular nu. of thalamus
Lat. geniculate nu.
Sup. cerebellar peduncle
Trigeminal ganglion
Occipital b.
Middle cerebellar peduncle
Occipital b.
Rectus capitis ant. m.
Occipital condyle
CV I, lamina
CV I, pedicle

Sagittal -7 mm

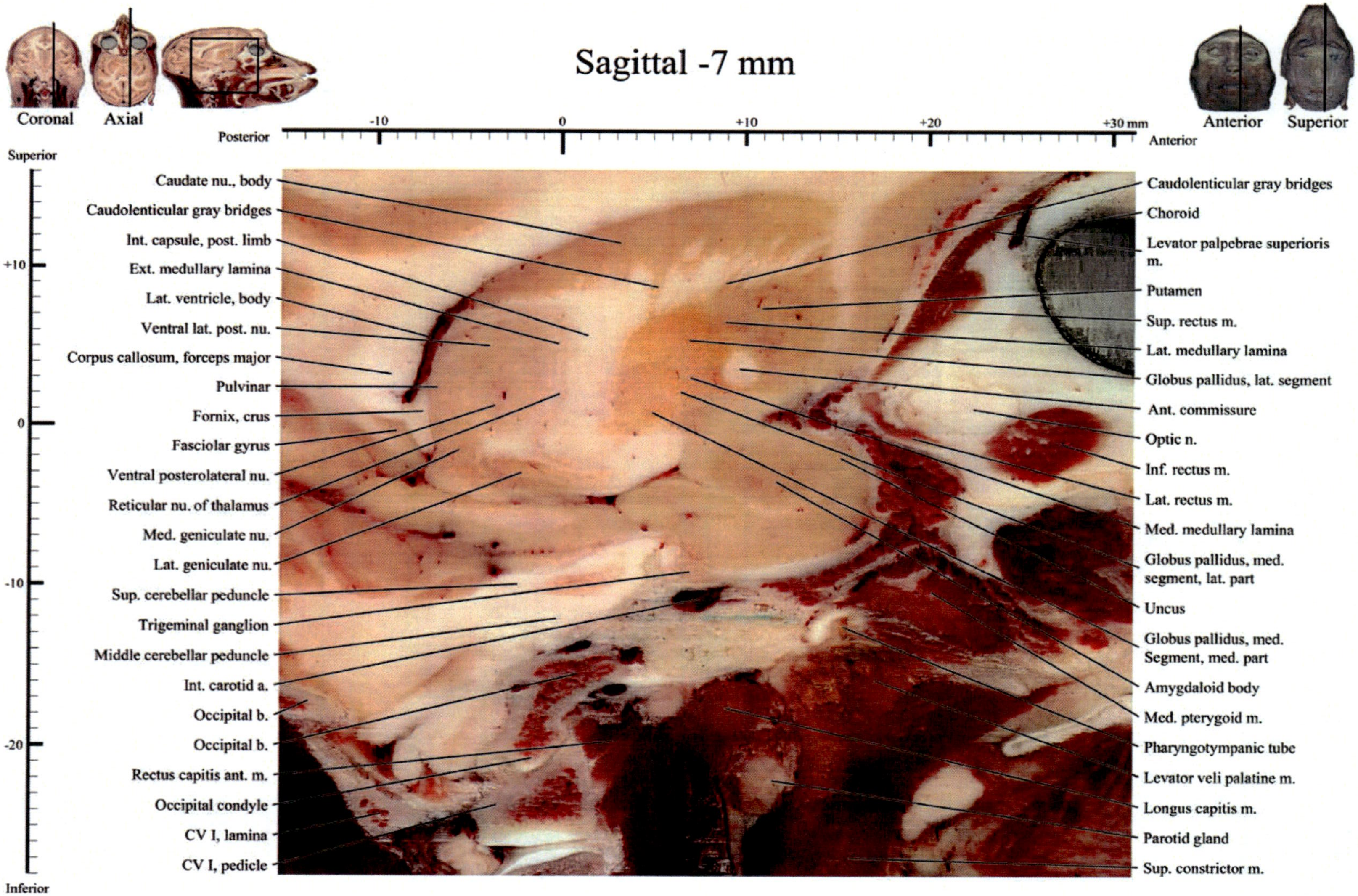

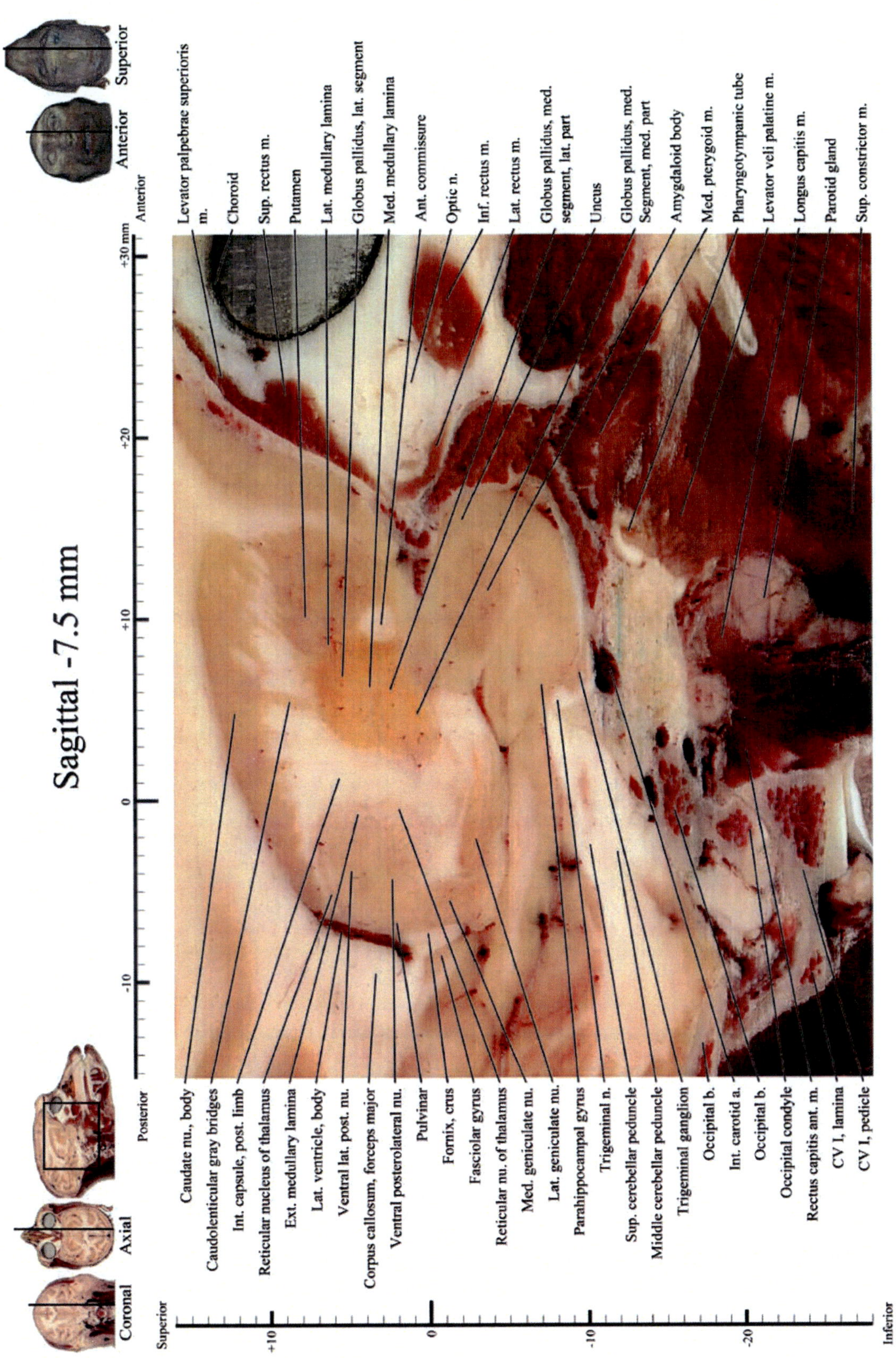
Sagittal -7.5 mm
Coronal
Axial
Anterior
Superior
Posterior
Inferior
+30 mm
+20
+10
0
-10
-20
Levator palpebrae superioris m.
Choroid
Sup. rectus m.
Putamen
Lat. medullary lamina
Globus pallidus, lat. segment
Med. medullary lamina
Ant. commissure
Optic n.
Inf. rectus m.
Lat. rectus m.
Globus pallidus, med. segment, lat. part
Uncus
Globus pallidus, med. Segment, med. part
Amygdaloid body
Med. pterygoid m.
Pharyngotympanic tube
Levator veli palatine m.
Longus capitis m.
Parotid gland
Sup. constrictor m.
Caudate nu., body
Caudolenticular gray bridges
Int. capsule, post. limb
Reticular nucleus of thalamus
Ext. medullary lamina
Lat. ventricle, body
Ventral lat. post. nu.
Corpus callosum, forceps major
Ventral posterolateral nu.
Pulvinar
Fornix, crus
Fasciolar gyrus
Reticular nu. of thalamus
Med. geniculate nu.
Lat. geniculate nu.
Parahippocampal gyrus
Trigeminal n.
Sup. cerebellar peduncle
Middle cerebellar peduncle
Trigeminal ganglion
Occipital b.
Int. carotid a.
Occipital b.
Occipital condyle
Rectus capitis ant. m.
CV I, lamina
CV I, pedicle

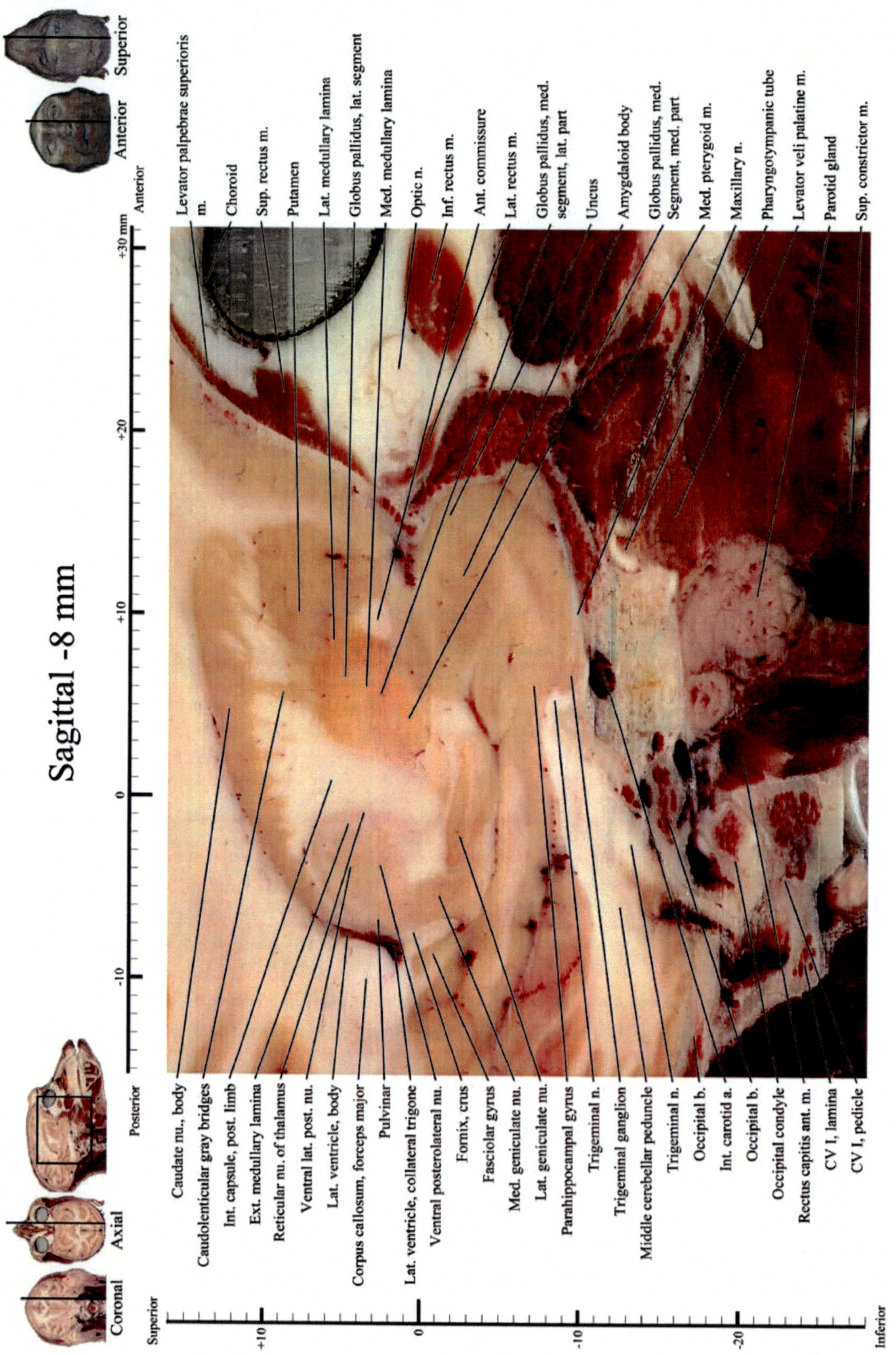
Sagittal -8 mm
Coronal
Axial
Anterior
Superior
Posterior
Anterior
+30 mm
+20
+10
0
-10
Superior
+10
0
-10
-20
Inferior
Levator palpebrae superioris m.
Choroid
Sup. rectus m.
Putamen
Lat. medullary lamina
Globus pallidus, lat. segment
Med. medullary lamina
Optic n.
Inf. rectus m.
Ant. commissure
Lat. rectus m.
Globus pallidus, med. segment, lat. part
Uncus
Amygdaloid body
Globus pallidus, med. Segment, med. part
Med. pterygoid m.
Maxillary n.
Pharyngotympanic tube
Levator veli palatine m.
Parotid gland
Sup. constrictor m.
Caudate nu., body
Caudolenticular gray bridges
Int. capsule, post. limb
Ext. medullary lamina
Reticular nu. of thalamus
Ventral lat. post. nu.
Lat. ventricle, body
Corpus callosum, forceps major
Pulvinar
Lat. ventricle, collateral trigone
Ventral posterolateral nu.
Fornix, crus
Fasciolar gyrus
Med. geniculate nu.
Lat. geniculate nu.
Parahippocampal gyrus
Trigeminal n.
Trigeminal ganglion
Middle cerebellar peduncle
Trigeminal n.
Occipital b.
Int. carotid a.
Occipital b.
Occipital condyle
Rectus capitis ant. m.
CV I, lamina
CV I, pedicle

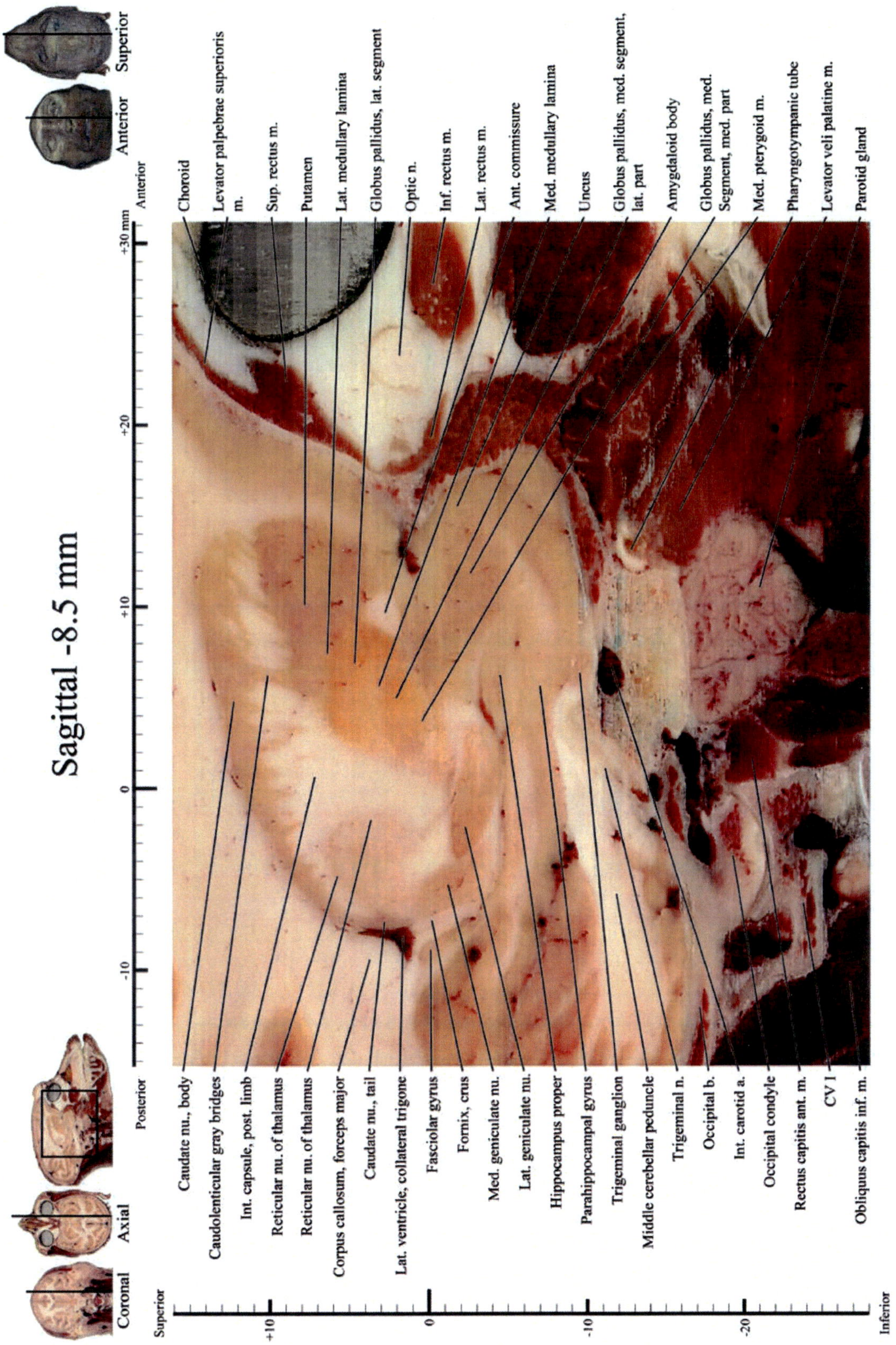

Sagittal -8.5 mm
Coronal
Axial
Anterior
Superior
Anterior
Posterior
Superior
Inferior
+30 mm
+20
+10
0
-10
+10
0
-10
-20
Choroid
Levator palpebrae superioris m.
Sup. rectus m.
Putamen
Lat. medullary lamina
Globus pallidus, lat. segment
Optic n.
Inf. rectus m.
Lat. rectus m.
Ant. commissure
Med. medullary lamina
Uncus
Globus pallidus, med. segment, lat. part
Amygdaloid body
Globus pallidus, med. Segment, med. part
Med. pterygoid m.
Pharyngotympanic tube
Levator veli palatine m.
Paroid gland
Caudate nu., body
Caudolenticular gray bridges
Int. capsule, post. limb
Reticular nu. of thalamus
Reticular nu. of thalamus
Corpus callosum, forceps major
Caudate nu., tail
Lat. ventricle, collateral trigone
Fasciolar gyrus
Fornix, crus
Med. geniculate nu.
Lat. geniculate nu.
Hippocampus proper
Parahippocampal gyrus
Trigeminal ganglion
Middle cerebellar peduncle
Trigeminal n.
Occipital b.
Int. carotid a.
Occipital condyle
Rectus capitis ant. m.
CV I
Obliquus capitis inf. m.

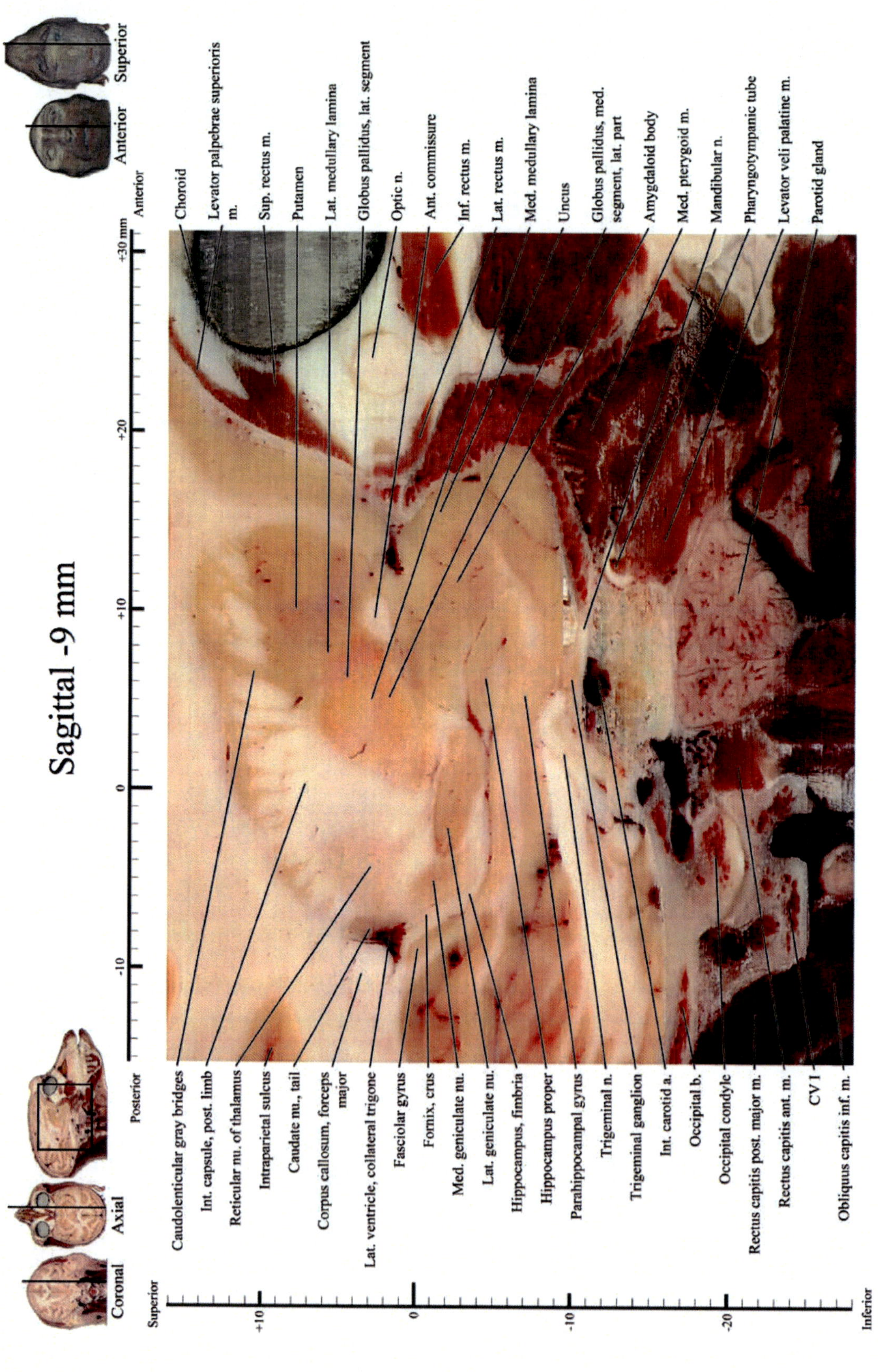
Sagittal -9 mm
Coronal
Axial
Anterior
Superior
Anterior
Posterior
+30 mm
+20
+10
0
-10
Superior
+10
0
-10
-20
Inferior
Choroid
Levator palpebrae superioris m.
Sup. rectus m.
Putamen
Lat. medullary lamina
Globus pallidus, lat. segment
Optic n.
Ant. commissure
Inf. rectus m.
Lat. rectus m.
Med. medullary lamina
Uncus
Globus pallidus, med. segment, lat. part
Amygdaloid body
Med. pterygoid m.
Mandibular n.
Pharyngotympanic tube
Levator veli palatine m.
Parotid gland
Caudolenticular gray bridges
Int. capsule, post. limb
Reticular nu. of thalamus
Intraparietal sulcus
Caudate nu., tail
Corpus callosum, forceps major
Lat. ventricle, collateral trigone
Fasciolar gyrus
Fornix, crus
Med. geniculate nu.
Lat. geniculate nu.
Hippocampus, fimbria
Hippocampus proper
Parahippocampal gyrus
Trigeminal n.
Trigeminal ganglion
Int. carotid a.
Occipital b.
Occipital condyle
Rectus capitis post. major m.
Rectus capitis ant. m.
CV I
Obliquus capitis inf. m.

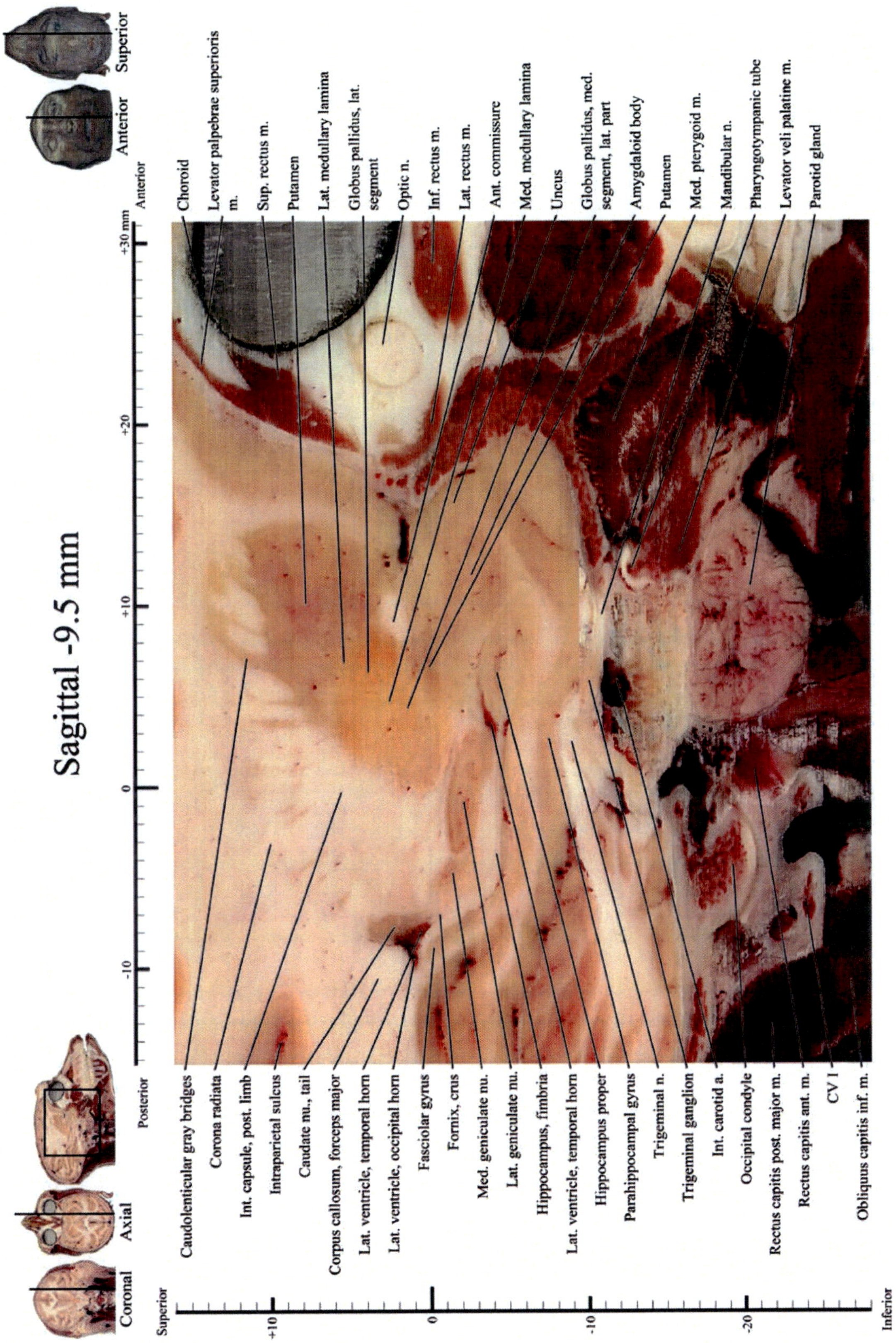

Sagittal -9.5 mm
Coronal
Axial
Anterior
Superior
Posterior
Inferior
+30 mm
+20
+10
0
-10
-20
Choroid
Levator palpebrae superioris m.
Sup. rectus m.
Putamen
Lat. medullary lamina
Globus pallidus, lat. segment
Optic n.
Inf. rectus m.
Lat. rectus m.
Ant. commissure
Med. medullary lamina
Uncus
Globus pallidus, med. segment, lat. part
Amygdaloid body
Putamen
Med. pterygoid m.
Mandibular n.
Pharyngotympanic tube
Levator veli palatine m.
Parotid gland
Caudolenticular gray bridges
Corona radiata
Int. capsule, post. limb
Intraparietal sulcus
Caudate nu., tail
Corpus callosum, forceps major
Lat. ventricle, temporal horn
Lat. ventricle, occipital horn
Fasciolar gyrus
Fornix, crus
Med. geniculate nu.
Lat. geniculate nu.
Hippocampus, fimbria
Lat. ventricle, temporal horn
Hippocampus proper
Parahippocampal gyrus
Trigeminal n.
Trigeminal ganglion
Int. carotid a.
Occipital condyle
Rectus capitis post. major m.
Rectus capitis ant. m.
CV I
Obliquus capitis inf. m.

Sagittal -10 mm

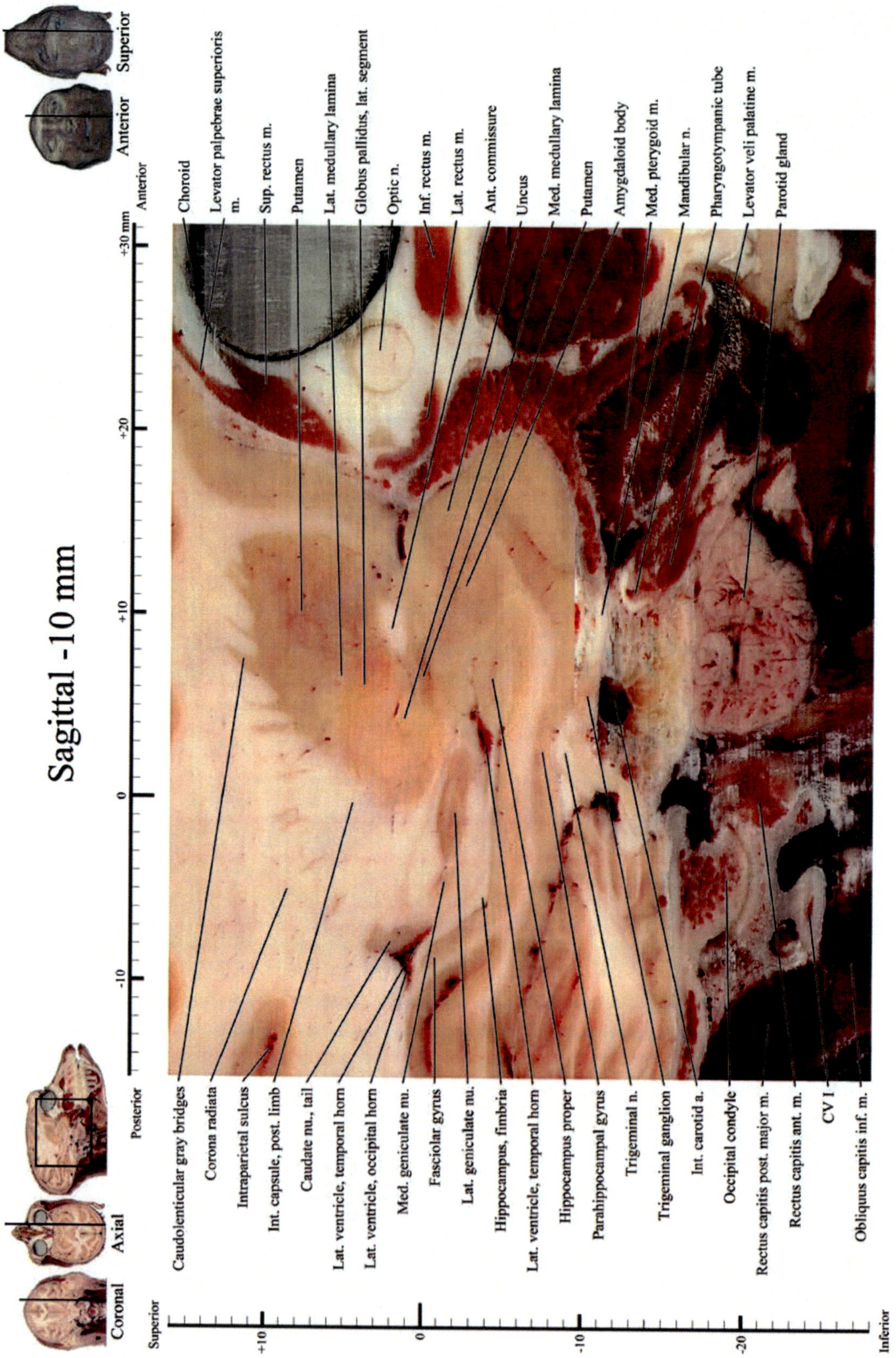

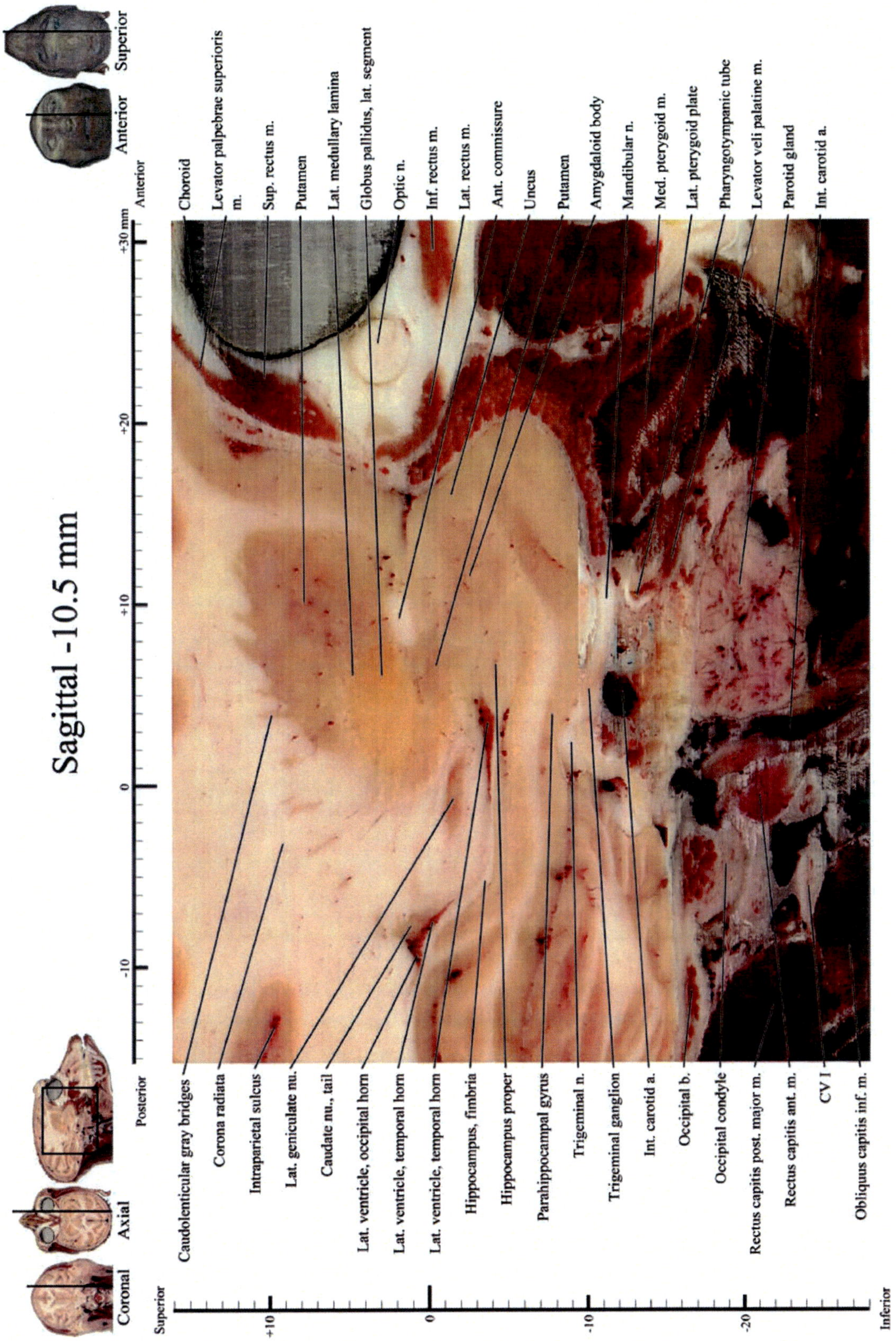
Sagittal -10.5 mm
Coronal
Axial
Anterior
Superior
Anterior
Posterior
+30 mm
+20
+10
0
-10
Superior
Inferior
+10
0
-10
-20
Caudolenticular gray bridges
Corona radiata
Intraparietal sulcus
Lat. geniculate nu.
Caudate nu., tail
Lat. ventricle, occipital horn
Lat. ventricle, temporal horn
Lat. ventricle, temporal horn
Hippocampus, fimbria
Hippocampus proper
Parahippocampal gyrus
Trigeminal n.
Trigeminal ganglion
Int. carotid a.
Occipital b.
Occipital condyle
Rectus capitis post. major m.
Rectus capitis ant. m.
CV I
Obliquus capitis inf. m.
Choroid
Levator palpebrae superioris m.
Sup. rectus m.
Putamen
Lat. medullary lamina
Globus pallidus, lat. segment
Optic n.
Inf. rectus m.
Lat. rectus m.
Ant. commissure
Uncus
Putamen
Amygdaloid body
Mandibular n.
Med. pterygoid m.
Lat. pterygoid plate
Pharyngotympanic tube
Levator veli palatine m.
Parotid gland
Int. carotid a.

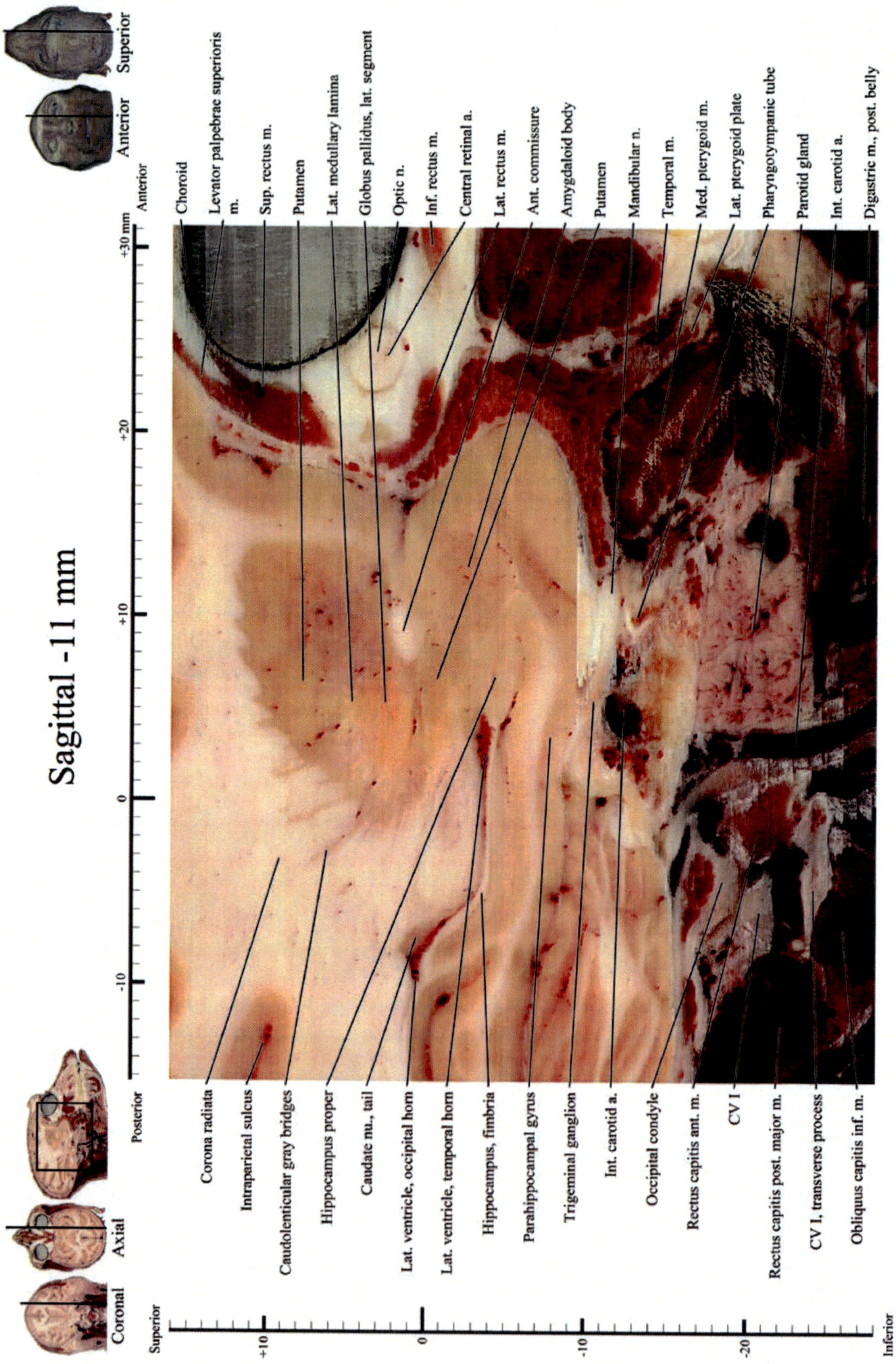
Sagittal -11 mm
Coronal
Axial
Anterior
Superior
Posterior
Anterior
Inferior
+30 mm
+20
+10
0
-10
+10
0
-10
-20
Choroid
Levator palpebrae superioris m.
Sup. rectus m.
Putamen
Lat. medullary lamina
Globus pallidus, lat. segment
Optic n.
Inf. rectus m.
Central retinal a.
Lat. rectus m.
Ant. commissure
Amygdaloid body
Putamen
Mandibular n.
Temporal m.
Med. pterygoid m.
Lat. pterygoid plate
Pharyngotympanic tube
Parotid gland
Int. carotid a.
Digastric m., post. belly
Corona radiata
Intraparietal sulcus
Caudolenticular gray bridges
Hippocampus proper
Caudate nu., tail
Lat. ventricle, occipital horn
Lat. ventricle, temporal horn
Hippocampus, fimbria
Parahippocampal gyrus
Trigeminal ganglion
Int. carotid a.
Occipital condyle
Rectus capitis ant. m.
CV I
Rectus capitis post. major m.
CV I, transverse process
Obliquus capitis inf. m.

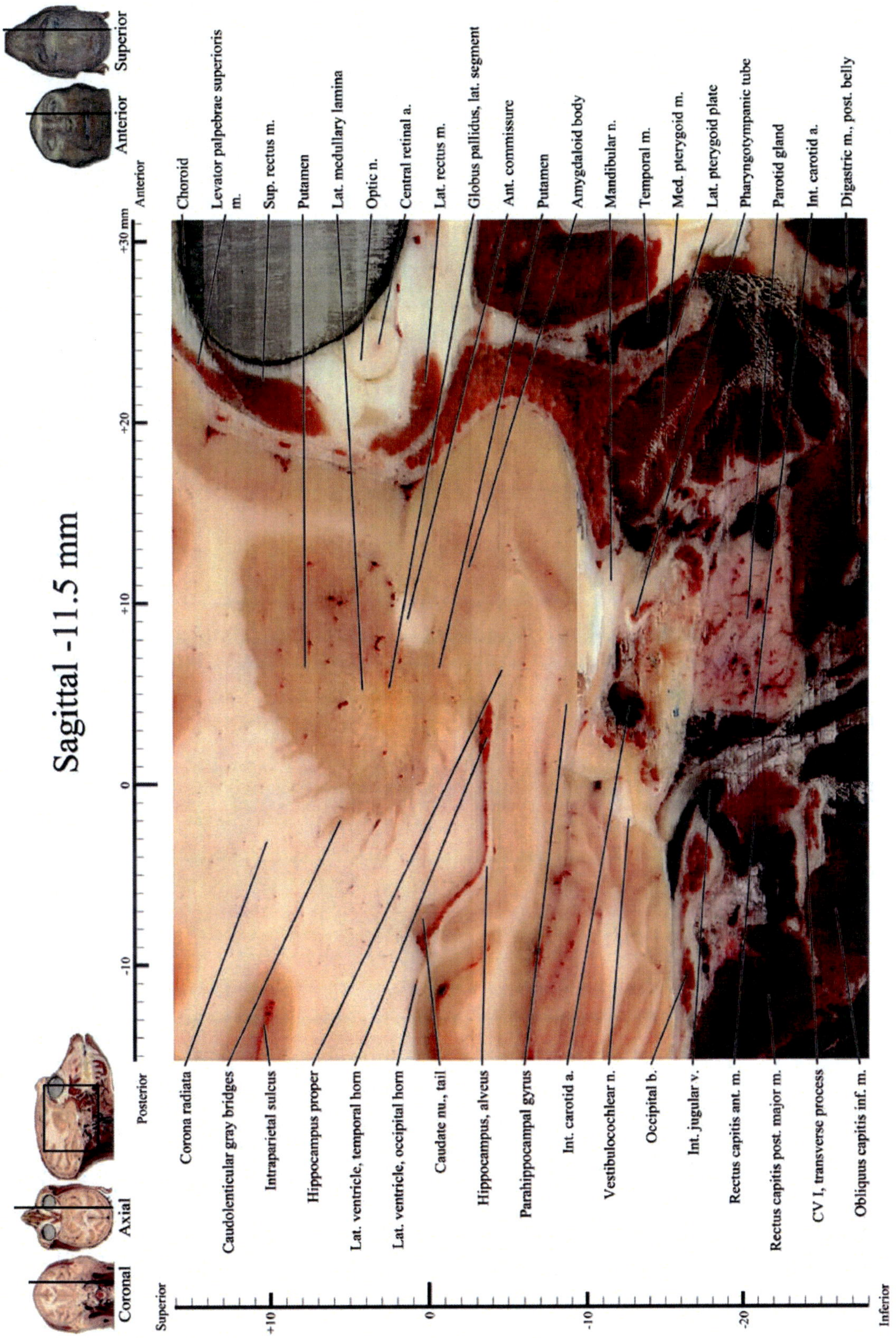
Sagittal -11.5 mm
Coronal
Axial
Superior
Anterior
Posterior
Inferior
+30 mm
+20
+10
0
-10
-20
Choroid
Levator palpebrae superioris m.
Sup. rectus m.
Putamen
Lat. medullary lamina
Optic n.
Central retinal a.
Lat. rectus m.
Globus pallidus, lat. segment
Ant. commissure
Putamen
Amygdaloid body
Mandibular n.
Temporal m.
Med. pterygoid m.
Lat. pterygoid plate
Pharyngotympanic tube
Parotid gland
Int. carotid a.
Digastric m., post. belly
Corona radiata
Caudolenticular gray bridges
Intraparietal sulcus
Hippocampus proper
Lat. ventricle, temporal horn
Lat. ventricle, occipital horn
Caudate nu., tail
Hippocampus, alveus
Parahippocampal gyrus
Int. carotid a.
Vestibulocochlear n.
Occipital b.
Int. jugular v.
Rectus capitis ant. m.
Rectus capitis post. major m.
CV I, transverse process
Obliquus capitis inf. m.

Sagittal -12 mm

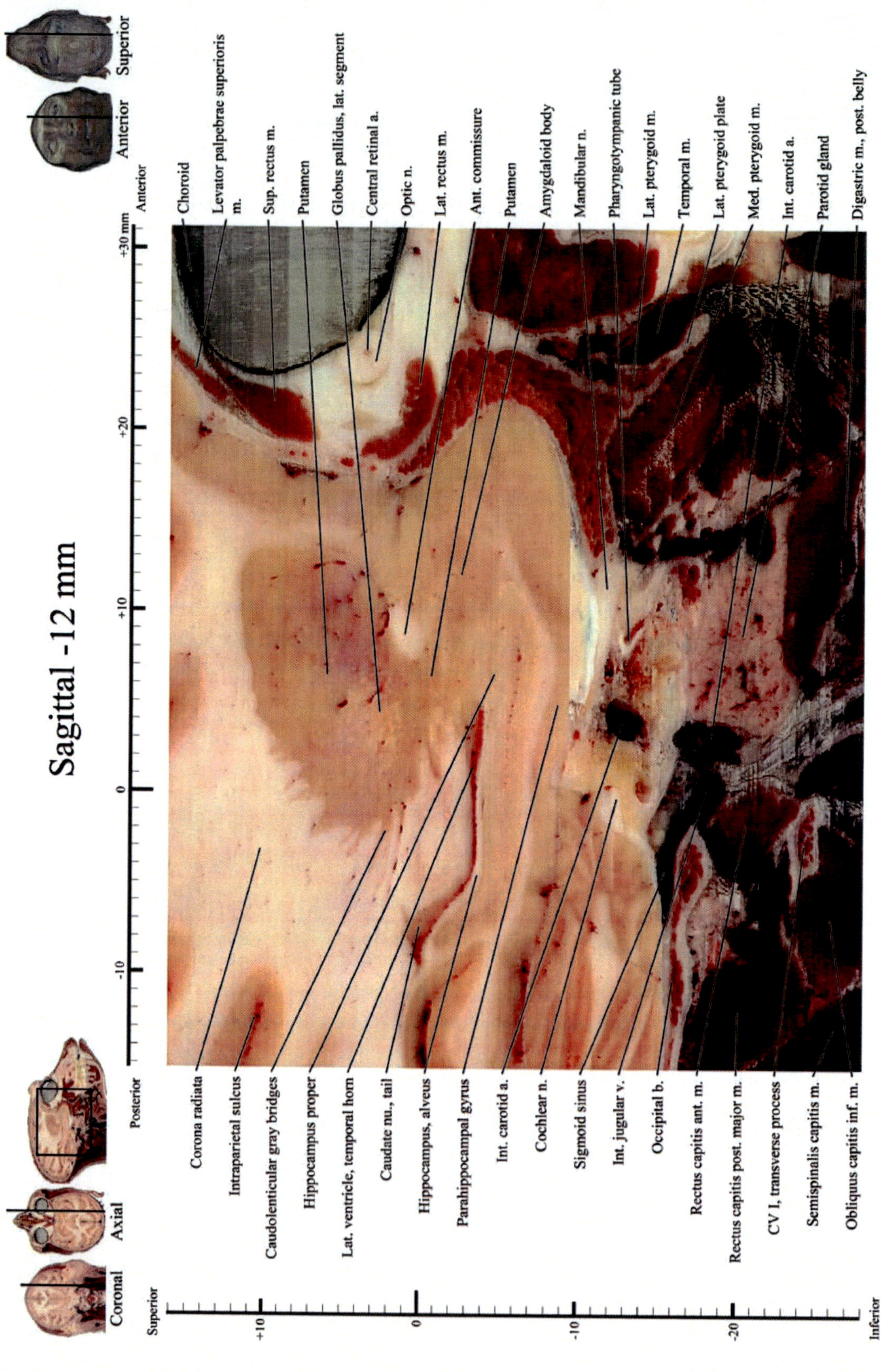

Sagittal -12.5 mm

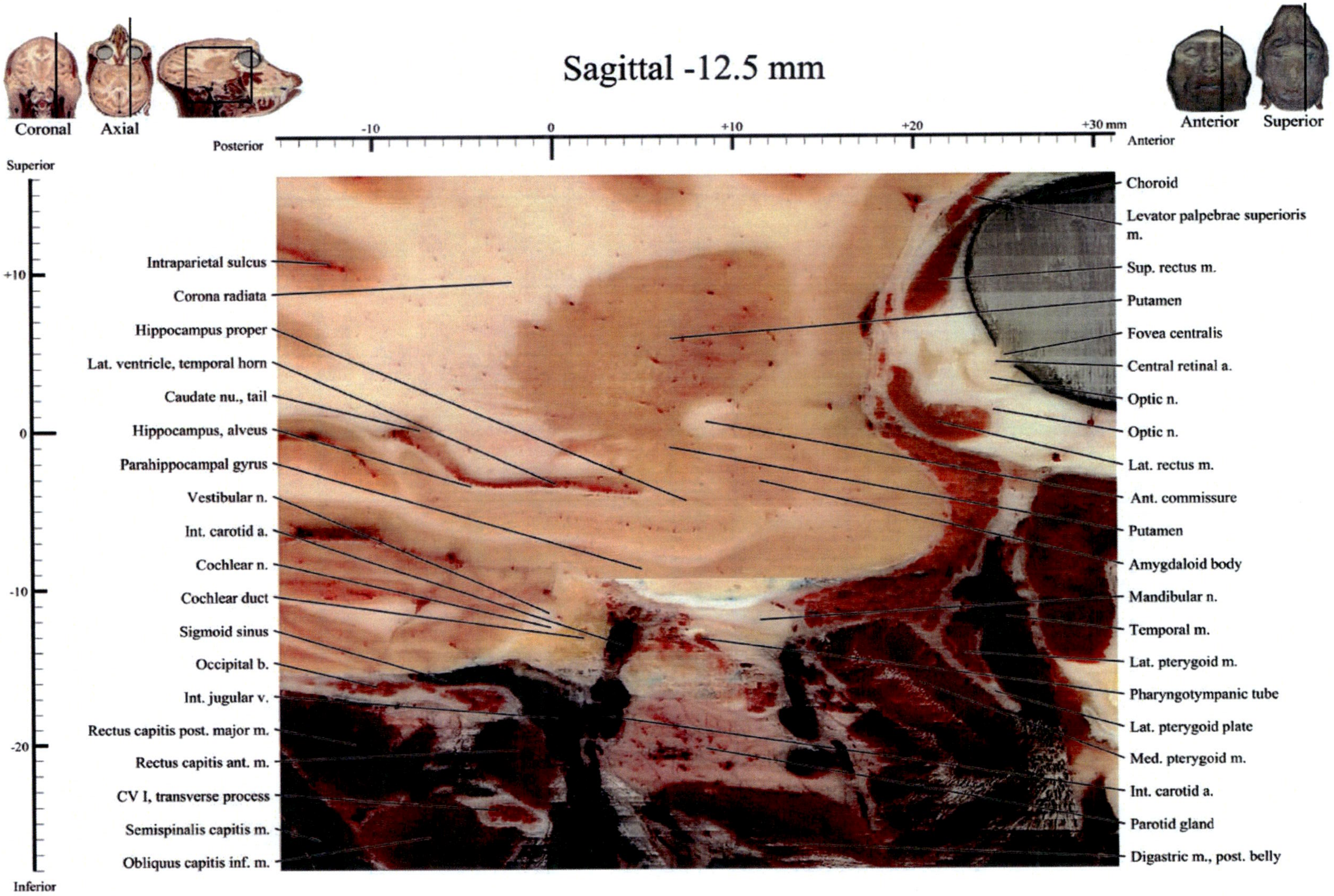

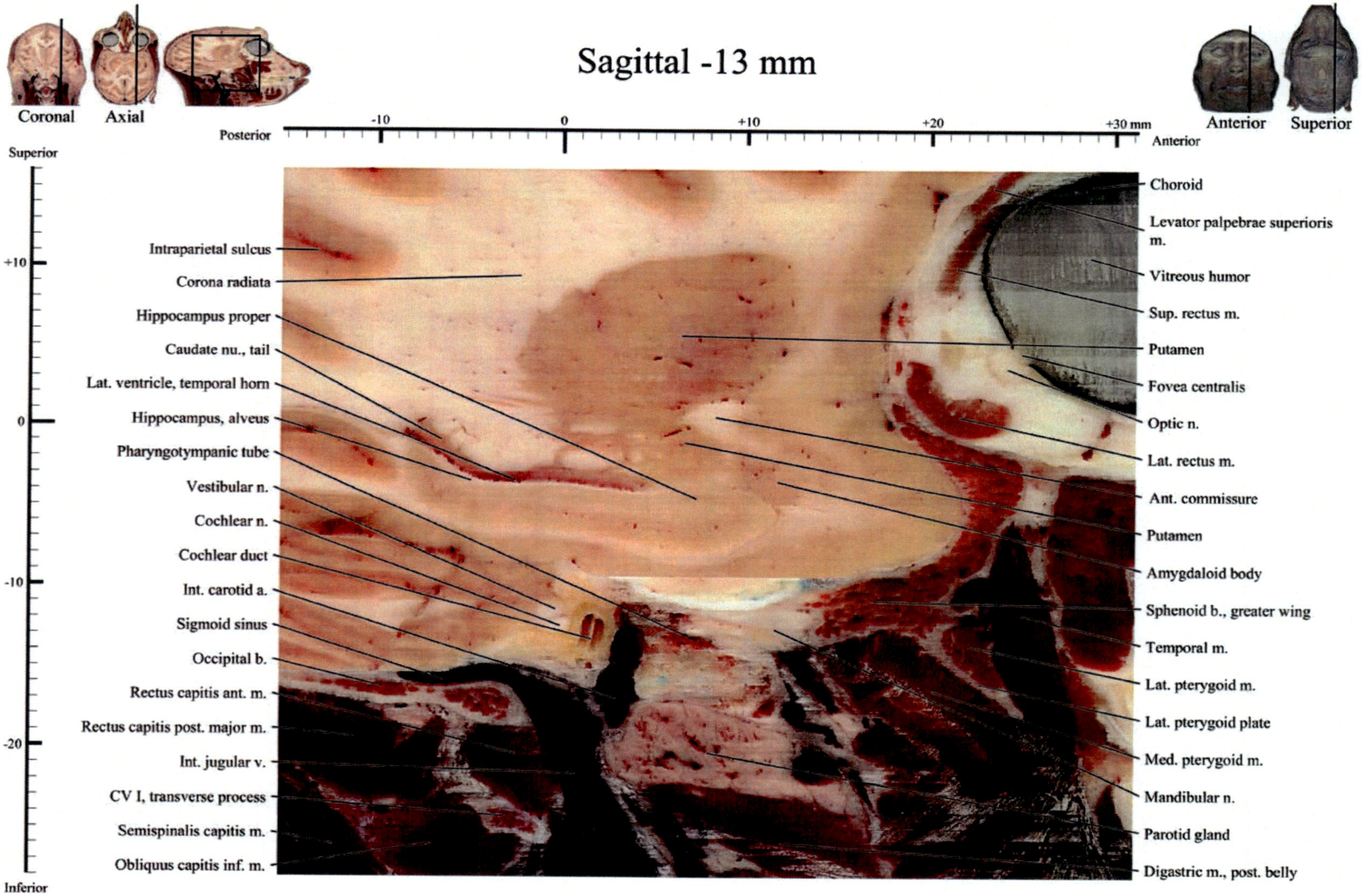
Sagittal -13 mm
Coronal
Axial
Anterior
Superior
Posterior
Anterior
-10
0
+10
+20
+30 mm
Superior
Inferior
+10
0
-10
-20
Intraparietal sulcus
Corona radiata
Hippocampus proper
Caudate nu., tail
Lat. ventricle, temporal horn
Hippocampus, alveus
Pharyngotympanic tube
Vestibular n.
Cochlear n.
Cochlear duct
Int. carotid a.
Sigmoid sinus
Occipital b.
Rectus capitis ant. m.
Rectus capitis post. major m.
Int. jugular v.
CV I, transverse process
Semispinalis capitis m.
Obliquus capitis inf. m.
Choroid
Levator palpebrae superioris m.
Vitreous humor
Sup. rectus m.
Putamen
Fovea centralis
Optic n.
Lat. rectus m.
Ant. commissure
Putamen
Amygdaloid body
Sphenoid b., greater wing
Temporal m.
Lat. pterygoid m.
Lat. pterygoid plate
Med. pterygoid m.
Mandibular n.
Parotid gland
Digastric m., post. belly

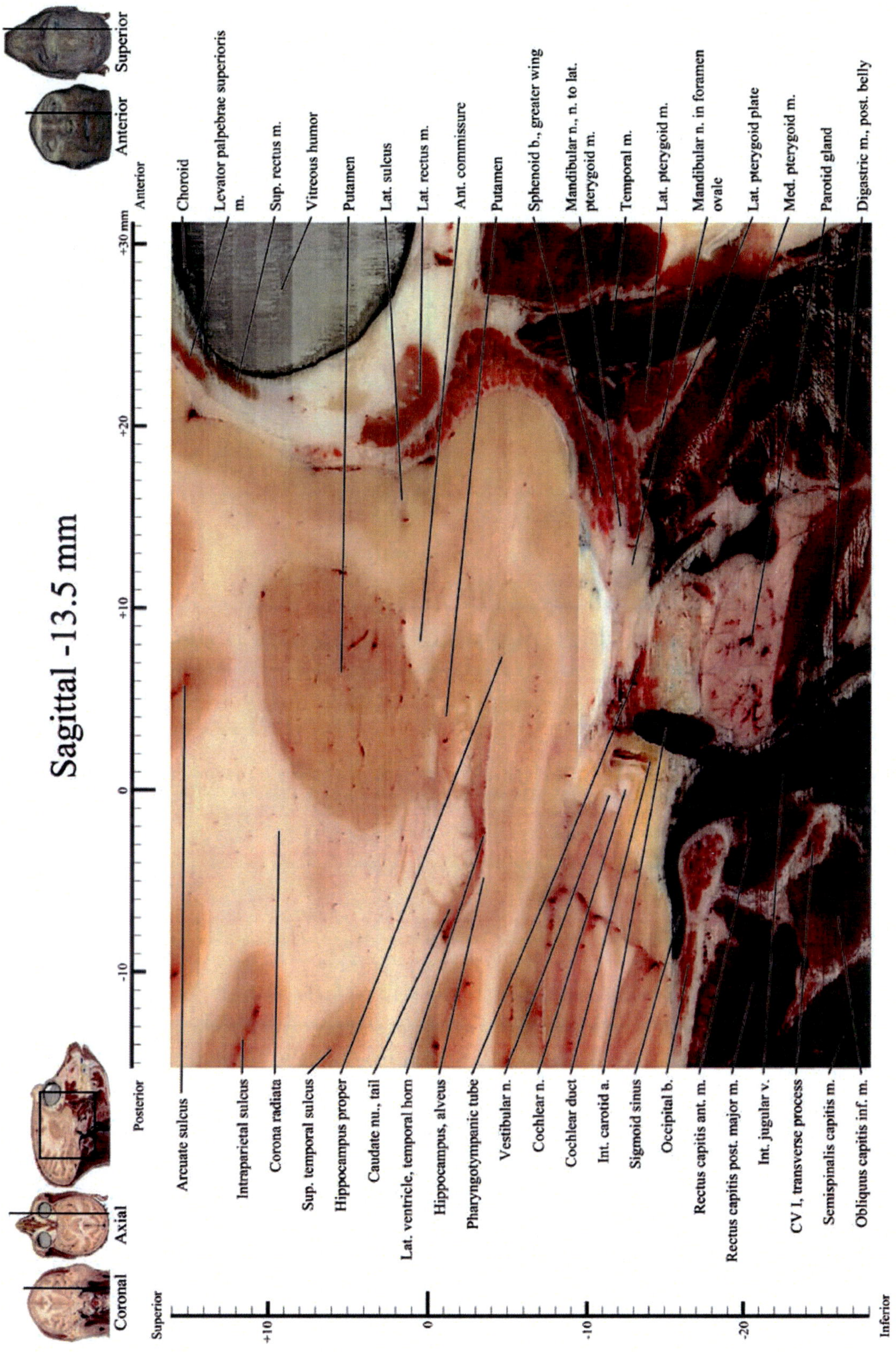
Sagittal -13.5 mm
Coronal
Axial
Anterior
Superior
Anterior
Posterior
Superior
Inferior
+30 mm
+20
+10
0
-10
+10
0
-10
-20
Choroid
Levator palpebrae superioris m.
Sup. rectus m.
Vitreous humor
Putamen
Lat. sulcus
Lat. rectus m.
Ant. commissure
Putamen
Sphenoid b., greater wing
Mandibular n., n. to lat. pterygoid m.
Temporal m.
Lat. pterygoid m.
Mandibular n. in foramen ovale
Lat. pterygoid plate
Med. pterygoid m.
Parotid gland
Digastric m., post. belly
Arcuate sulcus
Intraparietal sulcus
Corona radiata
Sup. temporal sulcus
Hippocampus proper
Caudate nu., tail
Lat. ventricle, temporal horn
Hippocampus, alveus
Pharyngotympanic tube
Vestibular n.
Cochlear n.
Cochlear duct
Int. carotid a.
Sigmoid sinus
Occipital b.
Rectus capitis ant. m.
Rectus capitis post. major m.
Int. jugular v.
CV I, transverse process
Semispinalis capitis m.
Obliquus capitis inf. m.

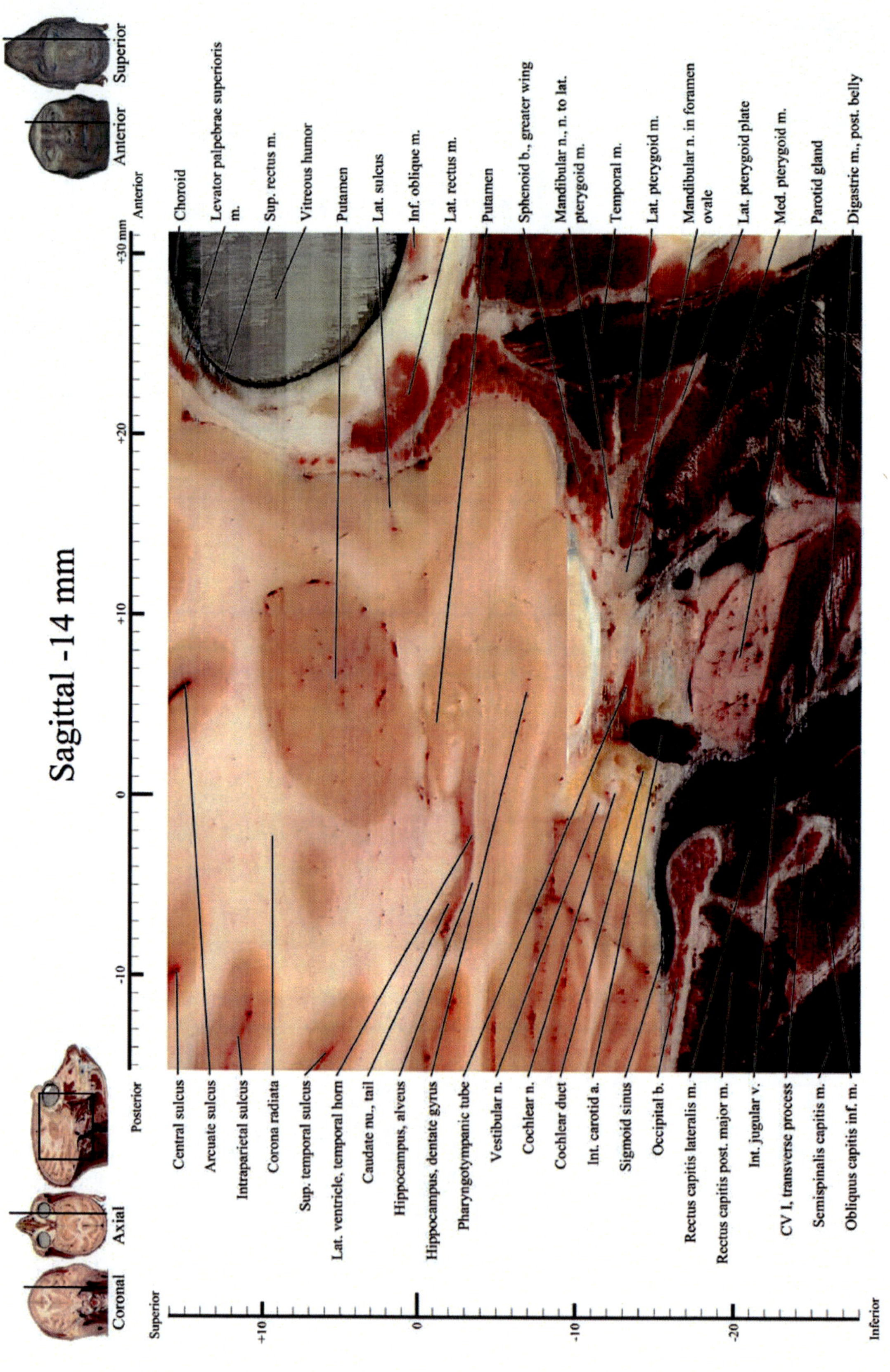
Sagittal -14 mm
Coronal
Axial
Anterior
Superior
Posterior
Inferior
+30 mm
+20
+10
0
-10
Choroid
Levator palpebrae superioris m.
Sup. rectus m.
Vitreous humor
Putamen
Lat. sulcus
Inf. oblique m.
Lat. rectus m.
Putamen
Sphenoid b., greater wing
Mandibular n., n. to lat. pterygoid m.
Temporal m.
Lat. pterygoid m.
Mandibular n. in foramen ovale
Lat. pterygoid plate
Med. pterygoid m.
Parotid gland
Digastric m., post. belly
Central sulcus
Arcuate sulcus
Intraparietal sulcus
Corona radiata
Sup. temporal sulcus
Lat. ventricle, temporal horn
Caudate nu., tail
Hippocampus, alveus
Hippocampus, dentate gyrus
Pharyngotympanic tube
Vestibular n.
Cochlear n.
Cochlear duct
Int. carotid a.
Sigmoid sinus
Occipital b.
Rectus capitis lateralis m.
Rectus capitis post. major m.
Int. jugular v.
CV I, transverse process
Semispinalis capitis m.
Obliquus capitis inf. m.
-20

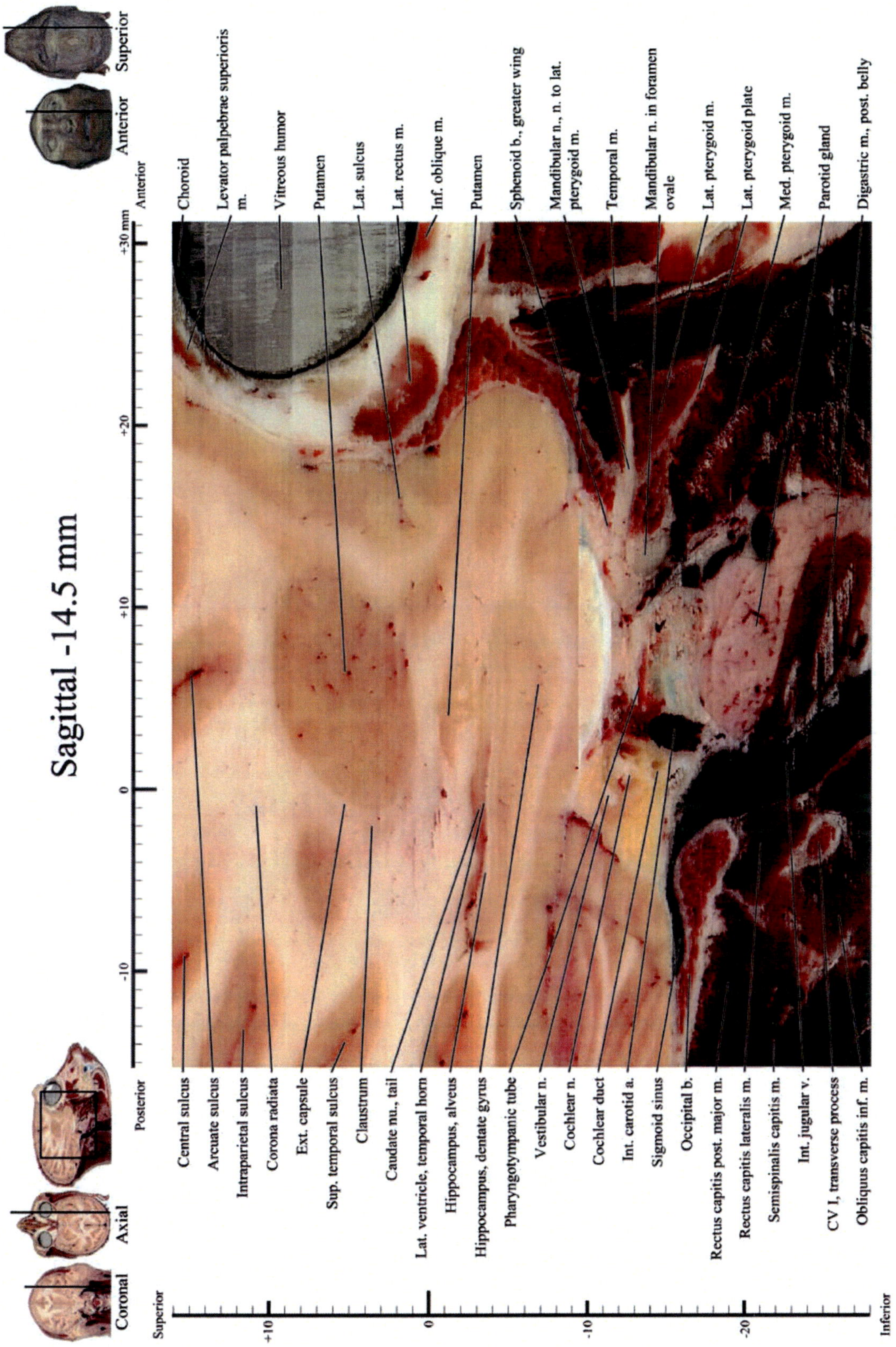
Sagittal -14.5 mm
Coronal
Axial
Anterior
Superior
Anterior
Posterior
Superior
Inferior
+30 mm
+20
+10
0
-10
+10
0
-10
-20
Choroid
Levator palpebrae superioris m.
Vitreous humor
Putamen
Lat. sulcus
Lat. rectus m.
Inf. oblique m.
Putamen
Sphenoid b., greater wing
Mandibular n., n. to lat. pterygoid m.
Temporal m.
Mandibular n. in foramen ovale
Lat. pterygoid m.
Lat. pterygoid plate
Med. pterygoid m.
Parotid gland
Digastric m., post. belly
Central sulcus
Arcuate sulcus
Intraparietal sulcus
Corona radiata
Ext. capsule
Sup. temporal sulcus
Claustrum
Caudate nu., tail
Lat. ventricle, temporal horn
Hippocampus, alveus
Hippocampus, dentate gyrus
Pharyngotympanic tube
Vestibular n.
Cochlear n.
Cochlear duct
Int. carotid a.
Sigmoid sinus
Occipital b.
Rectus capitis post. major m.
Rectus capitis lateralis m.
Semispinalis capitis m.
Int. jugular v.
CV I, transverse process
Obliquus capitis inf. m.

Sagittal -15 mm

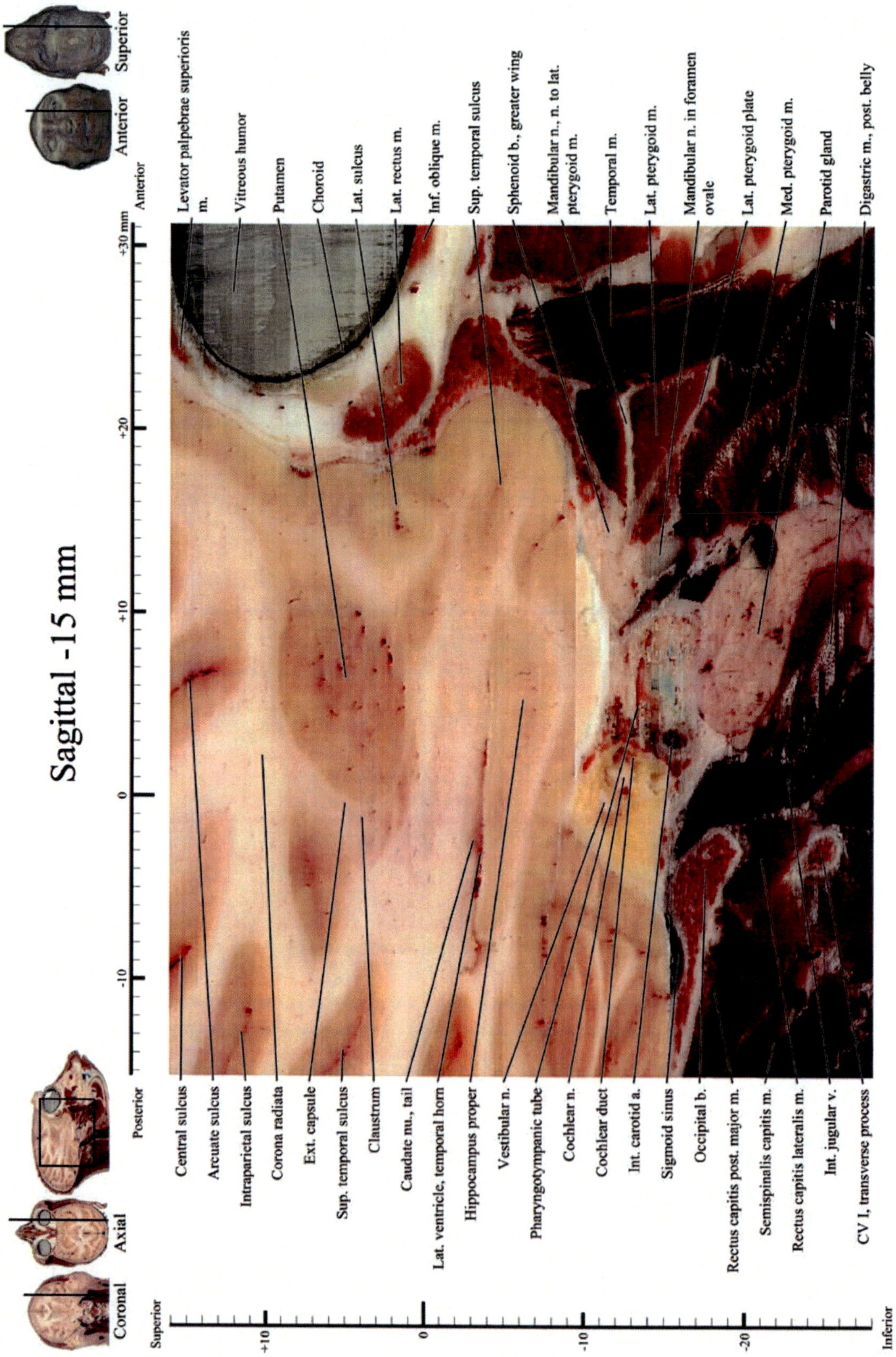

Sagittal -15.5 mm

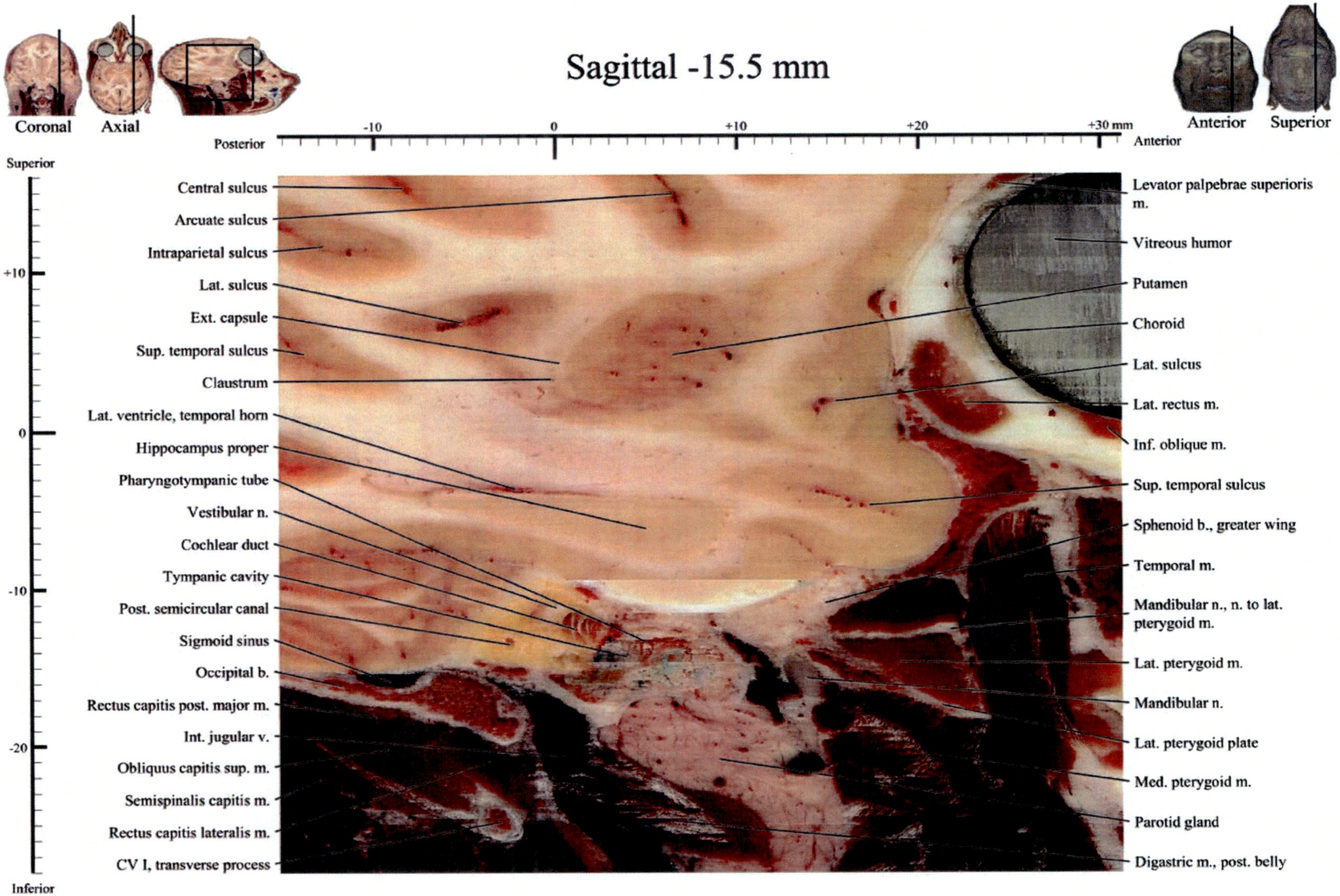

Sagittal -16 mm

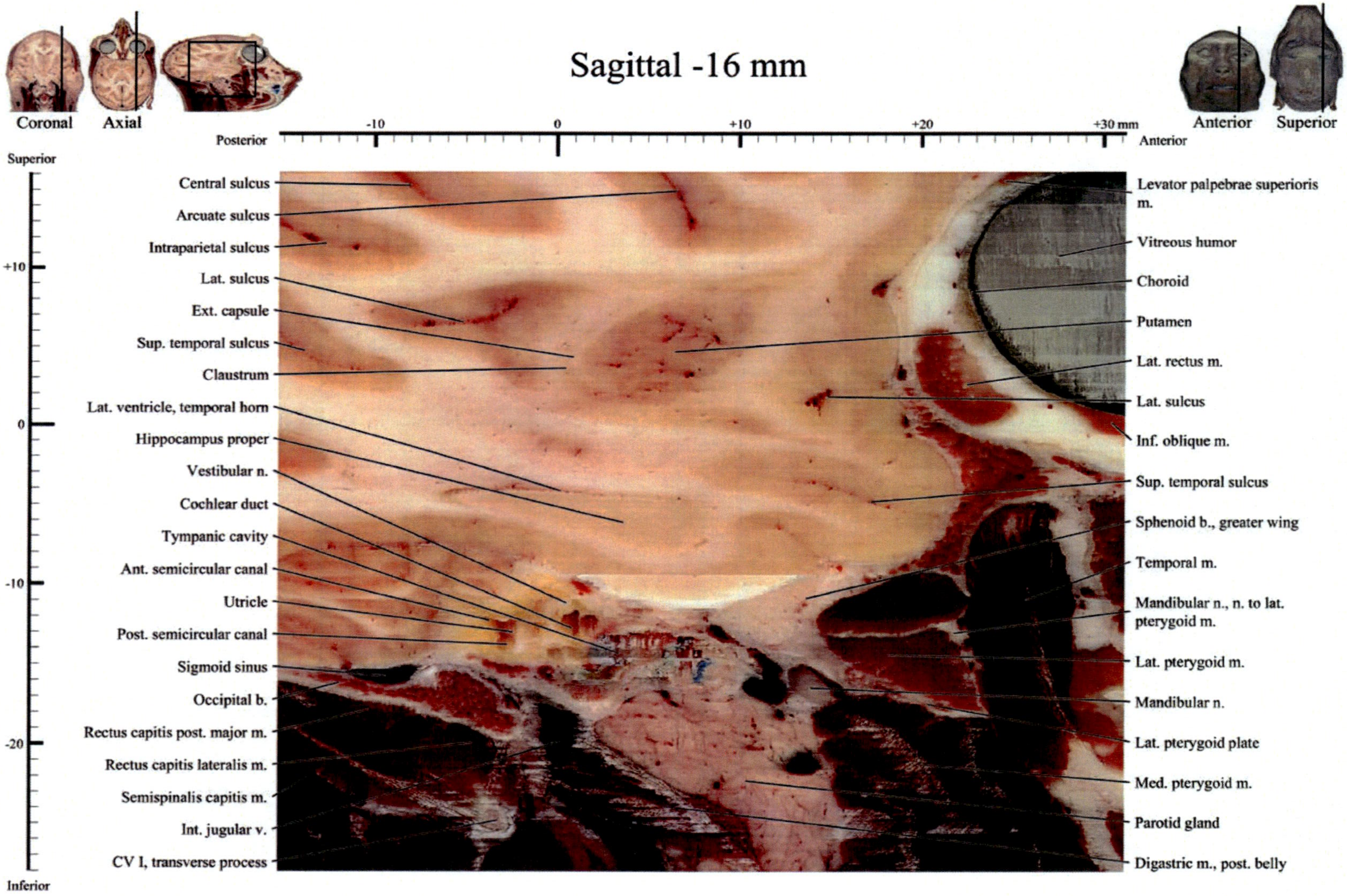

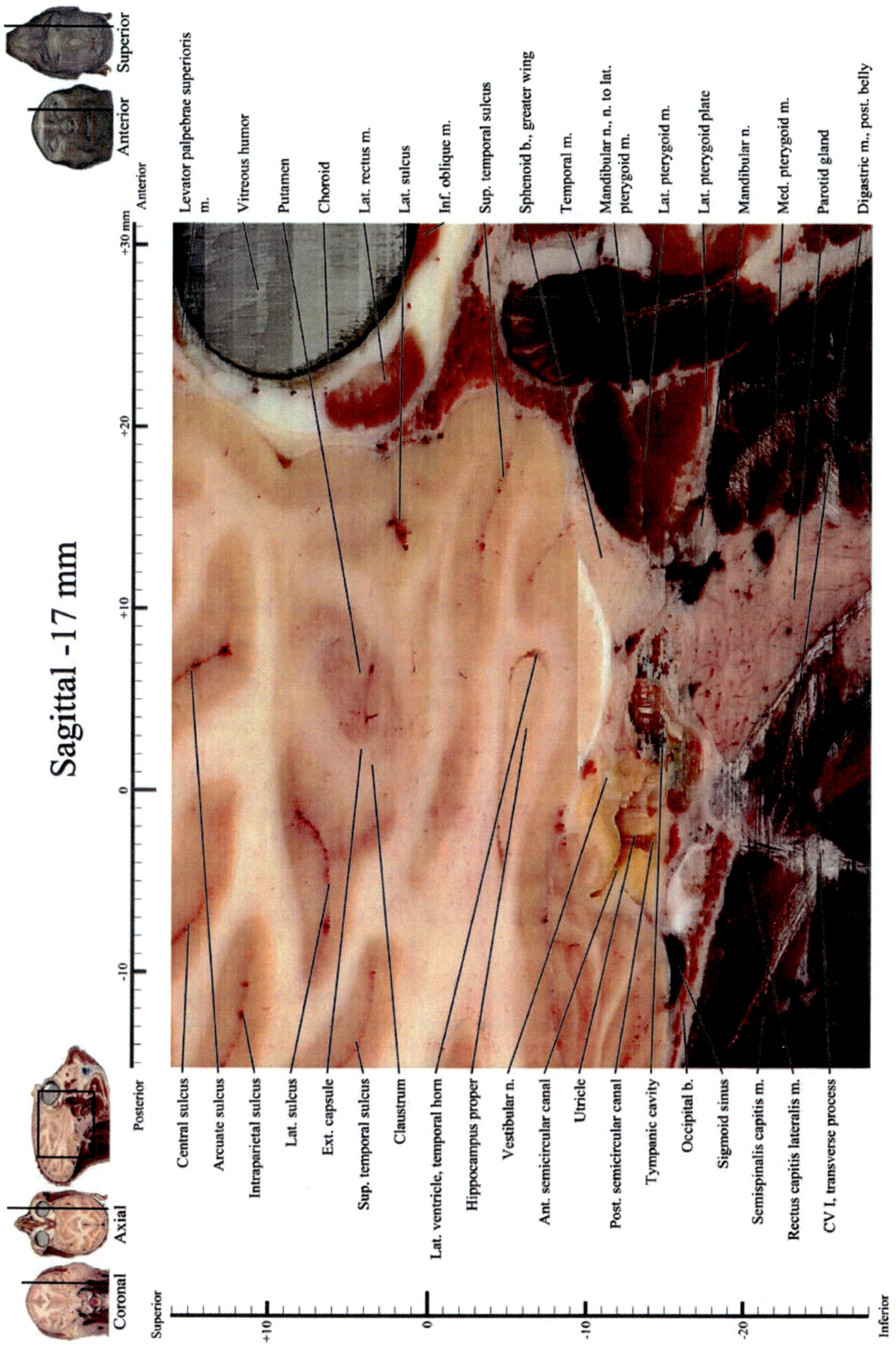
Sagittal -17 mm
Coronal
Axial
Posterior
Anterior
Superior
Anterior
+30 mm
+20
+10
0
-10
Superior
+10
0
-10
-20
Inferior
Levator palpebrae superioris m.
Vitreous humor
Putamen
Choroid
Lat. rectus m.
Lat. sulcus
Inf. oblique m.
Sup. temporal sulcus
Sphenoid b., greater wing
Temporal m.
Mandibular n., n. to lat. pterygoid m.
Lat. pterygoid m.
Lat. pterygoid plate
Mandibular n.
Med. pterygoid m.
Parotid gland
Digastric m., post. belly
Central sulcus
Arcuate sulcus
Intraparietal sulcus
Lat. sulcus
Ext. capsule
Sup. temporal sulcus
Claustrum
Lat. ventricle, temporal horn
Hippocampus proper
Vestibular n.
Ant. semicircular canal
Utricle
Post. semicircular canal
Tympanic cavity
Occipital b.
Sigmoid sinus
Semispinalis capitis m.
Rectus capitis lateralis m.
CV I, transverse process

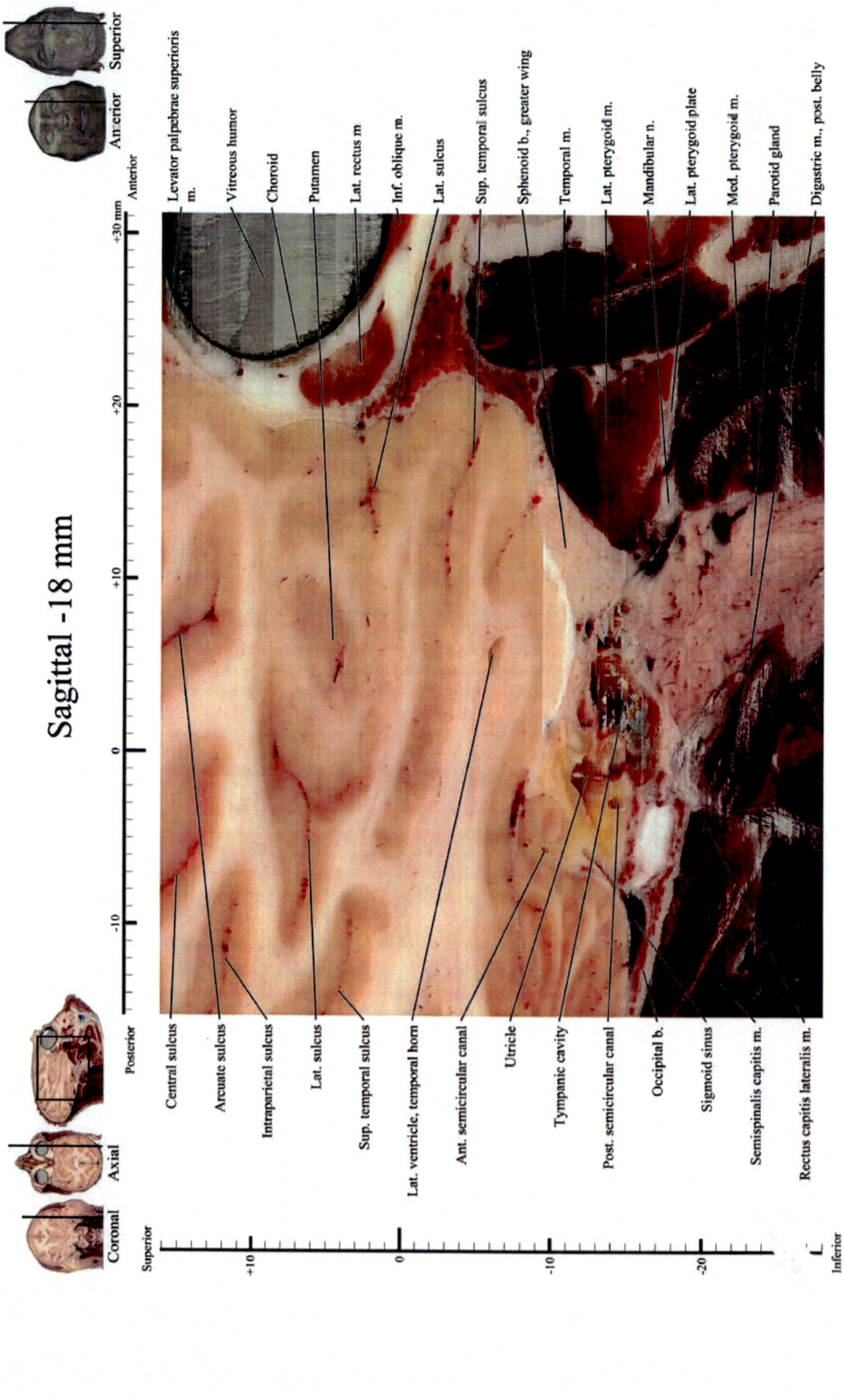
Sagittal -18 mm
Coronal
Axial
Anterior
Superior
Superior
+10
0
-10
-20
Inferior
Anterior
Posterior
+30 mm
+20
+10
0
-10
Levator palpebrae superioris m.
Vitreous humor
Choroid
Putamen
Lat. rectus m
Inf. oblique m.
Lat. sulcus
Sup. temporal sulcus
Sphenoid b., greater wing
Temporal m.
Lat. pterygoid m.
Mandibular n.
Lat. pterygoid plate
Med. pterygoid m.
Parotid gland
Digastric m., post. belly
Central sulcus
Arcuate sulcus
Intraparietal sulcus
Lat. sulcus
Sup. temporal sulcus
Lat. ventricle, temporal horn
Ant. semicircular canal
Utricle
Tympanic cavity
Post. semicircular canal
Occipital b.
Sigmoid sinus
Semispinalis capitis m.
Rectus capitis lateralis m.